MODERN CONCEPT OF COMMUNICATION SYSTEMS

MODERN CONCEPT OF COMMUNICATION SYSTEMS

B.M.K. Prasad
Principal
Dronacharya College of Engineering
Gurgaon

Amrit Bawankan
Larson and Toubro Ltd.
ECC Division, Gurgaon

Aparna Bawankan
Dronacharya College of Engineering
Greater Noida (U.P.)

KHANNA BOOK PUBLISHING CO. (P) LTD.

::: *Publisher of Engineering and Computer Books* :::

Head Office : 4C/4344, Ansari Road, Darya Ganj, New Delhi-110002
Phone: 011-2324 44 47-48 Fax: 011-2324 44 47

Regd. Office : 1694-95, Nai Sarak, Delhi-110006
Phone: 011-2328 32 11 Fax: 011-2325 66 58

E-Mail : sales@khannapublishers.com, contact@khannabooks.com

Website : www.khannapublishers.com

MODERN CONCEPT OF COMMUNICATION SYSTEMS

B. M. K. Prasad, Amrit Bawankan, Aparna Bawankan

ISBN : 978-93-82609-50-6

First Edition : 2014

Published by:
KHANNA BOOK PUBLISHING CO. (P) LTD.
4C/4344, Ansari Road, Daryaganj, New Delhi-110 002
Phone: 011-2324 44 47-48
E-mail: sales@khannapublishers.com
Website:www.khannapublishers.com

Cover Design & Layout by
Akriti Print & Graphics, Delhi

Printed in India by
S.P.S. Printers & Binders, Delhi

Dedicated To

Ram C. Prasad
C. Lal Prasad
G. Devi

Who have always encouraged to serve the society and the nation whole heartedly without delay and without any interest.

Preface

The proposed book Modern Concepts of Communication Systems is considered as one of the most respected and authentic texts in the field of communication systems and engineering. Its profound mathematical treatment, clarity of concepts, rich set of assignment problems and of course, novelty in approach defined in many way how analog or digital communication are taught in engineering colleges. The book is covering almost all the parts of syllabus of M.D. University, Rohtak; Gautam Buddha University, Lucknow; Mahamaya University, Noida; old UPT University, K.U., Kurukshetra etc.

Communication systems are a core subject for undergraduate courses in Electronics and Communications, and Computer Engineering. It is also offered as a special subject for undergraduate courses in Electrical and Instrumentation Engineering, and for postgraduate courses in Physics, Electronics and Computer Science. This book is primarily intended to serve as a textbook for undergraduate courses in Electronics and Communication Engineering.

The primary objective has been to make this book more student friendly with more illustrations, solved examples etc., so that it can be taken up for self study. Each chapter in this book has been completely reorganised and begins with an introduction that gently introduces the topics and links them with previous chapters as per the requirement.

Chapter one aims to give **an introduction** to modern communication systems through a brief overview. It is also intended to develop necessary background on signal and its spectral information. The discussion includes characterization of periodic signal by Fourier series and aperiodic signal by Fourier transform and finally representation of a signal in vector space. Fourier analysis brings out the frequency information of a signal which in communication context is very important. Reference to necessary Fourier series and Fourier transform properties are made at appropriate places.

One of the basic problems of Communication engineering is the design and analysis of systems which will allow many individual messages to be transmitted simultaneously over a single communication channel. A method by which such multiple transmission, called multiplexing, may be achieved consists of translating each message to different positions in frequency spectrum. This frequency multiplexing needs use of a carrier, with certain attribute of it varying according to message signal. In chapter two **amplitude modulation** has been discussed which is a fundamental requisite of communication systems. It is defined as the process by which some characteristics of a signal are called carrier. This is varies in accordance with instantaneous value of another signal called modulating signal.

In the amplitude modulation, the modulator output is a high frequency carrier signal with varying amplitude, the variation containing information of the modulating message. In chapter three, we discuss **modulation systems** in which the modulator output is of constant amplitude. Here the message signal is superimposed on the carrier by varying its frequency and phase, together called **angle**. Following the qualitative and quantitative presentation of underlying principles, we discuss a variety of modulation and demodulation techniques.

Chapter four is devoted to the **signal analysis**. In this chapter it has been considered that the signal is already in electrical form. A signal is generally given in time domain. Time domain specifies that what is the amplitude of the signal at a particular time instant.

Noise is an unwanted signal that corrupts a desired message signal. In general, noise may be predictable, in nature. In chapter five, it has been described how noise affects reception of amplitude modulated message. It is investigated signal side band (SSB), double side band suppressed carrier (DSB-SC) and double side band with carrier (DSB-C) in detail.

In chapter six, we discuss first pass band transmission of **digital signal** and then take up inter channel and inter symbol interference, way to reduce bandwidth of digital signal, Nyquist criterion and related issues.

In chapter seven **pulse modulation** has been discussed in detail. In pulse modulation, the carrier is no longer a continuous signal but consists of a pulse train, some parameter of which is varied in accordance with the instantaneous value of the following signal.

Acknowledgement

We are thankful to Prof. (Dr.) V.K. Rohtagi, Bhabha Atomic Research Centre, Mumbai; Prof. S.P. Verma, Patna University, Patna; Prof. (Dr.) Rana Khan, Jamia Milia Islamia University, New Delhi; Prof. P.K. Mukherjee, Delhi University, Delhi; Prof. (Dr.) O.P. Mehta, Ex-Principal, DCE, Gurgaon; Prof. R.C. Narula Ex-Vice-Principal, DCE, Gurgaon; Prof. Suresh Chandra, Patna University, Patna; Mr. Surender Kumar, IFS; Mr. Mahesh Kumar, Prof. Amar Kumar Mishra, Prof. (Dr.) Swambhu Kumar, Prof. R.K. Mishra, Prof. S.S. Mahto, Tribhuwan University, Nepal; Mr. S.P. Ghosh, DST, New Delhi for their guidance encouragement and constant support from every angle.

We are also thankful to Dr. Bachcha Prasad, Medical Officer (Rtd.) Govt. of Bihar, Mr. Lakhi Prasad, Dr. Vinay Kumar, Mr. Raj Kumar, Ms. Neelam, Mr. Sanjay Kumar, Mr. Ajay Kumar, Mr. Raju Kumar, Dr. Dhirendra Kumar, Dr. Surendra Prasad.

We are thankful to Mr. Om Prakash. Ms. Roop Kumari, Mr. Yamuna Prasad, Ms. Raj Kumari, Mr. Krishan Mohan Verma, Ms. Phool Kumari, Mr. Sushil Krishan Gore, RBI, Lucknow, Prof. Radha Krishan Gore, Deoria; Mr. Manish Mohan Gore, CSIR, New Delhi; Mr. Rajneesh Mohan Gore, Ms. Kiran Prasad, Mr. Ashok Kumar, Ms. Pushpa, Mr. Ajay Bowankan, Mr. Basu Bowankan, Ms. Divya Bowankan for their always constant support and encouragement in the family.

Special thanks to Mr. Puneet Khanna ..... of Khanna Publication Pvt. Ltd., New Delhi and their team members ....

Contents

DEDICATION *v*

PREFACE *vii*

ACKNOWLEDGEMENT *ix*

1. COMMUNICATION SYSTEMS **1—30**

1.1 Introduction 1
1.2 Process of Communication Systems with Elements 2
 1.2.1 Information Source 2
 1.2.2 Input Transducer 3
 1.2.3 Transmitter 3
 1.2.4 The channel and the Noise 3
 1.2.5 Receiver 4
 1.2.6 Destination 4
1.3 Information Sources 4
 1.3.1 Speech 4
 1.3.2 Television 5
 1.3.3 Fascimile 6
 1.3.4 Personal Computers 6
1.4 Communication Channels 7
 1.4.1 Telephone Channels 7
 1.4.2 Optical Fibre 8
 1.4.3 Mobile Radio Channels 9
 1.4.4 Satellite Channel 9
1.5 Baseband and Passband Signals 10
1.6 Representation of Communication Signals and Communication Systems 11
 1.6.1 Periodic and Non-periodic Signals 11
 1.6.2 Deterministic and Random Signals 11
 1.6.3 Energy and Power Signals 11
1.7 Classification of Communication 13
 1.7.1 Line Communication 13
 1.7.2 Wireless or Radio Communication 13
1.8 The Modulation Process 13
 1.8.1 Types of Modulation 13
 1.8.2 Need of Modulation/Benefits of Modulation 14
1.9 Basic Communication Resource 15

1.10 Information Capacity Theorem 15
1.11 Analog and Digital Signals 16
1.12 Conversion of Analog Signal to Digital Signals 17
1.13 Communications: Analog and Digital 18
1.13.1 Analog Communication 18
1.13.2 Digital Communication 19
1.14 Model of a Digital Communication Systems 19
1.15 Discrete Information Source 20
1.15.1 Source Alphabet 20
1.15.2 Symbol Rate 20
1.15.3 Source Alphabet Probabilities 20
1.15.4 Probabilistic Dependence of Symbols in a Sequence 20
1.16 Source Encoder and Decoder 20
1.16.1 Block Size 21
1.16.2 Codeword length 21
1.16.3 Average Data Rate 21
1.16.4 Efficiency of the Encoder 21
1.17 Channel Encoder and Decoder 21
1.18 Digital Modulators and Demodulators 22
1.19 Communication Channel 23
1.19.1 Signal Attenuation 23
1.19.2 Multipath Distortion 24
1.19.3 Additive Noise Interference 24
1.19.4 Amplitude and Phase Distortion 24
1.20 Advantage and Disadvantage of Digital Communication 24
1.20.1 Advantages 24
1.20.2 Disadvantages 24
1.21 Fundamental Limitations of Communication Systems 24
1.21.1 Noise Limitations 24
1.21.2 Equipment Limitation 25
1.21.3 Bandwidth Limitations 25
1.22 Bandwidth Requirements 26
1.23 Since wave and Fourier Series Review 26
1.24 Frequency Spectra of Non-sinusoidal Waves 29

2. AMPLITUDE MODULATION 31—160

2.1 Introduction 31
2.2 Continuous Wave or Analog Modulation 31
2.3 Pulse Modulation System 31

2.4	Multiplexing	31
	2.4.1 Frequency Division Multiplexing (FDM)	32
	2.4.2 Time-Division Multiplexing (TDM)	32
2.5	Amplitude Modulation	33
	2.5.1 Few Points about Amplitude Modulation (AM)	34
	2.5.2 Spectrum of AM Wave or Frequency-Domain Representation	34
2.6	Modulation Index, (m_a)	38
	2.6.1 Overmodulation	38
	2.6.2 Amplitude Modulation with $m_a < 1$	39
	2.6.3 Amplitude Modulation with $m_a > 1$	39
2.7	Single Tone Amplitude Modulation (AM)	39
2.8	Power Content in AM Wave	41
	2.8.1 Carrier Power	41
	2.8.2 Sideband Power	42
2.9	Transmission Efficiency of Amplitude Modulated Signal	43
2.10	Power of a Single Tone Amplitude Modulated Signal	43
2.11	Current Calculation for Single-Tone Amplitude Modulation	44
2.12	Power Content in Multiple-Tone Amplitude Modulation	45
	2.12.1 Total or Net Modulation Index for Multiple-Tone Modulation	47
2.13	Generation of Amplitude Modulation	47
	2.13.1 Low-level Amplitude Modulation	48
	2.13.2 High-level Amplitude Modulation	48
	2.13.3 Square Law Diode Modulation	48
2.14	Collector Modulation method	49
	2.14.1 Operating Principle	50
	2.14.2 Mathematical Analysis	51
2.15	Demodulation of AM Waves	51
	2.15.1 Square-law Detector	51
	2.15.2 Linear Diode or Envelope Detector	53
	2.15.3 Operating Principle	53
2.16	Double Sideband Suppressed Carrier (DSB-SC) System	54
	2.16.1 Few Points	56
2.17	Generation of DSB-SC Signal	57
	2.17.1 The Balanced Modulator	57
	2.17.2 Ring Modulator	59
2.18	Demodulation of DSB-SC Signals	61
	2.18.1 Synchronous Detection Method	61
	2.18.2 Working Principle	62
	2.18.3 Synchronous or Coherent Detection	63
	2.18.4 Envelope Detection after suitable Carrier Re-insertion	63

2.19 Quadrative—Amplitude Modulation (QAM) 64
2.20 Effect of Phase and Frequency Errors in Synchronous Detection 66
2.21 Carrier Acquisition in DSB-SC System or Synchronous/ Synchronization Techniques in DSB-SC System 67
2.21.1 Pilot Carrier 67
2.21.2 Costa's Receiver 68
2.21.3 Operating Principle 68
2.21.4 Limitation 69
2.21.5 Squaring loop 69
2.22 Single Sideband Suppressed—Carrier (SSB-SC) Modulation 70
2.22.1 Time-Domain Description of the SSB-SC Wave 71
2.22.2 SSB-SC Wave with Single-tone Modulating Signal 71
2.23 Hilbert Transform 72
2.23.1 Properties of a Hilbert transform are as under 75
2.23.2 Pre-envelope or Analytic Signal 75
2.23.3 SSB-SC for a General Modulating Signal 76
2.24 Generation of SSB-SC Signals 78
2.24.1 Frequency Discrimination Method 78
2.24.2 Limitations 78
2.25 Phase Discrimination or Phasing Method 80
2.25.1 Detection of SSB-SC Signals 82
2.25.2 Effect of Phase and Frequency Error 83
2.25.3 Phasor Diagram and Waveform of SSB Signals 85
2.25.4 Waveform of SSB with Large Carrier 86
2.25.5 Demodulation of the SSB Signal with Large Carrier 87
2.25.6 Compatible Single Sideband 87
2.25.7 Vestigial-Sideband Modulation (VSB) 88
2.26 Generation and Detection of VSB Signal 88
2.26.1 Filter Method 88
2.26.2 Filter Characteristics 88
2.26.3 Phase-Discrimination Method for Generation of VSB-SC Wave 90
2.26.4 Time Domain Description of VSB Wave 90
2.26.5 In Phase Component $f_c(t)$ 92
2.26.6 Quadrature-Phase Component $f_s(t)$ 92
2.27 Detection of VSB Signal with Large Carrier 93
2.28 Amplitude Modulation with Large Carrier (AM) 94
2.29 Envelope of the AM Wave 95
2.29.1 Bandwidth 95
2.29.2 Modulation Index 96
2.30 Overmodulation and Envelope Distortion 96
2.30.1 AM with $m_a < 1$ 96
2.30.2 AM with $m_a \approx 1$ 96

2.30.3 Overmodulated Signal with $m_a > 1$ 97
2.30.4 Partially Suppressed Carrier AM 97
2.30.5 Phasor Representation of an AM Signal 97
2.31 Experimental Determination of Modulation 98
2.31.1 Derivation 99
2.32 Superposition Theorem of Spectrum 99
2.32.1 Linear Modulation 100
2.32.2 Power Content in AM Signal 101
2.32.3 Carrier Power 101
2.32.4 Sideband Power 101
2.33 Transmission Efficiency of AM Signal 102
2.34 Transmission Efficiency in Suppressed Carrier Systems 103
2.34.1 Generation of AM Waves 103
2.34.2 Chopper-type Switching Modulator 103
2.34.3 Circuit Diagram 104
2.34.4 Modulators Using Amplifying Devices 104
2.34.5 Principle of Operation 106
2.34.6 Overmodulation 107
2.34.7 Collector Circuit Efficiency 107
2.35 AM Generation Using Non-linear Circuit 108
2.35.1 Balance Modulator 109
2.35.2 Advantage of a Balanced Modulator Over a Simple Non-linear Circuit 111
2.36 Modulation of AM Waves 111
2.36.1 Square Law Detectors 111
2.36.2 Distortion 112
2.36.3 Linear Diode Detector 112
2.36.4 Operation 113
2.36.5 Choice of Time Constant R-C 114
2.37 Rate of Decay of Envelope 114
2.37.1 Rate of Discharge of Capacitor 115
2.38 AM Transmitters and Receivers 116
2.38.1 Low-level Modulation 117
2.38.2 Higher power level modulation 117
2.39 Neutralization 120
2.39.1 AM Receiver 120
2.39.2 Interception 120
2.39.3 Selection 120
2.39.4 R.F. Amplification 120
2.39.5 Detection 120
2.39.6 Audio Amplification 121
2.39.7 Reproduction 121

2.39.8 TRF Receiver 121
2.39.9 Features of a Receiver 121
2.39.10 Selectivity 121
2.39.11 Sensitivity 122
2.39.12 Fidelity 122
2.39.13 Superheterodyne Receiver 122
2.39.14 Advantages of a Superheterodyne Receiver 122
2.39.15 Sensitivity 123
2.39.16 Selectivity 123
2.39.17 Detector Design 123
2.39.18 Advantages of TRF Receiver 123
2.40 Constituent Stages of a Superheterodyne Receiver 124
2.40.1 RF Amplifier 124
2.40.2 Image Signal 124
2.40.3 Mixer Stage 124
2.40.4 Tracking and Alignment 126
2.40.5 Local Oscillator 126
2.41 IF Amplifier 128
2.41.1 Choice of Intermediate Frequency 128
2.41.2 Second Detector 129
2.41.3 Audio Amplifier 129
2.41.4 Noise in Amplitude Modulated Systems 130
2.41.5 Calculation of Figure of Merit for Various Communication Systems 130
2.41.6 Channel Noise is Additive 130
2.41.7 Channel Noise is White and Gaussian 130
2.41.8 Bandpass Noise at the Input of Detector 131
2.41.9 The Input Noise Power N 131
2.41.10 Noise Calculation for Various AM Systems 131
2.41.11 DSB-SC 131
2.41.12 Noise Power 132
2.41.13 Signal Power 133
2.41.14 SSB-SC 135
2.41.15 Noise Power 135
2.41.16 Signal Power 137
2.41.17 Amplitude Modulation System (Envelop Detector) 138
2.41.18 Output Power 138
2.41.19 Small-Noise Case 139
2.41.20 Output Signal Power So 139
2.41.21 Output Noise Power 140

2.42 Synchronous Detection of AM System 141
2.42.1 Large Noise Case 141
2.42.2 Threshold Effect in An Envelope Detector 143
2.42.3 Comparison of Various AM System 143
2.42.4 Receiver End 143
2.42.5 Transmitter End 143
2.42.6 Generation of Modulated Signals 143
2.42.7 Bandwidth 144
2.42.8 Selective Fading 144
2.42.9 Transmission Efficiency 144
2.42.10 Noise Performance 144
2.42.11 Non-Linear Distortion 144
2.43 Frequency Division Multiplexing (FDM) 145

3. ANGLE MODULATION 161—244
3.1 Introduction 161
3.2 Concept of Angle Modulation 161
3.2.1 Mathematical Representation 162
3.2.2 Types of Angle Modulation 163
3.2.3 Frequency Deviation 165
3.3 Relationship between Phase Modulation (PM) and Frequency Modulation (FM) 166
3.3.1 Single Tone Frequency Modulation 167
3.4 Mathematical Expression for Single-tone Frequency Modulation 167
3.5 Phasor Representation of Angle Modulation 168
3.6 Types of Frequency Modulation (FM) 169
3.6.1 Narrowband Frequency Modulation 169
3.6.2 Conclusion 170
3.6.3 Generation of Narrowband FM 170
3.6.4 Expression for Single-Tone Narrowband FM 171
3.6.5 Wideband FM 172
3.7 Effect of Variation of Modulation Index mf On the Spectrum of FM Signal 174
3.8 Transmission Bandwidth of FM Signal 175
3.8.1 Universal Curve 176
3.8.2 Carson's Rule 176
3.9 FM Bandwidth for an Arbitrary Modulating Signal X(f) 177
3.10 Narrowband FM Versus Wideband FM 177
3.11 Multiple Frequency Modulation 178
3.12 Square Wave Modulation 180
3.13 Linear and Nonlinear Modulation 182
3.13.1 Linearization of FM signal for Small Modulation Index 182

3.14 Phase Modulation: An Analytic View 183
3.14.1 Translation Bandwidth of PM 184
3.15 Comparison of Angle Modulation Waves and Amplitude Modulated Wave 184
3.16 Comparison of Frequency Modulation and Amplitude Modulation 185
3.16.1 Advantages of FM over AM 185
3.16.2 Disadvantages of FM Over AM 186
3.17 FM Generation 186
3.17.1 The Direct Method or Parameter Variation Method 186
3.17.2 Varactor Diode Method For FM Generation 187
3.18 Drawbacks of Direct Method for FM Generation 188
3.19 The Indirect or Armstrong Method of FM Generation 188
3.20 Practical Armstrong Method for FM Generation 189
3.21 FM Demodulators or Detectors 190
3.22 Types of FM Demodulators 190
3.22.1 Slope Detector 190
3.22.2 Phase Difference Detectors 191
3.22.3 Slope Detector 191
3.22.4 Balanced Slope Detector 192
3.22.5 Disadvantages of the Balanced Slope Detector 193
3.23 Phase Difference Discriminators: 193
3.23.1 Foster-Seeley Discriminator 193
3.23.2 At resonance 194
3.23.3 Off resonance 195
3.23.4 Disadvantage 196
3.24 Ratio Detector 196
3.24.1 Merits of the Ratio Detector 197
3.24.2 Demerits of the Ratio Detector 197
3.25 FM Receiver 198
3.26 RF Amplifier 198
3.27 Frequency Mixer and Local Oscillator 198
3.28 IF Amplifier 198
3.29 Limiter 198
3.30 FM Detector 199
3.31 AF Amplifier and Speaker 199
3.32 Noise In Angle Modulated Systems Frequency Modulation 199
3.32.1 Signal Power 200
3.32.2 Noise Power 200
3.32.3 Small Noise Case 201
3.32.4 Noise-quieting Effect 203
3.32.5 Figure of Merit γ 203
3.32.6 Single-tone Modulator 203

3.32.7 Noise Performance of FM as Compared to AM 204
3.32.8 Phasor Representation of FM Noise 206
3.32.9 Threshold Effect in FM 207
3.33 Generation of Spikes and Triplet Pulses 207
3.33.1 Pulse Triplet Generation 207
3.33.2 Spike Generation 208
3.33.3 Noise Energy at the output of Baseband Filter 210
3.33.4 Spike Duration 211
3.33.5 Frequency of Spike Occurrence (Unmodulated Carrier) 211
3.33.6 Output Signal to Noise Ratio (So/No) 212
3.33.7 Effect of Modulation 213
3.34 Comparison of Threshold in Analog Modulated Systems 213
3.34.1 Comparison between AM-Systems 214
3.34.2 Comparison Between AM and FM 214
3.34.3 Comparison Between FM Systems of Different Bandwidths 214
3.35 Noise in Phase Modulation 215
3.35.1 Figure of Merit (γ_{PM}) 215
3.35.2 Comparison of Noise Performance in Phase Modulation and Frequency Modulation 217
3.35.3 Comparison Between AM and FM 218
3.36 Threshold Improvement in Discriminators 219
3.36.1 Pre-emphasis and De-emphasis circuits 219
3.36.2 Improvement Factor IR 223
3.36.3 Frequency Modulation with Feedback (FMFB) 224
3.37 Voltage Controlled Oscillator (VCO) 225
3.38 Output of the Discriminator with and Without Feedback 225
3.39 Bandwidth of FMFB 226
3.40 Threshold Extension in FMFB 227
3.41 Phase Locked Loop FM Demodulator 228
3.41.1 Initial Equilibrium 228
3.41.2 FM Demodulation 229
3.41.3 Phase Error $\psi_e(t)$ 230
3.41.4 First Order PLL 231
3.42 Solution of Differential Equation 233
3.42.1 Operating Range of PLL 233
3.42.2 Bandwidth 234
3.42.3 Instability in PLL 235
3.42.4 Limitation of the First Order PLL 235
3.42.5 The Second Order PLL 235
3.42.6 (S/N) ratio at the Output of a PLL 237

4.	**SIGNAL ANALYSIS**	**245—280**
4.1	Introduction	245
4.2	Classification of Signal	245
	4.2.1 Causal and Noncausal Signal	246
	4.2.2 Real and Complex Signal	246
	4.2.3 Analog and Digital Signal	248
	4.2.4 Periodic and Aperiodic Signals	248
	4.2.5 Even and Odd Signals	249
	4.2.6 Energy and Power Signal	250
	4.2.7 Deterministic and Random Signals	250
	4.2.8 Continuous-time and Discrete-time Signals	251
	4.2.9 Singularity Functions	252
4.3	Signals Representation	254
	4.3.1 Time-domain Representation	255
	4.3.2 Frequency-domain representation	255
4.4	Fourier Series and Dirichlet's Conditions	256
	4.4.1 Trigonometric Fourier Series	256
4.5	Polar Fourier Series Representation	257
4.6	Complex Fourier Exponentially Series	258
	4.6.1 Concept of Negative Frequency	260
4.7	Deviation of Trigonometric Fourier Series from Complex Exponential Series	260
	4.7.1 Fourier Series Representation of a Periodic Function Over the Entire Interval $(-\alpha < t < \alpha)$	261
4.8	Properties of Continuous-Time Fourier Series (CTFS)	262
	4.8.1 Linearity	262
	4.8.2 Time Shifting Property	262
	4.8.3 Frequency Shifting Property	262
	4.8.4 Time Scaling Property	263
	4.8.5 Time-Differentiation	263
	4.8.6 Time-Convolution	263
	4.8.7 Multiplication Property	264
	4.8.8 Symmetry Properties	264
	4.8.9 The Fourier Transform	264
4.9	Analysis of a Non-periodic Function Over Entire Interval	265
	4.9.1 Merits of Fourier Transform	266
	4.9.2 Limitations of Fourier Transform	267
	4.9.3 Existence of the Fourier Transform	267
	4.9.4 Continuous Spectrum	267
	4.9.5 Fourier Transform Involving Impulse Function	268
	4.9.6 Singularity Function	268
	4.9.7 Sampling Square Function	272
	4.9.8 Shifting Property of the Delta Function	272

4.10 Properties of Fourier Transform 273
4.10.1 Lineary Property 273
4.10.2 Time-Scalling Property 273
4.10.3 Duality or Symmetry Property 275
4.10.4 Time Shifting Product 276
4.10.5 Frequency Shifting Property 276
4.10.6 Area Under the Curve 277
4.10.7 Time Differentiation and Integration 278
4.10.8 Conjugate Functions 279
4.10.9 Frequency Differentiation, Dual of Time Differentiation 279

5. NOISE 281—322

5.1 Introduction 281
5.2 Sources of Noise 281
5.2.1 External Noise 281
5.2.2 Atmospheric Noise 282
5.2.3 Extraterrestrial Noise 282
5.2.4 Industrial Noise 283
5.2.5 Internal Noise 283
5.3 Voltage and Current Models of a Noisy Resistor 286
5.4 Addition of Noise Due to Several Sources in Series 288
5.5 Addition of Noise due to Several Sources in Parallel 289
5.6 Addition of Noise Due to Several Amplifiers in Series 289
5.7 Spectral Densities 290
5.8 Derivation of Thermal Noise Expressions Using Power Spectrum Density 292
5.8.1 Generalised Nyquist Theorem for Thermal Noise 293
5.8.2 Calculation of Thermal Noise for a Single Noise Source 294
5.8.3 Thermal Noise in RC Circuit 295
5.9 Calculation of Thermal Noise for Multiple Noise Sources 297
5.9.1 Superposition of Power Spectrum Density 297
5.9.2 White Noise 297
5.9.3 Equivalent Noise Bandwidth 298
5.9.4 Noise Equivalent Bandwidth for Low-Pass Filter (LPF) 301
5.9.5 Noise in Reactive Circuits 302
5.9.6 Signal to Noise Ratio 304
5.9.7 Noise Figure 304
5.9.8 Noise Temperature 305
5.9.9 Calculation of Noise Figure 307
5.9.10 Noise Figure from Equivalent Noise Resistance 309
5.9.11 Experimental Determination of Noise Figure 311
5.10 Noise Figure Determination for Cascaded Stages of Amplifiers 313

6. **PULSE MODULATION** **323—348**
6.1 Introduction 323
6.2 Pulse Amplitude Modulation 324
6.2.1 Natural Sampling 324
6.2.2 Flat-top Sampling 326
6.3 Spectrum of a Flat-top Sampled Signal 327
6.4 A PAM Modulator Circuit 327
6.5 Demodulation of PAM Signals 327
6.5.1 Using an Equalizer 327
6.5.2 Using Holding Circuit 329
6.6 A PAM Demodulator Circuit 330
6.7 Time Division Multiplexed PAM System 330
6.8 Comparison Between FDM and TDM 332
6.9 Cross-talked due to HF Cutoff of the Channel 333
6.10 Cross-talk Due to LF Cutoff of the Channel 335
6.11 Transmission of PAM Signals 335
6.12 Bandwidth of PAM Signals 336
6.13 S/N Ratio of PAM System 337
6.14 Pulse Time Modulation 337
6.15 Generation of PTM Signals 338
6.15.1 Indirect Method 338
6.15.2 Direct Method 339
6.16 APWM Modulator Circuit 339
6.17 A PPM Modulator Circuit 340
6.18 Demodulation of PTM Signals 340
6.19 A PWM Demodulator Circuit 341
6.20 A PPM Demodulator Circuit 341
6.21 Bandwidth of PTM Signals 342
6.22 S/N Ratio of PTM Systems 345

7. **DIGITAL MODULATION** **349—391**
7.1 Introduction 349
7.2 Types of Digital Modulation 349
7.2.1 Coherent Digital Modulation 349
7.2.2 Non-Coherent Digital Modulation 349
7.3 Coherent Binary Amplitude Shift Keying 350
7.4 Signal Space Diagram of ask 350
7.5 Binary Phase Shift Keying (BPSK) 351
7.6 Generation of BPSK Signal 352

7.7	Preception of BPSK Signal	352
7.8	The Spectrum of BPSK Signals	355
7.9	A Geometrical Representation for BPSK Signals	357
7.10	Bandwidth for BPSK Signal	358
7.11	Error Probability of BPSK Signal	359
	7.11.1 Employing Coherent	359
7.12	Drawback of BPSK	360
7.13	Binary Frequency Shift Keying (BFSK)	360
7.14	Generation of BFSK	361
7.15	The Spectrum of BFSK Signal	362
7.16	Bandwidth of BFSK Signal	364
7.17	Detection of BFSK	364
7.18	Geometrical Representation Orthogonal BFSK	365
7.19	Distance between Signal Points	366
7.20	Geometrical Representation of Non-orthogonal BFSK Signals	366
7.21	Probability of Error, P_e of BFSK Signals	367
7.22	Advantages and Disadvantages of BFSK Signals	367
7.23	Non-coherent Binary Modulation Detection	367
7.24	Amplititude Shift Keying (ASK)	367
7.25	Non-coherent detection of FSK	368
7.26	Generation of DPSK	368
7.27	Detection of DPSK	369
7.28	Evaluation of Bandwidth of DPSK Signal	370
7.29	Error Probability of DPSK	370
7.30	Advantages and Disadvantages of DPSK	371
	7.30.1 Advantages	371
	7.30.2 Disadvntages	371
7.31	Quadrature phase shift Keying—QPSK	372
7.32	Generation of QPSK	372
7.33	Reception of QPSK or Detection of QPSK	374
7.34	Concept of carrier Synchronization in QPSK	376
7.35	Signal Space representation in QPSK Signals	377
7.36	Distance between Signal Points	379
7.37	Spectrum of QPSK Signal	379
7.38	Bandwidth of QPSK Signal	380
7.39	Probability of Error of QPSK system	381
7.40	Advantages of QPSK	383
7.41	Minimum Shift Keying, MSK	383

7.42	Signal Space representation of MSK and Distance between the Signal Points or Geometrical representation of MSK	387
7.43	Distance between Signal Points	387
7.44	Power Spectral Density, PSD, and Band width of MSK	388
7.45	Bandwidth Calculation of MSK	389
7.46	Generation of MSK	389
7.47	Reception of MSK or Detection of MSK	390
7.48	Advantages and Disadvantages of MSK as compared to WPSK	390
	7.48.1 Advantages	390
	7.48.2 Disadvantages	391
	INDEX	**392—396**

Communication System 1

1.1 INTRODUCTION

Imagine a world without television, radio, telephone, mobile, fax machines, internet, e-mail, automatic teller machines, credit or debit card, global positioning systems, radar, satellites, aicraft or ship guidance system etc. Most of us will be dreaded to think about it and consider evolution is in reverse gear and we are back to stone-age again. How do we then communicate with a man or machines? Only a few hundred years back we were dependent only on physical mode of communication, i.e. someone had to carry the message physically from one place to another. The speed of transmission was severely limited, cost huge and reliability of service poor—constrained by geopolitical situation on ground. How come the situation is different today? The development in electronic communication system by which signal moves with the speed of light is changing the face of human civilization. With a click of a mouse you find details of a place which is continents part; click your remote to switch to a favourite TV channel-see a live soccer match or how the people are rescued after tsunami; press few buttons and you are connected within seconds to your near or dear one visiting other side of the globe. Today all these are everyday reality and we perhaps cannot live without it.

The important questions is—how communication is made so simple, so close to our life? What is the technology behind it? Yes, we'll try to unfurl the mystery slowly steadily in this book. Our emphasis will be more on fundamental principles but adequate references on emerging trends will also be made. Before we get into detail, let us have a quick overview of important issues related to electronic's communication systems.

Communication is simply the process of conveying message at a distance or communication is the basic process of exchanging information.

The electronic equipments which are used for communication purpose, are called communication equipments. Different communication equipments when assembled together form a communication system.

Typical examples of communication system are line telephony and line telegraphy, radio telephony and line telegraphy, radio broadcasting, point-to-point communication and mobile communication, computer communication, radar communication, television broadcasting, radio telemetry, radio aids to navigation, radio aids to aircraft landing etc.

The earliest communication system namely line-telegraphy originated in eighteen fourties (1840s). In addition to this, line telephony come a few decades later whereas radio-communication could become possible in the beginning of twentieth century on invention of triode valve. Radio communication was further greatly improved during World War II. It became more widely used through the invention of transistor, integrated circuits (ICs) and other semi-conductor devices in the subsequent years. Also, in recent years, communication has become more widespread with the use of satellites and fibre optics. Today, there has been an increasing emphasis on the on the use of computers in communication.

1.2 PROCESS OF COMMUNICATION SYSTEMS WITH ELEMENTS

In the most fundamental sense, communication involves the transmission of information from one point to another through a succession of process as listed ahead:

(*i*) the generation of a thought pattern or image in the mind of an originator.

(*ii*) the description of that image, with a certain measure of precision, by a set of aural or visual symbols.

(*iii*) the encoding of these symbols in a form that is suitable for transmission over a physical medium of interest.

(*iv*) the transmission of the encoded symbol to the described destination.

(*v*) the decoding and reproduction of the original symbols.

(*vi*) the recreation of the original thought pattern image, with a definable degradation in quality, in the mind of a recipient.

Hence, as discussed above, the purpose of a communication system is to transmit an information-bearing signal, from a source, located at one point, to a user or destination, located at another point some distance away. As per the Fig. 1.1 the block diagram of a general communication system in which the different functional elements are represented by blocks. The essential components of a communication system are information source, input transducer, transmitter, communication channel, receiver and destination.

Now, we shall briefly discuss the functioning of these blocks.

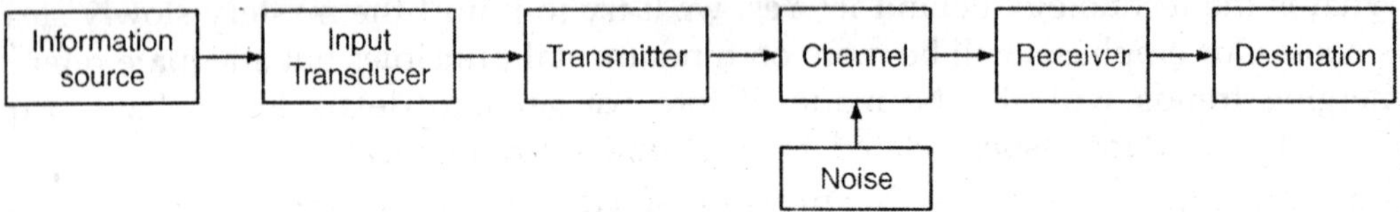

Fig. 1.1 *Block diagram of a communication system.*

1.2.1 Information Source

We know that a communication system serves to communicate a message or information. This message or information originates in the information source. In general, there can be various messages in the form of words, groups of words, code, symbols, sound signal etc. However, out of these messages, only the desired message is selected and conveyed or communicated.

In short, we can say that the function of information source is to produce required message which has to be transmitted.

1.2.2 Input Transducer

A transducer is a device which converts one form of energy into another form. The message from the information source may or may not be electrical in nature. In a case when the message produced by the information source is not electrical in nature, an input transducer is used to convert it into a time-varying electrical signal. For example, in case of radio-broadcasting, a microphone converts the information or message which is in the form of sound waves into corresponding electrical signal.

1.2.3 Transmitter

The function of the transmitter is to process the electrical signal from different aspects. For example in radio broadcasting the electrical signal obtained from sound signal, is processed to restrict it's range of audio frequencies (upto 5 kHz in amplitude modulation radio broadcast), and is often amplified. In wire telephony, no real processing is needed. However, in long-distance radio communication or broadcast, signal amplification is necessary before modulation. Modulation is the main function of the transmitter. In modulation, the message signal is superimposed upon the high-frequency carrier signal. In short, we can say that inside the transmitter, signal processings such as restriction of range of audio frequencies, amplification and modulation are achieved. All these processings of the message signal are done just to ease the transmission of the signal through the channel.

1.2.4 The channel and the Noise

The term channel means the medium through which the message travels from the transmitter to the receiver. In other words, we can say that the function of words, we can say that the function of the channel is to provide a physical connection between the transmitter and the receiver.

There are two types of channels, namely point-to-point channels and broadcast channels.

Examples of point-to-point channels are wirelines, microwave links and optical fibres. Wirelines operate by guided electromagnetic waves and they are used for local telephone transmission. In case of microwave links, the transmitted signal is radiated as an electromagnetic wave in free space. Microwave links are used in long-distance telephone transmission. An optical fiber is a low-less, well-controlled, guided optical medium. Optical fibres are used in optical communications. Although these three channels operate differently, they all provide a physical medium for the transmission of signals from one point to another point. Therefore, for these channels, the term point-to-point is used.

On the other hand, the broadcast channels provide a capability where several receiving stations can be reached simultaneously from a single transmitter. An example of a broadcast channel is a satellite in geostationary orbit, which covers about one-third of the earth's surface.

During the process of transmission and reception the signal gets distorted due to noise introduced in the system.

Noise is an unwanted signal which tend to interfere with the required signal. Noise signal is always random in character. Noise may interfere with signal at any point in a communication system. However, the noise has it's greatest effect on the signal in the channel.

1.2.5 Receiver

The main function of the receiver is to reproduce the message signal in electrical form from the distorted received signal. This reproduction of the original signal is accomplished by a process known as the demolition or detection. Demodulation is the reverse process of modulation carried out in transmitter.

1.2.6 Destination

Destination is the final stage which is used to convert an electrical message signal into it's original form. For example : in radio broadcasting, the destination is a loudspeaker which works as a transducer. i.e., it converts the electrical signal in the form of original sound signal.

1.3 INFORMATION SOURCES

As a matter of fact, the telecommunication environment is dominated by the following four important sources of information :

(*i*) speech

(*ii*) television

(*iii*) fascimile, and

(*iv*) personal computers.

A source of information may be characterised in terms of the signal which carries the information. Further, a signal is defined as a single values function of time that plays the role of the dependent variable. At every instant of time, the function has a unique value.

1.3.1 Speech

We know that speech in the primary method for human communication. Basically, the speech communication process involves the transfer of information from a speaker to a listener, which takes place in three successive stages as under :

1.3.1.1 Production

An intended message in the speaker's mind is represented by a speech signal that consists of a sounds (i.e., pressure waves) generated inside the speaker's mouth and whose arrangement is governed by the rules of language.

1.3.1.2 Propagation

The sound waves propagate through the air, reaching the listener's ears.

1.3.1.3 Perception

The incoming sounds are deciphered by the listener into a received message, and thus completing the chain of events that results in the transfer of information from the speaker to the listener.

In fact, the speech-production process may be viewed as a form of filtering, in which a sound source excites a vocal tract filter. The vocal tract consists of a tube of non-uniform cross-sectional are, beginning at the glottis (i.e., the opening between the vocal cords) and ending at the lips. Now, as the sound propagates along the vocal tract, the spectrum (i.e. the frequency content) is shaped by the frequency selectivity of the vocal tract. This effect is quite similar to the resonance phenomenon observed in organ pipes. At this stages, it may be noted that the hearing mechanism is highly sensitive to frequency. In addition to this, the type of communication system being considered has an important bearing on the bond of frequencies considered to be essential for the communication process. As an example, a frequency band from 300 to 3100 Hz is considered adequate for telephonic communication on a commercial basis.

1.3.2 Television

The second source of information, television (TV), refers to the transmission of pictures in motion by means of electrical signals. To accomplish this transmission, each complete picture has to be sequentially scanned. The scanning process is carried out in a TV camera. In a black-and-white TV, the camera contains optics designed to focus an image on a photocathode consisting of a large number of photosensitive elements.

The charge pattern so generated on the photo sensitive surface is scanned by an electron beam, and thus producing an output current which varies temporarily in accordance with the way in which the brightness of the original picture varies spatially from one point to another. This resulting output current is called a video signal.

The type of scanning used in television is a form of spatial sampling called raster scanning. The purpose of raster scanning is basically to convert a two-dimensional image intensity into a one-dimensional waveform.

Further, in television, a picture is divided into 525 lines, which constitute a frame. Now, each frame is decomposed into two interlaced fields, each one of which consists of 262.5 lines. for simplicity of presentation, we can refer to the two fields as I and II. In Fig. 1.2 scanning process has been illustrated.

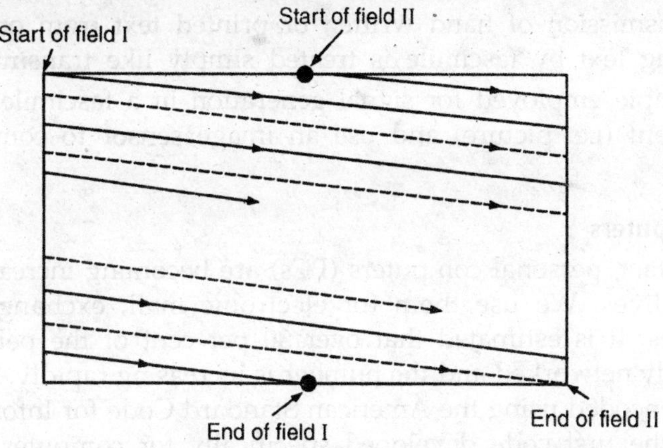

Fig. 1.2 *Illustration of interlaced answer scan.*

The lines of field I are depicted as solid lines, and the lines of field II are depicted as dashed lines. The figure also includes the start and end of each field. Field I is scanned first. The scanning spot of the TV camera moves with constant velocity across each line of the field from left to right. The image intensity at the centre of the spot is measured. Now, when the end of a particular line is reached, the scanning spot quickly flies back (in a horizontal direction) to the start of the next line down in the field. This flyback is known as the horizontal retrace. The scanning process is continued until the whole field has been covered. When this condition is reached, the scanning spot moves quickly (in a vertical direction) from the end of field I to the start of field II.

This second flyback is known as the vertical retrace. Field II is treated in the same fashion as field I. The time taken for each field to be scanned is $\frac{1}{60}$ second.

Correspondingly, the time taken for a frame or a complete picture to be scanned is $\frac{1}{30}$ second. With 525 lines in a frame, the line-scanning frequency equals 15.75 kHz. Hence, by flashing 30 still pictures per second on he display tube of the TV receiver, the human eye perceives them to be moving pictures. The effect is due to a phenomenon known as the persistence of vision.

Also, the reproduction quality of a TV picture is limited by the following two basic factors.

(*a*) the number of lines available in a raster scan, which limits resolution of the picture in the vertical direction.

(*b*) the channel bandwidth available for transmitting the video signal which limits resolution of the picture in the horizontal direction.

1.3.3 Fascimile

The purpose of the third source of information, fascimile (far) machine, is to transmit still pictures over a communication channel. Such a machine provides a highly popular facility for the transmission of hand written or printed text from one point to another. Further, transmitting text by fascimile is treated simply like transmitting a picture.

The basic principle employed for signal generation in a fascimile machine is to scan an original document (i.e. picture) and use an image sensor to convert the light to an electrical signal.

1.3.4 Personal Computers

As a matter of fact, personal computers (PCs) are becoming increasingly an important part of one daily lives. We use them for electronic mail, exchange of software, and sharing of resources. It is estimated that over 30 per cent of the personal computers in use today are already networked and the number is increasing rapidly. The text transmitted by a PC is usually encoded using the American Standard Code for Information Interchange (ASCII), which is the first code developed specifically for computer communications.

Each character in ASCII is represented by seven data bits constituting a unique binary pattern made up of 0s and 1s. Hence, a total of 2^7 = 128 different characters can be represented in ASCII. The characters are various lowercase and uppercase letters, numbers, special control symbols, and punctuation symbols commonly used such as @, $ and %. Some of the special "control" symbols such as BS (back space) and CR (carriage return), are used to control the printing of characters on a page. Other symbols, such as ENQ (enquiry) and ETB (end of transmission block), are used for communication purposes.

The seven data bits are ordered starting with the most significant bit b_7 down to the least significant bit b_1 which is clear from Fig. 1.3.

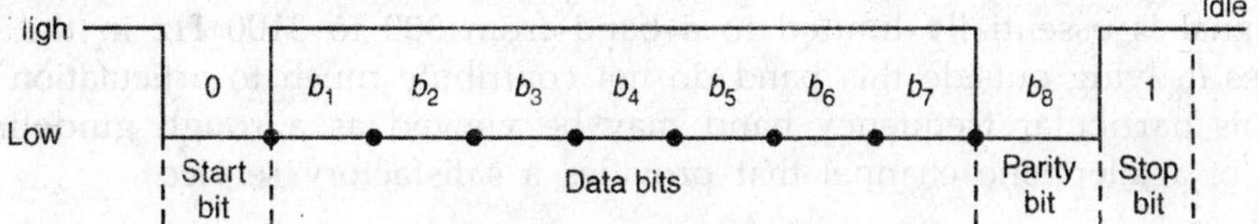

Fig. 1.3 *An ASCII format.*

At the end of the data bits, an extra bit b_8 is appended for the purpose of error detection. The error-detection bit is called a parity bit. A sequence of eight bits is referred to as a byte or an octet. The parity bit is set in such a way that the total number of 1s in each byte is odd for parity, or even for even parity.

Personal computers are often connected via their RS (recommended standard) – 232 ports. When ASCII data are transmitted through these ports, a start bit, set to 0, and one or more stop bits, set to 1, as depicted in Fig. 1.3, are added to provide character framing. When the transmission is idle, a long series of 1s is sent so as to keep the circuit connective alive.

1.4 COMMUNICATION CHANNELS

We have seen several sources of information which are responsible for the generation of message signals. These message signals need to be transmitted to the destination point. Now, let us discuss various communication channel (physical media) that carryout the transmission.

Basically, we shall consider following channels :

(*i*) telephone channels

(*ii*) optical fibre

(*iii*) mobile radio channels, and

(*iv*) satellite channels.

These four different channels have been because they play important roles in modern telecommunications environment.

1.4.1 Telephone Channels

A telephone network makes use of a switching mechanism. This switching mechanism is known as circuit switching and it is used to established an end-to-end communication link on a temporary basis. In fact, the primary purpose of the network is to ensure that the telephone transmission between a speaker at one end of the link and a listener at the other end, is an acceptable replacement for face-to-face conversation.

In this form of communication, the message source is the sound produced by the speaker's voice and the destination is the listener's ear.

However, it may be noted that the telephone channel supports only the transmission of electrical signals.

Because of this reason, appropriate transducer are used at the transmitting and receiving ends of the system.

Further, the telephone channel is essentially a linear, bandwidth-limited channel. A speech signal is essentially limited to a band from 300 to 3100 Hz in the sense that frequencies in lying outside this band do not contribute much to articulation efficiency. Hence, this particular frequency band may be viewed as a rough guideline for the passband of a telephone channel that provides a satisfactory service.

1.4.2 Optical Fibre

The second communication channel, an optical fibre, is a dielectric waveguide which transports light signals from one place to another just as a metallic wire pair or a co-axial cable, transports electrical signals.

An optical fibre consists of a central core within which the propagating electromagnetic field is confined and which is surrounded by a cladding layer, which is itself surrounded by a thin protective jacket. Basically, the core and cladding are both made of pure silica glass, whereas, the jacket is made of plastic. Optical fibers have unique characteristics that make them highly attractive as a transmission medium. Optical fibers offer the following unique advantage:

1.4.2.1 Enormous Potential Bandwidth

For optical fibers, enormous potential bandwidth results from the use of optical carrier frequencies around 2×10^{14} Hz. With such a high carrier frequency and a bandwidth roughly equal to 10 per cent of the carrier frequency, the theoretical bandwidth of a light wave system is around 2×10^{13} Hz, which is very large indeed.

1.4.2.2 Low Transmission Losses

Optical fibers offer low transmission losses as low as 0.2 dB/km.

1.4.2.3 Immunity to Electromagnetic Interference

In fact, immunity to electromagnetic interference is an inherent characteristic of an optical fiber which is viewed as a dielectric waveguide.

1.4.2.4 Small Size and Weight

For optical fibers, small size and weight are characterised by a diameter which is no greater than that of a human air.

1.4.2.5 Ruggedness and Flexibility

Optical fibers offer unique advantage of ruggedness and flexibility because of very high tensile strengths and the possibility of being bent or twisted without damage.

1.4.3 Mobile Radio Channels

The third communication channel, mobile radio channel, extends the capability of the public telecommunications network by introducing mobility into the network by virtue of it's ability to broadcast.

The term 'mobile radio' is usually meant to encompass terrestrial situations where a radio transmitter or receiver is capable of being moved, regardless of whether it actually moves or not. Basically, there is no 'line-of-sight' path for communication, rather, radio propagation takes place mainly by way of scattering from the surfaces of the surrounding buildings and by diffraction over and/or around them.

Therefore, the energy reaches the receiving antenna via more than one path. Therefore, in a mobile radio environment, we face a problem of multipath phenomenon in the sense that the various incoming radio waves reach their destination from different directions and with different time delays. Because of this, the received signal strength varies with location in a very complicated manner and therefore, a mobile radio channel may be viewed as a linear time-varying channel that is statistical in nature.

1.4.4 Satellite Channel

A satellite channel provides broad-area coverage in a continental as well as intercontinental sense. Moreover, access to remove areas not covered by conventional cable or fibre communications is also a distinct feature of satellites.

In almost all satellite communication system, the satellites are placed in geostationary orbit. For the orbit to be geostationary it has to satisfy two requirements. First, the orbit is geosynchronous, which requires the satellite to be at an altitude of 22,300 miles, a geosynchronous satellite orbits the earth in exactly 24 hours. Second, the satellite is placed in orbit directly above the equator on an eastward heading.

Viewed from earth, a satellite in geostationary orbit appears to be stationary in the sky.

Communications satellites in geostationary orbit offers the following unique system capabilities:

(*a*) broad-area coverage

(*b*) reliable transmission links

(*c*) wide transmission bandwidth

In terms of services, satellites can provide fixed point-to-point links extending over long distances and into remote areas, communication to mobile platforms (e.g., aircraft, ships), or broadcast capabilities.

In a typical satellite communication system, a message signal is transmitted from an earth station via an uplink to a satellite, amplified in a transponder on board the satellite. It is then retransmitted from the satellite via a downlink to another earth stations. In effect, the satellite acts as a powerful repeater in the sky. The most popular frequency band for satellite communication is 6 GHz for the uplink and 4 GHz for the downlink.

In fact, the use of this frequency band offers the following advantages :

(*a*) relatively inexpensive microwave equipment.

(*b*) low attenuation due to rain fall (rainfall is a primary atmospheric cause of signal loss).

(*c*) insignificant sky background noise.

In the $\frac{6}{4}$ GHz band, a typical satellite is assigned a 500 MHz bandwidth that is divided among 12 transponders on board the satellite. Each transponder, using approximately 36 MHz of the satellite bandwidth, corresponds to a specific radio channel.

A single transponder can carry at least one colour television signal, 1200 voice circuits, or digital data at a rate of 50 Mb/s.

As a matter of fact, communication channels may be classified in different ways as under:

(*a*) A channel may be linear or non-linear. A telephone channel is linear, whereas a satellite channel is usually (but not always) nonlinear.

(*b*) A channel may be time variant/varying. An optical fibre is time invariant whereas a mobile radio channel is varying.

(*c*) A channel may be bandwidth limited or power limited. A telephone channel is bandwidth limited whereas an optical fiber link and a satellite channel are both power limited.

1.5 BASEBAND AND PASSBAND SIGNALS

As discussed earlier, the message signal generated from the information source is known as baseband signal. This baseband signal may be a combination of two or more message signals. If the baseband signal is transmitted directly, then it is known as baseband transmission. This means that the baseband transmission does not used modulators and demodulators. The baseband signal may be both analog as well as digital.

The analog baseband signal varies continuously with time and has continuous amplitude. The digital baseband signal is discrete in both time and amplitude.

The baseband transmission is preferred at low frequencies and for short distances. However, Inter Symbol Interference (ISI) is the major problem associated with the baseband transmission.

Now, if the modulated signal is transmitted over the channel, it is known as bandpass or simply passband transmission. Whenever a modulating signal is impressed upon a carrier, the modulated signal is produced. This modulated signal has fixed band of frequencies around carrier frequencies. In fact, the nature of such a signal is bandpass type. Therefore, modulated signals are known as passband or bandpass signals.

An example, if a 1 kHz modulating signal is impressed upon a 100 kHz carrier signal by using amplitude modulation technique then two sideband will be at (100 – 1) i.e., 99 kHz and upper sideband will be at (100 + 1) i.e., 101 kHz.

Thus, the amplitude modulated signal will have frequencies from 99 kHz to 101 kHz. These frequencies are bandpass type. This means that the modulated signals are also bandpass i.e., passband type.

It may be noted that the bandpass transmission is generally used at high frequencies and for long distances. All television and satellite transmission are bandpass type. The bandpass transmission may be both analog and digital. If the digital signal modulates the carrier, then it is known as digital passband transmission.

1.6 REPRESENTATION OF COMMUNICATION SIGNALS AND COMMUNICATION SYSTEMS

In order to develop knowledge of the operation of a communication system, we require some mathematical tools for the representations of signals and system. As a matter of fact, the most useful method of signal representation depends upon the particular type of signal under consideration.

Therefore, depending upon the feature of interest, we may distinguish following three different classes of signals which arise in the study of communication systems :

(*i*) Periodic and non-periodic signals

(*ii*) Deterministic and random signals

(*iii*) Energy and power signals.

1.6.1 Periodic and Non-periodic Signals

A periodic signal $g(t)$ is a function of time which satisfies the following conditions:

$$g(t) = g(t + T_0) \text{ for all values of } t$$

where,

t = time, and

T_0 = constant

Now, the smallest value of T_0 for which this condition is satisfied is called the period of $g(t)$. Accordingly, the period T_0 defines the duration of one complete cycle of $g(t)$. Also, any signal for which above equation does not held for any T_0 is called a non-periodic or simply aperiodic signal.

1.6.2 Deterministic and Random Signals

A deterministic signal is a signal about which there is no uncertainty with respect to it's value at any time, past, present or future.

Accordingly, deterministic signals may be modelled as completely specified function of time.

On the other hand, a random, signal is a signal about which there is some degree of uncertainty before it actually occurs.

1.6.3 Energy and Power Signals

We already know that in electrical systems, a signal may represent a voltage or a current. Now, let us consider a voltage V(t) dissipated in this resistor may be defined as under:

$$p(t) = \frac{|V(t)|^2}{R}$$

or, $$p(t) = R|i(t)|^2$$

It may be noted that in both the above cases, the instantaneous power $p(t)$ is proportional to the squared amplitude of the signal. Further, for a resistor R equal to 1 ohm, equations both take on same mathematical form.

Accordingly, in signal analysis, it is customary to normalize the calculations by working with a 1 ohm resistor, so that, regardless of whether a given signal $g(t)$ represents a voltage or a current, we may express the instantaneous power associated with the signal simply as under:

$$p(t) = |g(t)|^2$$

Now, based upon above convention, the total energy of a signal $g(t)$ may be defined as

$$E = \lim_{T \to \infty} \int_{-T}^{T} |g(t)|^2\, dt$$

or, $$F = \int_{-\infty}^{\infty} |g(t)|^2\, dt$$

Correspondingly, the average power of a signal $g(t)$ is defined as

$$P = \lim_{T \to \infty} \frac{1}{2T} \int_{-T}^{T} |g(t)|^2\, dt$$

Now we can say that the signal $g(t)$ is an energy signal if and only if the total energy of the signal satisfies the condition

$$0 < E < \infty$$

Also, it can be conducted that the signal $g(t)$ is a power signal if and only if the average power of the signal satisfies the condition

$$0 < P < \infty$$

Theoretically, there are several possible methods for the representation of signals. However, practically, Fourier analysis, involving the resolution of signal into sinusoidal components, is most important among all other methods. Basically, this is a consequence of the well-known fact that the response of a system to sine-wave input is another sine-wave of the same frequency (but with a different phase and amplitude) under following two conditions:

(*i*) the system is linear in that it obeys the principle of superposition.

The means that if $y_1(t)$ and $y_2(t)$ denote the responses of a system to the input $x_1(t)$ and $x_2(t)$ respectively, the system is linear if the response to the composite input $a_1x_1(t) + a_1x_1(t)$ is equal to $a_1y_1(t) + a_2y_2(t)$, where a_1 and a_2 are arbitrary constants.

(*ii*) the system is time invariant

This means that if $y(t)$ is the response of a system to the input $x(t)$, the system is time invariant if the response to the time-shifted input $x(t - t_0)$ is equal to $y(t - t_0)$ where t_0 is constant.

1.7 CLASSIFICATION OF COMMUNICATION

Regarding the mode propagation, communication may be divided in the following two forms:

(*i*) Line Communication

(*ii*) Wireless or Radio Communication

1.7.1 Line Communication

In line communication, the medium of transmission is a pair of conductors called transmission line. This is also called as line channel. This means that in line communication, the transmitted and the receiver are connected through a wire or line. However, the installation and maintenance of a transmission line is not only costly and complex but also overcrowds the open space. Apart from this, it's message transmission capability is also limited.

1.7.2 Wireless or Radio Communication

In wireless or radio communication, a message is transmitted through open space by electromagnetic waves called as radio waves. Radio waves are radiated from the transmitter in open space through a device called antenna. A receiving antenna intercepts the radio waves at the receiver. All the radio, TV and satellite broadcasting are wireless or radio communication. The advantages of wireless communication are cost effectiveness, possible long distance communication and simplicity.

1.8 THE MODULATION PROCESS

As discussed earlier, the purpose of a communication system is to deliver a message signal from an information source in recognizable form to a user destination, with the source and the user being physical separated from each other. To do this, the transmitter modifies the message signal into a form which is suitable for transmission over channel. This modification is achieved by means of a process known as modulation, which involves varying some parameter of a carrier wave in accordance with the message signal. Also the receiver recreates the original message signal from a degraded version of the transmitted signal after propagation through the channel. This recreation is achieved by using a process called as demodulation. In fact, the demodulation is the reverse of the modulation process used in the transmitter. We can define modulation process in a more systematic form as under:

Modulation may be defined as the process by which some characteristic of a signal called carrier is varied in accordance with the instantaneous value of another signal called modulating signal. Signals containing information signals. This information bearing signal is also called baseband signal. The carrier frequency is greater than the modulating frequency. The signal resulting from the process of modulation is called modulated signal.

1.8.1 Types of Modulation

Modulation is basically of two types:

1.8.1.1 Continuous Wave Modulation

When the carrier wave is continuous in nature, the modulation process is known as continuous wave (CW) modulation or analog modulation. Examples of continuous wave modulation are Amplitude Modulation and Angle Modulation. When the amplitude of the carrier is varied in accordance with the message signal, it is known as amplitude modulation (AM). Also, when the angle of the carrier is varied according to the instantaneous value of the modulating signal, it is called angle modulation. Angle modulation may be further subdivided into Frequency Modulation (FM) and Phase Modulation (PM), in which the instantaneous frequency and phase of the carrier, respectively, are varied in accordance with the message signal.

1.8.1.2 Pulse Modulation

When the carrier wave is a pulse-type waveform, the modulation process is known as pulse modulation. In pulse modulation, the carrier consists of a periodic sequence of rectangular pulses. Pulse modulation can be of an analog or digital type.

In analog pulse modulation, the amplitude, duration or position of a pulse is varied in accordance with sample values of the message signal. The analog pulse modulation may be of following three types:

(*i*) pulse-amplitude modulation (PAM)

(*ii*) pulse-duration modulation (PDM)

(*iii*) pulse-position modulation (PPM)

On the otherhand, the digital form of pulse modulation is known as pulse-code modulation (PCM).

1.8.2 Need of Modulation/Benefits of Modulation

As discussed earlier, the message signal or baseband signal is used to modulate a high frequency carrier signal inside the transmitter. After modulation, the resulting modulated signal is transmitted with the help of an antenna which is connected at the output side of the transmitter. This modulated signal then travels down the channel to reach at the input of the receiver.

Now, one question can arise why we use modulation in communication system or what will happen if we transmit message signal or audio signal without modulation. The answer is that the modulation serves several purpose in communication system is discussed below:

1.8.2.1 Practical Values of Antenna

We know that in case when free space is used as a transmitting medium, messages are transmitted and received with the help of antenna. For efficient radiation and reception the, transmitting and receiving antennas must have lengths comparable to a quarter-wavelength of the frequency used.

In modulation process, low frequency or audio signal at radio stations are transmitted/translated to higher frequency spectrum i.e., radio frequency range. These higher radio frequency with the small wavelength act as carrier for the audio frequencies (i.e., modulating signal). Thus, the height of the antenna required is much reduced and becomes practical.

1.8.2.2 Reduction of noise

Noise is the major limitation of any communication. Although noise can not be eliminated completely, but with the help of several modulation schemes, the effect of noise can be minimized.

1.8.2.3 *To remove interference*

Another reason for not radiating modulating signal itself is that the frequency range of audio signal is from 20 Hz to 20 kHz. In radio-broadcasting, there are several radio stations. In case, there is no modulation, all these stations transmit audio or sound signals in the range of 20 Hz to 20 kHz. Due to this transmission over same range, the programmes of different stations will get mixed up.

Hence, in order to keep the various signals separate, it is necessary to translate or shift them to different portions of the electromagnetic spectrum. Thus each station is allocated a band of frequency. This also overcomes the drawback of poor radiation efficiency at low frequency.

1.9 BASIC COMMUNICATION RESOURCE

As a matter of fact, in a communication system, there are following two primary communication resources to be employed:

(*i*) transmitted power, and

(*ii*) channel bandwidth

The transmitted power refers to the average power of the transmitted signal.

The channel bandwidth is defined as the band of frequencies allocated for the transmission of the message signal.

Now, the most important system design objective is to use these two resources as efficiently as possible. In most communication channels, one resources may be considered more important than the other. Because of this, we may classify communication channels as power limited or band limited.

For example, the telephone circuit is a typical band limited channel, whereas a space communication link or a satellite channel is typically power limited channel.

A fundamental question that arises in the study of modulation schemes is the following:

With channel bandwidth and signal-to-noise ratio being the two principal parameters that are available to the designer of a communication system, which particular modulation scheme provides for their most efficient use? Or in other words we can say for a specified channel bandwidth, which modulation scheme requires the smallest signal to noise ratio for a prescribed level of system performance.

The answer to this important question lies with information capacity theorem by Shannon.

1.10 INFORMATION CAPACITY THEOREM

The information capacity theorem, given by Shannon, deals with a continuous channel. According to this theorem, channel bandwidth and signal-to-noise ratio are exchangeable in the sense that we may trade off one for the other for a prescribed system performance. The choice of one modulation scheme over another for the transmission of a message signal in the communication system is often dictated by the nature of this trade off. Infact, the interplay between channel bandwidth and signal-to-noise ratio, and the limitation that they impose on communication, highlighted by the information capacity theorem.

Let B denote the channel bandwidth, and SNR denote the received signal-to-noise ratio. The information capacity theorem states that ideally these two parameters are related by

$$C = B \log_2 (1 + SNR) \text{ b/s}$$

where C is the information capacity.

The information capacity is defined as the maximum rate at which information may be transmitted without error through the channel. It is measured in bits per second (b/s). Above equation clearly shows that for a prescribed information capacity, we may reduce the required SNR by increasing the channel bandwidth, hence the advantage of using a broad bandwidth to transmit messages. Moreover, above equation provides an idealized framework for comparing the noise performance of one modulation scheme against another.

1.11 ANALOG AND DIGITAL SIGNALS

The analog signal is that type of signal which varies smoothly and continuously with time. This means that analog signals are defined for every value of time and they take on continuous values in a given time interval. Thus, we can say that analog messages are characterized by data whose values vary over a continuous range. The signal as shown in Fig. 1.4 is an analog signal. In fact, for analog signal, the name derives from the fact that such a signal is analogous to the physical signal that it represents. The vast majority of signals in the world around us are analog. For example, the temperature or the atmospheric pressure of a certain location may very over a continuous range and may assume an infinite number of possible values. Similarly, a speech waveform is an analog signal since it has amplitudes that very over a continuous range.

An alternative form of signal representation is that of a sequence of numbers, each number representing the signal magnitude at an instant of time. The resulting signal is called a digital signal. Digital messages are constructed with a finite number of symbols. For example, the printed language consists of 26 letters, 10 numbers, a space and several punctuation marks. Therefore, any text is a digital message constructed from about 50 symbols.

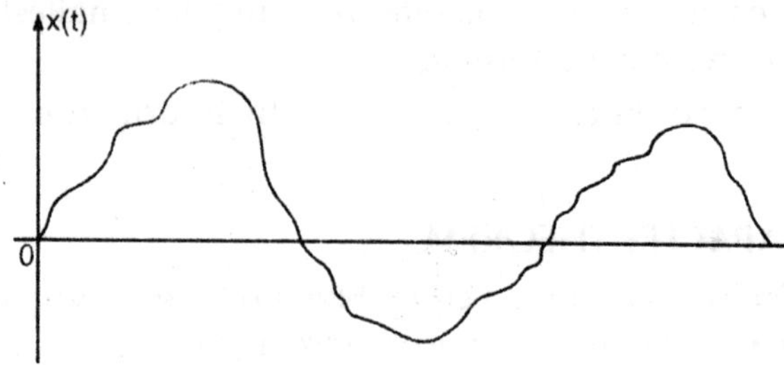

Fig. 1.4 *An analog signal.*

Now, since a digital signal is represented only by digits, therefore, we can use any number system to represent a digital signal. However, in practice, we generally use binary number system to represent a digital signal. In a binary system, each digit in the number takes on one of only two possible values, denoted 0 and 1. Correspondingly, the digital signals in binary system need have only two voltage levels which may be labeled low and high.

As Fig. 1.5 shows a digital signal. Observe that the waveform is a pulse train with 0 V representing as '0' signal or logical '0' and + 5 V representing logic '1'.

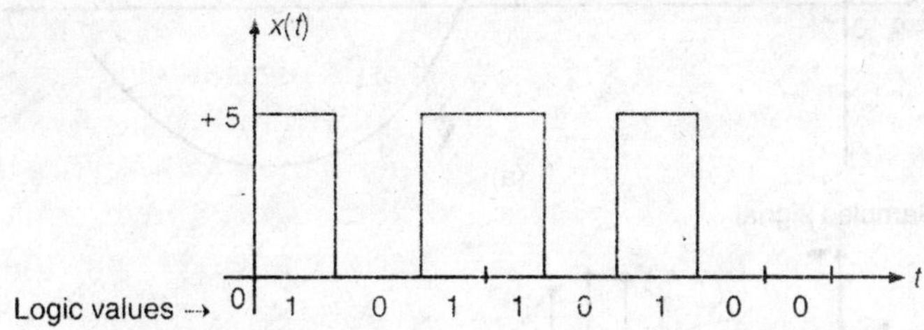

Fig. 1.5 *Variation of a binary signal with time.*

1.12 CONVERSION OF ANALOG SIGNAL TO DIGITAL SIGNALS

In communication systems, sometimes it happens that we are available with analog signal, however, we have to transmit a digital signal for a particular application. In such cases, we have to convert an analog signal into digital signal. This means that we have to convert a continuous time signal in the form of digits. To see how a signal can be converted from analog to digital form, let us consider an analog signal as shown in Fig. 1.6 (*a*). First of all, we get sample of this signal according to sampling theorem. For this purpose, we mark the time-instants to, t_1, t_2 and so on, at equal time intervals along the time axis. At each of these time-instants, the magnitude of the signal is measured and thus samples of the signal are taken. Fig. 1.6 (*b*) shows a representation of the signal of Fig. 1.6 (*a*) in terms of it's samples.

Now, we can say that the signal in Fig. 1.6 (*b*) is defined only at the sampling instants. This means that it no longer is a continuous function of time, but rather, it is a discrete-time signal. However, since the magnitude of each sample can take any value in a continuous range, the signal in Fig. 1.6 (*b*) is still an analog signal.

This difficulty is neatly resolved by a process known as quantization. In quantization, the total, amplitude range which the signal may occupy is divided into a number of standard levels.

As shown in Fig. 1.6 (*c*), amplitudes of the signal $x(t)$ lie in the range $(-m_p, m_p)$ which is partitioned into L intervals, each of magnitude $\Delta V = 2m_p/L$. Now, each sample is approximated or rounded off to the nearest quantized level as shown in figure. Since each sample is now approximated to one of the L numbers, therefore the information is digitized.

The quantized signal is an approximation of the original one. We can improve the accuracy of the quantized signal to any described degree simply by increasing the number of levels L.

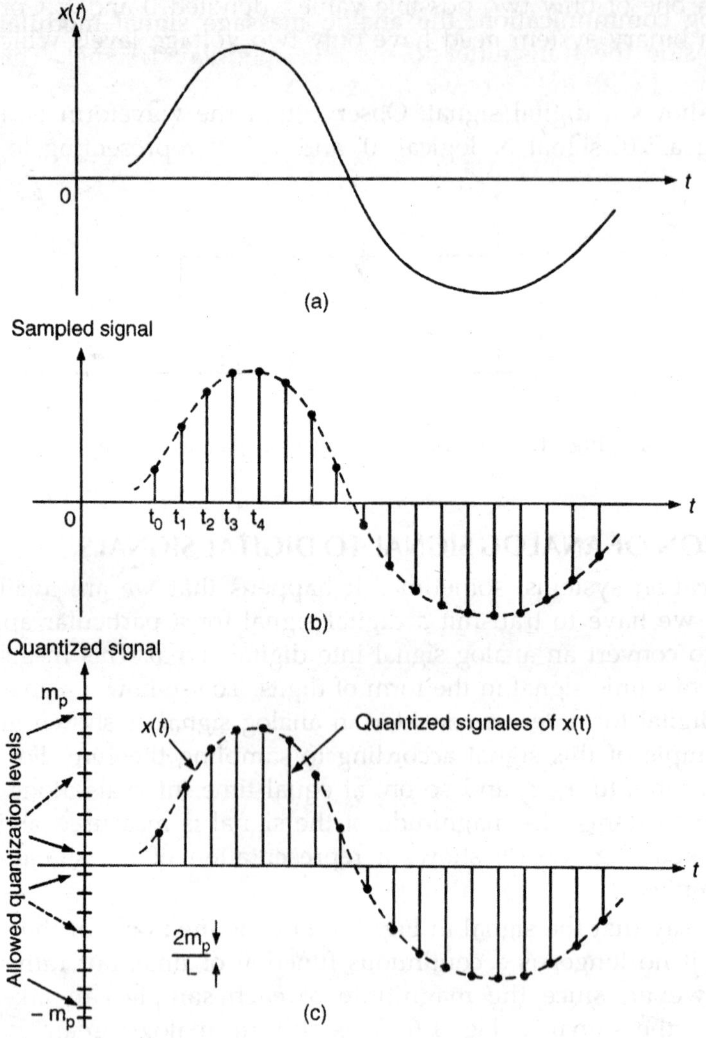

Fig. 1.6 (*a*) An analog signal, (*b*) Samples of analog signal, (*c*) Quantization

1.13 COMMUNICATIONS: ANALOG AND DIGITAL

Depending upon the message signal, communications may also be classified as:

(*i*) Analog communication

(*ii*) Digital communication

1.13.1 Analog Communication

Analog communication is that type of communication in which the message or information signal be transmitted is analog in nature. This means that in analog communication the modulating signal (i.e. baseband signal) is an analog signal. This analog message signal may be obtained from sources such as speech, video shooting etc.

In analog communication, the analog message signal modulates some high carrier frequency inside the transmitter to produce modulated signal. This modulated signal is then transmitted with the help of a transmitting antenna to travel through the transmission channel. At the receiver, this modulated signal is received and processed to recover the original message signal. Fig. 1.7 shows the block diagram of an analog communication system.

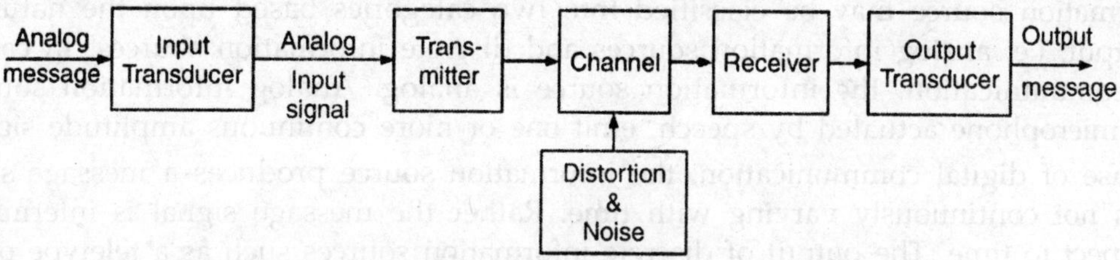

Fig. 1.7 Basic Analog Communication System

Presently all the AM, FM radio transmission and TV transmission are examples of analog communication.

1.13.2 Digital Communication

In digital communication, the message signal to be transmitted is digital in nature. This means that digital communication involves the transmission of information in digital form.

1.14 MODEL OF A DIGITAL COMMUNICATION SYSTEM

Fig. 1.8 illustrates the model of a digital communication system. The overall purpose of the system is to transmit the message or sequences of symbols coming out of a source to a destination point at as high a rate and accuracy as possible. The source and destination point are physically separated in space and a communication channel connects the source to the destination point.

The communication channel accepts electrical signals and the output of the channel is usually a smeared or distorted version of the input due to the non-ideal nature of the communication channel. In addition to this, the information-bearing signal is also corrupted by unpredictable electrical signals from both man-made and natural causes. Thus, the smearing and the noise introduce errors in the information being transmitted and limits the rate at which information can be communicated from the source of the destination.

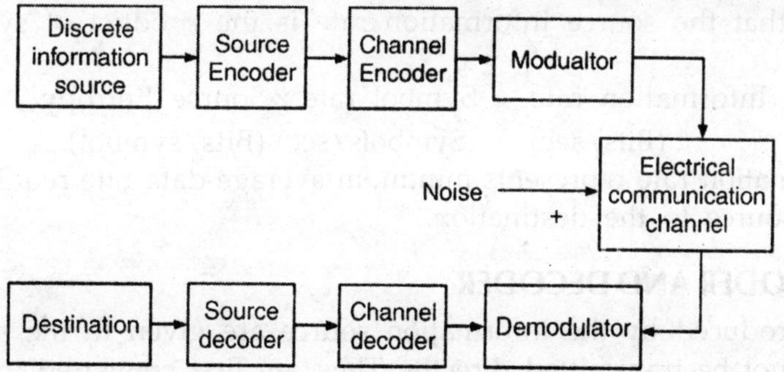

Fig. 1.8 Model of a Digital Communication System

The probability of incorrectly decoding a message symbol at the receiver is often used as a measure of performance of a digital communication system.

Now, let us have a detailed look at each of the functional blocks in a digital communication system.

1.15 DISCRETE INFORMATION SOURCE

Information source may be classified into two categories based upon the nature of their output, i.e. analog information sources and discrete information sources. In case of analog communication, the information source is analog. Analog information sources, such as microphone actuated by speech, emit one or more continuous amplitude signals.

In case of digital communication, the information source produces a message signal which is not continuously varying with time. Rather the message signal is intermittent with respect to time. The output of discrete information sources such as a teletype or the numerical output of a computer consists of a sequences of discrete symbols or letters. An analog information source may be transformed into a discrete information source through the process of sampling and quantizing. Discrete information sources are characterized by the following parameters.

1.15.1 Source Alphabet

These are the letters, digits or special characters available from the information source.

1.15.2 Symbol Rate

It is the rate at which the information source generates source alphabets. It is generally represented in symbols/sec unit.

1.15.3 Source Alphabet Probabilities

Each source alphabet from the source has independent occurrence rate in the sequence. Probability of the occurrence of each source alphabet can becomes one of the important property which is useful in digital communications.

1.15.4 Probabilistic Dependence of Symbols in a Sequence

The information carrying capacity of each source alphabet is different in a particular sequence. This parameters defines average information content of the symbols. The entropy of a source describes the average information content per symbol in long messages. Entropy may be defined in terms of bits per symbols. Bit is the abbreviation for a binary digit. This means that the source information rate is the product of symbol rate and source entropy, i.e.

$$\underset{\text{(Bits/sec)}}{\text{Information rate}} = \underset{\text{(Symbols/sec)}}{\text{Symbol rate}} \times \underset{\text{(Bits/symbol)}}{\text{Source Entropy}}$$

Thus, the information rate represents minimum average data rate required to transmit information from source to the destination.

1.16 SOURCE ENCODER AND DECODER

The symbols produced by the information source are given to the source encoder. These symbols cannot be transmitted directly. They are first converted into digital form

(1's and 0's) by the source encoder. Each binary '1' and '0' is known as a bit. The group of bits is called a codeword. The source encoder assigns codewords to the symbols. For each distinct symbol, there is an unique codeword.

The codeword can be of 4, 8, 16 or 32 bits length. As the number of bits are increased in each codeword, the symbols that may represented are also increased.

1.16.1 Block Size

Block size describes the maximum number of distinct codewords which can be represented by a source encoder. This depends on the number of bits in the codewords.

Example : The block size of 8 bits source encoder will be 2^8 bits source encoder will be 2^8 i.e. 256 codewords.

1.16.2 Codeword length

It is the number of bits used to represent each codeword.

Example : If 8 bits are assigned to each codeword, then the codeword length will be 8 bits.

1.16.3 Average Data Rate

It is the output bits per second from the source encoder. In fact, the source encoder assigns multiple number of bits to each input symbol. Hence, the data rate is generally higher than the symbol rate.

Example : If we consider that the symbols are given to the source encoder and the length of codeword is 8 bits, then the output data rate from the source encoder would be given as

$$\begin{aligned} \text{Date rate} &= \text{Symbol rate} \times \text{Codeword length} \\ &= 10 \times 8 \\ &= 80 \text{ bits/seconds} \end{aligned}$$

Also, since the information rate is the minimum number of bits per second needed to convey information from source to destination, therefore the optimum data rate is equal to the information rate. However, due to practical limitations, designing such type of source encoder is quite difficult. Thus, the average data rate is higher than the information rate and hence symbol rate also.

1.16.4 Efficiency of the Encoder

The efficiency of the encoder is the ratio of minimum source information rate to the actual output data rate of the source encoder.

At the receiver end, some sort of decoder is used to perform the reverse operation to that of source encoder. It converts the binary output of the channel decoder into a symbol sequence. Some decoders also use memory to store codewords. The decoders and the encoders can be synchronous or asynchronous.

1.17 CHANNEL ENCODER AND DECODER

After converting the message or information signal in the form of binary sequence by the source encoder, the signal is transmitted through the channel. The communication channel adds noise and interference to the signal being transmitted. Hence, errors are introduced in the binary sequence received at the receiver end.

Therefore, the errors are also introduced in the symbols generated from these binary codewords. Thus channel coding is done to avoid these types of errors. In fact, the channel encoder adds some redundant binary bits to the input sequence. Also, these redundant bits are always added with some properly defined logic.

It may be noted from Table 1.1 that each codeword at the output of channel encoder contains "even" number of 1's. Now, at the receiver end, if odd number of 1's are detected, then the receiver comes to know that there is an error in the received signal. The channel decoder at the receiver is thus able to reconstruct error free accurate bit sequence and reduce the effects of channel noise and distortion.

This means that the channel encoder and decoder serve to increase the reliability of a received signal. However the extra bits which are added by the channel encoders carry not information rather, they are used by the channel decoder to detect and correct errors if any.

Table 1.1

Output of source encoder			Bit to be added by the channel encoder for an even parity	Output of a channel encoder			
b_3	b_2	b_1	b_0	b_3	b_2	b_1	b_0
1	0	1	0	1	1	0	0
0	1	0	1	0	1	0	1
0	0	0	0	0	0	0	0
1	1	1	1	1	1	1	1
	⋮		⋮		⋮		
	⋮		⋮		⋮		

The coding and decoding operation at the encoder and decoder needs the memory and processing of binary data. However, in the modern time, due to use of microcontrollers and computers, the complexity of the encoders and decoders is much reduced.

A channel encoder must have the following important parameters:

(*i*) The coding rate that depends upon the redundant bits added by the channel encoder.

(*ii*) the coding method used.

(*iii*) Coding efficiency which is the ratio of data rate at the input to the data rate at the output of the encoder.

(*iv*) Error control capabilities.

(*v*) Feasibility of the encoder and decoder.

1.18 DIGITAL MODULATORS AND DEMODULATORS

If the modulating signal is digital then digital modulation techniques are used. The carrier signal used by digital modulators is always continuous sinusoidal wave of high frequency. In fact, the digital modulators map the input binary sequence of 1's and 0's to the analog signal waveforms.

For example, if one bit at a time is transmitted, then digital modulator signal is $s_1(t)$ to transmit binary '0' and $s_2(t)$ to transmit binary '1' from the Fig. 1.9.

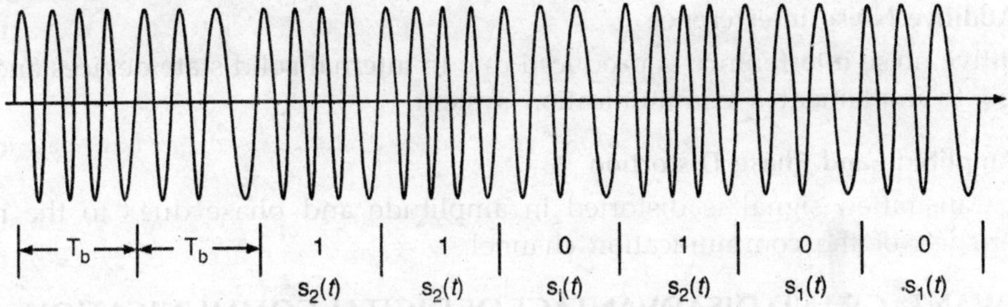

Fig. 1.9 *The output of a digital modulator.*

Here the signal $s_1(t)$ has low frequency compared to signal $s_2(t)$. Hence, here even though the modulated signal seems to be continuous, the modulation is discrete (in steps). This means that a signal carrier is converted into two waveforms $s_1(t)$ and $s_2(t)$ because of digital modulation.

Now, if the codeword consists of two bits and they are to be transmitted at a time, then there would be 2^2, i.e., 4 distinct symbols, i.e., codewords. Thus, these codewords will require four distinct waveforms for transmission purpose. Such types of modulators are known as M-ary modulators. Amplitude shift keying (ASK), phase shift keying (PSK), frequency shift keying (FSK), differential phase shift keying (DPSK) and minimum shift keying (MSK) are the examples of various digital modulators.

However, since these modulators use a continuous carrier wave, therefore they are also known as digital CW modulators.

At the receiver end, the digital demodulator converts the input modulated signal into the sequence of binary bits.

A digital modulation method must have following important parameters:

(i) Bandwidth needed to transmit the signal.

(ii) Probability of symbol or bit error.

(iii) Synchronous or asynchronous method of detection.

(iv) Complexity of implementations.

1.19 COMMUNICATION CHANNEL

The connection between transmitter and receiver is established through a communication channel. The communication can take place through wirelines, wireless or fiber optic channels. The other media such as optical disks, magnetic tapes and disks etc. may also be called as a communication channel since they can also carry data through them. However, it may be noted that each and every communication channel has some inherent problems. These are:

1.19.1 Signal Attenuation

The signal attenuation in channel occurs due to the internal resistance of the channel and fading of the signal.

1.19.2 Multipath Distortion

The multipath distortion occurs mostly in wireless communication channels.

1.19.3 Additive Noise Interference

Additive noise interference is produced due to internal solid state devices and resistors etc., used to implement a communication system.

1.19.4 Amplitude and Phase Distortion

The transmitted signal is distorted in amplitude and phase due to the non-linear characteristic's of the communication channel.

1.20 ADVANTAGE AND DISADVANTAGE OF DIGITAL COMMUNICATION

1.20.1 Advantages

1. In digital communication, the speech, video and other data may be merged and transmitted over a common channel using multiplexing.
2. Using data encryption, only permitted receivers may be allowed to detect the transmitted data. This property is of its most importance in military applications.
3. The digital communication systems are simpler and cheaper compared to analog communication systems because of the advances made in the IC technologies.
4. Since the transmitted signal is digital in nature, therefore, a large amount of noise interference may be tolerated.
5. Since in digital communication, channel coding is used, therefore the errors may be detected and corrected in the receivers.
6. Digital communication is adaptive to other advanced branches of data processing such as digital signal processing, image processing and data compression etc.

1.20.2 Disadvantages

Although digital communication offers, so many advantages as discussed above, it has some drawbacks also. However, the advantages of digital communication outweight disadvantages.

The disadvantages are as under:

1. Due to analog to digital conversion, the data rate becomes high. Therefore more transmission bandwidth is required for digital communication.
2. Digital communication needs synchronization in case of synchronous modulation.

1.21 FUNDAMENTAL LIMITATIONS OF COMMUNICATION SYSTEMS

While designing a communication system, an engineering generally faces several constraints or limitations. These are noise limitation, bandwidth limitation and equipment limitation.

1.21.1 Noise Limitations

The noise may be defined as an unwanted form of energy which tend to interface with the transmission and reception of the desired signals in a communication system. The noise cannot be eliminated completely. However, the effect of noise on desired signals can be minimised with the help of several techniques. Noise can be classified into two broad categories depending upon its source: external noise and internal noise. External noise is that type of noise whose sources are external to a communication system. The examples of external noise are atmospheric noise, galactic noise and industrial noise. On the other hand, internal noise is that type of noise whose sources are internal to a communication system. The examples of internal noise are thermal noise and shot noise. This type of noise is inevitable in every communication system and forms a basic limitation on transmission and reception of signals.

Infact, the noise limits are ability to identify the intended or desired message correctly and thus limits information transmission. Typical noise variations are measured in microvolts. If the signal variations are quite large in comparison, then the effect of noise is very small and may be ignored. Infact, in many practical communication systems, the effect of noise is generally unnoticeable. However, in long-distance communication system operating with a limited amount of signal power, the signal may be as small as noise or even smaller than the noise. Thus, in such cases, the presence of noise severally limits the capabilities of a communication system.

1.21.2 Equipment Limitation

The noise and bandwidth limitations dictate theoretically what can or cannot be achieved in terms of performance in a communication system. However, this theoretical limit may not be realised in a practical system due to equipment limitations.

1.21.3 Bandwidth Limitations

In addition to noise, the bandwidth limitation is another major constraint in a communication system. The frequency range or the band of frequency needed for a particular given transmission is known as **bandwidth**. This band of frequencies required for a particular transmission is also called channel. This band of frequencies or bandwidth for a particular transmission is always allocated by some Internation regulatory agencies. This type of regulation is essential to avoid interference among the signals having same frequency. But, for a given transmission, this allocated bandwidth may not be sufficient to convey the entire information.

Infact, the information theory states that the greater is the transmission bandwidth of a communication system, the more is the information that can be transmitted. Let us consider a simple example. Suppose one is listening to a music in an AM radio. The complete amount of information available to the human car is contained in a frequency range upto 15 kHz, i.e., musical information extends upto a frequency of 15 kHz. However, in AM radio the maximum modulating frequency is restricted upto 5 kHz and hence the maximum bandwidth of AM transmission is 10 kHz.

Therefore, an AM radio receiver cannot reproduce all the information contained in the music because this will require a bandwidth of 30 kHz. On the other hand, the bandwidth allocated to a FM transmission is about 200 kHz. Thus on FM receiver can easily reproduce the transmitted information without any distortion. This means that a FM system has a better fidelity than an AM system. Also, it may be observed in common life that one prefers to listen to a FM radio than an AM radio. Thus, we can conclude that bandwidth is a major fundamental limitation of a communication system.

1.22 BANDWIDTH REQUIREMENTS

It is reasonable to expect that the frequency range (i.e., bandwidth) required for a given transmission should depend on the bandwidth occupied by the modulation signals themselves. A high-fidelity audio signal requires a range of 50 to 15,000 Hz, but a bandwidth of 300 to 3400 Hz is adequatic for a telephone conversation. When a carrier has been similarly modulated with each, a greater bandwidth will be required for the high-fidelity (hi-fi) transmission. At this point, it is worth noting that the transmitted bandwidth need not be exactly the same as the bandwidth of the original signal, for reasons connected with the properties of the modulating systems.

Before trying to estimate the bandwidth of a modulated transmission, it is essential to know the bandwidth occupied by the modulating signal itself. If this consists of sinusoidal signals, there is no problem, and the occupied bandwidth will simply by the frequency range between the lowest and the highest sine-wave signal. However, if the modulating signals are non-sinusoidal, a much more complex situation results. Since such non-sinusoidal waves occur very frequently as modulating signals in communications.

1.23 SINCE WAVE AND FOURIER SERIES REVIEW

It is a very important in communication to have a basic understanding of a sine wave signal. Described mathematically in the time domain and in the frequency domain, this signal may be represented as follows:

$$V(t) = E_m \sin(2\pi ft + \phi)$$
$$= E_m \sin(\omega t + \phi) \quad ...(1.1)$$

where,

$V(t)$ = voltage as a function of time
E_m = peak voltage
sin = trigonometric sine function
f = frequency in hertz
ω = radian frequency ($\omega = 2\pi f$)
t = time
ϕ = phase

If the voltage waveform described by this expression were applied to the vertical input of an oscilloscope, a sine wave would be displayed on the CRT screen.]

The symbol f in equation 1.1 represents the frequency of the sine wave signal. Next we will review the Fourier series, which is used to express periodic time functions in the frequency domain, and the Fourier transform, which is used to express non-periodic time domain functions in he frequency domain.

To expand upon the topic of bandwidth requirements, we will define the terms of the expressions and provide examples so that these topic's can be clearly understood.

A periodic waveform has amplitude and repeats itself during a specific time period T. Some examples of waveform are sine, square, rectangular, triangular and smooth. Fig. 1.10 is an example of a rectangular wave, where A designates amplitude, T represents time, and τ indicates pulse width. This simplified review of the Fourier series is meant to reacquaint the student with the basic's.

The form for the Fourier series is follows:

$$f(t) = \frac{a_0}{2} + \sum_{n=1}^{\infty}\left[a_n \cos\left(\frac{2\pi nt}{T}\right) + b_n \sin\left(\frac{2\pi nt}{T}\right)\right] \quad ...(1.2)$$

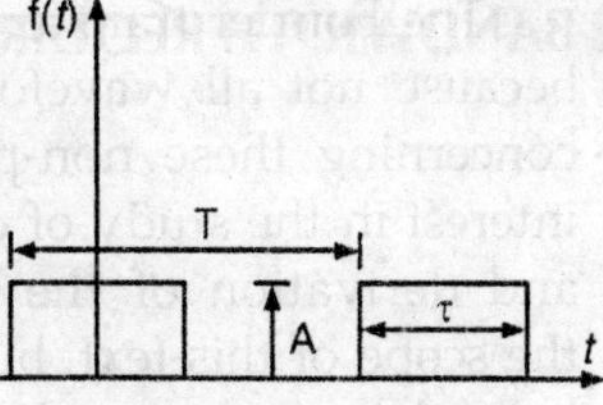

Fig. 1.10 *Rectangular wave.*

Each term is a simple mathematical symbol and shall be explained as follows:

$\sum_{n=1}^{\infty}$ = the sum of n term, in this case from 1 to infinity, where n takes on values of 1, 2, 3, 4, ...

a_0, a_n, b_n = the Fourier coefficients, determined by the type of waveform.

T = the period of the wave

$f(t)$ = an indication that the Fourier service as a function of time.

The expression will become clearer when the first four terms are illustrated:

$$f(t) = \left[\frac{a_0}{2}\right] + \left[a_1 \cos\left(\frac{2\pi t}{T}\right) + b_1 \sin\left(\frac{2\pi t}{T}\right)\right] + \left[a_2 \cos\left(\frac{4\pi t}{T}\right) + b_2 \sin\left(\frac{4\pi t}{T}\right)\right]$$

$$+\left[a_3 \cos\left(\frac{6\pi t}{T}\right) + b_3 \sin\left(\frac{6\pi t}{T}\right)\right] + ... \quad ...(1.3)$$

If we substitute ω_0 for $\frac{2\pi}{T}$ ($\omega_0 = 2\pi f_0 = 2\pi/T$) in equation 1.3, we can rewrite the Fourier series in random terms:

$$f(t) = \left[\frac{a_0}{2}\right] + [a_1 \cos \omega_0 t + b_1 \sin \omega_0 t] + [a_0 \cos 2\omega_0 t + b_2 \sin 2\omega_0 t]$$

$$+[a_3 \cos 3\omega_0 t + b_3 \sin 2\omega_0 t] + ... \quad ...(1.4)$$

Equation 1.4 supports the statement. The makeup of a square or rectangular wave is the sum of (harmonic) the sine wave components at various amplitudes.

The Fourier coefficients for the rectangular waveform in Fig. 1.10 as well as Fig. 1.11.

$$a_0 = \frac{2A\tau}{T}$$

$$a_n = \frac{2A\tau \sin(\pi n\tau/T)}{T(\pi n\tau/T)}$$

b_n = 0 because t = 0 (waveform is symmetrical)

The first four terms of this series for the rectangular waveform are:

$$f(t) = \left[\frac{A\tau}{T}\right] + \left[\frac{2A\tau}{T}\frac{\sin(2\pi\tau/T)}{(2\pi\tau/T)} \cos\left(\frac{2\pi t}{T}\right)\right] + \left[\frac{2A\tau}{T}\frac{\sin(2\pi\tau/T)}{(2\pi\tau/T)} \cos\left(\frac{4\pi t}{T}\right)\right]$$

$$+\left[\frac{2A\tau}{T}\frac{\sin(3\pi\tau/T)}{(3\pi\tau/T)} \cos\left(\frac{6\pi t}{T}\right)\right] + ... \quad ...(1.5)$$

The Fourier transform review material is included here because not all waveforms are periodic and information concerning these non-periodic waveforms are of greater interest in the study of communications. A complete study and derivation of the series and transform are beyond the scope of this text, but the student may find this review helpful in understanding these concepts.

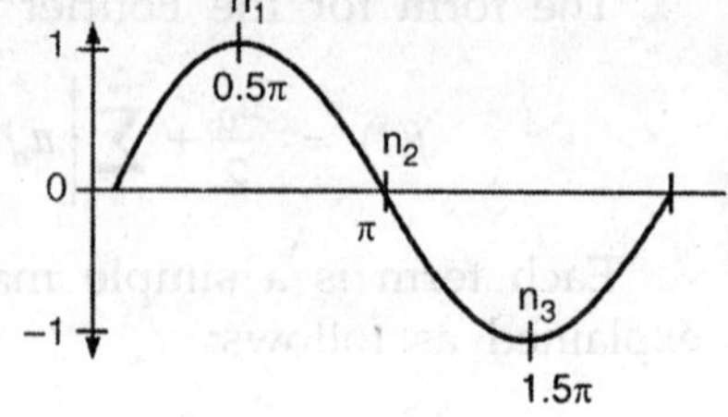

Fig. 1.11 *Sine in radius.*

Fig. 1.12 illustrates a single non-repetitive pulse. The transform for this pulse is:

$$F(\omega) = \frac{A\tau \sin(\omega\tau/2)}{(\omega\tau/2)} \quad ...(1.6)$$

$F(\omega)$ = Fourier transform

τ = pulse width

(ω) = radian frequency

A = amplitude in volts

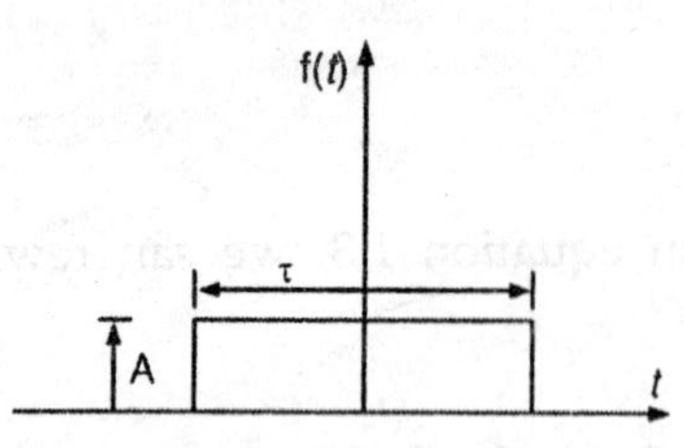

Fig. 1.12 *Single non-repetitive pulse.*

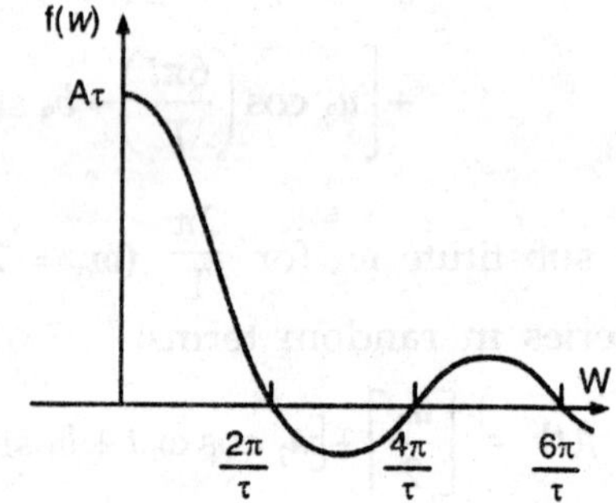

Fig. 1.13 *Fourier transform of a single pulse.*

SOME SOLVE EXAMPLES

EXAMPLE 1

Compute the first four terms in the Fourier series for 1 kHz rectangular waveform with a poise width of 500 μ sec and an amplitude of 10 V.

Solution:

$$T = \text{time } (I_t) = 1 \times 10^{-3}$$

$$\tau = \text{pulse width} = 500 \times 10^{-6}$$

$$A = 10 \text{ V}$$

$$\frac{\tau}{T} = \frac{500 \times 10^{-6}}{1 \times 10^{3}} = 0.5$$

From equation 1.5, we have,

$$f(t) = [(10)(0.5)] + \left[(2)(10)(0.5)\frac{\sin 0.5\pi}{0.5\pi}\cos(2\pi \times 10^3)\right]$$

$$+\left[(2)(10)(0.5)\frac{\sin(\pi)}{\pi}\cos(4\pi\times 10^3 t)\right]$$

$$+\left[(2)(10)(0.5)\left(\frac{\sin(1.5\pi)}{1.5}\right)\cos(6\pi\times 10^3 t)\right]$$

$$f(t) = [5]+\left[\frac{(10.n_1)}{0.5\pi}\cos 2\pi\times 10^3 t\right]+\left[\frac{(10.n_2)}{\pi}\cos 4\pi\times 10^3 t\right]$$

$$+\left[\frac{(10.n_3)}{1.5\pi}\cos 6\pi\times 10^3 t\right]$$

$$f(t) = [5]+\left[6.3666\cos(2\pi\times 10^3 t)\right]+[0]$$

$$+\left[-2.122\cos(6\pi\times 10^3 t)\right]$$

Because this waveform is a symmetrical square waveform, it has components at (An_0) DC, and at (An_1) 1 kHz and (An_3) 3 kHz points, and at odd multiple thereafter. Sine in radians, $n_1 = 1$, $n_2 = 0$, $n_3 = -1$.

EXAMPLE 2

Evaluate a single pulse with an amplitude of 8 mV and a first zero crossing at 0.5 kHz.

Solution:

$$\text{First zero crossing point} = \omega = 2\pi f = \frac{2\pi}{\tau}$$

$$\tau = \frac{1}{f} = \frac{1}{0.5\times 10^3} = 2\times 10^{-3}$$

$$V_{max}\ \text{transform} = F(w)_{max} = At$$

$$A = \frac{F(\omega)_{max}}{\tau} = \frac{8\times 10^{-3}}{2\times 10^{-3}} = 4\text{ V}$$

The single pulse has a maximum voltage of 4 V and a duration of 2 s as Fig. 1.13.

1.24 FREQUENCY SPECTRA OF NON-SINUSOIDAL WAVES

If any sinusoidal waves, such as square waves, are to be transmitted by a communication system, then it is important to realize that each such wave may be broken down into it's component sine waves. The bandwidth required will therefore be considering greater than might have been expected if only the repetition rate of such a wave had been taken into account.

It may be shown that any non-sinusoidal, single-valued repetitive waveform consists of sine waves or cosine waves. The frequency of the lowest-frequency, or fundamental, sine wave is equal to the repetition rate of the non-sinusoidal waveform, and all other's are harmonic's of the fundamental. There are an infinite number of such harmonics. Some non-sine wave recurring at a rate of 200 times per second will consist of a 200 Hz fundamental sine wave, and harmonic's at 400, 600 and 800 Hz, and so on. For some waveform only the even harmonics will be present. As a general rule, it may be added that the higher the harmonic, the lower it's energy level, so that in bandwidth calculations the highest harmonics are often ignored.

The preceding statement may be verified in any one of three different ways. It may be proved mathematically by Fourier analysis. Graphical synthesis may be used. In this case adding the appropriate sine-wave components, taken from a formula derived by Fourier analysis, demonstrates the truth of the statement. An added advantage of this method is that it makes it possible for us to see the effect on the overall waveform because of the absence of some of the components.

Finally, the presence of the component sine waves in the correct proportions may be demonstrated with a wave analyzer, which is basically a high-gain tunable amplifier with a narrow bandpass, enabling it to tune to each component sine wave and measure it's amplitude.; Some formulas for frequently encountered non-sinusoidal waves are now given, and more may be found in handbooks. If the amplitude of the non-sinusoidal wave is A and its repetition rate is $\frac{W}{2\pi}$ per second, then it may be represented as follows:

Square wave

$$e = \frac{4A}{\pi}\left(\cos \omega t - \frac{1}{3}\cos 3\omega t + \frac{1}{5}\cos 5\omega t - \frac{1}{7}\cos 7\omega t + ...\right) \quad ...(1.7)$$

Triangular wave

$$e = \frac{4A}{\pi^2}\left(\cos \omega t + \frac{1}{9}\cos 3\omega t + \frac{1}{25}\cos 5\omega t + ...\right) \quad ...(1.8)$$

Sawtooth wave

$$e = \frac{2A}{\pi}\left(\sin \omega t - \frac{1}{2}\sin 2\omega t + \frac{1}{3}\sin 3\omega t - \frac{1}{4}\sin 4\omega t + ...\right) \quad ...(1.9)$$

In each case several of the harmonics will be required, in addition to the fundamental frequency, if the wave is to be represented adequately. This, of course, will greatly increase the required bandwidth.

❑❑❑

Amplitude Modulation 2

2.1 INTRODUCTION

Modulation is a fundamental requisite of communication systems. It is defined as the process by which some characteristic of a signal called carrier is varied in accordance with instantaneous value of another signal called modulating signal. Signals containing information or intelligence to be transmitted are referred to as modulating signals. This information bearing signal is also called baseband signal. The term baseband designates the band of frequencies representing the signal supplied by the source of information. The carrier frequency is greater than the modulating frequencies. The signal resulting from the process of modulation is called modulated signal.

2.2 CONTINUOUS WAVE OR ANALOG MODULATION

When the carrier wave is continuous in nature the modulation process is known as continuous wave (CW) modulation or analog modulation. The carrier wave is usually a sinusoidal signal. Amplitude modulation, and angle modulation are examples of this type of modulation.

2.3 PULSE MODULATION SYSTEM

When the carrier wave is a pulse-type waveform, the modulation process is known as pulse modulation. PAM, PWM, PCM, DM etc., are examples of this type of modulation.

2.4 MULTIPLEXING

It is a technique in which several message signal are combined into a composite signal for transmission over a common channel. In order to transmit a number of these signals over the same channel, the signals must be kept apart so that they do not interfere with each other, and hence they can be separated easily at the receiver end.

Multiplexing is of two types as under:

(*i*) Frequency division multiplexing (FDM)

(*ii*) Time-division multiplexing (TDM)

2.4.1 Frequency Division Multiplexing (FDM)

The FDM scheme is shown in given Fig. 2.1 with the simultaneous transmission of three message or baseband signals. The spectra of the message signals and the sum of the modulated carriers are indicated in the figure. DSB modulation is used in illustrating the spectra of figure. Any type of modulation can be used in FDM as long as the carrier spacing is sufficient to avoid spectra overlapping. However, the most widely used method of modulation is SSB modulation. At the receiving end of the channel the three modulated signals are separated by bandpass filters (BPFs) and then demodulated.

FDM is used in telephone system, telemetry, commercial broadcast, television, and communication networks. Commercial AM broadcast stations use carrier frequency spaced 10 kHz apart in the frequency range from 540 to 1600 kHz. This separation is not sufficient to avoid spectral overlap for AM with a reasonably high fidelity (50 Hz to 15 kHz) audio signal. Therefore, AM stations on adjacent carrier frequencies are placed geographically for apart to minimize interference. Commercial FM (Frequency Modulation) broadcast uses carrier frequencies spaced 200 kHz apart. In a long-distance telephone system, upto 600 or more voice signals (200 Hz to 3.2 kHz) are transmitted over a coaxial cable or microwave links by using SSB modulation with carrier frequencies spaced 4 kHz apart. In practice, the composite signal formed by spacing several signals in frequency may, in turn, be modulated by using another carrier frequency. In this case, the first carrier frequencies are often called sub-carriers.

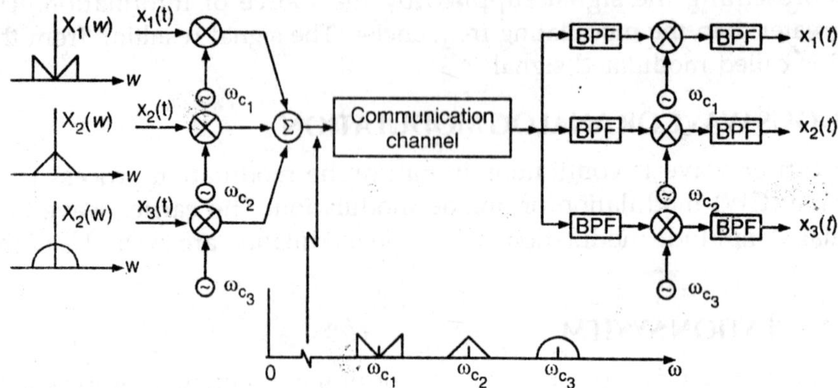

Fig. 2.1 *Frequency-division multiplexing*

2.4.2 Time-Division Multiplexing (TDM)

In Time-Division Multiplexing, the complete channel bandwidth is allotted to one user for a fixed time slot. As an example, if there are ten users, then every user can be given the time slot of one second. Thus, complete channel can be used by each user for one second time in every ten seconds. This techniques is suitable for digital signals. Because digital signals are transmitted intermittently and the time spacing between two successive digital codewords can be utilized by other signals. There is possibility of crosstalk in FDM whereas Intersymbol Interference is possible in TDM. These problems can be overcome by some special cares.

2.5 AMPLITUDE MODULATION

Amplitude Modulation may be defined as a system in which the maximum amplitude of the carrier wave is made proportional to the instantaneous value (amplitude) of the modulating or baseband signal.

Let us consider a sinusoidal carrier wave $c(t)$ a given as,

$$c(t) = A \cos \omega_c t \quad ...(2.1)$$

Here A is the maximum amplitude of the carrier wave and ω_c is the carrier frequency. For simplicity here we have assumed that the phase of the carrier wave is zero in eqn. 2.1.

Let us $x(t)$ denotes the modulating or baseband signal, then according to amplitude modulation, the maximum amplitude A of the carrier will have to be made proportional to the instantaneous amplitude of modulating signal $x(t)$.

The standard equation for amplitude modulated (AM) wave may be expressed as

$$s(t) = x(t) \cos \omega_c(t) + A \cos \omega_c t \quad ...(2.2)$$

or,

$$s(t) = \left[A + x(t)\right] \cos \omega_c t \quad ...(2.3)$$

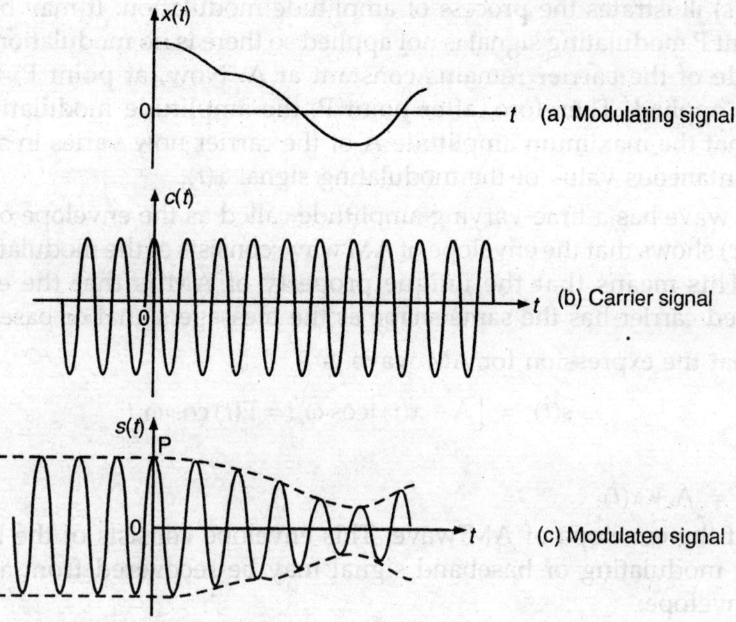

Fig. 2.2 *Illustration of amplitude modulation*

This figure shows the modulating signal or baseband signal, carrier signal and modulated signal.

2.5.1 Few Points about Amplitude Modulation (AM)

(*i*) It may be observed that eqn. (2.2) or eqn. (2.3) describes the time-domain behaviour of amplitude-modulated signal.

(*ii*) It may be noted that carrier signal is a fixed frequency signal having frequency ω_c. The modulating or baseband signal $x(t)$ contains the information or intelligence to be transmitted. In the process of amplitude modulation, this information is superimposed upon the carrier signal in the form of amplitude variations of the carrier signal. This means that the information to be transmitted, is now, contain in the amplitude variations of the carrier signal.

(*iii*) The resulting signal from the process of amplitude modulation is called amplitude modulated signal or simply AM wave.

(*iv*) In the process of amplitude modulation the frequency and phase of the carrier remain constant whereas the maximum amplitude varies according to the instantaneous value of the information signal.

(*v*) Eqn. (2.3) represents an amplitude modulated wave. This wave has a constant frequency ω_c and amplitude $A + x(t)$. This implies that the amplitude of the wave is changing around A in accordance with the value of the modulating signal $x(t)$. The frequency of the AM signal remains unchanged and is equal to ω_c.

(*vi*) Fig. 2.2 (c) illustrates the process of amplitude modulation. It may be observed that upto point P modulating signal is not applied so there is no modulation and maximum amplitude of the carrier remains constant at A. Now, at point P, the modulating signal is applied. Therefore, after point P, the amplitude modulation occurs. This means that the maximum amplitude A of the carrier now varies in accordance with the instantaneous value of the modulating signal $x(t)$.

(*vii*) The AM wave has a time-varying amplitude called as the envelope of the AM wave. Fig. 2.2 (c) shows that the envelope of AM wave consists of the modulating or baseband signal. This means that the unique property of AM is that the envelope of the modulated carrier has the same shape as the message signal or baseband signal.

We know that the expression for AM wave is

$$s(t) = \left[A + x(t)\right]\cos \omega_c t = E(t) \cos \omega_c t \quad ...(2.4)$$

where,

$$E(t) = A + x(t)$$

$E(t)$ is called the envelope of AM wave. This envelope consists of the baseband signal $x(t)$. Hence, the modulating or baseband signal may be recovered from an AM wave by detecting the envelope.

2.5.2 Spectrum of AM Wave or Frequency-Domain Representation

If $x(t)$ is modulating signal and carrier signal is given by the expression

$$c(t) = A \cos \omega_c t \quad ...(2.5)$$

Then the equation for AM wave will be

$$s(t) = x(t) \cos \omega_c t + A \cos \omega_c t \quad ...(2.6)$$

This equation describes the AM wave in time-domain. However, if we want to know the frequency description or frequencies present in AM wave, we will have to find it's spectrum or frequency domain representation. For this purpose, first we have to take the Fourier transform of AM wave.

Let $s(\omega)$ denote the Fourier transform of $s(t)$, $c(\omega)$ denotes the Fourier transform of $c(t)$ and $X(\omega)$ denotes the Fourier transform of $x(t)$.

Let the modulating signal $x(t)$ and it's Fourier transform $X(\omega)$ as shown in Fig. 2.3 (*a*)

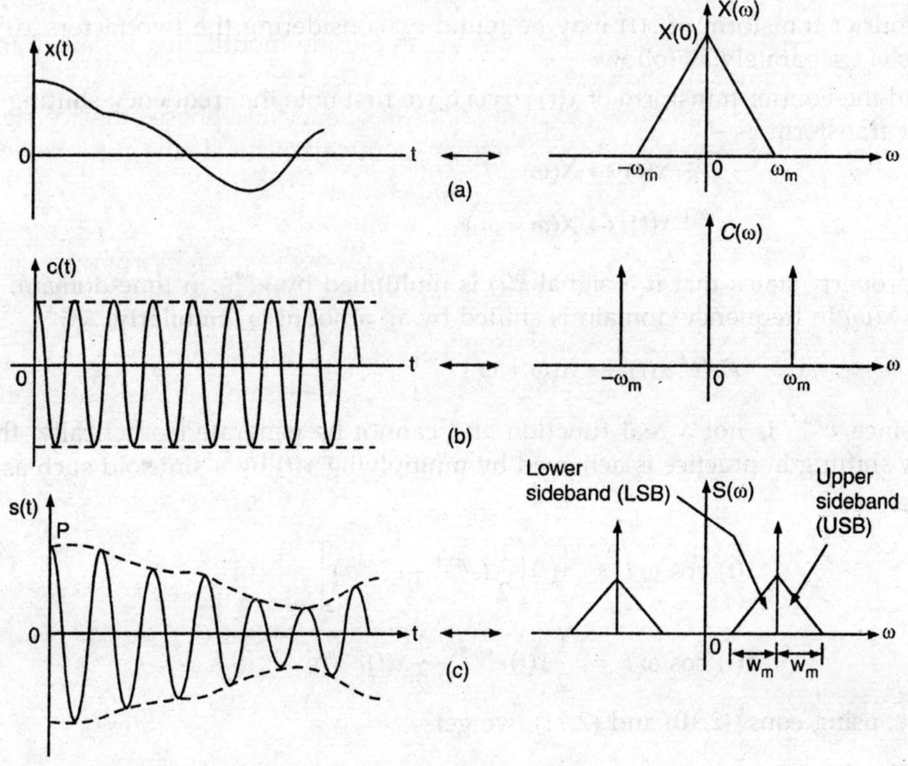

Fig. 2.3

Let the modulating signal or message signal $x(t)$ be band limited to the interval $-\omega_m \leq \omega \leq \omega_m$ as shown in above Fig. 2.3 (*a*). This means that the modulating signal does not have any frequency component outside the interval $(-\omega_m, \omega_m)$.

It may be noted at this point that the first figure shows, the modulating signal frequency ranges extend from $-\omega_m$ to ω_m i.e., it includes negative frequencies also from $-\omega_m$ to 0. Practically there is no meaning of negative frequency. In fact, the negative frequency is used for mathematical convenience only.

Hence, we can say that the modulating signal contains frequencies from 0 to ω_m or simply the bandwidth of modulating signal is ω_m. We know that the Fourier transform of a cosine signal $\cos \omega_c t$ consists of two impulse at ω_c and $-\omega_c$ as

$$\cos \omega_c t \leftrightarrow \pi[\delta(\omega + \omega_c) + \delta(\omega - \omega_c)] \quad ...(2.7)$$

Since the carrier signal is $c(t) = A \cos \omega_c t$, therefore,

$$A \cos \omega_c t \leftrightarrow A[\delta(\omega + \omega_c) + \delta(\omega - \omega_c)] \quad ...(2.8)$$

Fig. 2.3 (*b*) shows the carrier signal $A \cos \omega_c t$ and it's Fourier transform. Now, the AM wave is given as

$$s(t) = x(t) \cos \omega_c t + A \cos \omega_c t \quad ...(2.9)$$

To find the spectrum of AM wave, we take it's Fourier transform.

The Fourier transform of $s(t)$ may be found by considering the two factors $x(t) \cos \omega_c t$ and $A \cos \omega_c t$ separately as follow:

To find the Fourier transform of $x(t) \cos \omega_c t$, we first note the frequency-shifting theorem of Fourier transform as

If $$x(t) \leftrightarrow X(\omega)$$

then, $$e^{j\omega_c t}\, x(t) \leftrightarrow X(\omega - \omega_c) \quad ...(2.10)$$

This property states that if a signal $x(t)$ is multiplied by $e^{j\omega_c t}$ in time-domain, then it's spectrum $X(\omega)$ in frequency-domain is shifted by an amount ω_c. Similarly,

$$e^{-j\omega_c t}\, x(t) \leftrightarrow X(\omega + \omega_c) \quad ...(2.11)$$

But, since $e^{j\omega_c t}$ is not a real function and cannot be generated practically, therefore, frequency shifting in practice is achieved by multiplying $x(t)$ by a sinusoid such as $\cos \omega_c t$.

Therefore,

$$x(t) \cos \omega_c t = x(t)\left[\frac{1}{2}(e^{j\omega_c t} + e^{-j\omega_c t})\right] \quad ...(2.12)$$

$$x(t) \cos \omega_c t = \frac{1}{2}x(t)e^{j\omega_c t} + \frac{1}{2}x(t)e^{-j\omega_c t} \quad ...(2.13)$$

Hence, using eqns. (2.10) and (2.11), we get

$$x(t) \cos \omega_c t \leftrightarrow \frac{1}{2}[X(\omega - \omega_c) + X(\omega + \omega_c)] \quad ...(2.14)$$

This means that the multiplication of a signal $x(t)$ by a sinusoidal of frequency ω_c shifts the spectrum $X(\omega)$ by $\pm\, \omega_c$.

The Fourier transform of the second factor $A \cos \omega_c t$ will be as in eqn. (2.8)

$$A \cos \omega_c t \leftrightarrow \pi A[\delta(\omega + \omega_c) + \delta(\omega - \omega_c)] \quad ...(2.15)$$

This means that Fourier transform of $A \cos \omega_c t$ consists of two impulses at $\pm\, \omega_c$.

Therefore, the Fourier transform of AM wave given eqn. (2.6) will be the sum of eqns. (2.14) and (2.15).

Thus,

$$s(\omega) = \frac{1}{2}[X(\omega-\omega_c)+X(\omega+\omega_c)]+\pi A[\delta(\omega+\omega_c)+\delta(\omega-\omega_c)] \quad ...(2.16)$$

Above equation for Amplitude Modulated wave contains two factors. The first factor given as $\frac{1}{2}[X(\omega-\omega_c)+X(\omega+\omega_c)]$ represents the spectrum of original or baseband signal shifted in the positive as well as in the negative direction by the factor ω_c.

On the other hand, the second factor given as $\pi A[\delta(\omega+\omega_c)+\delta(\omega-\omega_c)]$ represents the presence of carrier signal i.e., two impulses each having strength equal to πA.

Thus the spectrum of modulated signal contains shifted spectrum of modulating signal and the spectrum of carrier signal as shown in Fig. 2.3 (*c*).

The following points may be observed from above figure:

(*i*) For positive frequencies, a portion of the spectrum of AM wave si lying above the carrier frequency ω_c. This band of frequency which is lying above the carrier frequency ω_c is known as the upper sideband (USB) whereas the symmetrical portion below carrier frequency ω_c is known as lower sideband (LSB). For negative frequencies, the upper sideband (USB) is represented by the portion of the spectrum below-ω_c and the lower sideband by the symmetrical portion above-ω_c. Two sidebands have been shown in Fig. 2.3 (*c*).

(*ii*) We generally keep $\omega_c > \omega_m$ which ensures that the two sidebands do not overlap each other.

(*iii*) At this point, it may be noted that the negative frequencies appeared in spectrum analysis is due to the exponential representation of Fourier transform. These negative frequencies are used for mathematical convenience. For a real function, these negative frequencies have no meaning. For general purpose, it is sufficient to consider only the positive frequency region while treating the negative frequency region as a replica of the positive frequencies. This means that to calculate the BW of a signal, we consider only positive side.

(*iv*) From figure, it is obvious that for positive side, the highest frequency component present in the spectrum of AM wave is $\omega_c + \omega_m$ and the lowest frequency component is $\omega_c - \omega_m$.

In fact, the difference between these two extreme frequencies is equal to the bandwidth of the AM wave.

Therefore,

Bandwidth, $B = (\omega_c + \omega_m) - (\omega_c - \omega_m)$

$$B = 2\omega_m$$

Thus, it is clear that the bandwidth of the amplitude modulated wave is twice the highest frequency present in the baseband or modulating signal.

2.6 MODULATION INDEX, (m_a)

Modulation Index is defined as the measure of extent of amplitude variation about an unmodulated maximum carrier. It is represented by m_a.

Mathematically,

$$m_a = \frac{|x(t)|_{max}}{\text{Maximum carrier amplitude}}$$

or,

$$\text{Modulation Index,} \quad m_a = \frac{|x(t)|_{max}}{A} \qquad \text{...(2.17)}$$

where,

$|x(t)|_{max}$ = the maximum amplitude of modulating signal, and

A = maximum amplitude of carrier signal.

The modulation index is also known as depth of modulation, degree of modulation or modulation factor.

Also, the absolute values of m_a multiplied by 100 is known as percentage modulation.

2.6.1 Overmodulation

We know that modulation index is given as

$$m_a = \frac{|x(t)|_{max}}{A} \qquad \text{...(2.18)}$$

where,

'A' is the maximum amplitude of the carrier signal.

The baseband or modulating signal will be preserved in the envelope of the AM signal only if we have

$$|x(t)|_{max} \leq A$$

i.e. modulation index m_a is less than or equal to unity.

In other words, the modulating signal is preserved in the envelope of AM signal only if the percentage modulation is less than or equal to 100 per cent.

On the other hand, if $m_a > 1$ or the percentage modulation is greater than 100, the baseband signal is not preserved in the envelope. If means that is this case, the baseband signal recovered from the envelope will be distorted. This type of distortion is called envelope distortion and the AM signal with $m_a > 1$ or $m_a > 100\%$ is called overmodulated signal.

Let us consider a baseband signal $x(t)$ and a carrier signal A cos $\omega_c t$ as shown in given Figs. 2.4 (*a*) and 2.4 (*b*). On Amplitude modulation, following two cases arise depending upon the amplitude of the baseband signal relative to the maximum amplitude of the carrier signal.

2.6.2 Amplitude Modulation with $m_a < 1$

In this case, the waveform of the AM signal as shown in Fig. 2.4 (c). Here the maximum amplitude of baseband signal is less than maximum carrier amplitude A.

$$\therefore \qquad |x(t)|_{max} < A$$

From figure, it may be observed that the envelope is not reaching the zero-amplitude axis of the AM waveform and so the baseband signal may be fully recovered from the envelop of AM wave.

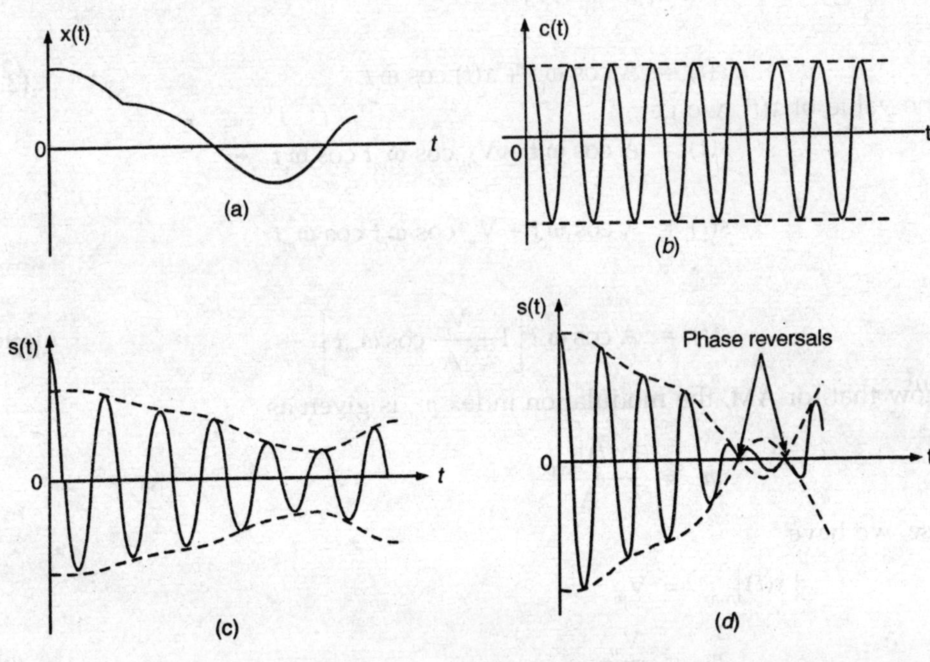

Fig. 2.4 (a)

2.6.3 Amplitude Modulation with $m_a > 1$

In this case, the waveform of the AM signal as shown in Fig. 2.4 (*d*). Here, the amplitude of a baseband signal exceeds maximum carrier amplitude.

$$\therefore \qquad |x(t)|_{max} < A$$

In this case,

the modulation index m_a, is more than 1 or 100%.

2.7 SINGLE TONE AMPLITUDE MODULATION (AM)

In this Modulation, we shall discuss amplitude modulation in which the modulating or baseband signal consists of only one (signal) frequency, i.e., modulation is done by a single frequency or tone. This type of amplitude modulation is known as Single Tone Amplitude Modulation.

Let us consider, a single tone modulating signal as

$$x(t) = V_m \cos \omega_m t \quad ...(2.19)$$

[$\because$ Signal $x(t)$ may be a voltage signal or a current signal. Here, we suppose that $x(t)$ is a voltage signal with maximum amplitude equal to V_m.]

Which contains a single frequency ω_m. Let the carrier signal be

$$c(t) = A \cos \omega_c t \quad ...(2.20)$$

We know that the general expression for AM signal is

$$s(t) = [A + x(t)] \cos \omega_c t$$

or,

$$s(t) = A \cos \omega_c t + x(t) \cos \omega_c t \quad ...(2.21)$$

Putting the value of $x(t)$, we get,

$$s(t) = A \cos \omega_c t + V_m \cos \omega_m t \cos \omega_c t$$

or,

$$s(t) = A \cos \omega_c t + V_m \cos \omega_c t \cos \omega_m t$$

or,

$$s(t) = A \cos \omega_c t \left[1 + \frac{V_m}{A} \cos \omega_m t\right] \quad ...(2.22)$$

But, we now that for AM, the modulation index m_a is given as

$$m_a = \frac{|x(t)|_{max}}{A}$$

In this case, we have

$$|x(t)|_{max} = V_m$$

$$\therefore \quad m_a = \frac{V_m}{A}$$

Putting this value of m_a in eqn. (2.22) we have,

$$s(t) = A \cos \omega_c t [1 + m_a \cos \omega_m t] \quad ...(2.23)$$

This is the desired expression for Single Tone Modulating Signal.

Eqn. (2.23) may further be simplified to observe the frequency components present in AM signal.

$$s(t) = A \cos \omega_c t [1 + m_a \cos \omega_m t]$$

or,

$$s(t) = A \cos \omega_c t + A.m_a \cos \omega_c t \cos \omega_m t$$

or,

$$s(t) = A \cos \omega_c t + \frac{A.m_a}{2} [2 \cos \omega_c t \cos \omega_m t]$$

$$= A\cos\omega_c t + \frac{A.m_a}{2}\left[\cos(\omega_c t + \omega_m) + \cos(\omega_c - \omega_m)t\right]$$

or,

$$s(t) = A\cos\omega_c t + \frac{A.m_a}{2}\cos(\omega_c + \omega_m) + \frac{A.m_a}{2}\cos(\omega_c - \omega_m)t \quad ...(2.24)$$

Eqn. (2.24) reveals that the AM signals has three frequency components as follow:

(*i*) carrier frequency ω_c having amplitude A.

(*ii*) upper sideband $(\omega_c + \omega_m)$ having amplitude $m_a.\frac{A}{2}$.

(*iii*) lower sideband $(\omega_c - \omega_m)$ having amplitude $m_a.\frac{A}{2}$.

With the help of these frequency components, we can plot the frequency-spectrum of single-tone amplitude modulated (AM) wave.

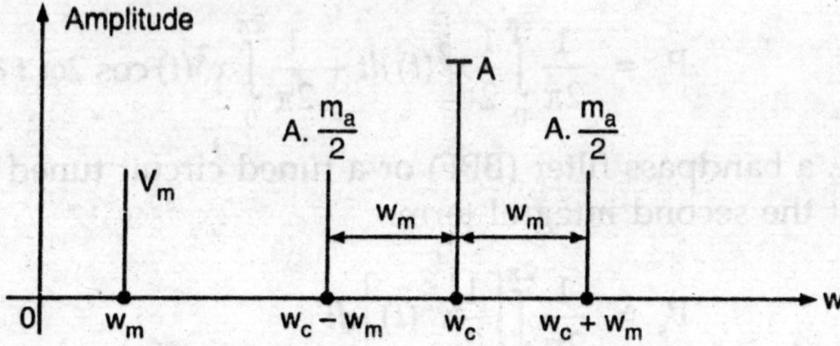

Fig. 2.4 (*b*) *Single-sided frequency spectrum of single-tone AM wave*

2.8 POWER CONTENT IN AM WAVE

It may be observed from the expression of AM wave that the carrier component of the amplitude modulated wave has the same amplitude as unmodulated carrier. In addition to carrier component, the modulated wave consists of two sideband components. It means that the modulated wave contains more power than the unmodulated carrier. However, since the amplitudes of two sidebands depends upon the modulation index, it may be anticipated that the total power of the amplitude modulated wave would depend upon the modulation index also.

We know that the general expression of AM wave is given as

$$s(t) = A\cos\omega_c t + x(t)\cos\omega_c t \quad ...(2.25)$$

The total power P of the AM wave is the sum of the carrier power P_c and sideband power P_c.

2.8.1 Carrier Power

The carrier power P_t is equal to the mean-square (ms) value of the carrier term $A\cos\omega_c t$, i.e.

$$P_c = \text{mean square value of } A\cos\omega_c t$$

$$P_c = \overline{[A\cos\omega_c t]^2} = \frac{1}{2\pi}\int_0^{2\pi} A^2\cos^2\omega_c t.dt = \frac{A^2}{2} \qquad \text{...(2.26)}$$

2.8.2 Sideband Power

The sideband power P_s is equal to the mean figure value of the sideband term $x(t)\cos\omega_c t$, i.e.,

$$P_s = \text{mean square value of } x(t)\cos\omega_c t$$

$$P_s = \overline{[x(t)\cos\omega_c t]^2} = \frac{1}{2\pi}\int_0^{2\pi} x^2(t)\cos^2\omega_c t\, dt$$

$$P_s = \frac{1}{2\pi}\int_0^{2\pi}\frac{1}{2}\left[2\cos^2\omega_c t\right]x^2(t)\,dt$$

or,

$$P_s = \frac{1}{2\pi}\int_0^{2\pi}\frac{1}{2}x^2(t)\,dt + \frac{1}{2\pi}\int_0^{2\pi} x^2(t)\cos 2\omega_c t\,dt \qquad \text{...(2.27)}$$

In AM generation, a bandpass filter (BPF) or a tuned circuit tuned to carrier frequency ω_c is used to filter out the second integral term.

$$\therefore \quad P_s = \frac{1}{2\pi}\int_0^{2\pi}\left[\frac{1}{2}x^2(t)\right]dt$$

or,

$$P_s = \text{mean square (ms) value of } \frac{1}{2}x^2(t)$$

or,

$$P_s = \frac{1}{2}\overline{x^2(t)} \qquad \text{...(2.28)}$$

The total sideband power P_s is due to the equal contributions of the upper and lower sidebands. Hence, the power carried by the upper and the lower sidebands will be

$$P_{s\,(LSB)} = P_{s\,(USB)} = \frac{P_s}{2} = \frac{1}{4}\overline{x^2 t} \qquad \text{...(2.29)}$$

Therefore,

The total power P_t of the AM signal is the sum of the carrier power P_c and sideband power P_s.

$$P_t = P_c + P_s = \frac{1}{2}A^2 + \frac{1}{2}\overline{x^2(t)}$$

$$P_t = \frac{1}{2}\left[A^2 + \overline{x^2(t)}\right] \quad \text{...(2.30)}$$

2.9 TRANSMISSION EFFICIENCY OF AMPLITUDE MODULATED SIGNAL

We know that,

The total modulated power of an AM signal is

$$P_t = P_c + P_s = \frac{1}{2}\left[A^2 + \overline{x^2(t)}\right] \quad \text{...(2.31)}$$

Out of this total power P_t, the useful message or baseband power is the power carried by the sidebands. The large carrier power P_c is a waste from the transmission point of view because it does not carry any information or message. This large carrier power P_c is transmitted alongwith the sideband power only for the convenient and cheap detection. **Hence, P_s is the only useful message power present in the AM wave.**

In AM wave, the amount of useful message power P_s may be expressed by a term known as Transmission Efficiency, η.

Hence transmission efficiency of AM wave may be defined as the percentage of total power contributed by the sidebands.

Transmission efficiency, η,

$$\eta = \frac{P_s}{P_t} \times 100 \quad \text{...(2.32)}$$

or,

$$\eta = \frac{\frac{1}{2}\overline{x^2(t)}}{\frac{1}{2}A^2 + \frac{1}{2}\overline{x^2(t)}} \times 100$$

or,

$$\eta = \frac{100\,\overline{x^2(t)}}{A^2 + \overline{x^2(t)}} \quad \text{...(2.33)}$$

The maximum transmission efficiency of the AM is only 33.33%. This implies that only one-third of the total power is carried by the sidebands and the rest two-third is wasted.

2.10 POWER OF A SINGLE TONE AMPLITUDE MODULATED SIGNAL

Let us consider that a carrier signal $A \cos \omega_c t$ is amplitude-modulated by a single-tone modulating signal $x(t) = V_m \cos \omega_m t$.

Then the unmodulated or carrier power

$$P_c = \text{mean square}$$

$$P_c = \overline{(A \cos \omega_c t)^2} = \frac{A^2}{2}$$

The sideband power P_s,

$$P_s = \frac{1}{2}\overline{x^2(t)}$$

$$P_s = \frac{1}{2}\overline{(V_m \cos \omega_m t)^2}$$

$$P_s = \frac{1}{2}\frac{V_m^2}{2} = \frac{1}{4}V_m^2 \quad ...(2.34)$$

Also,

We know that, the total modulated power P_t is the sum of P_c and P_s

$$\therefore \quad P_t = P_c + P_s$$

$$= \frac{A^2}{2} + \frac{1}{4}.V_m^2$$

$$P_t = \frac{A^2}{2}\left[1 + \frac{1}{2}\left(\frac{V_m}{A}\right)^2\right]$$

but,

$$\frac{V_m}{A} = \frac{\text{Maximum basebad amplitude}}{\text{Maximum carrier amplitude}} = m_a$$

$$= \text{Modulation index for AM}$$

Hence,

$$P_t = \frac{A^2}{2}\left[1 + \frac{1}{2}.m_a^2\right]$$

but,

$$\frac{A^2}{2} = P_c = \text{carrier power}$$

$$\therefore \quad P_t = P_c\left(1 + \frac{m_a^2}{2}\right)^2 \quad ...(2.35)$$

2.11 CURRENT CALCULATION FOR SINGLE-TONE AMPLITUDE MODULATION

In this case, the modulation index may be calculated from the values of unmodulated and modulated currents in the AM transmitter.

Let I_c be the r.m.s value of the carrier or unmodulated current and I_t be the r.m.s. value of the total or modulated current of an AM transmitter. Let R be the antenna resistance through which these currents flow.

Now, we know that for a single-tone modulation the power relation is expressed as,

$$P_t = P_c\left(1+\frac{m_a^2}{2}\right) \quad ...(2.36)$$

where,

P_t = total or modulated power

P_c = carrier or unmodulated power

m_a = modulation index

From eqn. (2.36), we can write,

$$\frac{P_t}{P_c} = 1+\frac{m_a^2}{2}$$

or,

$$\frac{I_t^2.R}{I_c^2.R} = 1+\frac{m_a^2}{2}$$

or,

$$\frac{I_t}{I_c} = \sqrt{1+\frac{m_a^2}{2}}$$

or,

$$I_t = I_c\sqrt{1+\frac{m_a^2}{2}} \quad ...(2.37)$$

2.12 POWER CONTENT IN MULTIPLE-TONE AMPLITUDE MODULATION

It is that type of modulation in which the modulating signal consists of more than one frequency components.

Let us consider that a carrier signal A cos $\omega_c t$ is modulated by a baseband or modulating signal $x(t)$ which is expressed as

$$x(t) = V_1 \cos \omega_1 t + V_2 \cos \omega_2 t + V_3 \cos \omega_3 t \quad ...(2.38)$$

We know that the general expression for AM wave is

$$s(t) = A \cos \omega_c t + x(t) \cos \omega_c t \quad ...(2.39)$$

Putting the value of $x(t)$, we get,

$$s(t) = A \cos \omega_c t + \left[V_1 \cos \omega_1 t + V_2 \cos \omega_2 t + V_3 \cos \omega_3 t\right]\cos \omega_c t$$

or,

$$s(t) = A\left[1+\frac{V_1}{A}\cos \omega_1 t+\frac{V_2}{A}\cos \omega_2 t+\frac{V_3}{A}\cos \omega_3 t\right]\cos \omega_2 t \quad ...(2.40)$$

But we know that,

$$\frac{V}{A} = \frac{\text{Maximum amplitude of modulating signal}}{\text{Maximum amplitud of carrier signal}}$$

$$\frac{V}{A} = \text{modulation index } m_a$$

Therefore,

$$s(t) = A\left[1 + m_1 \cos \omega_1 t + m_2 \cos \omega_2 t + m_3 \cos \omega_3 t\right] \cos \omega_c t \quad ...(2.41)$$

where,

$m_1 = \frac{V_1}{A}$, $m_2 = \frac{V_2}{A}$, $m_3 = \frac{V_3}{A}$, are the modulation indexes of the corresponding frequency components.

The expression for AM wave in eqn. (2.4) may further be expand as

$$s(t) = A \cos \omega_c t + m_1 A \cos \omega_c t \cos \omega_1 t + m_2 A \cos \omega_c t \cos \omega_2 t + m_3 A \cos \omega_c t \cos \omega_3 t$$

Now, we know that the total power in Amplitude Modulation is given as

$$P_t = \text{carrier power} + \text{sideband power}$$

$$P_t = P_c + P_s \quad ...(2.42)$$

The carrier power P_c is given as

$$P_c = (A \cos \omega_c t)^2 = \frac{A^2}{2}$$

Similarly, the sideband power is given as

$$P_s = \frac{1}{2}\overline{x^2(t)}$$

$$P_s = \frac{1}{2}\left[\overline{(V_1 \cos \omega_1 t)^2} + \overline{(V_2 \cos \omega_2 t)^2} + \overline{(V_3 \cos \omega_3 t)^2}\right] \quad ...(2.43)$$

But, we know that,

$$m_1 = \frac{V_1}{A} \text{ so that } V_1 = m_1 A$$

$$m_2 = \frac{V_2}{A} \text{ so that } V_2 = m_2 A$$

and

$$m_3 = \frac{V_2}{A} \text{ so that } V_3 = m_3 A$$

Putting these values of V_1, V_2 and V_3 in eqn. (2.43), we get,

$$P_s = \frac{1}{2}\left[\overline{(m_1 A \cos \omega_1 t)^2} + \overline{(m_2 A \cos \omega_2 t)^2} + \overline{(m_3 A \cos \omega_3 t)^2}\right]$$

$$P_s = \frac{1}{2}\left[\frac{m_1 A^2}{2} + \frac{m_2 A^2}{2} + \frac{m_3 A^2}{2}\right]$$

or,

$$P_s = \frac{1}{4}A^2\left[m_1^2 + m_2^2 + m_3^2\right] \quad ...(2.44)$$

Now putting the value of P_c and P_s in eqn. (2.41), we get

$$\Rightarrow \quad P_t = P_c + P_s$$

$$\Rightarrow \quad P_t = \frac{A^2}{2} + \frac{1}{4}A^2\left[m_1^2 + m_2^2 + m_3^2\right]$$

$$\Rightarrow \quad P_t = \frac{A^2}{2}\left[1 + \frac{1}{2}(m_1^2 + m_2^2 + m_3^2)\right]$$

$$\Rightarrow \quad P_t = P_c\left[1 + \frac{m_1^2 + m_2^2 + m_3^2}{2}\right]$$

This expression may be extended upto *n*-modulating terms, then,

$$P_t = P_c\left[1 + \frac{m_1^2}{2} + \frac{m_2^2}{2} + \frac{m_3^2}{2} + ... + \frac{m_n^2}{2}\right] \quad ...(2.45)$$

2.12.1 Total or Net Modulation Index for Multiple-Tone Modulation

Let us consider that m_t is the total or net modulation for a multiple-tone modulation. We know that for a multiple-tone modulation, the total power is expressed as

$$P_t = P_c\left[1 + \frac{m_1^2}{2} + \frac{m_2^2}{2} + \frac{m_3^2}{2} + ... + \frac{m_n^2}{2}\right] \quad ...(2.46)$$

where m_1, m_2, m_3 ... m_n are the modulation indexes for different modulating signals.

The power for AM wave is also expressed as,

$$P_t = P_c\left(1 + \frac{m_t^2}{2}\right) \quad ...(2.47)$$

Comparing eqns. (2.46) and (2.47), we have

or,

$$m_t = \sqrt{m_1^2 + m_2^2 + m_3^2 + ... + m_n^2} \quad ...(2.48)$$

This is the desired expression for the total or net modulation index.

2.13 GENERATION OF AMPLITUDE MODULATION

The device which is used to generate an amplitude modulated (AM) wave is known as Amplitude Modulator.

The methods of AM generation may be broadly classified as follow :

(*i*) Low-level AM Modulation

(*ii*) High-level AM Modulation

2.13.1 Low-level Amplitude Modulation

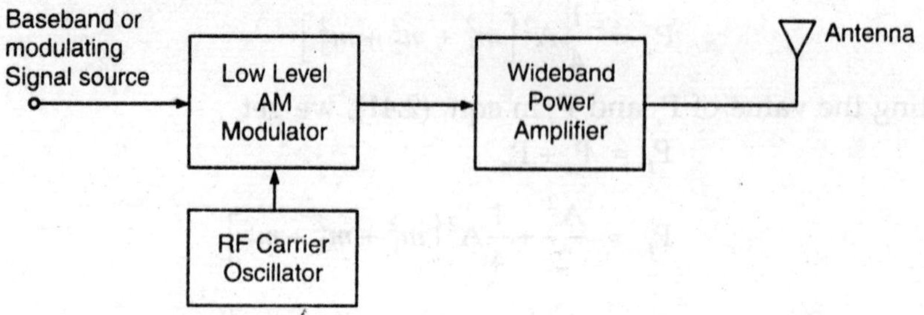

Fig. 2.5 (*a*) *Block Diagram for Low AM Modulation*

In a low-level amplitude modulation system, the modulation is done at low power level. At low power levels, a very small power is associated with the carrier signal and the modulating signal. Because of this, the output power of modulation is low. Therefore, the power amplifiers are required to boost the amplitude modulated signals upto the desired output level. From the above figure, it is clear that modulation is done at low power level. After this, the amplitude modulated signal is applied to a wideband power amplifier. A wideband power amplifier is used just to preserve the sidebands of the modulated signals. Amplitude modulated system, employing modulation at low power levels are also called low-level amplitude modulation transmitters.

2.13.2 High-level Amplitude Modulation

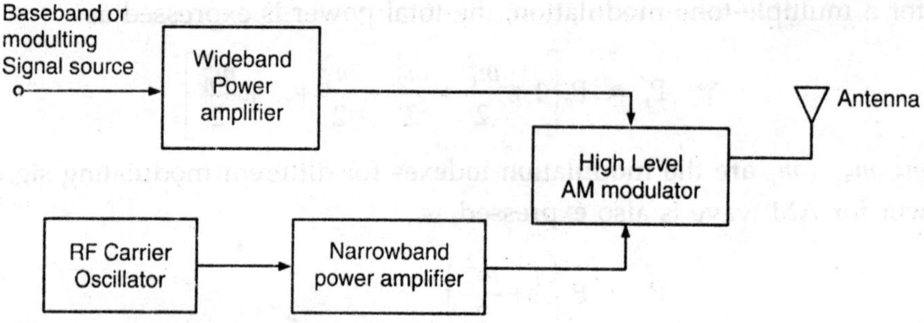

Fig. 2.5 (*b*) *Block diagram for high-level AM Modulation*

In a high-level amplitude modulation system, the modulation is done at high power level. Therefore, to produce amplitude modulation signal and the carrier signal are first power amplified and then applied to AM high-level modulator. For modulating signal, the wideband power amplifier is required just to preserve all the frequency components present in modulating signal. On the other hand, for carrier signal, the narrow band power amplifier is required because it is a fixed frequency signal. The collector modulation method is the example of high-level modulation.

2.13.3 Square Law Diode Modulation

This method is suited at low voltage levels because of the fact that current-voltage characteristics of a diode is highly non-linear particularly in the low voltage region.

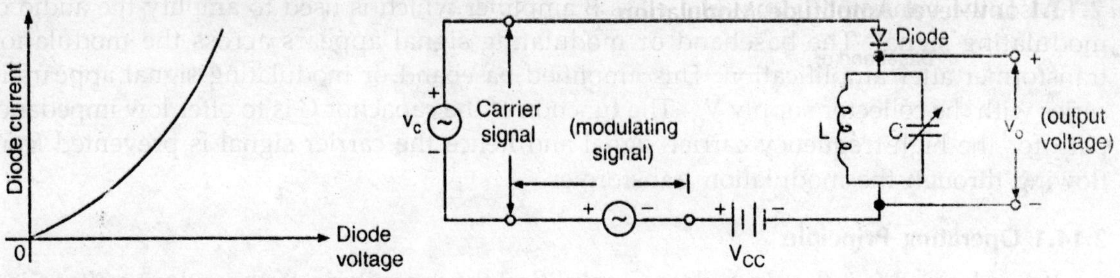

Fig. 2.6 *Current-voltage characteristic of a diode* **Fig. 2.7** *Square-low diode modulation*

It may be observed from the above figure that carrier and modulating signals are applied across the diode. A d.c. battery V_{cc} is connected across the diode to get a fixed operating point on the *v-i* characteristics of diode. The working of this circuit may be explained by considering the fact when two different frequencies are passed through a non-linear device, the process of amplitude modulation takes place. Hence, when carrier and modulating frequencies are applied at the input of diode, then different frequency terms appear at the output of diode. These different frequency terms are applied across a tuned circuit which is tuned to the carrier frequency and has a narrow bandwidth just to pass to sidebands the carrier and reject other frequencies. Hence at the output of tuned circuit, carrier and two sidebands are obtained, i.e., Amplitude Modulated wave is produced.

2.14 COLLECTOR MODULATION METHOD

Collector modulation method is a very popular method for AM generation. Here, the transistor T_1 makes a radio frequency (RF) class-C amplifier. At the base of T_1, the carrier signal is applied. V_{cc} makes the collector supply used for biasing purpose.

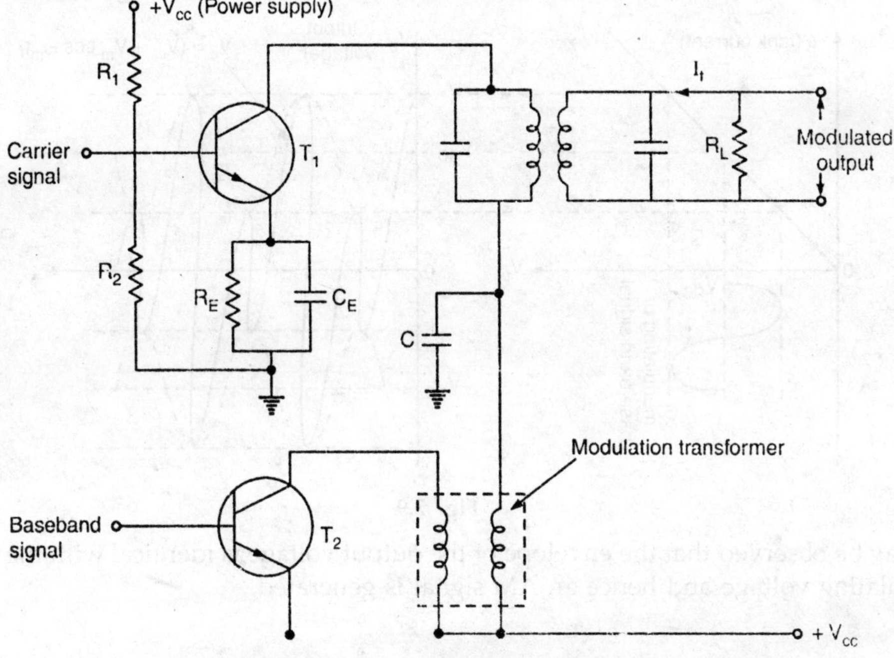

Fig. 2.8 *Illustration of collector modulation method*

Also, the transistor T_2 makes a class B amplifier which is used to amplify the audio or modulating signal. The baseband or modulating signal appears across the modulation transformer after amplification. This amplified baseband or modulating signal appears in series with the collector supply V_{cc}. The function of the capacitor C is to offer low impedance path for the high-frequency carrier signal and hence the carrier signal is prevented from flowing through the modulation transformer.

2.14.1 Operating Principle

It is a known fact that in a class-C amplifier, the magnitude of the output voltage is a definite fraction of or at the most equal to the supply voltage V_{cc}. In addition to this, a linear relationship exists between the output tank current T_t and the variable supply Voltage V_c. This means that in class C amplifier, the output voltage will be an exact replica of the input voltage waveform and the magnitude of the output voltage will be an approximately equal to the carrier supply voltage V_{cc}.

Now if R is the resistance of the output tank circuit at resonance, then the magnitude of the output voltage is given as

$$RI_t \cong V_{cc}$$

Therefore, the unmodulated carrier is amplified by class C modulated amplifier using transistor T_1 and its magnitude will remain constant at V_{cc} since there appears no voltage across the modulating transformer in the absence of baseband or modulating voltage.

But now if a baseband or modulating voltage $v_m = V_m \cos \omega_m t$ appears across the modulating transformer, this signal will be added to the carrier supply voltage V_{cc}. This results in a quite slow variation in carrier supply voltage V_{cc}. This type of slow variation in carrier supply voltage changes the magnitude of the carrier signal voltage at the output of the modulated class C amplifier as shown in given below figure.

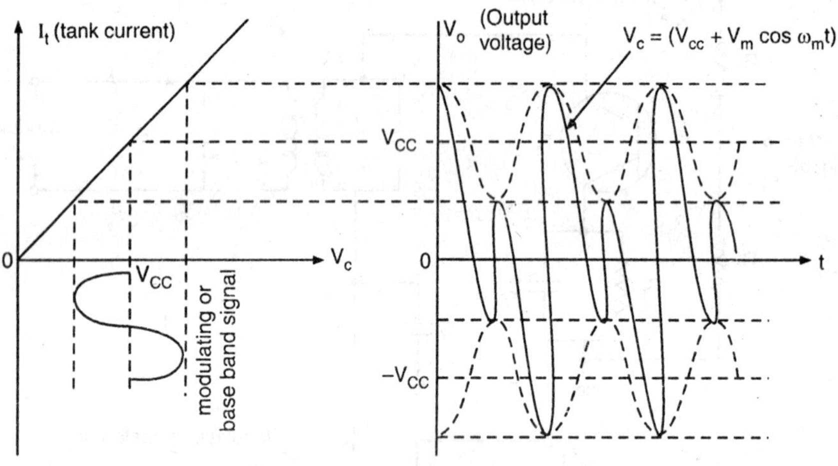

Fig. 2.9

It may be observed that the envelope of the output voltage is identical with the baseband or modulating voltage and hence an AM signal is generated.

2.14.2 Mathematical Analysis

The slowly changing carrier supply voltage V_c may be expressed as,

$$V_c = V_{cc} + v_m$$

or,

$$V_c = V_{cc} + V_m \cos \omega_m t \qquad ...(2.49)$$

But we know that for AM, the modulation index m_a is given by

$$m_a = \frac{\text{Maximum modulating voltage}}{\text{Maximum carrier voltage}}$$

In this case,

$$m_a = \frac{V_m}{V_{cc}} \qquad ...(2.50)$$

$$\therefore \quad V_c = V_{cc} + m_a.V_{cc} \cos \omega_m t$$

$$V_c = V_{cc}(1 + m_a \cos \omega_m t) \qquad ...(2.51)$$

Let the carrier voltage be

$$v_c = V_{cc} \cos \omega_c t$$

Then the modulated output voltage will be

$$v_0 = V_c \cos \omega_c t \qquad ...(2.52)$$

Putting the value of V_c in above equation from eqn. (2.51), we get,

$$v_0 = V_{cc} (1 + m_a \cos \omega_m t) \cos \omega_c t$$

This is the required expression for AM wave.

2.15 DEMODULATION OF AM WAVES

The process of extracting a modulating or baseband signal from the modulated signal is called Demodulation or Detection. The devices used for demodulation or detection are called demodulators or detectors. For amplitude modulation, demodulators are categorized as:

(*i*) Square-law Detectors

(*ii*) Envelop Detectors

AM signal with large carrier are detected by using the envelope detector. The envelope detector uses the circuit which extracts the envelope of the AM wave. Infact, the envelope of the AM wave is the baseband or modulating signal. But a low-level amplitude modulated signal can only be detected by using square-law detectors in which a device operating in the non-linear region is used to detect the modulating signal.

2.15.1 Square-law Detector

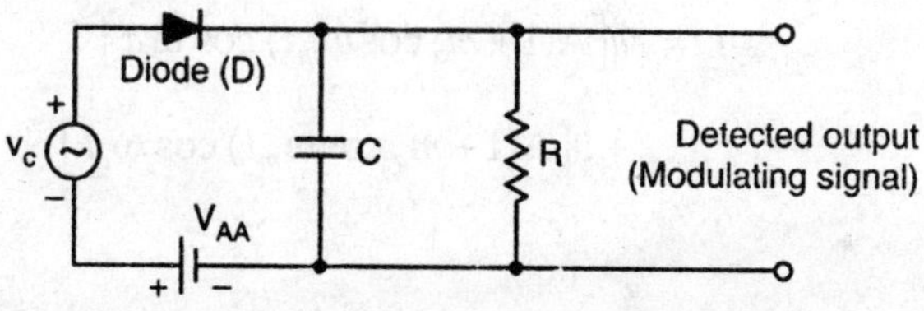

Fig. 2.10 *Basic circuit of square law diode detector*

The square law detector circuit is used for detecting modulated signal of small magnitude so that the operating region may be restricted to the non-linear portion of the V-I characteristics of the device. Above figure shows the circuit of a square-law detector. It may be observed that the circuit is very similar to the square-law modulator. The only difference lies in the filter circuit. In a square law modulator, the filter used is a bandpass filter whereas in a square law detector, a low-pass filter is used.

In the circuit, the dc supply voltage V_{AA} is used to get the fixed operating point in the non-linear portion of the diode V-I characteristic. Since, the operation is limited to the non-linear region of the diode characteristics, the lower half-portion of the modulated waveform is compressed. This produces envelope applied distortion. Due to this, the average value of the diode-current is no longer constant, rather it varies with time.

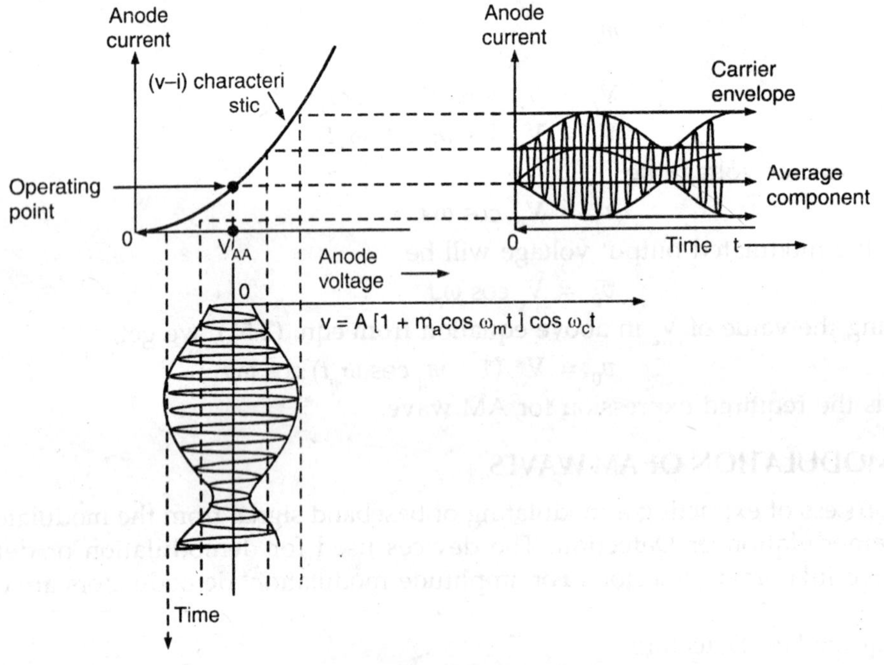

Fig. 2.11

This distorted output diode current is expressed by the non-linear *v-i* relationship as

$$i = av + bv^2 \quad \text{...(2.53)}$$

Here, v is the input modulated voltage.

We know that AM wave is expressed as

$$v = A(1 + m_a \cos \omega_m t) \cos \omega_c t \quad \text{...(2.54)}$$

Substituting, the value of v in eqn. (2.53), we get,

$$i = a\left[A(1 + m_a \cos \omega_m t) \cos \omega_c t\right] + b\left[A(1 + m_a \cos \omega_m t) \cos \omega_c t\right]^2 \quad \text{...(2.55)}$$

Now, if above expression is expanded, then we may observed the presence of terms of frequencies like $2\omega_c$, $2(\omega_c \pm \omega_m)$, ω_m and $2\omega_m$ besides the input frequency terms.

Hence, this device current i containing all these frequency terms is passed through a low-pass filter which allows to pass the frequency below or upto modulating frequency ω_m and rejects the other higher frequency components. Therefore, the modulating or baseband signal with frequency ω_m is recovered from the input modulated signal.

2.15.2 Linear Diode or Envelope Detector

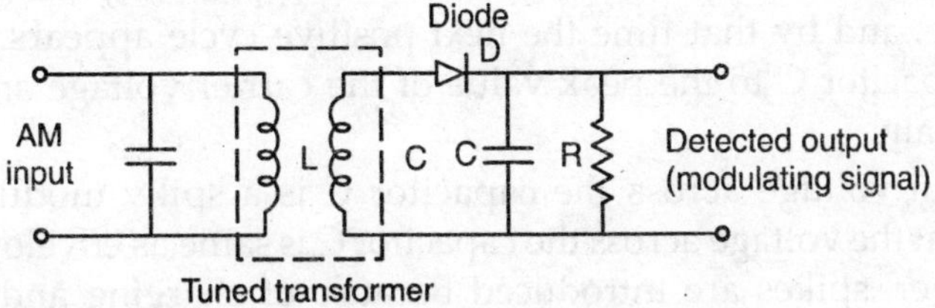

Fig. 2.12 *A linear diode detector*

It is a known fact that a diode operating in a linear region of it's V-I characteristics can extract the envelope of an AM wave. This type of detector is known as envelope detector or a linear detector. Envelope detector is most popular in commercial receiver circuits since it is very simple and is not expensive, also at the same time, it gives satisfactory performance for the reception of broadcasting programmes. In the input portion of the circuit, the tuned transformer provides perfect tuning at the desired carrier frequency, R-C network is the time-constant network. If the magnitude of the modulated signal at the input of the detector is 1 volt or more, the operation takes place in the linear portion of the V-I characteristic's of diode.

2.15.3 Operating Principle

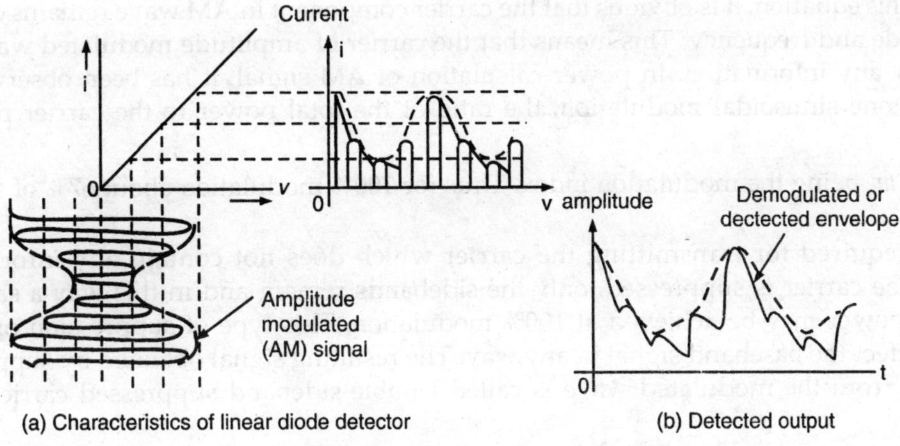

Fig. 2.13

First of all, let us assume that capacitor is absent in the circuit. In this case, the detector circuit will work as a half-wave rectifier. Therefore, the output waveform would be a half-rectified modulated signal as shown in the above figure. Now let us consider that the capacitor is introduced in the circuit. For the positive half-cycle, the diode conducts and the capacitor is charged to the peak value of the carrier voltage. However, for a negative half-cycle, the diode is reverse-biased and does not conduct. This means that the input carrier voltage is disconnected from the R-C circuit. Therefore, the capacitor starts discharging through the resistance R with a time-constant τ = RC. If the time-constant τ = RC is suitably chosen, the voltage across the capacitor C will not fall appreciably during the small period of negative half-cycle, and by that time the next positive cycle appears. This positive cycle again charges the capacitor C to the peak value of the carrier voltage and thus this process repeats again and again.

Hence, the output voltage across the capacitor C is a spiky modulating or baseband signal. This means that the voltage across the capacitor C is same as envelope of the modulated carrier signal, however, spikes are introduced because of charging and discharging of the capacitor C.

(*b*) figure illustrates the resulting detected modulating or baseband signal. We can reduce discharges negligibly amount by keeping the time constant RC large so that the capacitor C discharges negligible small. However, the large value of τ = RC produces another problem known as diagonal clipping. Thus, we cannot increase the time-constant beyond a certain limit. Therefore, the time-constant is an important consideration for envelope detector.

2.16 DOUBLE SIDEBAND SUPPRESSED CARRIER (DSB-SC) SYSTEM

The equation of AM wave in it's simplest from, i.e., single-tone modulation is expressed as

$$s(t) = A\cos\omega_c t + A.\frac{m_a}{2}\cos(\omega_c + \omega_m)t + A.\frac{m_a}{2}\cos(\omega_c - \omega_m)t \qquad ...(2.56)$$

From this equation, it is obvious that the carrier component in AM wave remains constant in amplitude and frequency. This means that the carrier of amplitude modulated wave does not convey any information. In power calculation of AM signal, it has been observed that for single-tone sinusoidal modulation, the ratio of the total power to the carrier power is $\left(1+\frac{m_a^2}{2}\right)$, m_a being the modulation index. Thus for 100% modulation about 67% of the total power is required for transmitting the carrier which does not contain any information. Hence, if the carrier is suppressed, only the sidebands remain and in this way a saving of two-third power may be achieved at 100% modulation. This type of suppression of carrier does not affect the baseband signal in any way. The resulting signal obtained by suppressing the carrier from the modulated wave is called Double-sideband suppressed carrier (DSB-SC) system.

Thus in a double-sideband suppressed carrier modulation there is no carrier signal only sidebands are present.

We know that the frequency-shifting of Fourier Transform is given as

If

$$x(t) \leftrightarrow X(\omega)$$

then

$$e^{j\omega_c t}\, x(t) \leftrightarrow X(\omega - \omega_c) \quad ...(2.57)$$

This property states that if a signal $x(t)$ is multiplied by $e^{j\omega_c t}$ in time-domain then its spectrum $X(\omega)$ in frequency-domain is shifted by an amount ω_c.

Similarly,

$$e^{-j\omega_c t}\, x(t) \leftrightarrow X(\omega + \omega_c) \quad ...(2.58)$$

But, since $e^{j\omega_c t}$ is not a real function and cannot be generated practically, therefore, frequency shifting in practice is achieved by multiplying $x(t)$ by a sinusoidal such as $\cos \omega_c t$.

Hence,

$$\Rightarrow \quad x(t) \cos \omega_c t = x(t).\frac{1}{2}(e^{j\omega_c t} + e^{-j\omega_c t})$$

$$\Rightarrow \quad x(t) \cos \omega_c t = \frac{1}{2}x(t)\, e^{j\omega_c t} + \frac{1}{2}x(t)\, e^{-j\omega_c t} \quad ...(2.59)$$

Using frequency-shifting property in equation (2.59), we get,

$$x(t) \cos \omega_c t \leftrightarrow \frac{1}{2}\left[X(\omega - \omega_c) + X(\omega + \omega_c)\right] \quad ...(2.60)$$

This means that the multiplication of a signal $x(t)$ by a sinusoidal of frequency ω_c shifts the spectrum $X(\omega)$ by $\pm\, \omega_c$.

Now if $x(t)$ is taken as modulating or baseband signal and $\cos \omega_c t$ is taken as carrier signal, then $x(t) \cos \omega_c t$ represents the modulated signal. Further, the Fourier transform of this modulated signal is given by the eqn. (2.60). This equation shows that the spectrum of modulated signal contains only shifted spectrum of signal and there is no carrier components. However, we know that the modulated signal, which contains no carrier but two sidebands is called Double-sideband Suppressed Carrier (DSB-SC) modulation.

Therefore, a DSB-SC signal is obtained by simply multiplying modulating signal $x(t)$ with carrier signal $\cos \omega_c t$. This is achieved by a product modulator.

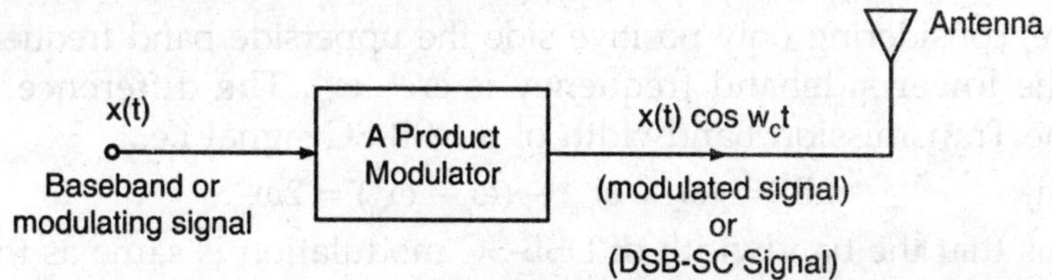

Fig. 2.14

This is the block diagram of a product modulator (↑).

Fig. 2.15 shows the modulating signal and its Fourier transform, carrier signal and its Fourier transform and DSB-SC signal its frequency spectrum.

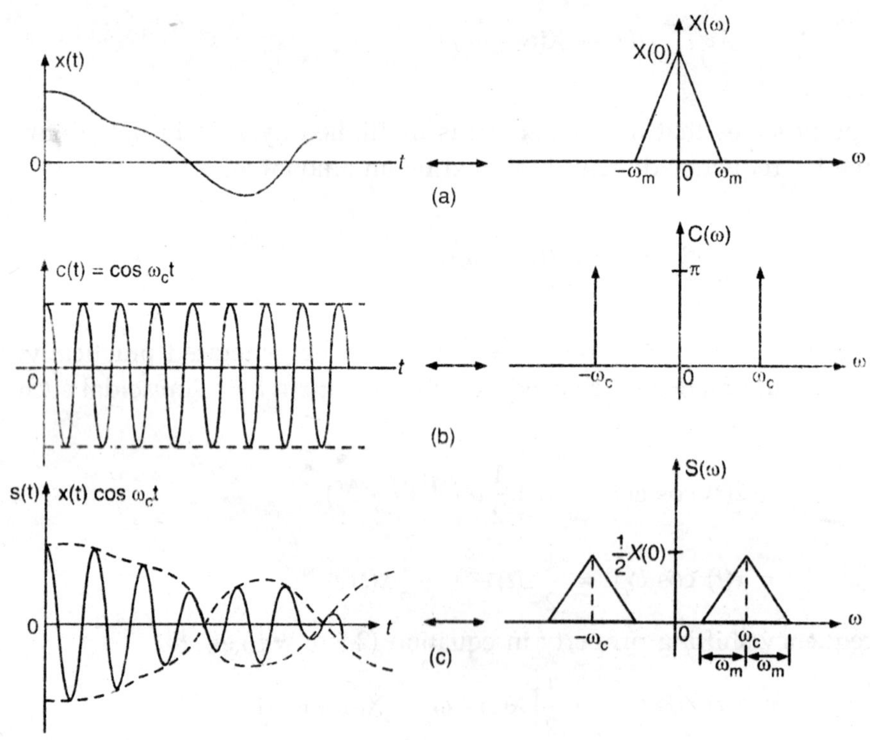

Fig. 2.15

2.16.1 Few Points

(*i*) The DSB-SC signal exhibits phase-reversal at zero crossings. Because of this, the envelope of a DSB-SC modulated signal is different from the message signal. This is unlike the case of an AM wave.

(*ii*) It is clear from the above figure that the impulses at $\pm \omega_c$ are missing which means that the carrier term is suppressed in the spectrum and only two sideband terms, USB and LSB are left. Therefore, it is called Double-sideband Suppressed Carrier (DSB-SC) system.

(*iii*) From figure, considering only positive side the upperside band frequencies is $\omega_c + \omega_m$ whereas the lower sideband frequency is $\omega_c - \omega_m$. The difference of these two is equal to the Transmission Bandwidth of a DSB-SC signal i.e.,

Bandwidth, $$B = (\omega_c + \omega_m) - (\omega_c - \omega_m) = 2\omega_m.$$

It is obvious that the bandwidth of DSB-SC modulation is same as that of generated AM waves.

2.17 GENERATION OF DSB-SC SIGNAL

The expression for DSB-SC signal is given as

$$s(t) = x(t) \cos \omega_c t$$

where,

$$x(t) = \text{Baseband signal}$$
$$\cos \omega_c t = \text{Carrier signal}$$

From the expression, we see that a DSB-SC signal is basically the product of the modulating or baseband signal and the carrier signal. Unfortunately, a single electronic device cannot generate a DSB-SC signal. As, we already discussed earlier, a circuit to achieve the generation of a DSB-SC signal is called a product modulator. In this section, we shall discuss two types of product modulator, namely the Balanced Modulator and the Ring Modulator.

2.17.1 The Balanced Modulator

We know that a non-linear resistance or a non-linear device may be used to produce Amplitude Modulation, i.e., one carrier and two sidebands. However, a DSB-SC signal contains only two sidebands. Thus, if two non-linear devices such as diodes, transistors etc., are connected in a balanced mode so as to suppress the carriers of each other, then only sidebands are left, i.e., a DSB-SC signal is generated.

Therefore, a balanced modulator may be defined as a circuit in which two non-linear devices are connected in a balanced mode to produce a DSB-SC signal.

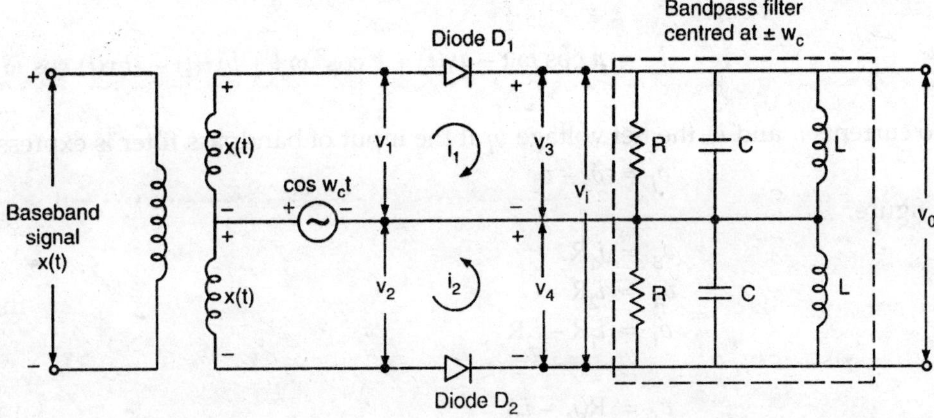

Fig. 2.16 *Balanced Modulator using Diodes*

This figure shows a balanced modulator circuit using two diodes. A modulating signal $x(t)$ is applied to the two diodes through a centre-tapped transformer with the carrier signal $\cos \omega_c t$.

A non-linear v-i relationship is given as

$$i = av + bv^2 \qquad ...(2.61)$$

Here, we have neglected the higher power terms.

In above expression, v is the input voltage applied across a non-linear device and i is the current through the non-linear device.

From figure,

$$v_1 = \cos \omega_c t + x(t) \quad ...(2.62)$$

$$v_2 = \cos \omega_c t - x(t) \quad ...(2.63)$$

For diode D_1, the non-linear v-i relationship becomes,

$$i_1 = av_1 + bv_1^2 \quad ...(2.64)$$

Similarly, for diode D_2, the non-linear v-i relationship becomes,

$$i_2 = av_2 + bv_2^2 \quad ...(2.65)$$

In the expression of current i_1, substituting the value of v_1, we get,

$$i_1 = a[\cos \omega_c t + x(t)] + b[\cos \omega_c t + x(t)]^2$$

or,

$$i_1 = a \cos \omega_c t + ax(t) + b[\cos^2 \omega_c t + x^2(t) + 2x(t) \cos \omega_c t]$$

or,

$$i_1 = a \cos \omega_c t + ax(t) + b \cos^2 \omega_c t + bx^2(t) + 2bx(t) \cos \omega_c t \quad ...(2.66)$$

Similarly, in the expression of current i_1, substituting the value of v_2, we get,

$$i_2 = a[\cos \omega_c t - x(t)] + b[\cos \omega_c t - x(t)]^2$$

or,

$$i_2 = a \cos \omega_c t - ax(t) + b \cos^2 \omega_c t + bx^2(t) - 2bx(t) \cos \omega_c t \quad ...(2.67)$$

Due to currents i_1 and i_2, the net voltage v_i at the input of bandpass filter is expressed as,

$$v_i = v_3 - v_4 \quad ...(2.68)$$

but from figure,

$$v_3 = i_1 R$$

and,

$$v_4 = i_2 R$$

$\therefore$

$$v_i = i_1 R - i_2 R$$

or,

$$v_i = R(i_1 - i_2)$$

In above equations, substituting the values of i_1 and i_2 from eqns. (2.66) and (2.67), we get,

$$v_i = R[2ax(t) + 4bx(t) \cos \omega_c t]$$

or,

$$v_i = 2R[ax(t) + 2bx(t) \cos \omega_c t]$$

This voltage filter is that type of filter which allows to pass a band of frequencies. Here, since the bandpass filter is centred around $\pm\, \omega_c$, it will pass a narrowband of frequencies centred at $\pm\, \omega_c$ with a small bandwith of $2\omega_m$ to preserve the sidebands.

Therefore, the output of BPF centred around $\pm\, \omega_c$ is given by

$$v_o = 4a\text{R}\ x(t) \cos \omega_c t$$

$$v_o = 4a\text{R}\ x(t) \cos \omega_c t$$

This is the expression for a DSB-SC signal.

2.17.2 Ring Modulator

Ring modulator is another product modulator, which is used to generate DSB-SC signal.

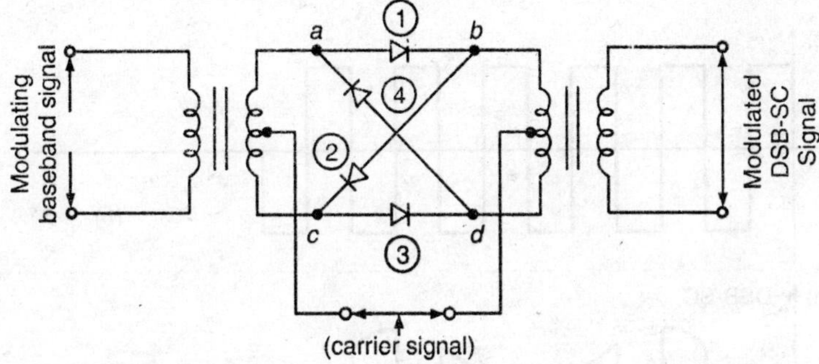

Fig. 2.17 *Circuit diagram of a ring modulator*

In a ring modulator circuit, four diodes are connected in the form of a ring in which all four diodes point in the same manner. All the four diodes in ring are controlled by a square wave carrier signal $c(t)$ of frequency f_c applied through a centre-tapped transformed.

In case, when diodes are ideal and transformer are perfectly balanced, the two outer diodes are switched on if the carrier signal is positive whereas the two inner diodes are switched off and thus presenting very high impedance as shown in given Fig. 2.18 (*a*). Under this condition, the modulator multiplies the modulating signal $x(t)$ by + 1.

Now, in case when carrier signal is negative, the situation becomes reversed as show in given below Fig. 2.18 (*b*). In this case, the modulator multiplies the modulating signal by – 1.

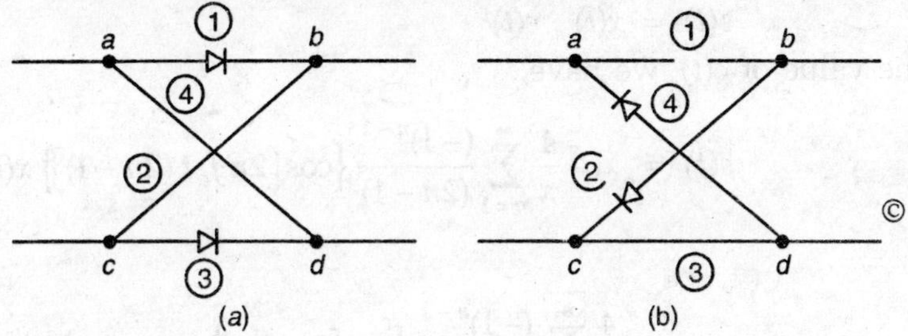

Fig. 2.18 *(a) Illustration of condition when the diodes (1) and (3) are switched ON and diodes (2) and (4) are switched OFF.*

(b) Illustration of condition when the diodes (2) and (4) are switched ON and diodes (1) and (3) switched OFF.

Hence, the ring modulator is a product modulator for a square wave carrier and modulating signal.

The below Fig. 2.19 illustrates the modulating signal, square wave carrier signal and the modulated (DSB-SC) signal.

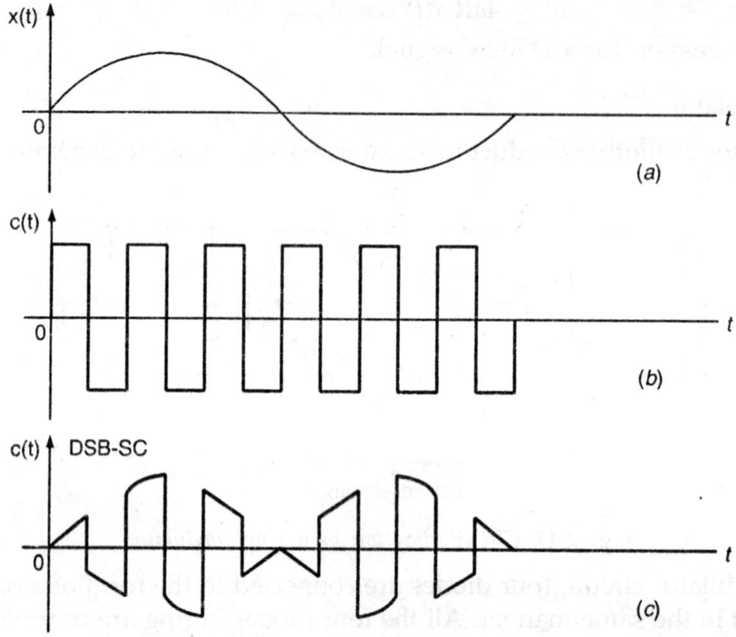

Fig. 2.19 *(a) Sinusoidal modulating wave (b) Square wave carrier (c) Modulated wave (DSB-SC) output of the Ring Modulator*

The square wave carrier may be represented in Fourier series as under:

$$c(t) = \frac{4}{\pi}\sum_{\pi}^{\infty}\frac{(-1)^{n-1}}{(2n-1)}\{\cos[2\pi f_c t(2n-1)]\}$$

We have,

$$s(t) = x(t) \,.\, c(t)$$

Substituting the value of $c(t)$, we have,

$$\Rightarrow \quad s(t) = x(t)\frac{4}{\pi}\sum_{n=1}^{\infty}\frac{(-1)^{n-1}}{(2n-1)}\{\cos[2\pi f_c t(2n-1)]\}\, x(t)$$

or,

$$\Rightarrow \quad s(t) = \frac{4}{\pi}\sum_{n=1}^{\infty}\frac{(-1)^{n-1}}{(2n-1)}\{\cos[2\pi f_c t(2n-1)]\}\, x(t)$$

Thus, with the help of above mathematical analysis, it may be verified that the output from ring modulator does not have any component at carrier frequency. Hence, the modulated output does not have any component at carrier frequency. Hence, the modulated output contains only product terms.

A ring modulator is also known as a Double-Balanced modulator since it is balanced with respect to the baseband signal as well as the square wave carrier signal.

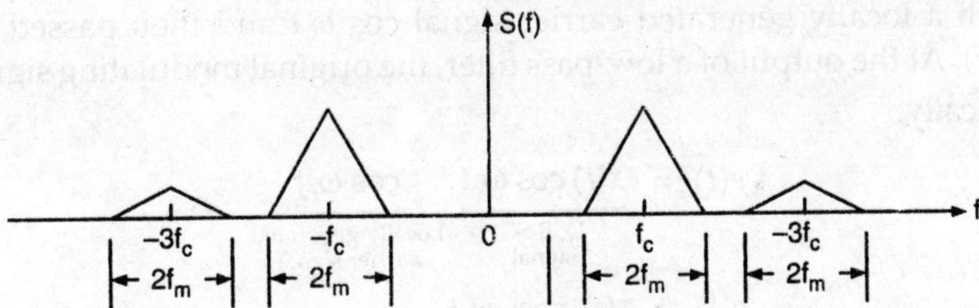

Fig. 2.20 *Spectrum of the DSB-BC output of the ring modulator*

The frequency spectrum of the ring modulator output contains sidebands around each of the odd harmonics of the square wave carrier signal. We have assumed that the modulating or baseband signal is bandlimited to $-f_m \leq f \leq f_m$. The desired sideband around the carrier frequency f_c may be selected by using a bandpass filter (BPF) having center frequency ω_c and bandwidth $2\,f_m$.

Above Fig. 2.20 we observed that to avoid overlapping of sidebands we must have $f_c > f_m$.

2.18 DEMODULATION OF DSB-SC SIGNALS

The DSB-SC signal may be demodulated by following two methods:

(*i*) Synchronous Detection Method

(*ii*) Using envelope detector after carrier reinsertion.

2.18.1 Synchronous Detection Method

We know that the DSB-SC system is used at the transmitter end to shift the modulating signal to a higher carrier frequency $\pm\,\omega_c$. Now, this modulated (DSB-SC) signal is transmitted from the transmitter and it reaches the receiver through a transmission medium. At the receiver end, the original modulating signal $x(t)$ is recovered from the modulated (DSB-SC) signal. This can be achieved by simply retranslating the baseband or modulating signal from a higher spectrum, centred at $\pm\,\omega_c$, to the original spectrum. This process of retranslation is called Demodulation or Detection. Hence, the original or baseband signal is recovered from the modulated signal by the detection process.

A method of DSB-SC detection is known as Synchronous Detection. The given Fig. 2.21 shows the block diagram of synchronous detection method.

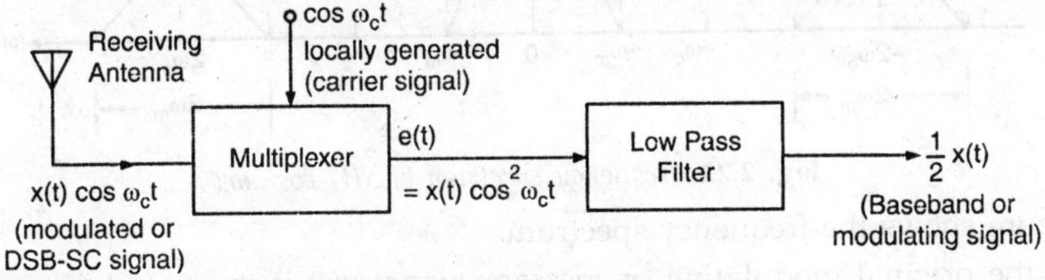

Fig. 2.21 *Synchronous detection method*

2.18.2 Working Principle

In synchronous detection method, the received modulated or DSB-SC signal is first multiplied with a locally generated carrier signal $\cos \omega_c t$ and then passed through a low-pass filter (LPF). At the output of a low-pass filter, the original modulating signal is recovered.

Mathematically,

$$e(t) = \underbrace{x(t)\cos\omega_c t}_{\text{DSB-SC signal}} . \underbrace{\cos\omega_c t}_{\text{Locally generated carrier signal}} \quad ...(2.69)$$

$$\Rightarrow \quad e(t) = x(t)\cos^2\omega_c t$$

$$= \frac{1}{2}x(t)\left[2\cos^2\omega_c t\right]$$

$$\Rightarrow \quad e(t) = \frac{1}{2}x(t)\left[1+\cos 2\omega_c t\right]$$

$$= \frac{1}{2}x(t) + \frac{1}{2}x(t)\cos 2\omega_c t \quad ...(2.70)$$

Now, it may be observed that when multiplied signal $e(t)$ is passed through a low-pass filter, then the term $\frac{1}{2}x(t)\cos 2\omega_c t$, centred at $\pm 2\omega_c$ is suppressed by low-pass filter and thus at the output of low-pass filter, the original modulating signal $\frac{1}{2}x(t)$ is obtained.

We may get the frequency-spectrum of multiplied signal $e(f)$ using Fourier Transform as

$$x(t)\cos^2\omega_c t \leftrightarrow \frac{1}{2}X(\omega) + \frac{1}{4}\left[X(\omega+2\omega_c) + X(\omega-2\omega_c)\right] \quad ...(2.71)$$

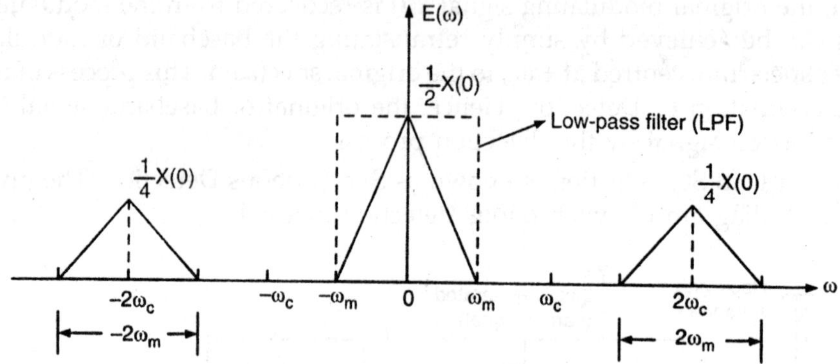

Fig. 2.22 *Frequency spectrum of $x(t)\cos^2\omega_c t$*

This figure shows the frequency spectrum.

Hence, the original modulating or message signal $x(t)$ is recovered from the DSB-SC signal.

2.18.3 Synchronous or Coherent Detection

From the previous topics, we observed that the detection process for DSB-SC requires a local oscillator at the receiver end. The frequency and phase of the locally generated carrier signal and the carrier signal at the transmitter carrier must be identical. This means that the local oscillator signal must be exactly coherent or synchronised with the carrier signal at the transmitter, both in frequency and phase, otherwise the detected signal would get distorted. Therefore, this method of recovery is called synchronous detection. Thus, the demerits of the synchronous detection is that if requires an additional system at the receiver to ensure that the locally generated carrier is synchronized with the transmitter carrier making the receiver complex and costly.

2.18.4 Envelope Detection after suitable Carrier Re-insertion

The other possible method of demodulating DSB-SC signal is by inserting a carrier generated at the receiver end with the help of a local oscillator. However, the phase and the frequency of the re-inserted carrier must be properly synchronized with those at the transmitter end in order to avoid distortion. We know that if we insert a sufficient carrier of same frequency and phase to DSB-SC signal, it converts DSB-SC signal into a conventional AM wave. Now, this AM wave is demodulated by an envelope detector. However, phase and frequency errors will result in similar type of distortion as obtained in coherent detection.

Let us consider that the received DSB-SC signal is expressed by

$$s(t) = c(t)\ .\ x(t)$$

or,

$$s(t) = A\cos(2\pi f_c t)\ x(t) \qquad \text{...(2.72)}$$

Let us assume, A = 1, in eqn. (2.72). Then the equation becomes,

$$s(t) = \cos(2\pi f_c t)\ x(t) \qquad \text{...(2.73)}$$

The inserted carrier at the receiver will be

$$c'(t) = A\cos(2\pi f_c t + \phi)$$

where ϕ = amount of phase discrepancy.

Then the resulting signal will be

$$r(t) = s(t) + \text{re-inserted carrier at the receiver}$$

or,

$$r(t) = s(t)\ .\ c'(t) \qquad \text{...(2.74)}$$

Substituting eqns. (2.72) and (2.73) in eqn. (2.74), we get,

$$r(t) = s(t) + c'(t)$$

$$r(t) = \cos(2\pi f_c t)\ x(t) + A\cos(2\pi f_c t + \phi)$$

or,

$$r(t) = x(t)\cos(2\pi f_c t) + A\cos(2\pi f_c t)\cos\phi - A\sin(2\pi f_c t)\sin\phi$$

or,

$$r(t) = \left[x(t) + A\cos\phi\right]\cos(2\pi f_c t) - (A\sin\phi)\sin(2\pi f_c t)$$

or,

$$r(t) = e(t)\cos\left[(2\pi f_c t) + \theta(t)\right] \quad ...(2.75)$$

where,

$$e(t) = \sqrt{\left[A + x(t)\right]^2 - 2A\,x(t)\left[1 - \cos\phi\right]}$$

and,

$$\theta(t) = \tan^{-1}\left[\frac{A\sin\theta}{x(t) + A\cos\phi}\right]$$

Now, from this expression,

$$r(t) = e(t)\left\{\cos\left[(2\pi f_c t) + \theta(t)\right]\right\}$$

it may be observed that $e(t)$ is the envelope of the resulting signal $r(t)$.

Also, if we take $\phi = 0$, then the envelope will be given by

$$e(t) = A + x(t)$$

Hence, modulating signal $x(t)$ can be recovered from $r(t)$ using an envelope detector since the $r(t)$ is basically a conventional AM wave given by

$$r(t) = \left[A + x(t)\right]\cos 2\pi f_c t \quad ...(2.76)$$

This is however possible only when $\left[A + x(t)\right] > 0$ for all values of t.

It is possible only when the modulation index m is less than unity.

If $\phi \neq 0$, then the phase error exists between the two carriers. It is given as

$$e(t) = A\left[1 + \frac{2x(t)}{A}\cos\phi + \left\{\frac{x(t)}{A}\right\}^2\right]^{1/2} \quad ...(2.77)$$

If $A \gg |x(t)|$, then, we have,

$$e(t) \cong A + x(t)\cos\phi \quad ...(2.78)$$

The desired signal output will thus be $x(t)\cos\phi$. If $\phi = 0$ and there is a difference in frequency Δf between the two oscillators, then the envelope of the resulting signal $r(t)$ will be given as,

$$e(t) = A + x(t)\cos\left[2\pi\,\Delta f t\right] \quad \text{for} \quad A \gg |x(t)| \quad ...(2.79)$$

The envelope can be detected by an envelope detector but here distortion will be identical to coherent detection.

2.19 QUADRATIVE—AMPLITUDE MODULATION (QAM)

This modulation scheme is also called Quadrative-Carrier Multiplexing. This modulation scheme enables two DSB-SC modulated signals to occupy the same transmission bandwidth and therefore it allows for the separation of the two message signals at the receiver output. It is therefore known as a bandwidth-Conservation scheme.

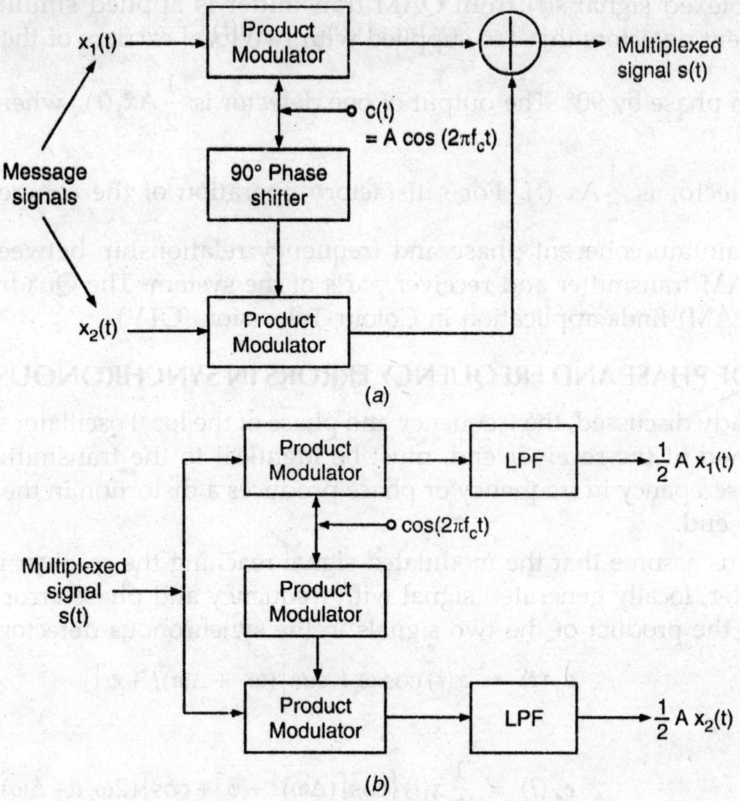

Fig. 2.23 *Illustration of Quadrative Amplitude (QAM) system*
(a) QAM transmitter (b) QAM receiver

From the above figure, it shows the block diagram of Quadrature Amplitude Modulation (QAM) system. Figure (*a*) shows the QAM transmitter and Fig. (*b*) shows the QAM receiver.

The QAM transmitter consists of two separate balanced modulators (BM) which are supplied with two carrier waves of the same frequency but differing in phase by 90°. The output of the two balanced modulators are added in the adder and transmitted. The transmitted signal is thus given by

$$s(t) = x_1(t) \text{ A} \cos (2\pi f_c t) + x_2(t) \text{ A} \sin (2\pi f_c t)$$

or,

$$s(t) = \text{A}x_1(t) \text{ A} \cos (2\pi f_c t) + \text{A}x_2(t) \sin (2\pi f_c t) \quad ...(2.80)$$

where,

$x_1(t)$ and $x_2(t)$ are the two different message signals applied to the product modulators. Both $x_1(t)$ and $x_2(t)$ are band-limited in the interval $- f_m \leq f \leq f_m$, then $s(t)$ will occupy a bandwidth of $2f_m$. This bandwidth $2f_m$ is centred at the carrier frequency f_e, where f_m is the bandwidth of message signal $x(t)$ or $x_2(t)$.

Hence, the multiplexed signal consists of the in-phase component $\text{A}x_1(t)$ and the quadrature phase component $- \text{A}x_2(t)$.

The multiplexed signal $s(t)$ from QAM transmitter is applied simultaneously to two separate coherent detectors that are supplied with two local carriers of the same frequency, but differing in phase by 90°. The output of one detector is $\frac{1}{2}Ax_1(t)$, whereas the output of the second detector is $\frac{1}{2}Ax_2(t)$. For satisfactory operation of the coherent detector, it is essential to maintain coherent phase and frequency relationship between the oscillators used in the QAM transmitter and receiver parts of the system. The Quadrature Amplitude Modulation (QAM) finds application in Colour Television (CTV).

2.20 EFFECT OF PHASE AND FREQUENCY ERRORS IN SYNCHRONOUS DETECTION

As we already discussed, the frequency and phase of the local oscillator signal in coherent detection method at the receiver end, must be identical to the transmitted carrier signal. Any kind of discrepancy in frequency or phase produces a distortion in the detection output at the receiver end.

In this, let us assume that the modulated signal reaching the receiver is denoted as $x(t)$ $\cos \omega_c t$. Consider, locally generated signal with frequency and phase error equal to $\Delta\omega$ and ϕ respectively, the product of the two signals in the synchronous detector provides

$$e_d(t) = x(t) \cos \omega_c t . \cos\left[(\omega_c + \Delta\omega)t + \phi\right]$$

or,

$$e_d(t) = \frac{1}{2} x(t)\left\{\cos\left[(\Delta\omega)t + \phi\right] + \cos\left[(2\omega_c t + \Delta\omega)t + \phi\right]\right\} \quad ...(2.81)$$

Now, when the signal is allowed to pass through a low-pass filter (LPF) having a cut-off frequency ω_m, the terms centered around $\pm 2\omega_c$ are filtered out and the filter output is given as

$$e_o(t) = \frac{1}{2} x(t) \cos\left[(\Delta\omega)t + \phi\right] \quad ...(2.82)$$

The baseband signal $x(t)$ is multiplied by a slow-time varying function.

$\cos\left[(\Delta\omega)t + \phi\right]$ which distorts the message signal $x(t)$.

Now, let us consider the following special cases:

(*i*) When the frequency error Dw and phase error f are both zero, then last equation (eqn. (2.82) gives

$$e_o(t) = \frac{1}{2} x(t)$$

This means that there is no distortion in the detected output signal.

(*ii*) When there is only the phase error,

$\Delta\omega = 0$ and $\phi \neq 0$

In this case, eqn. (2.82) gives

$$e_o(t) = \frac{1}{2}x(t)\cos\phi$$

This shows that the output signal is multiplied by cos ϕ. When ϕ is time-independent, there is no distortion, rather, there is only attenuation. The output will be maximum when $\phi = 0$ and minimum when $\phi = 90°$. But, in general ϕ randomly varies with respect to the time due to random variation of propagation media. This causes undesirable distortion in the detected output.

It may be noted that the detected output is zero when $\phi = 90°$. This is known as a quadrature null effect since the signal is zero when the local carrier is in-phase quadrature with the transmitted carrier signal.

(*iii*) When there is only the frequency error, i.e.,

$\Delta\omega \neq 0$ but $\phi = 0$

In this case, eqn. (2.81) gives

$$e_o(t) = \frac{1}{2}x(t)\cos(\Delta\omega)t$$

Here, the multiplying factor cos $(\Delta\omega)t$ is time-dependent and produces distortion in the detected output signal. The error $\Delta\omega$ is usually small and thus a message signal $x(t)$ is multiplied by a slow varying sinusoidal signal. This is a more serious distortion. Therefore, a frequency error must be avoided.

(*iv*) When both errors are non-zero, i.e.

$\Delta\omega \neq 0$ and $\phi \neq 0$

In this case, eqn. (2.82), itself provides the detected output signal. Also, in this case, the constant phase error provides attenuation and the frequency error produces distortion in the detected output signal.

Hence, we get an attenuated and distorted output signal at the receiver end.

2.21 CARRIER ACQUISITION IN DSB-SC SYSTEM OR SYNCHRONOUS/ SYNCHRONIZATION TECHNIQUES IN DSB-SC SYSTEM

Some important synchronization techniques are given below:

2.21.1 Pilot Carrier

A small amount of carrier signal known as pilot carrier is transmitted alongwith the modulated signal from the transmitter. This small amount of carrier signal is called pilot carrier. This pilot carrier, separated at the receiver by an approximate filter, is amplified, and is used to phase lock the locally generated carrier signal at the receiver. The phase locking provides synchronizations. This system, where a weak carrier is transmitted alongwith the DSB-SC signal is also referred to as partially suppressed carrier system as the carrier is not totally suppressed. The process in which a large carrier is transmitted alongwith DSB-SC signal is known as amplitude modulation. This has been already discussed. The large carrier simplifies the reception system. The DSB-SC with partially suppressed carrier is equivalent to an over modulated AM signal.

2.21.2 Costa's Receiver

This system used for synchronous detection of DSB-SC signal has been shown in below Fig. 2.24.

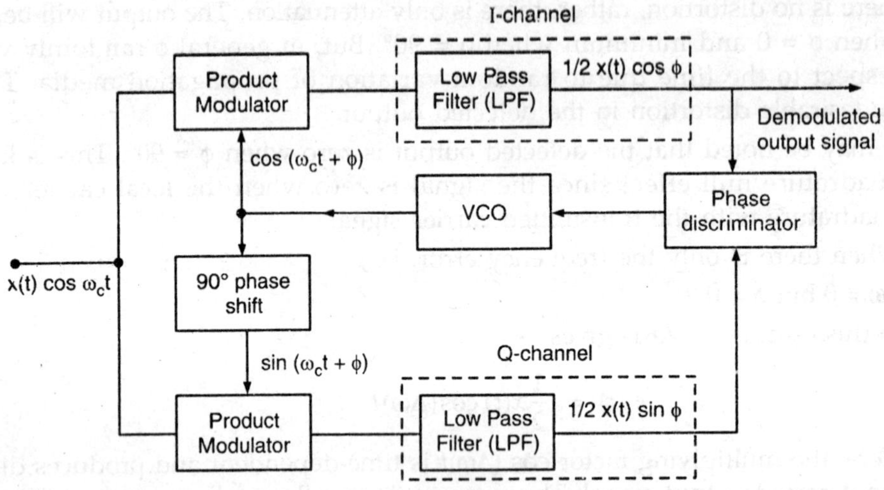

Fig. 2.24 *A Costa's receiver*

This system has two synchronous detectors-one detector is fed with a locally generated carrier signal which is in phase with the transmitted carrier signal. This detector circuit is called inphase coherent detector or I-channel. The other synchronous detector employs a local carrier which is in phase quadrature with the transmitted carrier signal and is called Quadrature phase coherent detector or Q-channel. On combining, the two detectors constitute a negative feedback system which synchronizes the local carrier signal with the transmitted carrier signal.

Figure shows (↑) a Costa's receiver.

2.21.3 Operating Principle

To start with, let us assume that the local carrier signal is synchronized with the transmitted carrier signal and $\phi = 0$. From the figure, the output of the I-channel is the desired modulating signal (since $\cos \phi = 1$), but the output of the Q-channel is zero (since $\sin \phi = 0$) because of the quadrature null effect. Now, assuming that the local oscillator frequency drifts slightly i.e., ϕ is oscillator frequency drifts slightly i.e., ϕ is a very small non-zero quantity, I-channel output will be almost unchanged, but Q-channel output now is not a zero, rather some signal would appear at its output and is proportional to $\sin \phi$. Thus, the output of the Q-channel,

→ is proportional to ϕ

→ would have a polarity same as the I-channel for one direction of phase shift in local oscillator, whereas, the polarity would be opposite to I-channel for the other direction of phase shift.

The phase discriminator provides a d.c. control signal which may be used to correct local oscillator phase error. The local oscillator is a voltage controlled oscillator (VCO). It's frequency may be adjusted by an error control d.c. signal.

2.21.4 Limitation

The Costa's receiver cases phase control when there is no modulation, i.e., $x(t) = 0$. The phase control reestablishes itself on the reappearance of modulation. However, the reestablishment is so fast that distortion is not perceptible in case of voice communication.

2.21.5 Squaring loop

In this method, the received signal is squared by a squaring circuit as shown in given figure.

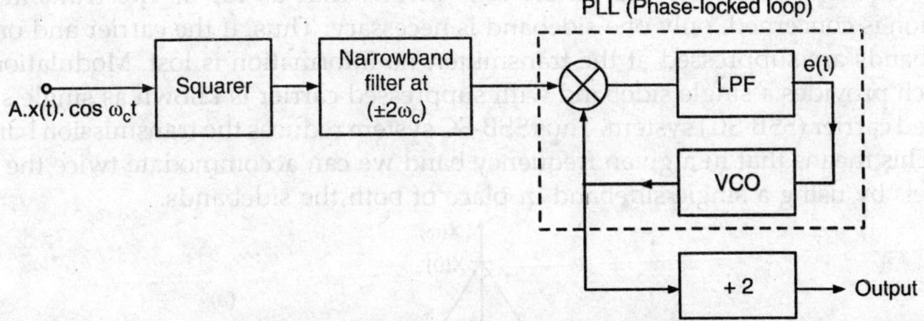

Fig. 2.25 *A squaring circuit for synchronization*

The output of the squarer will be given as

$$[A.x(t)\cos\omega_c t]^2 = A^2x^2(t)\cos^2\omega_c t$$

For simplicity let us assume that $x(t)$ is a single tone sinusoid denoted as $\cos\omega_m t$. i.e.,

$$x(t) = \cos\omega_m t$$

then the output of the squarer becomes,

$$[A\cos\omega_c t.\cos\omega_m t]^2 = A^2\cos^2\omega_m t\cos^2\omega_c t$$

$$= \frac{A^2}{4}(1+\cos 2\omega_m)(1+\cos 2\omega_c t)$$

$$= \frac{A^2}{4}[1+\cos 2\omega_m t+\cos 2\omega_c t+\cos 2\omega_c t\cos 2\omega_m t] \quad ...(2.83)$$

The term $\cos 2\omega_c t$ can be obtained by using a narrowband filter centred at $\pm 2\omega_c$. This frequency $\pm 2\omega_c$ is kept constant by tracking through a phase locked loop (PLL). The PLL uses negative feedback technique to provide a constant frequency signal, $\cos 2\omega_c t$.

Any drift in frequency is corrected by an error signals $e(t)$, generated at the output of the lowpass filter of PLL.

The VCO output is frequency divided by 2, to yield a synchronized local carrier of frequencies ω_c. This local carrier signal is used in synchronous detector. The frequency division can be accomplished by using a bistable multivibrator.

2.22 SINGLE SIDEBAND SUPPRESSED—CARRIER (SSB-SC) MODULATION

We have already discussed in previous section that amplitude modulation and double-sideband suppressed carrier (DSB-SC) modulation are wasteful of bandwidth since they both need a transmission bandwidth equal to twice the message signal bandwidth. In either case one half of the transmission bandwidth is occupied by the upper sideband of the modulated signal whereas the other half is occupied by the lower sideband. However, the lower and upper sidebands are uniquely related to each other by virtue of their symmetry about the carrier frequency, i.e., if amplitude and phase spectra of either sidebands is given, we can uniquely determine the other. This means that as far as the transmission of information is concerned, only one sideband is necessary. Thus, if the carrier and one of the two sidebands are suppressed at the transmitter, no information is lost. Modulation of this type which provides a single sideband with suppressed carrier is known as single sideband suppressed carrier (SSB-SC) system. Thus SSB-SC system reduces the transmission bandwidth by half. This means that in a given frequency band we can accommodate twice the number of channels by using a single sideband in place of both the sidebands.

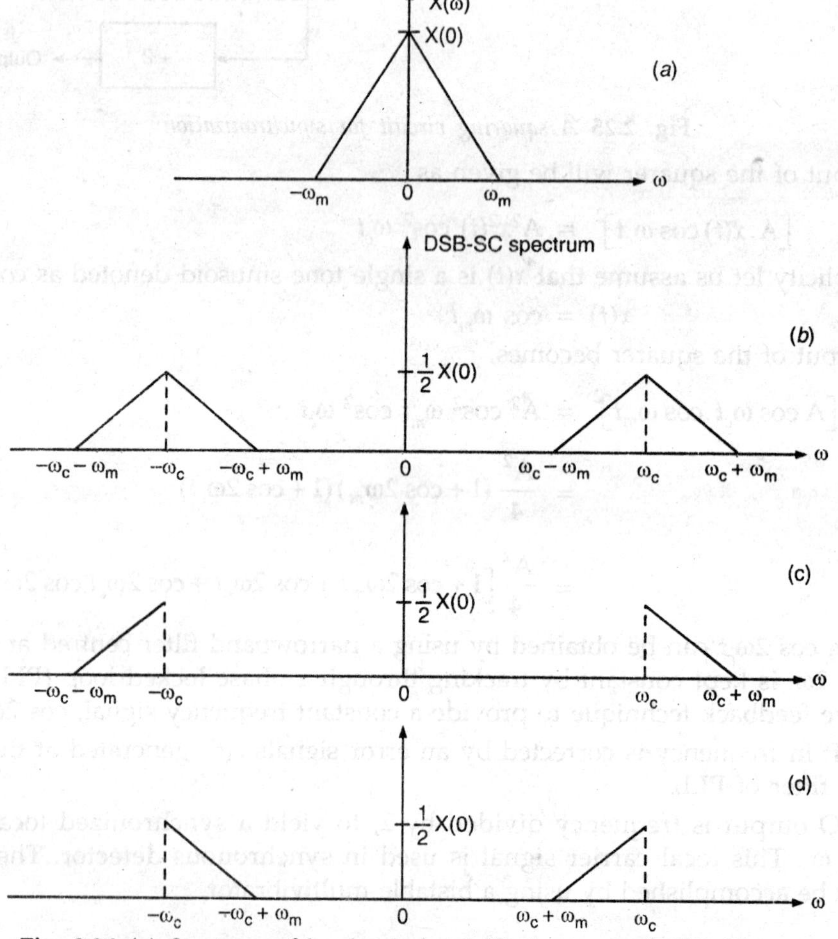

Fig. 2.26 *(a) Spectrum of baseband signal (b) Spectrum of DSB-SC wave (c) Spectrum of SSB-SC wave with the upper sideband transmitted (d) Spectrum of SSB-SC wave with the lower sideband transmitted.*

2.22.1 Time-Domain Description of the SSB-SC Wave

We can derive an expression to represent the SSB wave in time-domain considering a special case of single-tone modulating signal. We can derive a general expression for SSB wave using the concept of a pre-envelope.

2.22.2 SSB-SC Wave with Single-tone Modulating Signal

Let us consider a single-tone modulating signal as

$$x(t) = \cos \omega_m t \qquad ...(2.84)$$

The frequency spectrum of this modulating signal consist of two impulses located at $\omega = \pm\, \omega_c$. If this modulating signal, modulates a carrier signal $\cos \omega_c t$, then the resulting spectrum of DSB-SC signal will be shown in the Fig. (2.27) (*b*).

To get the SSB-SC waveform, we will have to eliminate one of the two sidebands.

Fig. 2.27 (*c*) shows the spectrum of SSB-SC wave with lower sidebands. From the figure, it is clear that this spectrum corresponds to a time-domain signal $\cos (\omega_c - \omega_m)t$, because the frequency spectrum of cosine function contains two impulses in the frequency domain. This means that a SSB-SC wave with lower sideband may be expressed as

$$\cos (\omega_c + \omega_m)t = \cos \omega_m t \cos \omega_c t + \sin \omega_m t \sin \omega_c t \qquad ...(2.85)$$

In the same manner, the expression for the single-tone SSB-SC wave with upper sideband may be expressed as

$$\cos (\omega_c + \omega_m)t = \cos \omega_m t \cos \omega_c t - \sin \omega_m t \sin \omega_c t \qquad ...(2.86)$$

Both these eqns. (2.85) and (2.86) may be combined

$$s(t)_{SSB} = \cos \omega_m t \cos \omega_c t \pm \sin \omega_m t \sin \omega_c t \qquad ...(2.87)$$

Here, the (+ve) sign represents the lower sideband and (–ve) sign represented the upper sideband. The terms $\sin \omega_c t$ and $\omega_c t$ may be written as

$$\sin \omega_m t = \cos\left(\omega_c t - \frac{\pi}{2}\right) \qquad ...(2.88)$$

$$\sin \omega_m t = \cos\left(\omega_c t - \frac{\pi}{2}\right) \qquad ...(2.89)$$

This means that the sine terms may be obtained using the corresponding cosine terms simply by giving a phase shift of $\left(-\frac{\pi}{2}\right)$.

The expression in eqn. (2.89) represents the SSB-SC wave for the case of single-tone modulation, however, it provides a way to get a general expression for SSB-SC wave.

In the expression of eqn. (2.88), the term $\sin \omega_m t$ is obtained by given by giving a phase shift of $\left(-\frac{\pi}{2}\right)$ to the modulating frequency $\cos \omega_m t$.

Similarly, in a general modulating signal $x(t)$, if all the frequency components are shifted by $\left(-\frac{\pi}{2}\right)$, it may lead to a general expression of SSB-SC signal.

Now, it may be noted that $x(t)$ may be expressed as a continuous sum of sinusoidal.

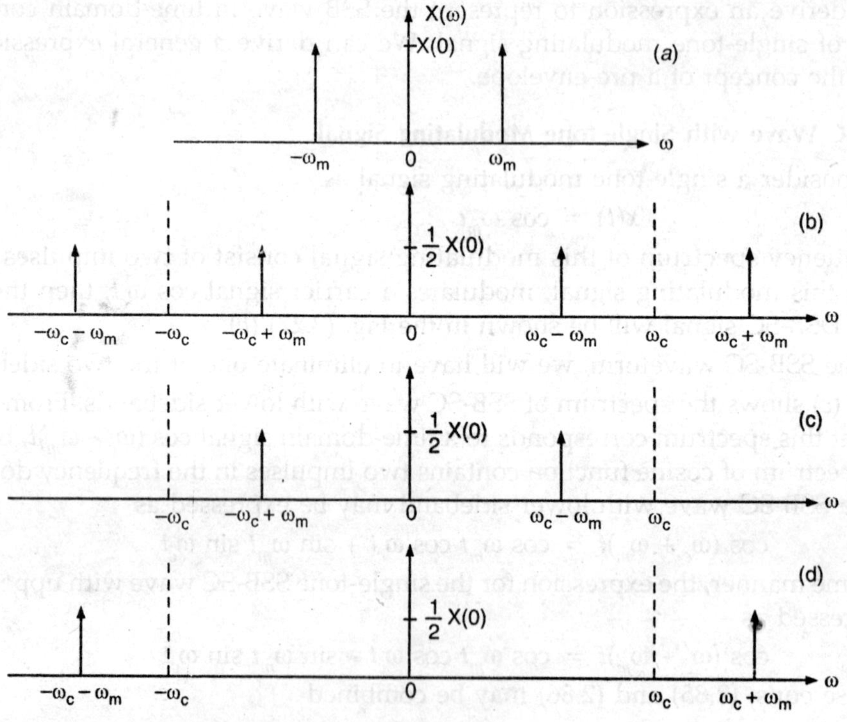

Fig. 2.27 *(a) Frequency spectrum of modulating signal (b) Frequency spectrum of DSB-SC wave (c) Spectrum of SSB-SC with lower sideband, (d) Spectrum of SSB-SC with upper sideband*

Therefore, expression in eqn. (2.90) may be extended for a SSB-SC signal modulated by a general modulating signal $x(t)$ as expressed below:

$$s(t)_{SSB} = x(t) \cos \omega_c t \pm \sin \omega_c t \qquad ...(2.90)$$

where $x_h(t)$ is a signal obtained by shifting the phase of every component present in $x(t)$ by $\left(-\frac{\pi}{2}\right)$.

Similarly, to eqn. (2.87), (+ve) sign corresponds to the lower sideband and (–ve) sign corresponds to the upper sideband.

2.23 HILBERT TRANSFORM

It can be seen that $f_\lambda(t)$ obtained by giving $\left(-\frac{\pi}{2}\right)$ phase shift to each frequency component present in $f(t)$, actually represents the Hilbert Transform of $f(t)$ i.e. $f_\lambda(t)$ is Hilbert Transform of $f(t)$ is defined as

$$f_\lambda(t) = \frac{1}{\pi} f(t) * \frac{1}{t}$$

$$= \frac{1}{\pi} \int_{-\infty}^{+\infty} \frac{f(\tau)}{t-\tau}\, d\tau \qquad ...(2.91)$$

The inverse Hilbert Transform is defined as

$$f(t) = -\frac{1}{\pi} \int_{-\infty}^{+\infty} \frac{f_\lambda(\tau)}{t-\tau}\, d\tau \qquad ...(2.92)$$

EXAMPLE 1

Show that if every frequency component of a signal $f(t)$ is shifted by $\frac{\pi}{2}$, the resultant signal $f_\lambda(t)$ is the Hilbert Transform of $f(t)$.

Solution: The situation may be regarded as though the signal $f(t)$ is passed through a phase shifting system H(ω), and the output is $f_\lambda(t)$ as shown in Fig. 2.28.

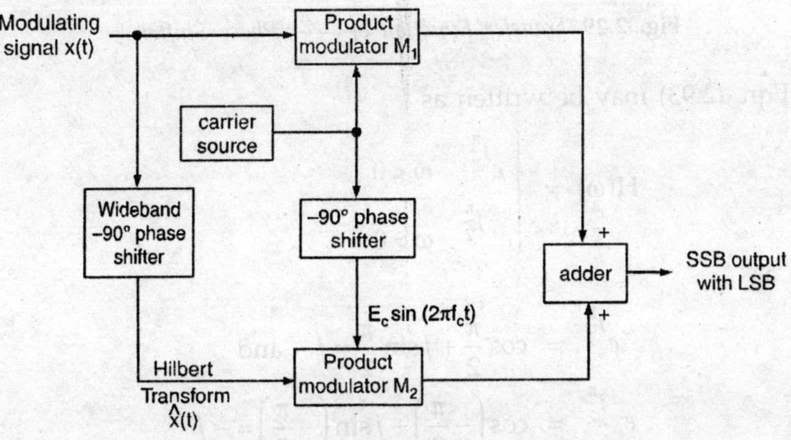

Fig. 2.28 *A Phase Shifting System*

The characteristics of the system are specified as:

(*i*) The magnitude of the frequency component present in $f(t)$ remains unchanged when it is passed through the system, i.e., H(ω) = 1, and

(*ii*) The phase of the positive frequency components is shifted by $-\frac{\pi}{2}$. Since the phase spectrum θ(ω) has an odd symmetry, the phase of the negative frequency components is shifted by $+\frac{\pi}{2}$. H(ω) and θ(ω) are plotted in Fig. 2.31.

The transfer function is given as

$$H(\omega) = |H(\omega)|\, e^{j\theta(\omega)}$$

$$= 1\, e^{j\theta(\omega)} \qquad ...(2.93)$$

It is evident from Fig. 2.29 that

$$\theta(\omega) = \begin{cases} +\dfrac{\pi}{2}, & \omega < 0 \text{ (i.e., negative frequencies)} \\ -\dfrac{\pi}{2}, & \omega > 0 \text{ (i.e., positive frequencies)} \end{cases}$$

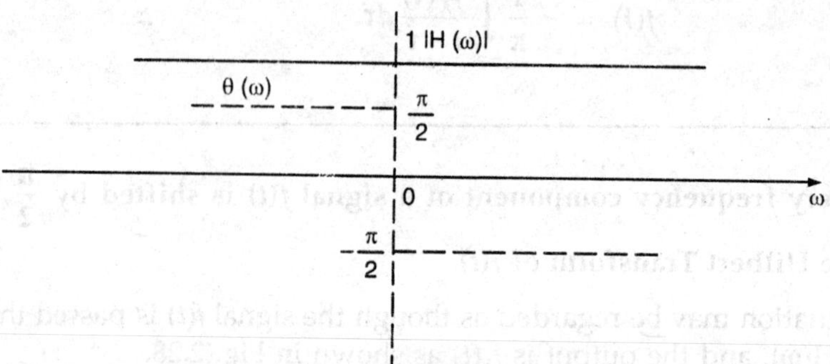

Fig. 2.29 *Transfer Function of* $-\dfrac{\pi}{2}$ *Phase Shifter*

Therefore, Eqn. (2.93) may be written as

$$H(\omega) = \begin{cases} e^{j\frac{\pi}{2}} & \omega < 0 \\ e^{-j\frac{\pi}{2}} & \omega > 0 \end{cases}$$

It is known as

$$e^{j\frac{\pi}{2}} = \cos\frac{\pi}{2} + j\sin\frac{\pi}{2} = j \quad \text{and}$$

$$e^{-j\frac{\pi}{2}} = \cos\left(-\frac{\pi}{2}\right) + j\sin\left(-\frac{\pi}{2}\right) = -j$$

Hence, H(ω) becomes

$$\frac{H(\omega)}{j} = \begin{cases} 1 & \omega < 0 \\ -1 & \omega > 0 \end{cases}$$

$$= -\operatorname{sgn}(\omega)$$

or

$$H(\omega) = -j\operatorname{sgn}(\omega) \qquad \text{...(2.94)}$$

The response $F_\lambda(\omega)$ of the system is related to input F(ω), as

$$F_\lambda(\omega) = F(\omega)\, H(\omega)$$

where

$$f(t) \leftrightarrow F(\omega) \text{ and } f_\lambda(t) \leftrightarrow F_\lambda(\omega)$$

Substituting H(ω) from Eqn. (2.94)

$$F_\lambda(\omega) = -jF(\omega)\operatorname{sgn}(\omega) \qquad \text{...(2.95)}$$

Taking the inverse Fourier Transform of both sides of Eqn. (2.95)

$$f_\lambda(t) = F^{-1}[-jF(\omega)\operatorname{sgn}(\omega)]$$

The time domain of sgn (ω) is given by

$$\frac{1}{\pi t} \leftrightarrow \text{sgn}\ (\omega)$$

Using time convolution theorem,

$$f_\lambda(t) = \frac{1}{\pi}\left[f(t) * \frac{1}{t}\right]$$

$$= \frac{1}{\pi}\int_{-\infty}^{\infty} \frac{f(\tau)}{t-\tau}\,d\tau$$

Which is the Hilbert Transform of $f(t)$. Some applications of the Hilbert Transform are as:

(*i*) generation of SSB signals

(*ii*) design of minimum phase type filters

(*iii*) representation of bandpass signals

2.23.1 Properties of a Hilbert transform are as under

1. A signal $f(t)$ and its Hilbert transform $f_\lambda(t)$ have the same energy density spectrum.
2. A signal $f(t)$ and its Hilbert transform $f_\lambda(t)$ have the same auto correlation function.
3. A signal $f(t)$ and its Hilbert transform $f_\lambda(t)$ are mutually orthogonal, i.e.,

$$\int_{-\infty}^{\infty} f(t)\, f_\lambda(d) = 0$$

4. If $f_\lambda(t)$ is a Hilbert transform of $f(t)$, then the Hilbert transform of $f_\lambda(t)$ is $-f(t)$, that is, if

$$\text{H}[f(t)] = f_\lambda(t)$$

then

$$\text{H}[f_\lambda(t)] = -f(t)$$

where H denotes the Hilbert Transform.

2.23.2 Pre-envelope or Analytic Signal

The concept of fibre-envelope, also known as analytic function is useful in deriving the general expression is of the SSB-SC signal. The pre-envelope of a real valued signal $f(t)$ is defined as

$$f_p(t) = f(t) + j f_\lambda(t) \qquad \text{...(2.96)}$$

where $f_\lambda(t)$ is a Hilbert transform of $f(t)$. Obviously, the pre-envelope $f_p(t)$ is a complex-valued signal. The real part of $f_p(t)$, and the imaginary part is its Hilbert transform $f_\lambda(t)$. The complex conjugate of the pre-envelop denoted by $f_p(t)$ is given as

$$f_p^* = f(t) - j f_\lambda(t) \qquad \text{...(2.97)}$$

2.23.3 SSB-SC for a General Modulating Signal

We have already derived an expression for a generalized SSB-SC signal by extending the concept of SSB-SC for a single-tone modulation. We will now derive the same expression, as Eqn. 2.96, by using the concept of pre-envelope, or analytic signal.

Pre-envelope of a function $f(t)$ as defined as Eqn. (2.96) is reproduced as follows

$$f_p(t) = f(t) + j f_\lambda(t)$$

The Fourier Transform of $f_p(t)$ is the sum of the Fourier Transform of $f(t)$ and $f_\lambda(t)$,

$$F[f_p(t)] = F[f(t)] + jF[f_\lambda(t)]$$

or

$$F_p(\omega) = F(\omega) + j[-jF(\omega)\,\text{sgn}\,(\omega)]$$

$$= F(\omega) + F(\omega)\,\text{sgn}\,(\omega) \qquad \text{...(2.98)}$$

It is known that,

$$\text{sgn}\,(\omega) = \begin{cases} 1 & \text{for } \omega > 0 \\ -1 & \text{for } \omega < 0 \end{cases}$$

Hence

$$F_p(\omega) = \begin{cases} 2F(\omega), & \omega > 0 \\ 0, & \omega < 0 \end{cases} \qquad \text{...(2.99)}$$

Eqns. 2.98 and 2.99 show F(ω) and $F_p(\omega)$ respectively. It is evident from Fig. 2.30 (*b*) that $F_p(\omega)$ vanishes for negative frequencies.

Similarly, we can find the Fourier transform of $f_p^*(t)$ defined by Eqn. (2.97),

$$f_p^*(\omega) = F(\omega) - j[-jF(\omega)\,\text{sgn}\,(\omega)]$$

$$= F(\omega) - F(\omega)\,\text{sgn}\,(\omega) \qquad \text{...(2.100)}$$

Which may be written as,

$$f_p^*(\omega) = \begin{cases} 0, & \omega > 0 \\ 2F(\omega), & \omega < 0 \end{cases} \qquad \text{...(2.101)}$$

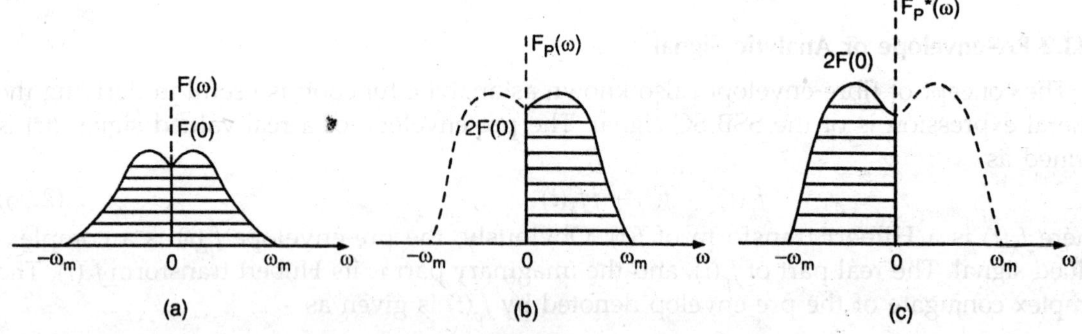

Fig. 2.30

This is plotted in Fig. 2.30(*c*). Evidently, $F_p(\omega)$ vanishes for positive frequencies.

Now, take an SSB-SC wake consisting of only the lower sidebands of a general modulating signal $f(t)$ as shown in Fig. 2.31. We observe the following points in Fig. 2.31.

(*i*) The right-hand position of the figure represents a spectrum of $\frac{1}{4} f_p^*(t)\, e^{j\omega_c t}$. This is equivalent to shifting the spectrum $f_p^*(\omega)$ towards right by ω_c.

(*ii*) Similarly, left-hand portion of Fig. 2.31 represents the spectrum of

$$\frac{1}{4} f_p(t)\, e^{-j\omega_c t}.$$

Therefore, Fig. 2.31 represents the spectrum of a combined signal $f_p(t)\, e^{+j\omega_c t} + f_p(t)\, e^{-j\omega_c t}$.

Hence, the time domain representation of the SSB-SC spectrum shown in Fig. 2.31 gives as

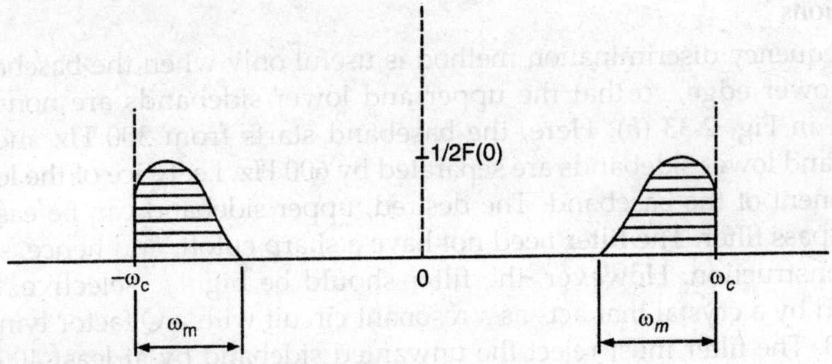

Fig. 2.31 *SSB-SC Spectrum Consisting of Lower Sidebands*

$$f_{SSB}(t) = \frac{1}{4}\left[f_p^*(t)\, e^{j\omega_c t} + f_p(t)\, e^{-j\omega_c t}\right] \qquad ...(2.102)$$

Substituting in terms of Hilbert transform from Eqns. (2.96) and (2.97), we have

$$\phi_{SSB}(t) = \frac{1}{4}[f(t) - j f_\lambda(t)]\, e^{j\omega_c t} + \frac{1}{2}[f(t) + j f_\lambda(t)]\, e^{-j\omega_c t}$$

$$= \frac{1}{2} f(t)\left[\frac{e^{j\omega_c t} + e^{-j\omega_c t}}{2}\right] + \frac{j}{2} f_\lambda(t)\left[\frac{e^{-j\omega_c t} + e^{j\omega_c t}}{2}\right]$$

$$= \frac{1}{2}[f(t)\cos\omega_c t + f_\lambda(t)\sin\omega_c t]$$

Which is the time domain description of the SSB-SC signal consisting of only the lower sidebands. Similarly, we can derive an expression for an SSB-SC signal consisting of the upper sidebands,

$$\phi_{SSB}(t) = \frac{1}{2}[f(t)\cos\omega_c t - f_\lambda(t)\sin\omega_c t]$$

Therefore, the time domain description of the SSB-SC wave is represents by the expression

$$\phi_{SSB}(t) = f(t)\cos\omega_c t \pm f_\lambda(t)\sin\omega_c(t)$$

Where + and – sign corresponds to the lower sidebands and upper sidebands, respectively.

2.24 GENERATION OF SSB-SC SIGNALS

SSB-SC wave can be generated by two methods:

(*i*) frequency discrimination method, or filter method;

(*ii*) phase discrimination method, or phase-shift method or phasing method.

2.24.1 Frequency Discrimination Method

First a DSB-SC signal is generated by using a simple product modulator, or a balanced modulator, and then of the sidebands is filtered out by an appropriate bandpass filter. The schematic diagram is shown in Fig. 2.32. The design of bandpass filter is very critical and puts some limitations on the modulating and carrier frequencies.

2.24.2 Limitations

1. The frequency discrimination method is useful only when the baseband is restricted at its lower edge, so that the upper and lower sidebands are non-overlapping as shown in Fig. 2.33 (*b*). Here, the baseband starts from 300 Hz and therefore, the upper and lower sidebands are separated by 600 Hz, i.e. twice of the lowest frequency component of the baseband. The desired, upper sideband can be easily obtained by a bandpass filter. The filter need not have a sharp cutoff, and hence, simple in design and construction. However, the filter should be highly selective. Such filters are realized by a crystal that acts as a resonant circuit with a Q factor lying between 1000 to 2000. The filter must reject the unwanted sideband by at least 40 dB as shown in Fig. 2.33 (*b*). When the baseband is not restricted at its lower edge and contains very low frequencies, the upper and lower sideband may overlap each other, as shown in Fig. 2.33 (*a*). The baseband starts from d.c., and hence there is no separation between the two sidebands; rather they touch each other at ω_c. Therefore, a sharp cutoff bandpass filter is needed to separate the desired upper sideband from the lower one. It is impossible to realize such an ideal filter.

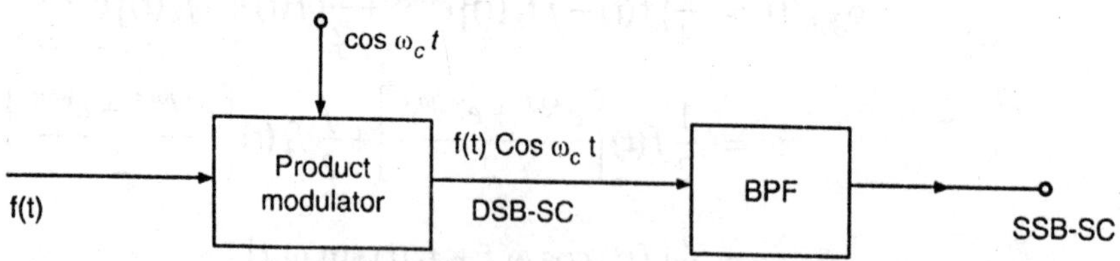

Fig. 2.32 *Filter Method for Generating SSB-SC*

The filter method is used for speech communication where lowest spectral component is 70 Hz, and it can be taken as 300 Hz without affecting the intelligibility of the speech. However, the system is not useful for video communication where the baseband starts from as low as d.c.

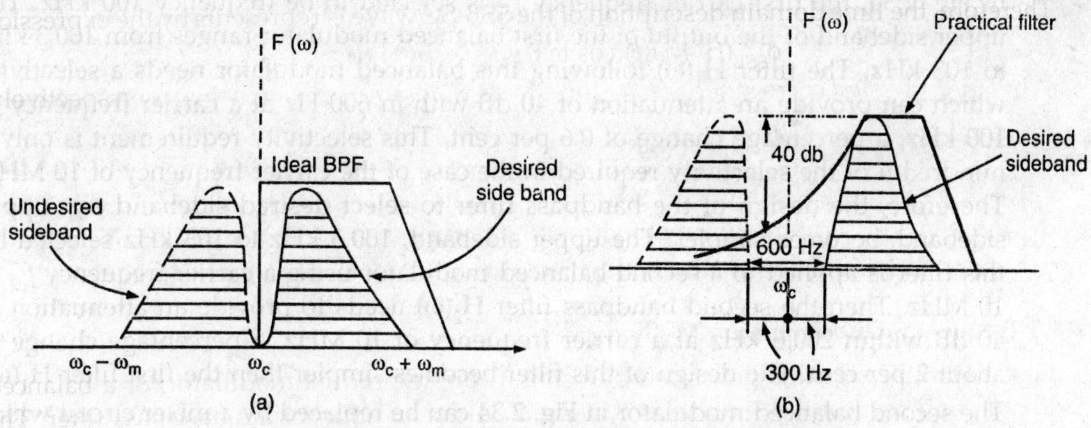

Fig. 2.33 *Filter Requirement for SSB-SC Generation*
(a) Overlapping Sidebands (b) Non-overlapping Sidebands

2. The frequencies discrimination method has another restriction that the baseband should be appropriately related to the carrier frequency. The design of the bandpass filter becomes very difficult if the carrier frequency is much higher than the bandwidth of the baseband signal. Let us consider an example in which a voice signal with a baseband ranging from 300 Hz to 300 kHz modulates a carrier wave of 10 MHz. The bandpass filter requires a selectivity that can provide an attenuation of 40 dB within 600 Hz at a carrier frequency of 10 MHz, a percentage frequency change of only 0.006 per cent. The design and construction of such a sharp selective filter is very complex and difficult. This problem is solved by performing the translation of the baseband signal to final carrier frequency in several stages. This approach is illustrated in Fig. 2.34. Which involves two translation stages.

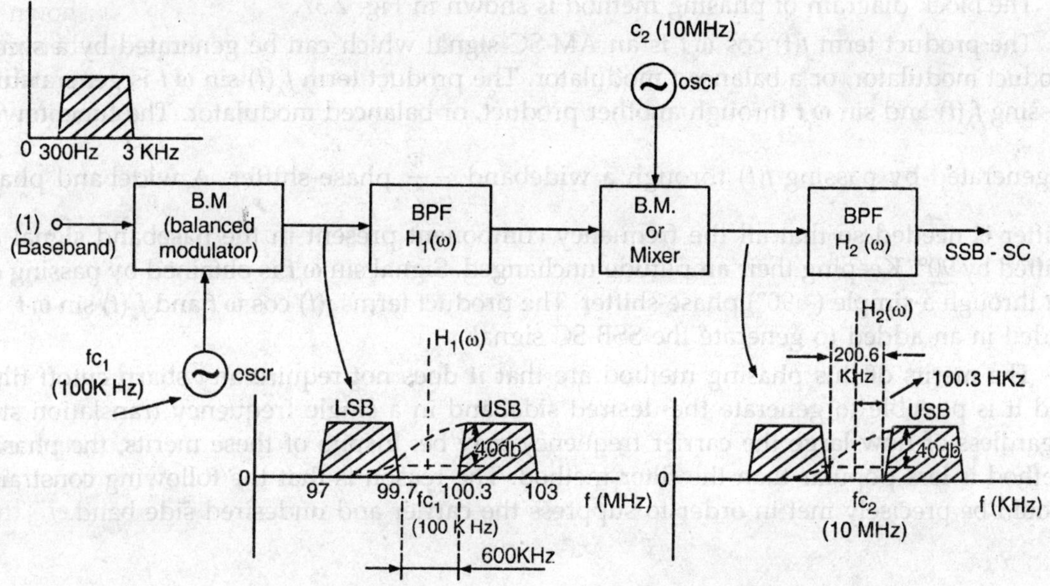

Fig. 2.34 *SSB-SC with Two Translation Stages*

In Fig. 2.34, the first carrier frequency f_{c1} is selected to be frequency 100 kHz. The upper sideband of the output of the first balanced modulator ranges from 100.3 kHz to 103 kHz. The filter $H_1(\omega)$ following this balanced modulator needs a selectivity which can provide an attenuation of 40 dB with in 600 Hz at a carrier frequency of 100 kHz, a percentage change of 0.6 per cent. This selectivity requirement is only a hundredth of the selectivity required in the case of the carrier frequency of 10 MHz. Therefore, the design of the bandpass filter to select desired sideband e.g. upper sideband, becomes simpler. The upper sideband, 100.3 kHz to 103 kHz selected by this filter is applied to a second balanced modulator using a carrier frequency f_{c2} of 10 MHz. Then the second bandpass filter $H_2(\omega)$ needs to provide an attenuation of 40 dB within 200.6 kHz at a carrier frequency of 10 MHz, a percentage change of about 2 per cent. The design of this filter becomes simpler then the first filter $H_1(\omega)$. The second balanced modulator in Fig. 2.34 can be replaced by a mixer circuit which is a simpler frequency translation device. A mixer, however, has the disadvantage that it produces not only the desired sum and difference frequencies, but also the undesired input frequencies. But this problem is not so serious because the input frequencies, including carrier frequency are well out of the passband of the second filter $H_2(\omega)$. Thus, while generating SSB, if the carrier is of the MHz order, or more, the frequency translation should be achieved in more than stage for convenience in appropriate filter design.

2.25 PHASE DISCRIMINATION OR PHASING METHOD

In this method, the time domain description of the SSB-SC wave given in equation

$$\phi_{SSB}(t) = f(t)\cos\omega_c t \pm f_\lambda(t)\sin\omega_c t$$

is used. A lower sideband SSB-SC signal is given by the expression

$$\phi_{SSB}(t) = f(t)\cos\omega_c t + f_\lambda(t)\sin\omega_c t$$

The block diagram of phasing method is shown in Fig. 2.37.

The product term $f(t)\cos\omega_c t$ is an AM-SC signal which can be generated by a simple product modulator, or a balanced modulator. The product term $f_\lambda(t)\sin\omega_c t$ is generated by passing $f_\lambda(t)$ and $\sin\omega_c t$ through another product, or balanced modulator. The function $f_\lambda(t)$ is generated by passing $f(t)$ through a wideband $-\frac{\pi}{2}$ phase-shifter. A wideband phase-shifter is needed so that all the frequency component present in the baseband signal are shifted by 90°. Keeping their amplitude unchanged. Signal $\sin\omega_c t$ is obtained by passing $\cos\omega_c t$ through a simple (– 90°) phase-shifter. The product terms $f(t)\cos\omega_c t$ and $f_\lambda(t)\sin\omega_c t$, are added in an added to generate the SSB-SC signal.

The merits of this phasing method are that it does not require any sharp cutoff filter, and it is possible to generate the desired sideband in a single frequency translation step, regardless of how large the carrier frequency may be. Inspite of these merits, the phasing method is less popular than the filter method. The reason is that the following constraints should be precisely met in order to suppress the carrier and undesired side band.

(i) Each balanced modulator need to be carefully balanced in order to suppress the carrier.

(ii) In practice, it is difficult to design a wideband phase shifting network, particularly where the baseband extends over many octaves. This network is needed to generate Hilbert transform of $f(t)$.

(iii) Each modulator should have equal sensitivity to the baseband signal.

(iv) The carrier phase shifting network must provide an exact 90° phase shift at the carrier frequency.

The most difficult problem arises while designing a wideband phase shifting network to generate Hilbert transform. The wideband constant phase shift is obtained over a given frequency range by including a phase-shifting network in each path of modulation, i.e. before each of the two balanced modulators.

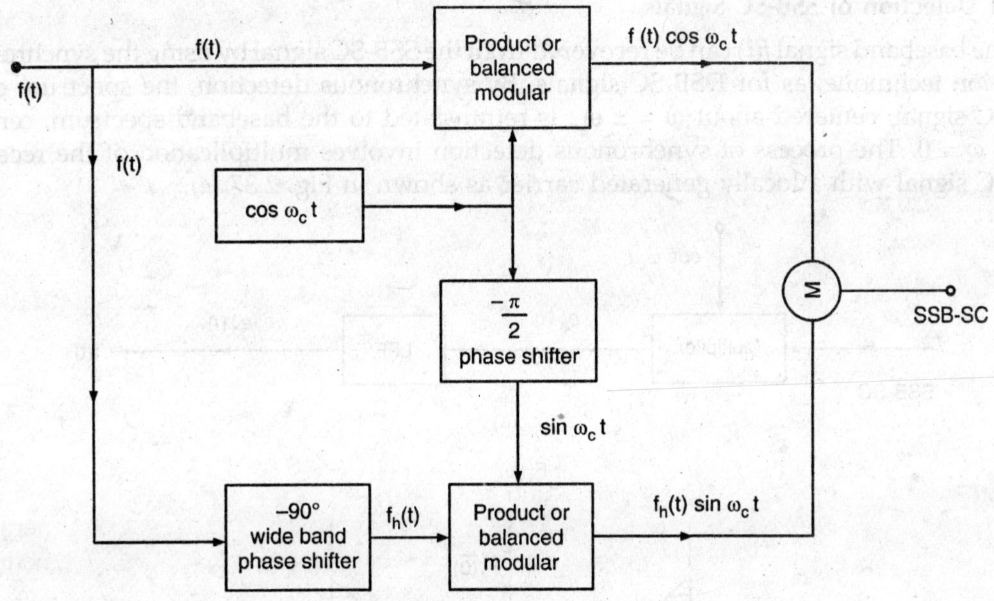

Fig. 2.35 *Phase Shift Method for SSB-SC Generation*

This type of arrangement is shown in Fig. 2.36. The two phase shifting networks are denoted by α and β. The phase-shifts α and β are related by

$$\beta - \alpha = \frac{\pi}{2}$$

The phase-shifts α and β are adjusted to provide a constant phase-shift over a given frequency range within the specified tolerance.

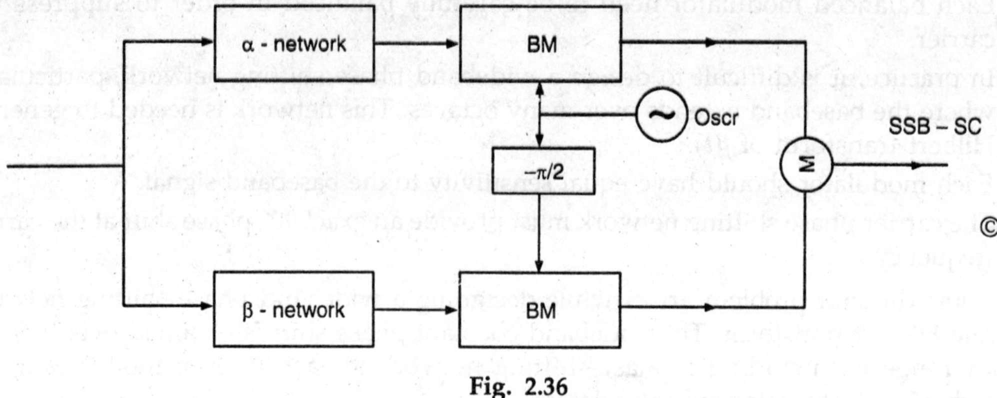

Fig. 2.36

2.25.1 Detection of SSB-SC Signals

The baseband signal $f(t)$ can be recovered from the SSB-SC signal by using the synchronous detection technique, as for DSB-SC signals. By synchronous detection, the spectrum of an SSB-SC signal, centered about $\omega = \pm\, \omega_c$, is retranslated to the baseband spectrum, centred about $\omega = 0$. The process of synchronous detection involves multiplication of the received SSB-SC signal with a locally generated carrier as shown in Fig. 2.37 (*a*).

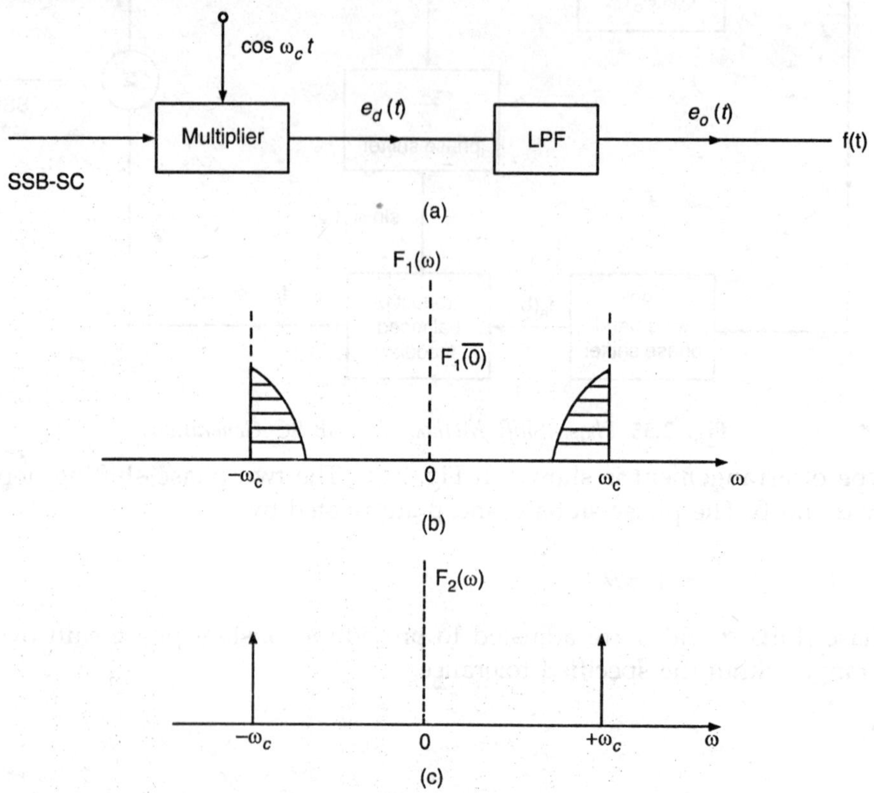

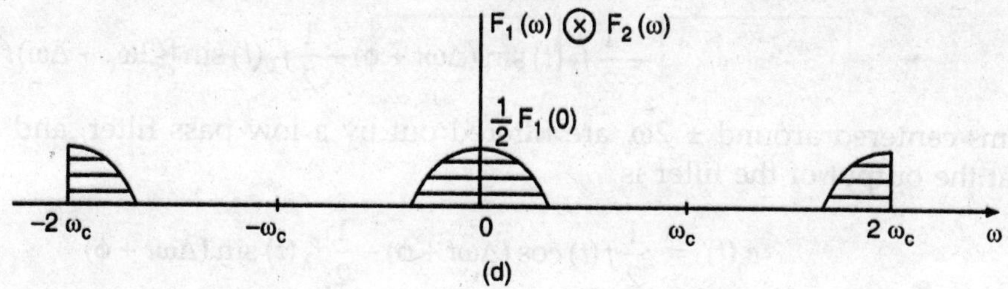

Fig. 2.37 *Detection of SSB-SC Signal (a) Synchronous Detector, (b) SSB-SC (lower sideband) (c) Spectrum of cos $\omega_c t$, (d) Convolution of Figures (b) and (c)*

The output of the multiplier is,

$$e_d(t) = \phi_{SSB}(t) \cos \omega_c t$$

$$= [f(t) \cos \omega_c t \pm f_\lambda(t) \sin \omega_c t] \cos \omega_c t$$

$$= f(t) \cos^2 \omega_c t \pm f_\lambda(t) \sin \omega_c t \cos \omega_c t$$

$$= \frac{1}{2} f(t) \{1 + \cos 2\omega_c t\} \pm \frac{1}{2} f_\lambda(t) \sin 2\omega_c t$$

$$= \frac{1}{2} f(t) + \frac{1}{2}[f(t) \cos 2\omega_c t \pm f_\lambda(t) \sin 2\omega_c t] \quad \text{...(2.103)}$$

When $e_d(t)$ is passed through a low pass filter, the terms centered about $\pm 2\omega_c$ are filtered out and we get, at the output of detector, e_o given by

$$e_o(t) = \frac{1}{2} f(t) \quad \text{...(2.104)}$$

The frequency domain explanation is evident from Fig. 2.37. Multiplication of the SSB signal with cos $\omega_c t$ in the time domain is equivalent to convolution of their spectra, as shown in Fig. 2.37 (*d*). The component centered around $\pm 2\omega_c$ is filtered out, whereas, the massaged signal centered at $\omega = 0$ appears at the output. The synchronous detection can be achieved by a ring demodulator circuit, or a modulator circuit using non-linear devices.

2.25.2 Effect of Phase and Frequency Error

As in DSB-SC signals, the synchronous detection of SSB-SC signals also needs phase and frequency synchronization. A detected output will be distorted when there is a discrepancy in phase and frequency of the locally generated carrier and transmitted carrier. Distortion due to discrepancy in frequency is somewhat same in both DSB-SC and SSB-SC detection, however, distortion because of phase discrepancy is more serious in the latter.

Let us presume that the locally generated carrier has a frequency error of $\Delta\omega$, and a phase-error of ϕ, then the output of multiplier in Fig. 2.37 (*a*) is given as

$$e_d(t) = [f(t) \cos \omega_c t + f_\lambda(t) \sin \omega_c t][\cos \{(\omega_c + \Delta\omega)t + \phi\}]$$

$$= \frac{1}{2} f(t) \cos(\Delta\omega t + \phi) + \frac{1}{2} f(t) \cos[(2\omega_c + \Delta\omega)t + \phi]$$

$$-\frac{1}{2}f_\lambda(t)\sin(\Delta\omega t+\phi)+\frac{1}{2}f_\lambda(t)\sin[(2\omega_c+\Delta\omega)t+\phi]$$

The terms centered around $\pm\, 2\omega_c$ are filtered out by a low pass filter, and the signal appearing at the output of the filter is

$$e_o(t) = \frac{1}{2}f(t)\cos(\Delta\omega t+\phi)-\frac{1}{2}f_\lambda(t)\sin(\Delta\omega t+\phi) \qquad ...(2.105)$$

Consider the following cases:

(*i*) When there is no phase and frequency error, ($\Delta\omega = 0$, $\phi = 0$) then

$$e_o(t) = \frac{1}{2}f(t) \text{ (without any distortion)}$$

(*ii*) When phase error is zero, and frequency error, $\Delta\omega$ then from (2.105)

$$e_o(t) = \frac{1}{2}f(t)\cos(\Delta\omega)t-\frac{1}{2}f_\lambda(t)\sin(\Delta\omega)t$$

The signal is slow variant with time causing a distortion similar to that observed in DSB-SC detection. The amplitude fluctuation at a rate of $\Delta\omega$ is known as Warble and is tolerable upto $\Delta\omega = 30$ Hz.

(*iii*) When the frequency error is zero and phase error is equal to ϕ, then

$$e_o(t) = \frac{1}{2}[f(t)\cos\phi - f_\lambda(t)\sin\phi] \qquad ...(2.106)$$

Since $f(t)$ and $f_\lambda(t)$ contain the same frequency spectrum they cannot be separated. Now let us see that the undesired term $f_\lambda(t)\sin\phi$ introduces a phase distortion, shifting every component of $f(t)$ in phase by ϕ radians. The Fourier transform of Eqn. (2.106) yields

$$E_o(\omega) = \frac{1}{2}[F(\omega)\cos\phi - F_\lambda(\omega)\sin\phi] \qquad ...(2.107)$$

Substituting this value of $f_\lambda(t)$ from Eqn. (2.107) and recalling the definition of sgn (ω) we can write

$$E_o(\omega) = \begin{cases} \frac{1}{2}F(\omega)e^{j\phi}, & \omega > 0 \\ \frac{1}{2}F(\omega)e^{-j\phi}, & \omega < 0 \end{cases} \qquad ...(2.108)$$

Multiplication of $e^{j\phi}$ in frequency domain is equivalent to phase shift of ϕ in time domain. Hence Eqn. (2.108) shows that every component of $f(t)$ is shifted in phase by ϕ radians, causing phase distortion in the detected output. In DSB-SC, the phase error introduced only attenuation, and not distortion. The phase distortion does not seriously affect voice communication, as it may change only quality of speech and not its intelligibility. This is because the human ear is less sensitive to phase distortion, distortion being known as Donald Duck Voice effect. However, phase distortion is intolerable in video signal and music transmission.

2.25.3 Phasor Diagram and Waveform of SSB Signals

Now let us consider a single-tone modulating signal $E_1 \cos \omega_1 t$ that modulates a carrier $E_c \cos \omega_c t$ to generate an SSB-SC signal. The resulting signal is given as, considering lower sideband

$$\phi_{SSB}(t) = \frac{1}{2} E_1 E_c \cos(\omega_c - \omega_1)t \quad ...(2.109)$$

The signal represents a single sinusoidal with an amplitude $\frac{1}{2}E_1E_c$ and a frequency $(\omega_c - \omega)$. Therefore, the phasor diagram consists of a single phasor, and there will be no amplitude fluctuation in the modulated wave. This is obvious from the phasor diagram in Fig. 2.38 (*a*), and the waveform drawn in Fig. 2.38 (*b*). It is also from the waveform in that both amplitude and frequency of the wave are constant, i.e. therefore is no amplitude modulation.

Therefore, an SSB-SC signal modulated by a single tone modulating signal does not carry any useful information. However, the baseband signal appearing in actual systems are multiple-tone rather than single tone. Consider a two tone baseband given by

$$f(t) = E_1 \cos \omega_1 t + E_2 \cos \omega_2 t$$

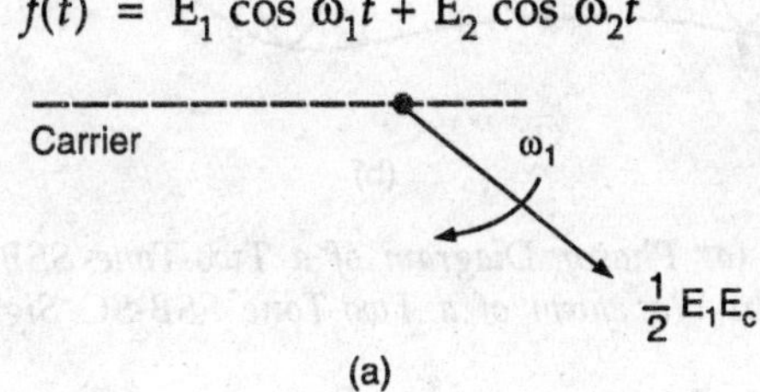

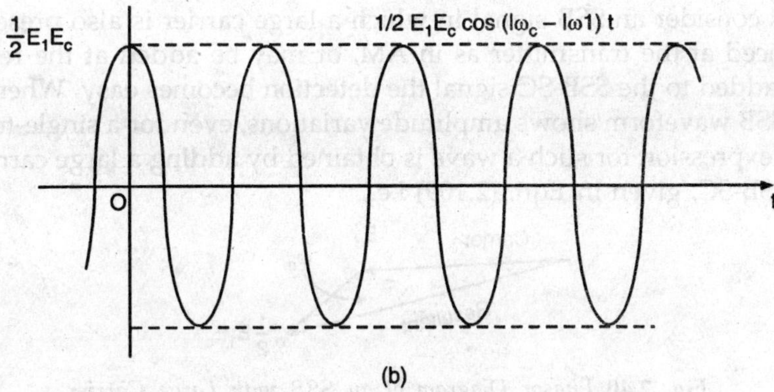

Fig. 2.38 *(a) Phasor Diagram of a Single-Tone SSB-SC Signal, (b) Waveform of a Single-Tone SSB-SC Signal*

Fig. 2.39 (*a*) shows the phasor diagram of an SSB-SC signal modulated by the baseband signal $f(t)$. The resultant phasor varies with time as the two phasor rotate. The amplitude variation in the resulting wave form is shown in Fig. 2.39 (*b*). Thus the multiple tone SSB-SC signal shows amplitude variation.

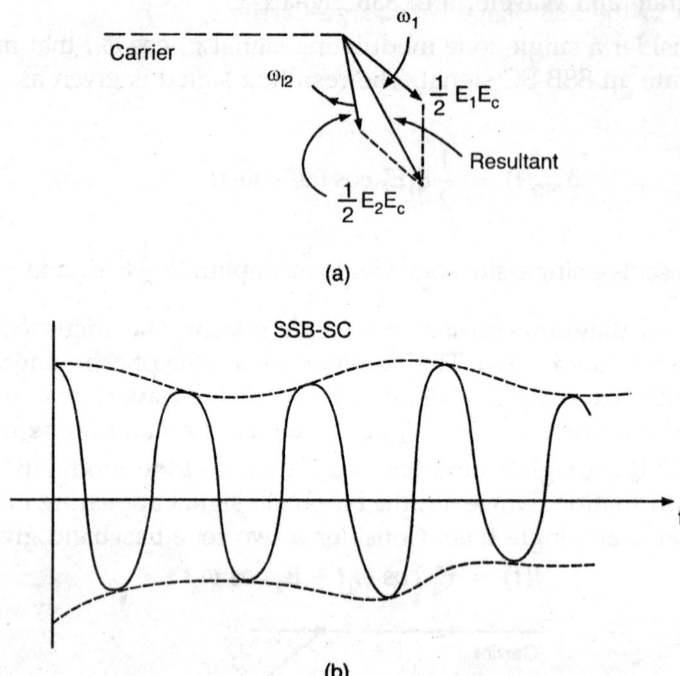

Fig. 2.39 *(a) Phasor Diagram of a Two-Tone SSB-SC Signal, (b) Waveform of a Two-Tone SSB-SC Signal*

2.25.4 Waveform of SSB with Large Carrier

Now, let us consider an SSB signal in which a large carrier is also present. This carrier may be introduced at the transmitter as in AM, or may be added at the receiver. When a large carrier is added to the SSB-SC signal the detection becomes easy. When a large carrier is present, the SSB waveform shows amplitude variations, even for a single-tone modulating frequency. The expression for such a wave is obtained by adding a large carrier terms in the expression of SSB-SC, given in Eqn. (2.109) i.e.,

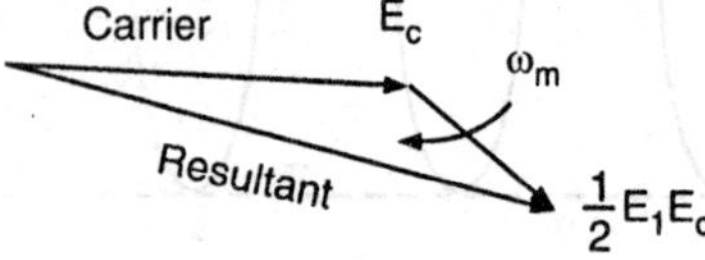

Fig. 2.40 *Phasor Diagram of an SSB with Large Carrier*

$$\phi_{SSB}(t) = \frac{1}{2}E_1E_c \cos(\omega_c - \omega_1)t + E_c \cos\omega_c t \quad E_c > E_1 \qquad ...(2.110)$$

The phasor diagram is shown in Fig. 2.40. SSB-SC waveform is formed by superposition of two waveforms, lower sideband large carrier term.

2.25.5 Demodulation of the SSB Signal with Large Carrier

When a large carrier is introduced, a synchronous detector is not necessary for recovering the baseband signal. An envelope detector is simpler and cheaper than synchronous detector. An expression for a SSB-signal with a large carrier A cos $\omega_c t$ can be written as

$$\phi_{SSB}(t) = f(t) \cos \omega_c t + f_\lambda(t) \sin \omega_c t + A \cos \omega_c t \qquad ...(2.111)$$

Here it should be kept in mind that the carrier is added to SSB-SC signal and not multiplied with it as a synchronous detection. This technique is called carrier reinsertion technique. Eqn. (2.111) can be written as

$$\phi_{SSB}(t) = [A + f(t)] \cos \omega_c t + f_\lambda(t) \sin \omega_c t$$

$$= e(t) \cos (\omega_c t + \theta)$$

where $e(t) = \sqrt{[A + f(t)]^2 + f_\lambda^2(t)}$ is the envelope of the wave, and $\theta = -\tan^{-1}\left[\frac{f_\lambda(t)}{A + f(t)}\right]$ is the phase of the wave. When this SSB signal is applied to an envelope detector, the output of the detector will be the envelope $e(t)$:

$$e(t) = \left[\{A + f(t)\}^2 + f_\lambda^2(t)\right]^{1/2}$$

$$= A\left[1 + \frac{2f(t)}{A} + \frac{f^2(t)}{A^2} + \frac{f_\lambda^2(t)}{A^2}\right]^{1/2}$$

Which can be expanded in the form a Binomial series, given by

$$e(t) = A\left[1 + \frac{2f(t)}{2A} + \text{higher terms}\right.$$

Since $A \gg |f(t)|$ and $|f_\lambda(t)|$, higher terms may be neglected, then

$$e(t) \simeq A + f(t)$$

Thus the output of the envelope detector is close to the desired baseband $f(t)$. The carrier reinsertion technique is also useful for the detection of DSB-SC signals.

The large carrier can be added either at the transmitter or at the receiver. When the carrier is added at the receiver, the synchronization problem remains as it is, and this technique has no advantage. When carrier is added at the transmitter, this made has the advantage of both SSB and AM system. It needs only half the bandwidth, at the same time, detection becomes simple.

2.25.6 Compatible Single Sideband

An SSB signal can be generated in which the carrier is suppressed, even then it can be detected with an envelope detector, simple diode detector. Such a signal is compatible for reception using a commercial AM radio receiver. The signal with this characteristic is termed as compatible single sideband (CSSB). However, such a signal involve complex signal processing, and the system is presently impractical for commercial application.

2.25.7 Vestigial-Sideband Modulation (VSB)

The major advantage of the SSB-SC modulation is that it reduced the bandwidth requirement to half as compared to DSB-SC modulation. But SSB-SC signals are relatively difficult to generate due to the difficulty in isolating the desired sideband. The required filter must have a very sharp cutoff characteristic, particularly when the baseband signal contains extremely low frequencies i.e., television and telegraphic signals. This difficulty is over come by a scheme known as vestigial sideband modulation, which is a compromise between SSB-SC and DSB-SC modulation.

In vestigial sideband modulation (VSB) the desired sideband is allowed to pass completely. Whereas just a small portion, called trace or vestige, of the undesired sideband is also allowed, as shown in Fig. 2.41. The transmitted vestige of the undesired sideband compensate for the loss of the wanted upper sideband.

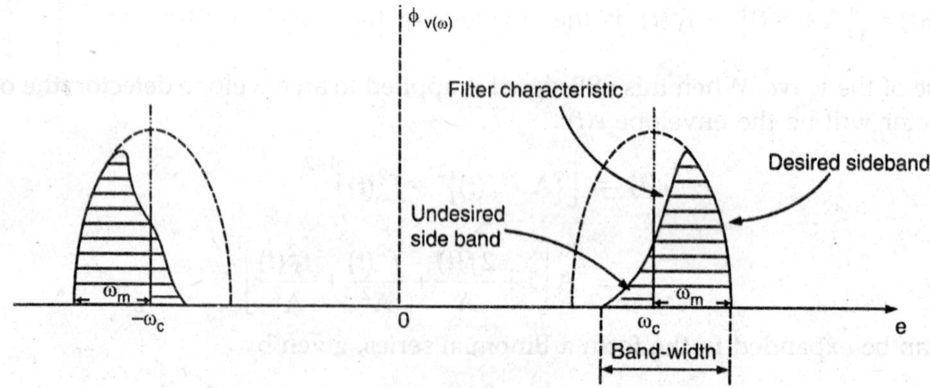

Fig. 2.41 *Spectrum of a Vestigial Sideband Signal*

It is evident from Fig. 2.41 that the filter required read not have a sharp cutoff, which is an advantage of the VSB system. However, as compared to SSB-SC, the bandwidth of the VSB becomes larger, although it remains much smaller than the USB-SC signal. Thus, VSB is a compromise between DSB-SC and SSB-SC signals. VSB is used in television for transmission of picture signals.

2.26 GENERATION AND DETECTION OF VSB SIGNAL

2.26.1 Filter Method

A VSB-SC signal can be generated by passing a DSB-SC signal through an appropriate filter as in Fig. 2.42 (*a*). The system is similar to DSB-SC except that the filter characteristic should be appropriate. The filter characteristic desired for VSB can be derived by analysing the demodulation technique of the VSB-SC signal which is nothing but the synchronous detection shown in Fig. 2.42 (*b*).

2.26.2 Filter Characteristics

Let us find the characteristics of the filter with a transfer function H(ω) shown in Fig. 2.43 (*a*) that may generate a VSB-SC signal from a DSB-SC signal. A DSB-SC signal has a spectrum given by

$$\phi_{SSB}(\omega) = \frac{1}{2}F(\omega+\omega_c)+\frac{1}{2}F(\omega-\omega_c)$$

The output of the filter H(ω) will be given by

$$\phi_v(\omega) = \frac{1}{2}H(\omega)[F(\omega+\omega_c)+F(\omega-\omega_c)] \quad ...(2.112)$$

where $\phi_v(t) = \phi_v(\omega)$

This signal $\phi_v(t)$ is passed through a product modulator for synchronous detection as shown in Fig. 2.42 (*b*). The output of this product modulator denoted by $e_d(t)$ will be given as

$$e_d(t) = \phi_v(t)\cos\omega_c(t)$$

or

$$e_d(t) \leftrightarrow \frac{1}{2}[\phi_v(\omega+\omega_c)+\phi_v(\omega-\omega_c)]$$

Substituting the value of $\phi_v(\omega)$ from Eqn. (2.112), we get

$$e_d(t) \leftrightarrow \frac{1}{2}\Big[\{F(\omega+2\omega_c)+F(\omega)\}H(\omega+\omega_c) + \{F(\omega)+F(\omega-2\omega_c)\}H(\omega-\omega_c)\Big]$$

This signal $e_d(t)$ is passed through a low pass filter and, hence, the components centred around $\pm 2\omega_c$ are filtered out. Thus, the output of LPF

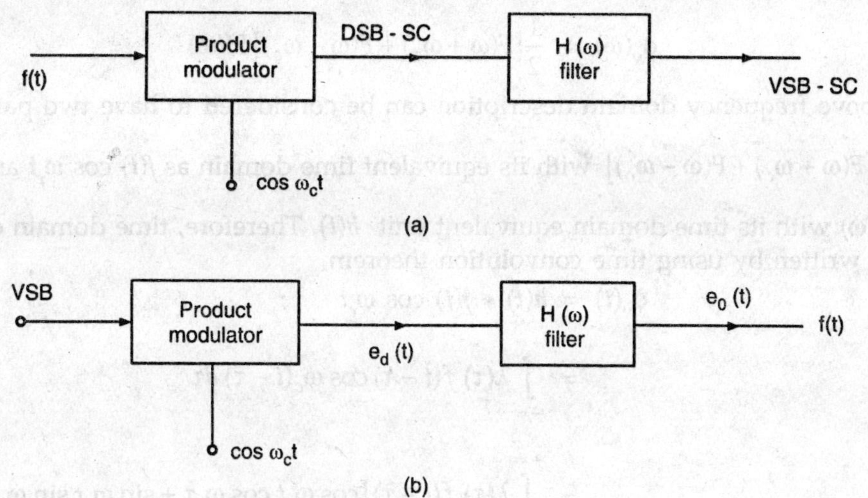

Fig. 2.42 *(a) VSB-SC Modulator (filter method); (b) VSB-SC Synchronous Detector*

of the synchronous detector will be given as

$$e_o(t) = \frac{1}{4}F(\omega)\{H(\omega+\omega_c)+H(\omega-\omega_c)\} \quad ...(2.113)$$

for getting a distortionless reception, the output $e_o(t)$ should be given as

$$e_o(t) = C_1 F(\omega) \qquad ...(2.114)$$

where C_1 is a constant.

When comparing Eqn. (2.113) with Eqn. (2.114) we get the following condition for a frequency range $|\omega| < \omega_m$, since the baseband is limited to ω_m

$$H(\omega + \omega_c) + H(\omega - \omega_c) = C \text{ (constant)} \qquad |\omega| < \omega_m \qquad ...(2.115)$$

The terms $H(\omega + \omega_c)$, and $H(\omega - \omega_c)$, represent $H(\omega)$ shifted $-\omega_c$ and $+\omega_c$, respectively, as shown in Fig. 2.43 (*b*) and (*c*). The sum these shifted spectrum should be constant over the frequency range $|\omega| < \omega_m$ as shown in Fig. 2.43 (*d*). It is obvious from Fig. 2.43 (*e*) that $H(\omega + \omega_c) + H(\omega - \omega_c)$ will be constant over $|\omega| < \omega_m$ only if the cutoff characteristic of filter $H(\omega)$ has a complementary symmetry, odd symmetry around the carrier frequency.

2.26.3 Phase-Discrimination Method for Generation of VSB-SC Wave

Like SSB-SC signals, VSB-SC signals also can be generated by using the phase discrimination method. The principle of this method will become clear once we determine the time domain discrimination of the VSB-SC wave.

2.26.4 Time Domain Description of VSB Wave

The frequency domain description of the VSB signal is given by Eqn. 2.112 as follows:

$$\phi_v(\omega) = \frac{1}{2}[F(\omega + \omega_c) + F(\omega - \omega_c)]\,H(\omega)$$

The above frequency domain description can be considered to have two parts:

(*i*) $\frac{1}{2}[F(\omega + \omega_c) + F(\omega - \omega_c)]$ with its equivalent time domain as $f(t) \cos \omega_c t$ and

(*ii*) $H(\omega)$ with its time domain equivalent unit $h(t)$. Therefore, time domain of $\phi_v(t)$ can be written by using time convolution theorem,

$$\phi_v(t) = h(t) * f(t) \cos \omega_c t \qquad ...(2.116)$$

$$= \int_{-\infty}^{+\infty} \lambda(\tau)\, f(t - \tau) \cos \omega_c (t - \tau)\, d\tau$$

$$= \int_{-\infty}^{+\infty} \lambda(\tau)\, f(t - \tau) [\cos \omega_c t \cos \omega_c \tau + \sin \omega_c t \sin \omega_c \tau]$$

The above equation may be split into two parts:

$$\phi_v(t) = \cos \omega_c t \int_{-\infty}^{+\infty} \lambda(\tau) \cos \omega_c \tau\, f(t - \tau)\, d\tau$$

$$+ \sin \omega_c t \int_{-\infty}^{+\infty} \lambda(\tau) \sin \omega_c \tau f(t - \tau) \, d\tau$$

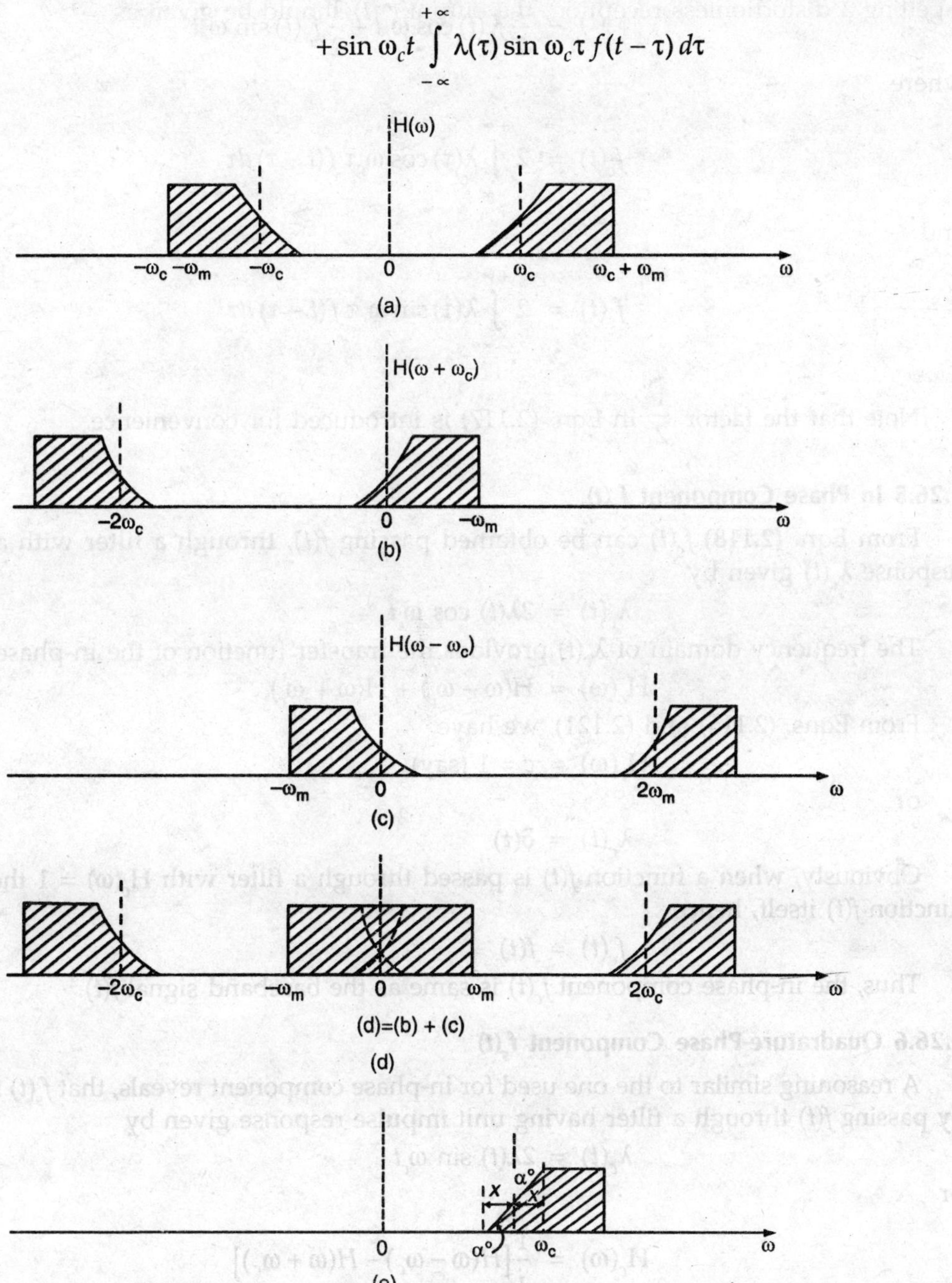

Fig. 2.43 *Filter Characteristics for VSB-SC Generation*

The equation can be written in a simplified form by defining the in-phase and the quadrature-phase components denoted by $f_c(t)$, $f_s(t)$, respectively. The simplified version is similar to the expression of the SSB-SC wave in time domain. Thus, the time domain description of VSB-SC signal is,

$$\phi_v(t) = \frac{1}{2} f_c(t) \cos \omega_c t + \frac{1}{2} f_s(t) \sin \omega_c t \qquad \text{...(2.117)}$$

where

$$f_c(t) = 2 \int_{-\infty}^{+\infty} \lambda(\tau) \cos \omega_c \tau \, f(t - \tau) \, d\tau \qquad \text{...(2.118)}$$

and

$$f_s(t) = 2 \int_{-\infty}^{+\infty} \lambda(\tau) \sin \omega_c \tau \, f(t - \tau) \, d\tau \qquad \text{...(2.119)}$$

Note that the factor $\frac{1}{2}$ in Eqn. (2.117) is introduced for convenience.

2.26.5 In Phase Component $f_c(t)$

From Eqn. (2.118) $f_c(t)$ can be obtained passing $f(t)$, through a filter with an impulse response $\lambda_c(t)$ given by

$$\lambda_c(t) = 2\lambda(t) \cos \omega_c t \qquad \text{...(2.120)}$$

The frequency domain of $\lambda_c(t)$ provides the transfer function of the in-phase filter,

$$H_c(\omega) = H(\omega - \omega_c) + H(\omega + \omega_c) \qquad \text{...(2.121)}$$

From Eqns. (2.115) and (2.121), we have

$$H_c(\omega) = c = 1 \text{ (say)}$$

or

$$\lambda_c(t) = \delta(t)$$

Obviously, when a function $f(t)$ is passed through a filter with $H_c(\omega) = 1$ the output is function $f(t)$ itself, hence

$$f_c(t) = f(t) \qquad \text{...(2.122)}$$

Thus, the in-phase component $f_c(t)$ is same as the baseband signal $f(t)$.

2.26.6 Quadrature-Phase Component $f_s(t)$

A reasoning similar to the one used for in-phase component reveals, that $f_s(t)$ is obtained by passing $f(t)$ through a filter having unit impulse response given by

$$\lambda_c(t) = 2\lambda(t) \sin \omega_c t$$

or

$$H_c(\omega) = \frac{1}{j}[H(\omega - \omega_c) - H(\omega + \omega_c)] \qquad \text{...(2.123)}$$

Thus, we need a filter with the transfer function $H_s(\omega)$ given by Eqn. (2.123) to get the quadrature component $f_s(t)$.

The time domain description of the VSB-SC signal is given by using Eqn. (2.117) and (2.122) as

$$\phi_v(t) = \frac{1}{2} f(t) \cos \omega_c t + \frac{1}{2} f_s(t) \sin \omega_c t \qquad ...(2.124)$$

Thus time domain description given in Eqn. (2.124) provides a basis for VSB signal generation as given as in Fig. 2.44. Note that DSB-SC and SSB-SC are two special case of VSB-SC wave defined by Eqn. (2.124). It becomes DSB-SC when $f_s(t)$ equals zero, and SSB-SC when $f_s(t) = f_\lambda(t)$.

The block diagram and Eqn. (2.124) corresponds to the VSB signal with a full upper sideband and a partial lower sideband. Likewise, Eqn. (2.124) will represent a VSB signal having a full lower sideband and a partial upper sideband if the plus sign, is replaced by a minus sign. Accordingly, the lower path signal in the block diagram of Fig. 2.44 will subtracted instead of added.

2.27 DETECTION OF VSB SIGNAL WITH LARGE CARRIER

Such signals can be detected by using an envelope detector. However, distortion is introduced due to the quadrature component $f_s(t)$. This can be seen by simple analysis as given below:

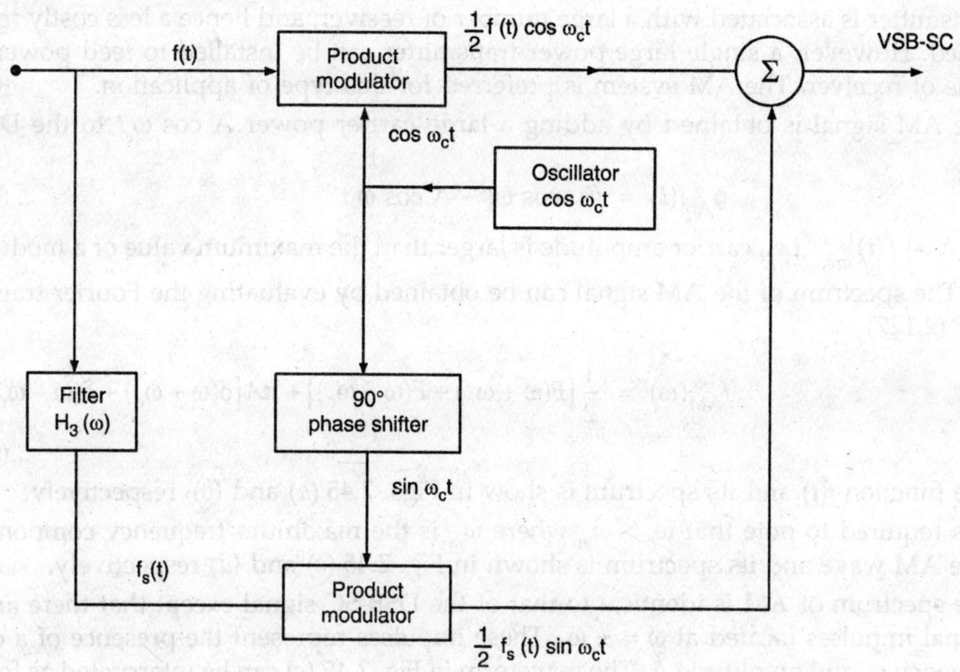

Fig. 2.44 *VSB-SC Generation (Phase-Discriminator Method)*

When a large carrier A cos $\omega_c t$ is added to the Eqn. (2.124) scaled by a factor (m_aA), the modified VSB at the input of the envelope detector is,

$$\phi_v(t) = \frac{1}{2} A m_a f(t) \cos \omega_c t + \frac{1}{2} A m_a f_s(t) \sin \omega_c t + A \sin \omega_c t \qquad ...(2.125)$$

The output of the envelope detector is the magnitude of the signal $\phi_v(t)$,

$$|\phi_v(t)| = A\left[1+\frac{1}{2}m_a f(t)\left[1+\left\{\frac{\frac{1}{2}m_a f_s(t)}{1+\frac{1}{2}m_a f(t)}\right\}\right]\right] \quad ...(2.126)$$

It is obvious from Eqn. (2.126) that quadrature component $f_s(t)$ produces distortion. This distortion can be minimized by: (*i*) reducing m_a; (*ii*) increasing the bandwidth of VSB to reduce $f_s(t)$. In commercial television broadcasting or telecasting, the vestigial sideband occupies a width of about one-sixth of a full sideband which keep the distortion within tolerable limit for $m_a = 1$.

2.28 AMPLITUDE MODULATION WITH LARGE CARRIER (AM)

The conventional amplitude modulation system used in broadcast systems consists of a large carrier power alongwith upper and lower sidebands, and referred as AM. The main advantage of this system is that the receiver is made simple and less expensive. In a broadcast system, transmitter has to transmit power for a large number of receivers. In other words, the transmitter is associated with a large number of receiver, and hence a less costly receiver is needed. However, a single large power transmitter can be installed to feed power for a multiple of receiver. The AM system is preferred for this type of application.

The AM signal is obtained by adding a large carrier power A cos $\omega_c t$ to the DSB-SC signal

$$\phi_{AM}(t) = f(t)\cos\omega t + A\cos\omega_c t \quad ...(2.127)$$

where $A > |f(t)|_{max}$ i.e., carrier amplitude is larger than the maximum value of a modulating signal. The spectrum of the AM signal can be obtained by evaluating the Fourier transform of Eqn. (2.127)

$$f_{AM}(\omega) = \frac{1}{2}[F(\omega+\omega_c)+F(\omega-\omega_c)] + \pi A[\delta(\omega+\omega_c)+\delta(\omega-\omega_c)] \quad ...(2.128)$$

The function $f(t)$ and its spectrum is show in Figs. 2.45 (*a*) and (*b*), respectively.

It is required to note that $\omega_c > \omega_m$ where ω_m is the maximum frequency component in $f(t)$. The **AM** wave and its spectrum is shown in Fig. 2.45 (*c*) and (*d*) respectively.

The spectrum of AM is identical to that of the DSB-SC signal except that there are two additional impulses located at $\omega = \pm\,\omega_c$. These impulses represent the presence of a carrier of frequency ω_c and amplitude A. The waveform in Fig. 2.45 (*c*) can be interpreted as follows:

The AM wave in Eqn. (2.127) can be written as

$$\phi_{AM}(t) = [A + f(t)]\cos\omega_c t \quad ...(2.129)$$

Eqn. (2.129) represents a wave of frequency and amplitude A + $f(e)$ Thus, the amplitude of the wave is changing around A in accordance with the value of the modulating signal $f(t)$. The frequency of the AM signal is unchanged and is equal to ω. The variation in the amplitude about A becomes obvious when comparing the carrier A cos $\omega_c t$ with the AM wave shown in Fig. 2.45(*c*).

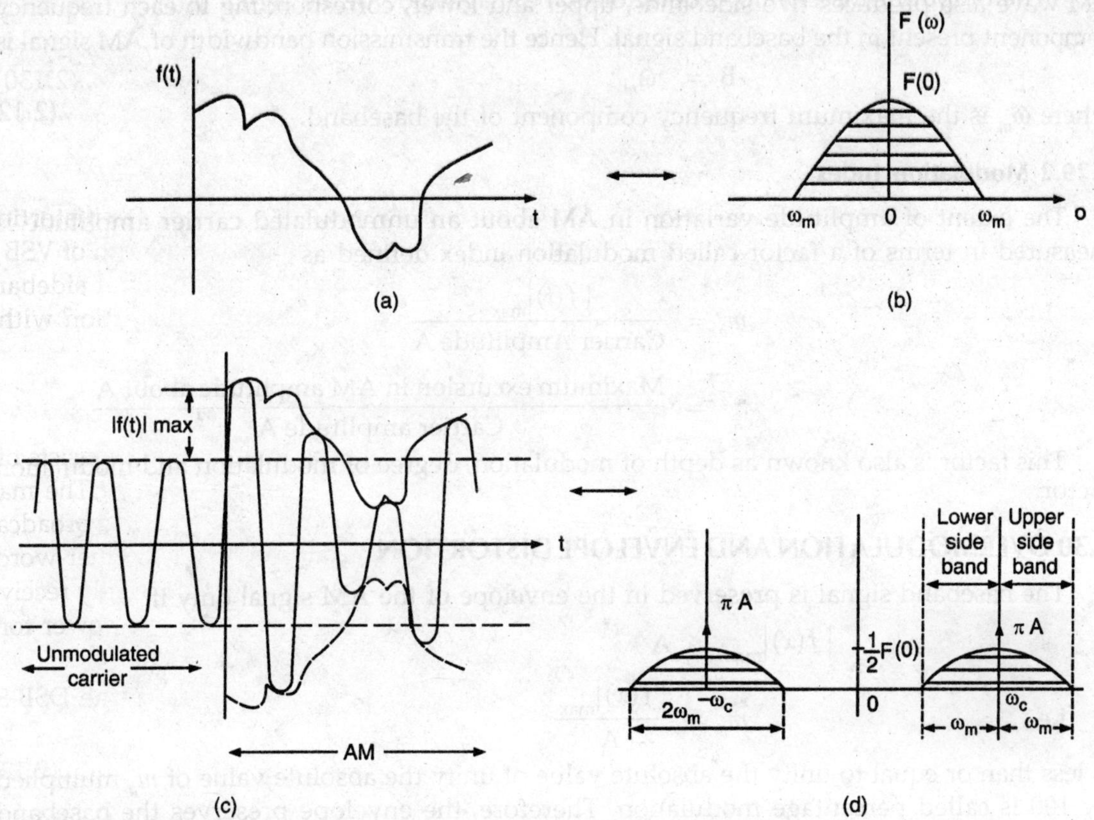

Fig. 2.45 *(a) Baseband Signal f(t); (b) Spectrum F(ω); (c) AM Wave; (d) Spectrum of AM wave*

2.29 ENVELOPE OF THE AM WAVE

The AM wave has a time varying amplitude called as the envelope of the AM wave. Fig. 2.45 (*c*) shows that the envelope of the wave consists of the baseband signal. The expression for the AM wave is given by Eqn. (2.129).

$$\phi_{AM} = [A + f(t)] \cos \omega_c t$$

$$= E(t) \cos \omega_c t$$

Where E(*t*) is the envelope of the AM wave. This envelope consists of the baseband signal *f*(*t*). Therefore, the baseband signal can be recovered from an AM wave by detecting the envelope. By using an envelope detector, the detection is less costly and simple. This is an important feature of the AM wave.

An envelope is defined as the magnitude of the pre-envelope of a function.

2.29.1 Bandwidth

The bandwidth of the AM wave is twice the bandwidth of the baseband signal. This is obvious from Fig. 2.45(*b*) and (*d*). The bandwidth is the same as that of DSB-SC signal. The

AM wave also produces two sidebands, upper and lower, corresponding to each frequency component present in the baseband signal. Hence the transmission bandwidth of AM signal is

$$B = 2\omega_m \quad ...(2.130)$$

where ω_m is the maximum frequency component of the baseband.

2.29.2 Modulation Index

The extent of amplitude variation in AM about an unmodulated carrier amplitude is measured in terms of a factor called modulation index defined as

$$m_a := \frac{|f(t)|_{max}}{\text{Carrier Amplitude A}}$$

$$= \frac{\text{Maximum excursion in AM amplitude about A}}{\text{Carrier amplitude A}}$$

This factor is also known as depth of modulation, degree of modulation and modulation factor.

2.30 OVERMODULATION AND ENVELOPE DISTORTION

The baseband signal is preserved in the envelope of the AM signal only if

$$|f(x)|_{max} \leq A$$

i.e.,

$$m_a \doteq \frac{|f(x)|_{max}}{A}$$

is less than or equal to unity the absolute value of unity the absolute value of m_a multiplied by 100 is called percentage modulation. Therefore, the envelope preserves the baseband signal only if the percentage modulation is less than, or equal to, 100 per cent. However, if $m_a > 1$ or the percentage modulation is greater than 100, the envelope does not preserve the baseband signal, rather, the baseband signal recovered from the envelope is distorted. This type of distortion is known as envelope distortion. An AM signal with $m_a > 100\%$ is known as overmodulated signal. Consider a baseband shown in Fig. 2.45 a modulating a carrier A cos $\omega_c t$. The following three cases arise depending on the amplitude of the baseband signal relative to carrier amplitude A.

2.30.1 AM with $m_a < 1$

The waveform of this type of signal is shown in Fig. 2.45 (*b*). Here, the amplitude of the baseband signal is less than A.

$$|f(x)|_{max} < A$$

As is evident from the waveform, the envelope is not reaching the zero-amplitude axis of the AM wave, and the baseband signal is fully preserved in the envelope of the AM wave. An envelop detector can recover the baseband signal without distortion.

2.30.2 AM with $m_a \approx 1$

The waveform is shown in Fig. 2.45 (*c*). Here, the amplitude of the baseband is equal to A,

$$|f(x)|_{max} = A$$

and hence, the waveform envelope just touches the zero-amplitude axis. The baseband signal remains preserved in the envelope and can be recovered by using an envelope detector. However, this is a limiting factor as even a small variation in the magnitude of $f(t)$ will cause envelope distortion.

2.30.3 Overmodulated Signal with $m_a > 1$

An overmodulated signal is shown in Fig. 2.48(*d*). Where, the modulation index is more than 100%. Hence,.

$$|f(t)|_{max} < A$$

i.e., the amplitude of a baseband signal exceeds carrier amplitude. Therefore, the portion of the envelope crosses the zero-amplitude axis, for both positive and negative excursions cancelling each other. If the resulting waveform is observed on a cathode ray oscilloscope, a trace will appear. The envelope and the baseband signal are not same. This is called envelope distortion. An envelope detector provides distorted baseband signal. A portion of the detected baseband signal is clipped.

Now it is clear that a baseband signal can be recovered from an AM signal by using a less costly and simple envelope detector, provided the percentage modulation does not exceed 100% i.e., the signal is not overmodulated. The limiting value of m_a = 100% is also avoided to be on safe side. However, an overmodulated signal can be recovered using a costly and as well as complex technique; synchronous detection.

2.30.4 Partially Suppressed Carrier AM

An overmodulated AM signal where the carrier amplitude A is less than the maximum value of a baseband signal and thus has a percentage modulation greater than 100% is equivalent to a partially suppressed carried AM. The carrier is not totally suppressed, rather it is presented in a smaller quantity. This type of AM signal is analogous to an AM-SC system with a pilot carrier.

2.30.5 Phasor Representation of an AM Signal

The relative amplitude an phase of various frequency components present in AM signal can be better understood by its phasor diagram. For convenience, we will consider a single-tone modulating signal.

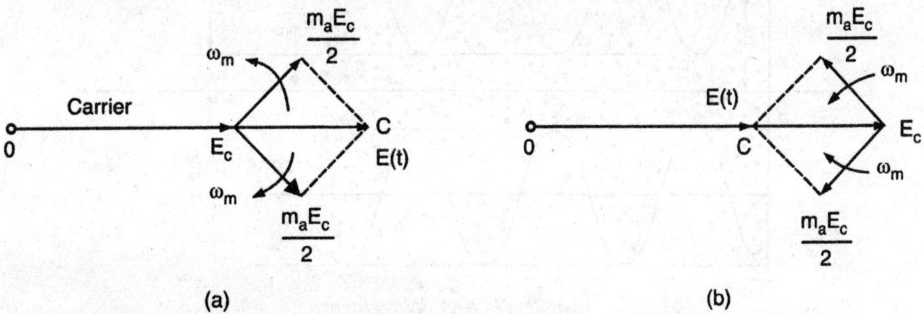

Fig. 2.46 *Phasor Diagram of AM Wave (a) Maximum Amplitude of the AM Wave (b) Minimum Amplitude of the AM wave*

The three frequency components present in AM speed given by Eqn.

$$e(t) = E_c \cos \omega_c t + \frac{m_a E_c}{2} \cos(\omega_c + \omega_m) + \frac{m_a E_c}{2} \cos(\omega_c - \omega_m) t$$

are represented by the three persons shown in Fig. 2.49. E_c represents the carrier phasor. The two sidebands $(\omega_c + \omega_m)$ and $(\omega_c - \omega_m)$ are represented by two phasors rotating in opposite directions with angular frequency ω_m. At any given time, the net amplitude, E(*t*), represented by phasor OC, depends on the position of the sideband phasors, with respect to the carrier phasor. Figures 2.46(*a*) and (*b*) illustrate two such situations causing maximum and minimum resultants respectively. The resultant of the two sideband phasors (OC) is always parallel to the carrier phasor.

The waveform of the AM signal can be interpreted as superposition of three waveforms: (*i*) carrier wave $E_c \cos \omega_c t$; (*ii*) upper sideband $\frac{m_a E_c}{2} \cos(\omega_c + \omega_m)t$; and (*iii*) lower sideband $\frac{m_a E_c}{2} \cos(\omega_c - \omega_m)t$. Note that each of three waveforms has a constant amplitude but the superposed waveform shows amplitude variation.

2.31 EXPERIMENTAL DETERMINATION OF MODULATION

The modulation index can be determined experimentally by observing AM waveform on a cathode ray oscilloscope (CRO). A single-tone modulating signal $E_m \cos \omega_m t$ modulating carrier $E_m \cos \omega_m t$ results in an AM wave as shown in Fig. 2.47. This waveform is applied to Y-defelecting plates of a CRO, as shown in Fig. 2.48. The trace is observed on the CRO screen. The maximum and minimum value of the envelope of the AM signal is measured with the help of the CRO. The modulation index is given by

$$m_a = \frac{E_{max} - E_{min}}{E_{max} + E_{min}} \quad \text{...(2.131)}$$

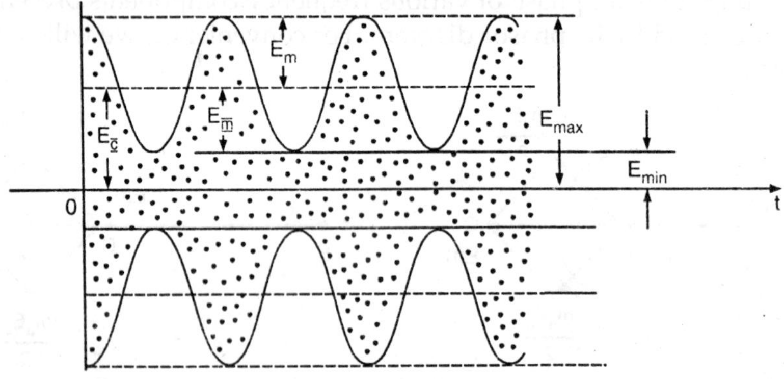

Fig. 2.47 *AM Waveform*

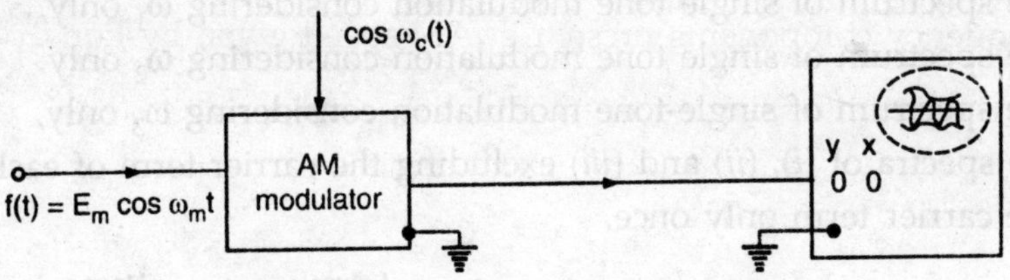

Fig. 2.48 *Measurement of m_a by CRO*

2.31.1 Derivation

Eqn. (2.131) is derived with the help of the waveform shown in Fig. 2.47.

It is obvious from the figure that

$$E_m = E_{max} - E_c = E_c - E_{min}$$

By definition,

$$m_a = \frac{E_m}{E_c}$$

This m_a may be written in two equivalent ways

$$m_a = \frac{E_{max} - E_c}{E_c} \qquad ...(2.132)$$

and

$$m_a = \frac{E_c - E_{max}}{E_c} \qquad ...(2.133)$$

By adding Eqns. (2.132) and (2.133), we

$$2m_a = \frac{E_{max} - E_{min}}{E_c}$$

or

$$m_a = \frac{E_{max} - E_{min}}{2E_c} \qquad ...(2.134)$$

Subtracting Eqn. (2.133) from Eqn. (2.132), we get

$$0 = E_{max} + E_{min} - 2E_c \qquad ...(2.135)$$

or

$$2E_c = E_{max} + E_{min} \qquad ...(2.136)$$

Substituting $2E_c$ from Eqn. (2.136) into Eqn. (2.134), we get

$$m_a = \frac{E_{max} - E_{min}}{E_{max} + E_{min}}$$

2.32 SUPERPOSITION THEOREM OF SPECTRUM

We can observe from the spectrum shown in Fig. 2.49 that the AM spectrum follows the superposition theorem. The theorem states that the sideband-spectrum of a multiple-tone AM signal is equal to the sum of the sideband-spectrum of the individual tone modulation, i.e., the multiple-tone AM spectrum can be obtained by steps given as:

(*i*) Find the spectrum of single tone modulation considering ω_1 only,
(*ii*) Find the spectrum of single tone modulation considering ω_2 only,
(*iii*) Find the spectrum of single-tone modulation considering ω_3 only,
(*iv*) Add the spectra of (*i*), (*ii*) and (*iii*) excluding the carrier term of each spectrum,
(*v*) Add the carrier term only once.

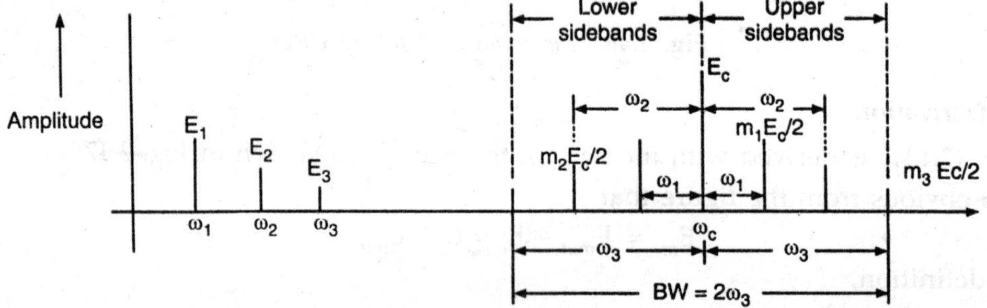

Fig. 2.49 *Line Spectrum of a Multiple-Tone AM*

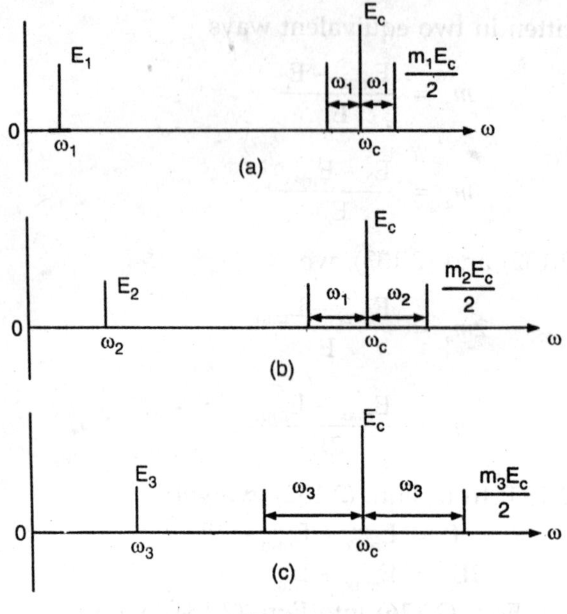

Fig. 2.50

This is shown graphically in Fig. 2.50. Fig. (*a*) shows the AM spectrum corresponding to ω_1 only. Similarly, Fig. (*b*) and (*c*) corresponding to AM spectrum for ω_2 and ω_3 respectively. The sum of the spectrum in Figs. (*a*), (*b*) and (*c*), and carrier terms provides the resulting spectrum shown in Fig. 2.49. Therefore, we can say that AM spectrum follows the superposition theorem of space.

2.32.1 Linear Modulation

The modulation system following the superposition theorem of spectra is known as linear modulation system. Thus, AM is linear modulation system. We will see later on that frequency modulation violates the law of superposition of spectra, and hence FM is a non-linear modulation system.

2.32.2 Power Content in AM Signal

The AM signal is represented by Eqn. (2.126) as reproduced below

$$\phi_{AM}(t) = A \cos \omega_c t + f(t) \cos \omega_c t$$

The total power of the AM wave is the sum of the carrier power P_c, and sideband power P_s. Remember that we are considering the normalized power, i.e., power dissipated in 1 ohms resistance.

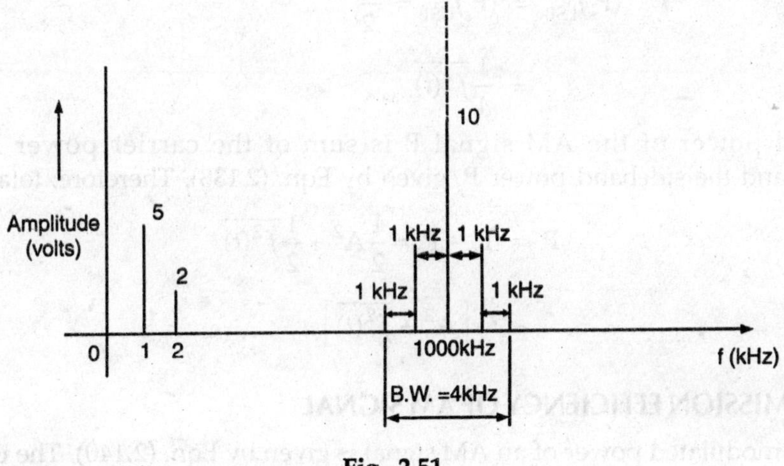

Fig. 2.51

2.32.3 Carrier Power

The carrier power P_c is the mean square value of the carrier term $A \cos \omega_c t$

$$P_c = \overline{[A \cos \omega_c t]^2} = \frac{A^2}{2} \qquad \text{...(2.137)}$$

2.32.4 Sideband Power

The sideband power P_s is the mean square value of the sideband $f(t) \cos \omega_c t$,

$$P_s = \overline{[f(t) \cos \omega_c t]^2}$$

$$= \frac{1}{2}\int_0^{2\pi} f^2(t) \cos^2 \omega_c t \, d(\omega t)$$

$$= \frac{1}{2\pi}\int_0^{2\pi} \frac{1}{2} f^2(t) [1 + \cos 2\omega_c t] \, d(\omega t)$$

$$= \frac{1}{2\pi}\int_0^{2\pi} \frac{1}{2} f^2(t) \, d(\omega t) + \frac{1}{2\pi}\int_0^{2\pi} \frac{1}{2} f^2(t) \cos 2\omega_c t \, d(\omega t)$$

The second integral-term is filtered out by the BPF centred around ω_c. Hence the, sideband power is obtained by solving the first integral.

$$P_s = \frac{1}{2}\overline{f^2(t)} \qquad ...(2.138)$$

This is half the baseband signal-power. The total sideband power P_s due to the equal contribution of the upper and lower sidebands. Hence, half of the P_s is contributed by the lower sidebands and the remaining half by the upper sideband. Therefore, power carried by the upper and the lower sidebands is,

$$(P_s)_{LSB} = (P_s)_{USB} = \frac{P_s}{2}$$

$$= \frac{1}{4}\overline{f^2(t)} \qquad ...(2.139)$$

The total power of the AM signal P is sum of the carrier power P_c defined by Eqn. (2.137), and the sideband power P_s given by Eqn. (2.138). Therefore, total AM power is

$$P = P_c + P_s = \frac{1}{2}A^2 + \frac{1}{2}\overline{f^2(t)}$$

$$= \frac{1}{2}\left[A^2 + \overline{f^2(t)}\right] \qquad ...(2.140)$$

2.33 TRANSMISSION EFFICIENCY OF AM SIGNAL

The total modulated power of an AM signal is given by Eqn. (2.140). The useful message, baseband, power out of this total power is the power carried by the sidebands, i.e., P_s. The large carrier power P_c does not carry any message, rather, this power is waste from the transmission point of view. However, this large carrier power P_c is transmitted alongwith the sideband power for convenient and cheap detection. Therefore, P_s is the only useful message power present in the AM wave. The amount of useful message power P_s present in AM wave is expressed by a term called transmission efficiency denoted by Eff. It is defined as percentage of total power contributed by the sideband.

Mathematically, the transmission efficiency is given by

$$(\text{Eff})_{AM} = \frac{P_s}{P} \times 100$$

$$= \frac{\frac{1}{2}\overline{f^2(t)}}{\frac{1}{2}A^2 + \frac{1}{2}\overline{f^2(t)}} \times 100$$

$$= \frac{100\overline{f^2(t)}}{A^2 + \overline{f^2(t)}} \qquad ...(2.141)$$

We can see that maximum transmission efficiency of the AM is 33.33%. This means, that only one-third of the total power is carried by the sidebands and the rest two third is a waste, and is transmitted only for a low cost reception system.

2.34 TRANSMISSION EFFICIENCY IN SUPPRESSED CARRIER SYSTEMS

In suppressed carrier systems, the carrier power is totally suppressed, and only the sidebands are transmitted. Therefore, the total AM power is carried by the sidebands and no power is contributed by the carrier, i.e. $P_c = 0$. Hence, the efficiency for a suppressed carrier (SC) system is

$$(\text{Eff})_{SC} = \frac{100 f^2(t)}{0 + f^2(t)} = 100 \text{ per cent}$$

Thus, the entire power in suppressed carrier systems is carried by sidebands, of course, at the cost of expensive and complex receiver.

2.34.1 Generation of AM Waves

The methods employed for generating AM-SC signals can also be used to generate AM signals, here AM refers to signals with large carriers. The methods for generation of AM waves are broadly divided into two parts. There are:

(*i*) using linear time invariant circuits, and

(*ii*) using non-linear circuits.

Circuits like switching or chopper modulator, collector modulator, drain modulator etc. are examples of the first category. Non-linear characteristics of devices, such as transistors, diodes, field effect transistors etc., are used in the second category. Now let us discuss some of the methods from both the categories.

2.34.2 Chopper-type Switching Modulator

The principle of this method is similar to that of AM-SC signal, except that chopping action is done one, a combined carrier plus message signal, i.e.,

$[A \cos \omega_c t + f(t)]$. It is noted that in the AM-SC generation only $f(t)$ was chopped.

The chopping is done at the rate of carrier frequency, and the chopping action is equivalent to multiplication of A cos $\omega_c t + f(t)$ with a periodic square signal $p(t)$ of frequency ω_c. The spectrum of the chopped signal is, therefore, obtained by convolving the spectrum of $[A + f(t)]$ with the spectrum of the square pulse, which is a sampling function. This is shown in Fig. 2.55.

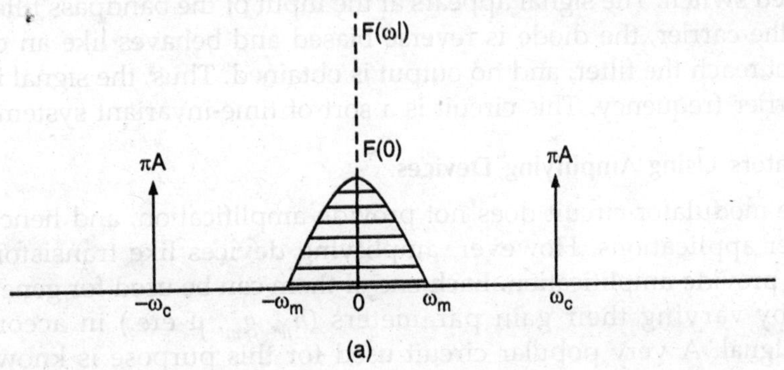

(a)

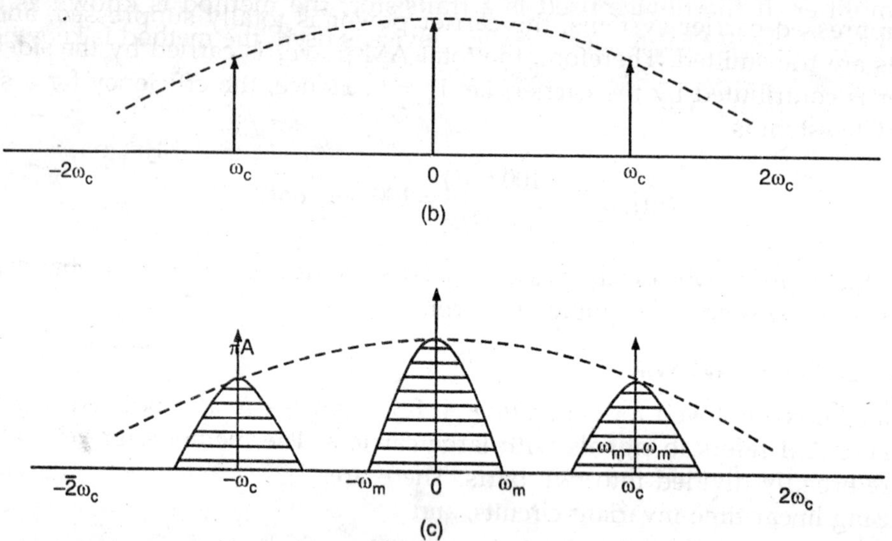

Fig. 2.52 *(a) Spectrum of A cos w_ct + f(t); (b) Spectrum of Square Signal; (c) Spectrum of Chopped Signal*

Fig. 2.52(*a*) shows the spectrum of $[A \cos \omega_c t + f(t)]$. The impulses of $\pm\omega_c$ correspond to the carrier A cos $\omega_c t$. Fig. 2.52(*b*) shows the spectrum of square signal $p(t)$ which is a sampling function. Fig. 2.52(*c*) represents the spectrum of the chopped signal, which is obtained by convolving Figs. 2.52 (*a*) and (*b*). When passed through a bandpass filter. The spectrum in Fig. 2.52(*c*) yields spectrum F(ω) shifted by $\pm\omega_c$, along with impulses at $\pm\omega_c$, which is the spectrum of an AM signal.

2.34.3 Circuit Diagram

The switching action can be obtained by a ring modulator used for AM-SC generation. A simple diode can perform the job of switching, provided $A \gg |f(t)|_{max}$. A simple diode chopper circuit used for AM generation is shown in Fig. 2.53.

The diode is forward biased for every positive half cycle of the carrier, and behaves like a short circuited switch. The signal appears at the input of the bandpass filter. For a negative half cycle of the carrier, the diode is reverse biased and behaves like an open switch. The signal does not reach the filter, and no output is obtained. Thus, the signal is chopped at the rate of the carrier frequency. This circuit is a sort of time-invariant system.

2.34.4 Modulators Using Amplifying Devices

The diode modulator circuit does not provide amplification, and hence, it can be used for low power applications. However, amplifying devices like transistors, FET, electron tubes etc. can provide amplification. Each one of them can be used for generating amplitude modulation by varying their gain parameters (h_{fe}, g_m, μ etc.) in accordance with the modulating signal. A very popular circuit used for this purpose is known as modulated

Class C amplifier. If the device used is a transistor, the method is known as collector modulation. Similarly, if FET, or electron tube device is used, the method is known as drain modulation, or plate modulation respectively. Fig. 2.54 shows a collector modulation circuit.

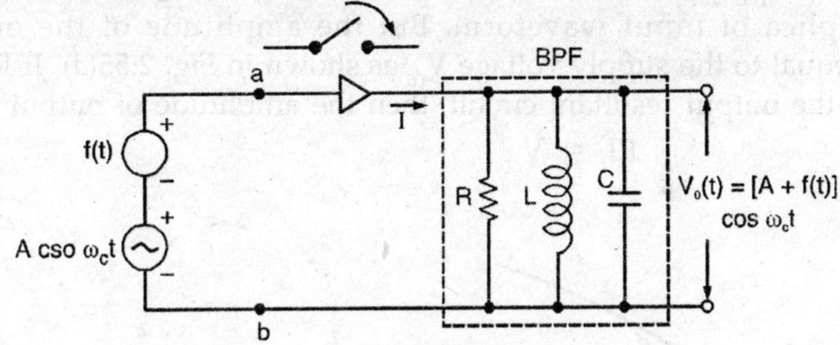

Fig. 2.53 *A Simple Diode Switching Modulator*

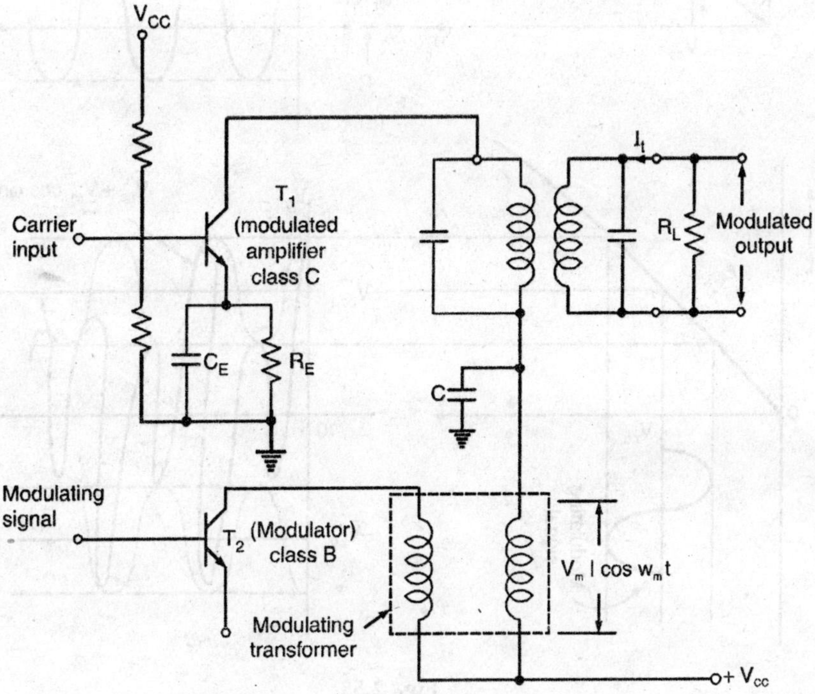

Fig. 2.54 *A Collector Modulation Circuit*

Transistor T_1 forms a radio frequency amplifier in Class C mode for higher efficiency. The carrier signal is applied at the base T_1. V_{cc} is the collector supply used for biasing. The transistor T_2 forms a Class B amplifier used amplify the baseband modulating signal. After amplification, the modulating signal appears across the modulating transformer. This modulating signal exists in series with the collect or supply, V_{cc}. Capacitor *c* offers low path for the carrier signal, and as such, the carrier is prevented from flowing through the modulation transform.

2.34.5 Principle of Operation

In a Class C amplifier, the amplitude of the output voltage is a definite fraction of or at the most equal to the supply voltage V_{cc}. A linear relation exists between the output lack current I_t and the supply variable voltage V_c, assume V_c to be variable supply. The output voltage is a replica of input waveform. But the amplitude of the output voltage is approximately equal to the supply voltage V_{cc} as shown in Fig. 2.55(*a*). If R is the resistance at resonance of the output resultant circuit, then the amplitude of output voltage will be

$$RI_t \simeq V_{cc}$$

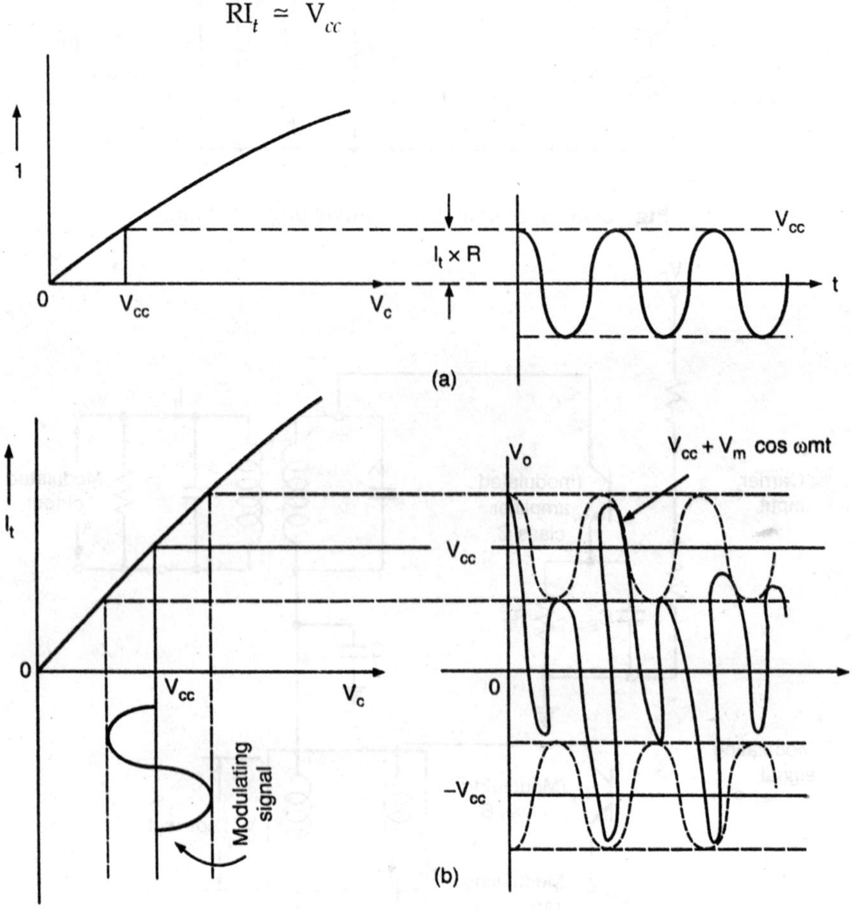

Fig. 2.55

Thus, the unmodulated carrier is amplified by the Class C modulated amplifier, and its amplitude remains constant equal to V_{cc}, as no voltage appears across the modulation transformer in the absence of modulating voltage. However, when a modulating voltage $V_m \cos \omega_m t$ appears across the modulation transformer, it is added with the supply voltage. The net effect is a slow variation in supply voltage V_{cc}; slow because as compared to the carrier frequency, the modulating frequency is small. This slow variation in supply voltage changes the amplitude of the carrier voltage at the output of the modulated Class C amplifier, as shown in Fig. 2.55(*b*). The envelope of the output voltage is identical with the modulating voltage and thus an AM wave is generated. This slowly varying supply voltage V_c is given by

$$V_c = V_{cc} + V_m \cos \omega_m t \quad ...(2.142)$$

$$= V_{cc}[1 + m_a \cos \omega_m t] \quad ...(2.143)$$

and the modulated output voltage V_o is given by

$$V_o = V_{cc}[1 + m_a \cos \omega_m t] \cos \omega_c t$$

where m_a the modulation index given by

$$m_a = \frac{V_m}{V_{cc}}$$

2.34.6 Overmodulation

When $V_m > V_{cc}$, the envelope crosses the zero axis and an envelope distortion occurs due to overmodulation as shown in Fig. 2.48(*d*).

2.34.7 Collector Circuit Efficiency

The modulated power delivered to the output load depends on the input supplied by supply voltage and power dissipation in the collector circuit. Out of the total power input in the collector circuit, only a part of it reaches the output load, the remaining power is lost in the collector resistance, and other dissipating components. Let us assume that

P_{in} = input power in the collector circuit

P_{out} = output power delivered to a load R_L and

P_d = Power dissipation in the collector circuit, then

$$P_d = P_{in} - P_{out}$$

$$= P_{in}\left(1 - \frac{P_{out}}{P_{in}}\right) \quad ...(2.144)$$

The ratio of output power and input power, in the collector circuit is defined as collector circuit efficiency denoted by η_c

$$\eta_c = \frac{P_{out}}{P_{in}} \quad ...(2.145)$$

Hence,

$$P_d = P_{in}(1 - \eta_c) \quad ...(2.146)$$

The power input in the collector circuit is the multiplication of a slow varying supply voltage

$$V_{cc}(1 + m_a \cos \omega_m t),$$

and the resulting collector current

where I_c is the average current in the absence of modulation. This multiplication provides instantaneous input power at any moment. The average input power is obtained by taking the average of this instantaneous power over a cycle. Thus,

$$P_{in} = \int_0^{2\pi} V_{cc}(1 + m_a \cos \omega_m t)\, I_c(1 + m_a \cos \omega_m t) d\omega x$$

$$= V_{cc} + I_c + \frac{m_a^2}{2} V_{cc} I_c$$

$$= V_{cc} I_c \left(1 + \frac{m_a^2}{2}\right) \quad \text{...(2.147)}$$

$$= P_{cc} \left(1 + \frac{m_a^2}{2}\right) \quad \text{...(2.148)}$$

where $P_{cc} = V_{cc} I_c$ is the power supplied by V_{cc}.

The total input power is composed of two parts: P_{cc} and $\frac{m_a^2}{2} P_{cc}$. The second part represents the sideband power introduced by the modulating amplifier in the collector circuit. Let us denote this power by P_m, i.e., $P_m = \frac{m_a^2}{2} P_{cc}$. This modulating power is introduced in the form of a varying power supply. When this modulating signal does not appear across the modulation transformer in the collector circuit, then $m_a = 0$, and the total input power is only P_{cc}. Thus P_{cc} is the unmodulated carrier power. Now substituting the value of P_{in} from Eqn. (2.147) into Eqn. (2.146), we get

$$P_d = V_{cc} I_c \left(1 + \frac{m_a^2}{2}\right)(1 - \eta_c) \quad \text{...(2.149)}$$

Thus, dissipation in the collector circuit increases with an increase in modulation index. This dissipation generates heat. In large power transmitters, if the modulation index is large, the necessary arrangement for cooling is made.

When the modulation index $m_a = 0$, the power dissipation, P_{d0} is,

$$P_{d0} = V_{cc} I_c (1 + \eta_c)$$

Eqn. (2.149) can be written in terms of P_{d0} as follows

$$P_d = P_{d0} \left(1 + \frac{m_a^2}{2}\right) \quad \text{...(2.150)}$$

From Eqn. (2.145) and (2.147) we get

$$P_{out} = \eta_c P_{in}$$

$$= \eta_c V_{cc} I_c \left(1 + \frac{m_a^2}{2}\right) \quad \text{...(2.151)}$$

2.35 AM GENERATION USING NON-LINEAR CIRCUIT

A simple diode can be used as a non-linear modulator by restricting its operation to non-linear region of its characteristic. The undesired frequency terms are filtered out by a bandpass filter. The non-linear circuit is exactly similar to the circuit shown earlier in Fig. 2.53 for AM generation using switching modulator. However, in this case, the total input to the diode $[f(t) + A \cos \omega_c t]$ is kept so small that the diode operates in a non-linear region of its V-I characteristics. Thus, this method is useful only for small signal amplitude modulation.

The input voltage to the diode is given by

$$e = f(t) + A \cos \omega_c t$$

The resulting current i will be given by

$$i = ae + be^2$$

$$= a[f(t) + A \cos \omega_c t] + b[f(t) + A \cos \omega_c t]$$

The expansion of this equation provides the presence of a number of frequency components, such as ω_c, $2\omega_c$, ω_m, $2\omega_c$, $(\omega_c \pm \omega_m)$, $(\omega_c \pm 2\omega_m)$ and so on. All these components, except the desired ones, i.e. ω_c and $\omega_0 \pm \omega_m$ are filtered out by a bandpass filter tuned to ω_c.

The diode modulator does not provide amplification. Also, a single diode is unable to balance out the undesired frequency completely. These limitations can be eliminated by using amplifying devices like transistor, FET, or an electron tube in a balanced mode.

2.35.1 Balance Modulator

A transistor balanced modulator circuit used for AM generation is shown in Fig. 2.56. The circuit is similar to the AM-SC generator except that feeding point of the carrier and the modulating signal are interchanged. The carrier voltage across the two windings of a center-tap transformer are equal, and opposite in phase i.e., $e_c = -e_c'$.

The input voltage to transistor T_1 is given by

$$e_{bc} = e_c + e_m$$

$$= E_c \cos \omega_c t + E_m \cos \omega_m t \quad ...(2.152)$$

Since, both e_c and e_m are in phase.

Similarly, the input voltage to transistor T_2 is given by

$$e_{bc}' = e_c' + e_m$$

$$= -E_c \cos \omega_c t + E_m \cos \omega_m t \quad ...(2.153)$$

This is because e_c' and e_c are in phase opposite i.e., $e_c = -e_c'$.

By the non-linearly relationship,

$$i_c = a_1 e_{bc} + a_2 e_{bc}^2$$

and

$$i_c' = a_1 e_{bc}' + a_2 e_{bc}'^2$$

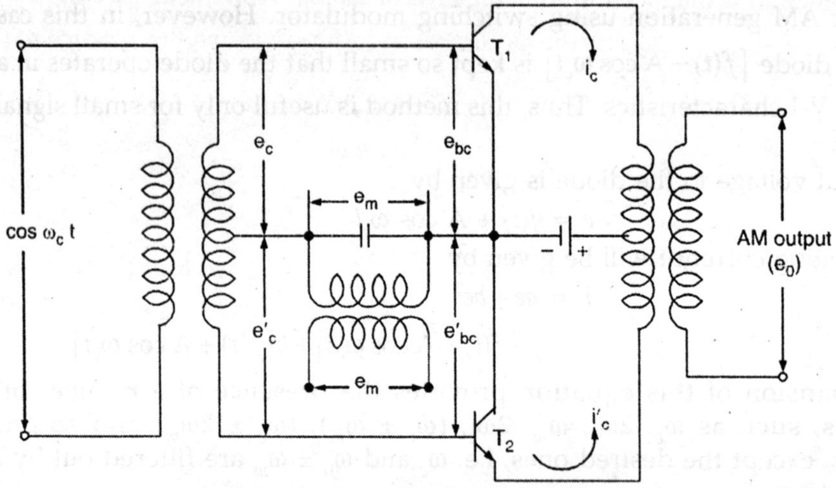

Fig. 2.56 *Balanced Modulator (AM)*

Substituting the value of e_{bc} and e'_{bc} from Eqns. (2.152) and (2.153), we get

$$i_c = a_1(E_c \cos \omega_c t + E_c \cos \omega_c t)$$

$$+ a_2[E_c \cos \omega_c t + E_m \cos \omega_m t]^2$$

$$= a_1[E_c \cos \omega_c t + E_m \cos \omega_m t] + a_2[E_c^2 \cos \omega_c t$$

$$+ E_m^2 \cos^2 \omega_m t + 2E_m E_c \cos \omega_c t \cos \omega_m t] \quad \text{...(2.154)}$$

Similarly,

$$i'_c = a_1[-E_c \cos \omega_c t + E_m \cos \omega_m t] + a_2[E_c^2 \cos^2 \omega_c t$$

$$+ E_m^2 \cos^2 \omega_m t - 2E_c E_m \cos \omega_c t \cos \omega_m t] \quad \text{...(2.155)}$$

The output AM voltage e_o is given as

$$e_o = K(i_c - i'_c) \quad \text{...(2.156)}$$

This is because the current i_c and i'_c flow in opposite directions in a tuned circuit. K is a constant depending on impedance and other circuit parameters. Substituting Eqns. (2.154) and (2.155) in Eqn. (2.156), we get

$$e_o = 2Ka_1E_c \cos \omega_c t + 4Ka_2E_cE_m \cos \omega_c t \cos \omega_m t$$

The other terms are balanced out, we can write,

$$e_o = 2KE_c a_1 \left[1 + \frac{2a_2E_m}{a_1} \cos \omega_m t\right] \cos \omega_c t$$

$$= 2Ka_1E_c[1 + m_a \cos \omega_m t] \cos \omega_c t \quad \text{...(2.156A)}$$

where $m_a = \dfrac{2a_2E_m}{a_1}$ is the modulation index.

2.35.2 Advantage of a Balanced Modulator Over a Simple Non-linear Circuit

In simple-non-linear circuits, the undesired non-linear terms, harmonics are eliminated by a bandpass filter. Hence, the bandpass filter must be carefully designed. But in a balanced modulator, the undesired non-linear terms are automatically balanced out and, at the output we get only the desired terms, so filter design is not so stringent.

2.36 MODULATION OF AM WAVES

The process of extracting a baseband modulating signal from the modulated signal is known as demodulation. As we know that AM signals with large carrier are detected by using the envelope detector. The envelope detector employs circuits that extract the envelope of the AM wave.

Evidently, the envelope of the AM wave is the baseband signal. However, a low-level modulated signal can be detected by using square law detectors to which a device operating in the non-linear region is used to detect the baseband signal. Thus, we can categorize the detectors into two parts:

(*i*) Square law detectors, and

(*ii*) Envelope detectors

2.36.1 Square Law Detectors

Square law detectors are used for detecting low-level modulated signals, say below 1 Volt, so that operating region of device-characteristic is restricted to non-linear region. The circuit is very similar to a square law non-linear modulator. The only difference is in filter circuits. In a detector, the filter is a low-pass filter instead of a bandpass filter. A simple square detector is shown in Fig. 2.57(*a*). The non-linear characteristic, modulated input voltage and resulting diode current waveforms are shown in Fig. 2.57(*b*). The dc source V_d is used to adjust the operating point. The operation is limited to the non-linear region due to which the lower half portion of the current waveform is compressed. This causes envelope distortion. The average value of the diode current does not remain constant, rather it varies with the time as shown in Fig. 2.57(*b*).

The distorted diode current is given by the non-linear, square law relation

$$i = a_1 e + a_2 e^2$$

where e is input modulated voltage.

$$e = E_c(1 + m_a \cos \omega_m t) \cos \omega_c t$$

Hence,

$$i = a_1 [E_c(1 + m_a \cos \omega_m t) \cos \omega_c t] + a_2 [E_c(1 + m_a \cos \omega_m t) \cos \omega_c t]^2 \quad ...(2.157)$$

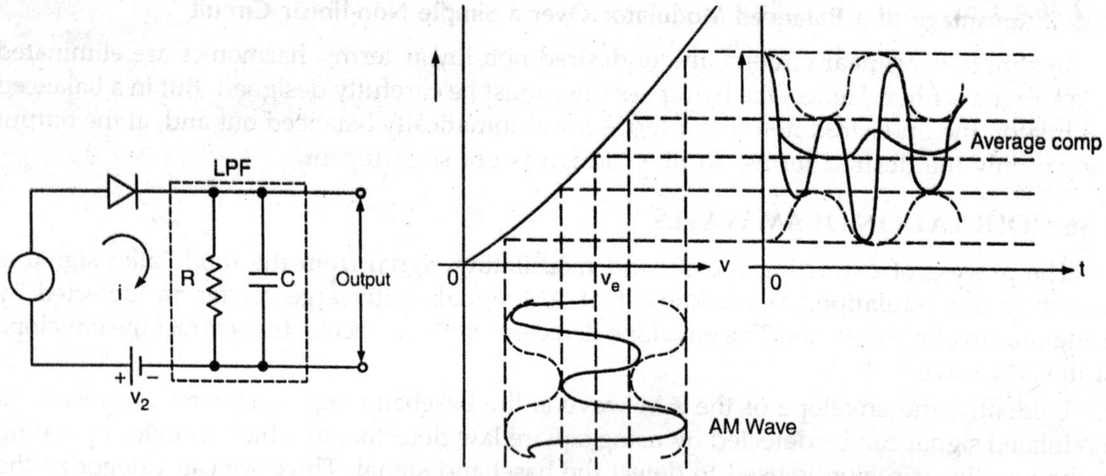

Fig. 2.57 *Square Law Detector (a) Circuit Diagram (b) Characteristics*

The expansion of Eqn. (2.157) reveals the presence of the components $2\omega_c$, $2(\omega_c \pm \omega_m)$, ω_m and $2\omega_m$ besides the input frequency terms. This current is passed through a low pass filter which allows to pass the below ω_m and suppresses the other higher components. Thus, the baseband signal with frequency ω_m is recovered.

2.36.2 Distortion

The non-linear characteristics of the diode produces additional frequency components. Frequencies centered about ω_c and $2\omega_c$ are easily suppressed by using a low pass filter, as they are far away from ω_m. But $2\omega_m$, is very close to ω_m and hence it cannot be totally suppressed by the low pass filter. Therefore, component $2\omega_m$ introduces distortion. This distortion term $2\omega_m$ cannot be completely eliminated and is always present as distortion. Thus, a square law detector cannot provide a distortionless AM detection. However, the degree of distortion can be reduced by keeping the magnitude of the component $2\omega_m$ much less than the amplitude of the desired component ω_m.

2.36.3 Linear Diode Detector

A diode operating in a linear region of its characteristic can extract the envelope of an AM wave. Such a detector is called envelope detector. This detector is extremely popular in commercial receiver circuits because it is very simple and less expensive, and at the same time provides satisfactory performance for the reception of broadcast programmes. The circuit diagram of a linear diode detector is known in Fig. 2.58(*a*). The tuned transformer provides perfect tuning at the desired carrier frequency. R-C forms the time constant network. When the modulated carrier at the input of detector is 1 Volt, or more, the operation takes place in the linear region of the diode characteristic. The idealized linear characteristic of the diode along with input voltage and output current waveforms are shown in Fig. 2.58(*b*).

2.36.4 Operation

Let us suppose that the capacitor is absent. The output waveform will be a half-rectified carrier wave as shown in Fig. 2.58(*b*). The diode will present a low resistance r_d, forward biased. The total resistance of the circuit becomes r_d + R. A current flowing in the diode circuit through this total resistance, will result in a half-rectified carrier voltage across resistance R.

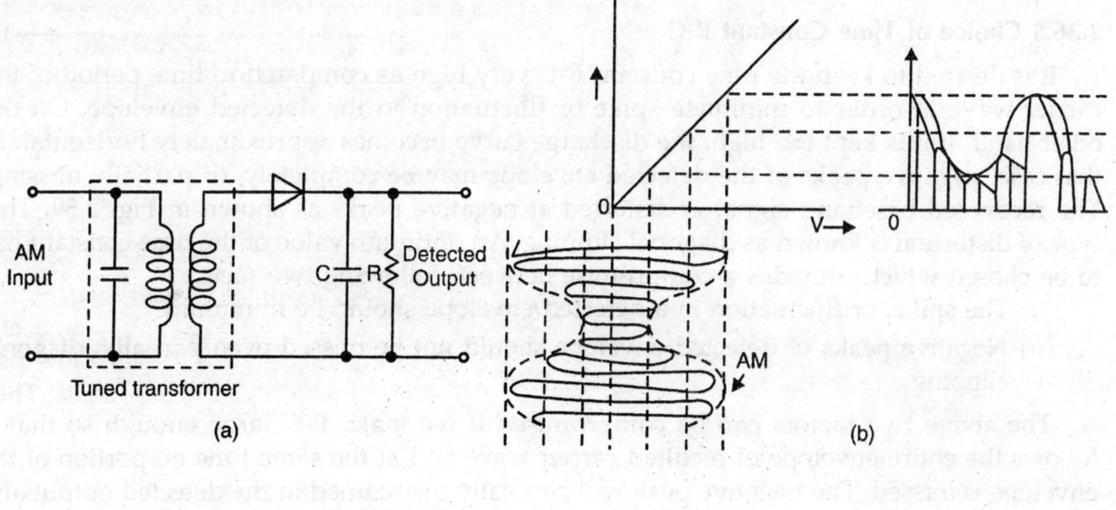

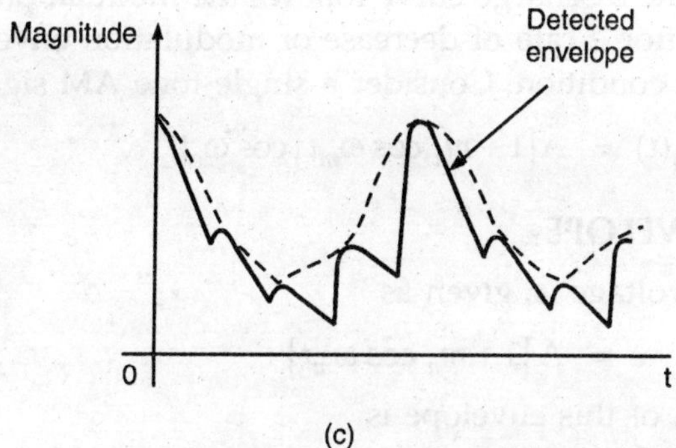

Fig. 2.58 *Linear Diode Detector (a) Circuit (b) Characteristics (c) Detected Output*

Now, let us assume that the capacitance has been introduced in the circuit. For the positive half-cycle of a rectified wave, the capacitor is charged to the peak value of the carrier voltage through the resistance (r_d + R). But for a negative half-cycle, the diode is reverse biased, and the carrier voltage is disconnected from the R-C circuit. So the capacitor starts discharging through the resistance R with a time constant τ = RC. If the time constant is properly chosen, the voltage across the capacitor does not fall appreciably during the small period of a negative half-cycle, and by that time the next positive cycle appears. This positive cycle further charges the capacitor to the peak value of the carrier voltage and the process continues. Thus, the voltage across the capacitor is just like a spiky baseband signal, i.e. voltage across the capacitor is same as envelope of the modulated carrier, but spike are introduced due to charging and discharging of the capacitor. The resulting detected baseband signal is shown in Fig. 2.58(*c*). The spikes can be reduced to a negligible amount by keeping the time constant and R-C large, so that the capacitor discharges negligible small. But the large value of R-C creates another problem called diagonal clipping, and hence, we cannot increase it beyond a certain limit. The choice of time constant is an important point for consideration.

2.36.5 Choice of Time Constant R-C

It is desired to keep the time constant R-C very high as compared to time period of the carrier wave in order to minimize spike or fluctuation in the detected envelope. On the other hand, if it is kept too high, the discharge curve becomes approximately horizontal. In that case, negative peaks of the detected envelope may be completely, or partially missing. The recovered baseband signal is distorted at negative peaks as shown in Fig. 2.59. This type of distortion is known as diagonal clipping. An optimum value of the time constant has to be chosen which provides a compromise between following two facts:

(*i*) The spike, or fluctuation in a detected envelope should be minimum.

(*ii*) Negative peaks of detected envelope should not be missed even partially, diagonal clipping.

The above two factors can be compromised if we make R-C large enough so that it follows the entire envelope of rectified carrier wave and at the same time no portion of the envelope is missed. The negative peak will be totally maintained in the detected output and no portion will be clipped, if the discharge curve follows the modulation envelope. For this, the rate of discharge of capacitor ≥ rate of decrease of modulation envelope. Let us derive a relation which satisfied this condition. Consider a single-tone AM signal given by

$$\phi_{AM}(t) = A[1 + m_a \cos \omega_m t] \cos \omega_c t$$

2.37 RATE OF DECAY OF ENVELOPE

The envelope of the AM voltage is, given as

$$e = A[1 + m_a \cos \omega_m t] \quad ...(2.158)$$

The rate of change, slope, of this envelope is

$$-\frac{de}{dt} = A m_a \, \omega_m \sin \omega_m t \quad ...(2.159)$$

The negative slope indicates the decay of the voltage.

The slope of an instant t_0 will be

$$-\left(\frac{de}{dt}\right)_{t=t_0} = A m_a \, \omega_m \sin \omega_m t_0 \quad ...(2.160)$$

This is the rate of decay of the envelope at an instant $t = t_0$

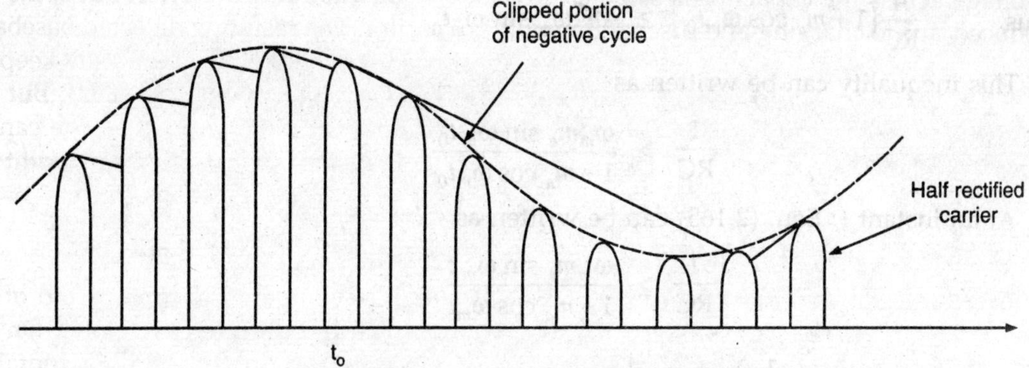

Fig. 2.59 *Diagonal Clipping*

2.37.1 Rate of Discharge of Capacitor

The capacitor C discharges exponentially. Let us assume that the capacitor starts discharge at $t = t_0$. Before the capacitor discharging at the instant, it has been charged to a voltage equal to the value of the envelope at $t = t_0$. This initial voltage across the capacitor is equal to the envelope voltage e at $t = t_0$, given by

$$e_o = (e)_{t=t_0} = A[1 + m_a \cos \omega_m t_0] \qquad ...(2.161)$$

The capacitor discharges exponentially from the initial voltage e_0. The capacitor voltage at any instant t is given as

$$e_{cap} = e_o e^{-\frac{t-t_0}{RC}} \qquad ...(2.162)$$

The rate of charge of capacitor voltage, slope of the discharge curve, is

$$-\frac{de_{cap}}{dt} = -\frac{d}{dt}\left[e_o e^{-\frac{t-t_0}{RC}}\right]$$

$$= \frac{e_o}{RC} e^{-\frac{(t-t_0)}{RC}} \qquad ...(2.163)$$

This rate of change at $t = t_0$ is given as

$$-\left(\frac{de_{cap}}{dt}\right)_{t=t_0} = \frac{e_0}{RC}$$

Substituting e_o from Eqn. (2.161),

$$-\left(\frac{de_{cap}}{dt}\right)_{t=t_0} = \frac{A}{RC}\left[1 + m_a \cos \omega_m t_0\right] \qquad ...(2.164)$$

To avoid diagonal clipping the slope of the discharge-curve at $t = t_0$ given by Eqn. (2.164) must be equal to, or greater than, the envelope-decay rate given in Eqn. (2.160).

Thus, $\frac{A}{RC}[1 + m_a \cos \omega_m t_0] \geq A m_a \omega_m \sin \omega_m t_0$

This inequality can be written as

$$\frac{1}{RC} \geq \frac{\omega_m m_a \sin \omega_m t_0}{1 + m_a \cos \omega_m t_0} \quad ...(2.165)$$

At an instant t, Eqn. (2.165) can be written as

$$\frac{1}{RC} \geq \frac{\omega_m m_a \sin \omega_m t}{1 + m_a \cos \omega_m t} \quad ...(166)$$

It is desired that $\left(\frac{1}{RC}\right)$ is always greater than, or equal to, the right hand side, RHS, of Eqn. (2.158). This can be achieved by maximizing the RHS. The condition for maximizing the RHS is obtained by equating the derivative of RHS of Eqn. (2.166) to zero.

$$\frac{d}{dt}\left[\frac{\omega_m m_a \sin \omega_m t}{1 + m_a \cos \omega_m t}\right] = 0$$

Solution of the above equation yields,

$$\cos \omega_m t = -m_a \quad ...(2.167)$$

or

$$\sin \omega_m t = \sqrt{1 - m_a^2} \quad ...(2.168)$$

Substituting these values of $\sin \omega_m t$ and $\cos \omega_m t$ in Eqn. (2.166), we get

$$\frac{1}{RC} \geq \omega_m \frac{m_a \sqrt{1 - m_a^2}}{1 - m_a^2}$$

or

$$\frac{1}{RC} \geq \frac{\omega_m m_a}{\sqrt{1 - m_a^2}} \quad ...(2.169)$$

If $m_a \ll 1$, small modulation index, then Eqn. (2.169) reduces to

$$\frac{1}{RC} \geq \omega_m m_a \quad ...(2.170)$$

Thus, Eqn. (2.169) provides the desired relation for obtaining optimum value of time constant RC in terms of modulation index, and modulating frequency.

2.38 AM TRANSMITTERS AND RECEIVERS

A transmitter not only performs the modulation process, but also raises the power level of a modulated signal to the desired extent for effective radiation. The AM transmitters are divided into two categories:

(*i*) Low-level Modulation

(*ii*) High-level Modulation

2.38.1 Low-level Modulation

The block diagram for this type of transmission is shown in Fig. 2.60. This system is used in low power transmitters. In this method, the modulation process takes place at a low power level, and the signal is then amplified by a Class B power amplifier to raise to the desired power level. The limitation of the system is lower efficiency of the Class B r.f. amplifier as compared to Class C, the latter cannot be used as it does not faithfully reproduce the modulated signal consisting of carrier and sideband terms. The amplifier should possess sufficient bandwidth to amplify these terms equally. The Class C amplifier have narrow bandwidth, which may cause sideband cutting, and thereby introduce distortions. The remedy is to increase the power level of the carrier before modulation to the desired extent, and then perform the modulation. This is done in another type of transmitter known as right-level transmitter.

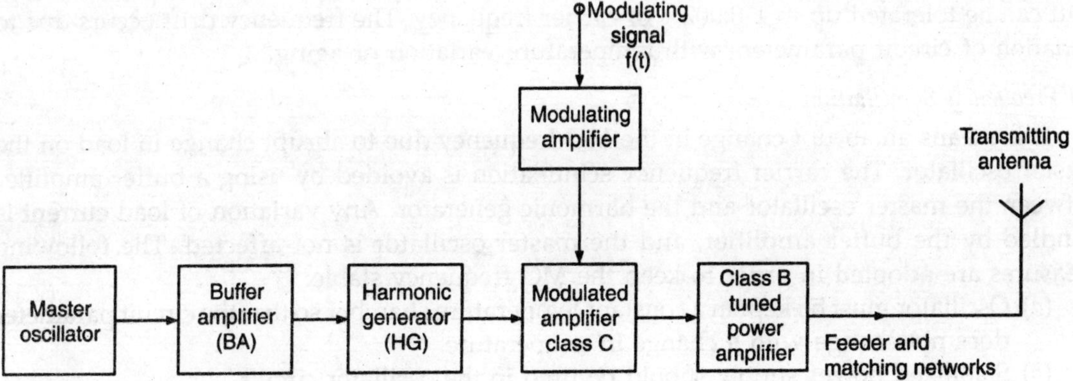

Fig. 2.60 *AM Transmitter Using Low Level Modulation*

2.38.2 Higher power level modulation

The block diagram for this type of transmitter is shown in Fig. 2.61. In high level power modulation, the carrier is first amplified by a Class C amplifier, and then modulation is done at the high power level. Here, the Class C amplifier have to amplify only the carrier frequency, and hence, can have high gain and efficiency. This type of transmitter is complex, and more expensive than the low level transmitters. It is, however suitable for high power transmission. Each constituent stage of the block diagrams is described briefly as follows:

2.38.2.1 *Master Oscillator (MO)*

The master oscillator generates a stable sub-harmonic carrier frequency, i.e., the fraction of a desired carrier frequency. This stable sub-harmonic oscillation is generated by using a crystal oscillator and then frequency is raised to the desired value by harmonic generators. The sub-harmonic frequency is used because stable crystal oscillators are easy to make at lower carrier frequencies. For example, if one wants 1 MHz of carrier frequency, it is better to make a crystal oscillator of 250 kHz which is the fourth sub-harmonic of 1 MHz. It is noted that a 250 kHz crystal oscillator will be much more stable than a 1 MHz crystal oscillator. So, the master oscillator generates a sub-harmonic of 250 kHz, and then it is increased four times by a harmonic generator to get the desired carrier frequency of 1 MHz. The limitation of the crystal oscillator is that it can provide only single frequency, and this frequency cannot be altered. This is not a problem in a broadcast transmitter which normally transmits only one carrier frequency.

Stability of MO Frequency

The carrier frequency generated by an MO ought to be very stable. Any change in the MO frequency will cause interference with other transmitting stations and the receiver will accept programmes from more than one transmitter. There are several reasons for the change in the frequency of a master oscillator. They are:

(i) Frequency Drift

This means a slow variation in frequency with time. The frequency drift permitted in medium wave transmitters is ± 20 Hz. Whereas, for shortwave and UHF transmitters the drift can be tolerated up to ± 0.002% of carrier frequency. The frequency drift occurs due to variation of circuit parameters with temperature variation or aging.

(ii) Frequency Scintillation

This means an abrupt change in the MO frequency due to abrupt change in load on the master oscillator. The carrier frequency scintillation is avoided by using a buffer amplifier between the master oscillator and the harmonic generator. Any variation of load current is handled by the buffer amplifier, and the master oscillator is not affected. The following measures are adopted in order to keep the MO frequency stable:

(*a*) Oscillator must be kept in a constant temperature chamber so that the circuit parameter does not change with a change in temperature.

(*b*) Stabilized power supply should be used in the oscillator circuit.

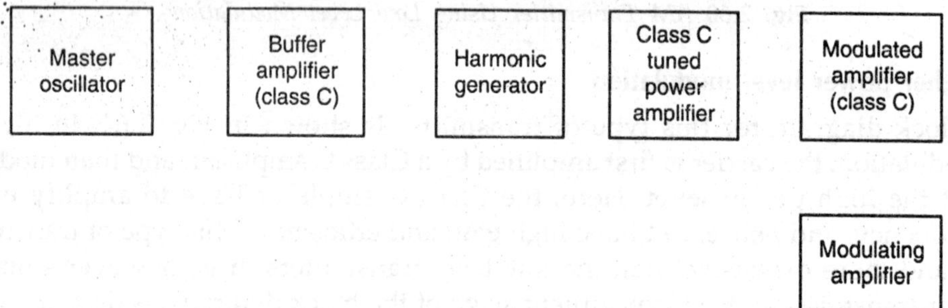

Fig. 2.61 *AM Transmitter Using High Level Modulation*

(*c*) Effective Q of the tuned circuit should be vary high.

(*d*) The active devices used in the oscillator circuit should have a high value of the ratio (g_m/C) where g_m is the mutual transconductance, and C is the device capacitance.

(*e*) Master oscillator should generate a subharmonic of the carrier frequency.

(*f*) Buffer amplifier must be used between the master oscillator and other stages.

2.38.2.2 *Buffer Amplifier*

This is a tuned amplifier providing a high input impedance at the master oscillator frequency. Any variation in load current does not affect the master oscillator due to this high input impedance of buffer amplifier at the operating frequency of the master oscillator. Thus, buffer amplifier isolates the MO from the succeeding stages, so that the loading effect may not change the frequency of the MO. A harmonic generator stage can work as a buffer, because its input and output tank circuits are tuned to different frequencies. If the current flow is large at a frequency in the output circuit, the current flow at input circuit will be small, as it is tuned to a different frequency, and thereby offers a high input impedance.

2.38.2.3 *Harmonic Generator*

It is an electronic circuit that generates harmonic of its input frequency. The principle of harmonic generation is the same as that of a non-linear modulator. When a signal is given to a non-linear circuit, it generates harmonic of input frequency. The desired harmonic is selected by a properly tuned circuit. One of the various possible circuit arrangement is shown in Fig. 2.62. The circuit uses a Class C tuned amplifier. It is well known that in the Class C operation the current flows in the form of pulses recurring with the same frequency as the input signal. These pulses contain harmonics of input frequency ω_c. This can be seen by Fourier analysis of the pulse train. If this pulse train is passed through a resonant circuit tuned to $n\omega_c$ the nth harmonic will be selected, and the others will be rejected. Thus, the output of the harmonic generator will have a frequency $n\omega_c$. The Class C operation is obtained by biasing the device much below cutoff.

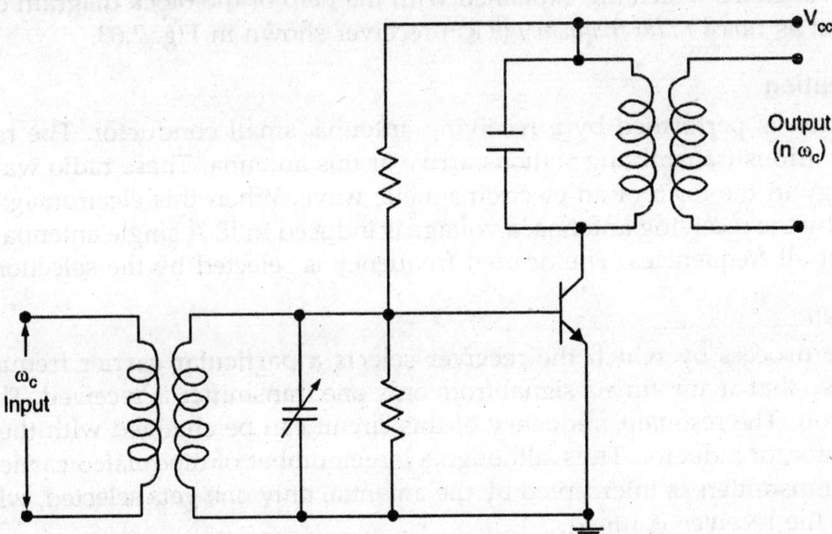

Fig. 2.62 *A Class C Harmonic Generator*

2.38.2.4 *Drive Amplifier Or Intermediate Power Amplifier*

One or more stages of a Class C tuned amplifier is used to increase the power level of a carrier signal to provide a large drive to the modulated Class C amplifier. The output of the harmonic generator provides a low power carrier signal. This power is amplified to raise the power to desired level to drive the find the final amplifier stage. The number of Class C stages depends on the desired power level, and also the gain of each stage. This stage is termed as Driver Amplifier or Intermediate Power Amplifier (IPA).

2.38.2.5 Modulation System

The collector modulation circuit is used for modulation in high power transmitters. The modulating amplifier is a Class A, or Class B, amplifier amplifying the baseband signal.

2.38.2.6 Feeder and Antenna

The transmitter power is fed to a transmitting antenna for effective radiation. The length of the antenna, a conductor, should be of the order of the wavelength for effective radiation. The antenna is normally located at a distance from the transmitter and hence power from the transmitter, is feed to the antenna through a properly designed transmission time called feeder. The impedance of the feeder line must be properly matched with the transmitter impedance at one end, and with the antenna impedance at the other end.

2.39 NEUTRALIZATION

Neutralization circuits are used in transmitters to eliminate the spurious oscillation caused by undesired feedback of energy through inter-electrode capacitance.

2.39.1 AM Receiver

A radio receiver is an electronic circuit that picks up a desired modulated radio frequency signal, and recovers the baseband signal from the modulated signal. A receiver performs the following functions which are explained with the help of the block diagram of a simple receiver known as *tuned radio frequency* (TRF) receiver shown in Fig. 2.63.

2.39.2 Interception

This function is performed by a receiving antenna, small conductor. The radio wave coming from various transmitting stations arrive at this antenna. These radio wave contain electrical energy in the form of an electromagnetic wave. When this electromagnetic wave is intercepted by the receiving antenna, a voltage is induced in it. A single antenna intercepts radio waves of all frequencies. The desired frequency is selected by the selection process.

2.39.3 Selection

This is the process by which the receiver selects a particular carrier frequency, and rejects others, so that at any time a signal from only one transmitter is received. This is done by a bank circuit. The resonant frequency of this circuit can be changed with the help of a variable capacitor, or inductor. Thus, although a large number of modulated carriers coming from various transmitters is intercepted by the antenna, only one gets selected, which is the one for which the receiver is tuned.

2.39.4 R.F. Amplification

The selected carrier, radio signal, is amplified by a Class C tuned amplifier. This is necessary to raise the carrier voltage level so that the linear diode detector following this stage may operate in linear region. The increased carrier voltage also suppresses the effect of noise.

2.39.5 Detection

Detection is the process of recovering a baseband signal from a modulated carrier. It has been discussed in detail earlier.

2.39.6 Audio Amplification

The detected audio signal is further amplified so that it can drive the speaker. This stage consists of a R-C coupled amplifier followed by a Class B push-pull power amplifier.

2.39.7 Reproduction

Reproduction is the process by which an electrical signal is converted into a desired physical message. In a commercial broadcast receiver, the output of the audio amplifier is fed into a speaker that produces sound according to the input audio signal.

2.39.8 TRF Receiver

A TRF receiver performs all the above functions and works satisfactorily at medium wave frequencies. At higher radio frequencies the performance of a TRF becomes poor. The perform of receiver is improved by a technique known as **heterodyning**. A receiver based on this technique is known as *superheterodyne receiver*. The performance of a receiver is judged from its various features such as selectivity, sensitivity and fidelity.

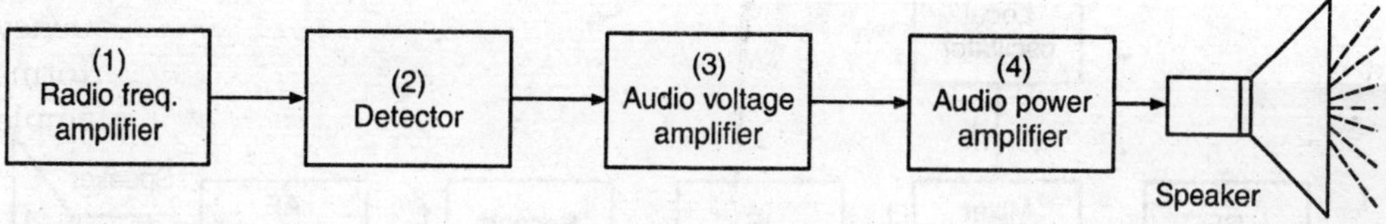

Fig. 2.63 *TRF Receiver*

2.39.9 Features of a Receiver

The various types of receivers can be compared in their performances by the following features

2.39.10 Selectivity

Selectivity is receivers' ability to distinguish between two adjacent carrier frequencies. By this feature it is decided how perfectly the receiver is able to select the desired carrier frequency and reject the others. Selectivity depends on the sharpness of the resonance curve of a tuned circuit involved in the receiver. The sharper the resonance curve, the better the selectivity. Better selectivity means that the receiver has greater capability to reject undesired signals.

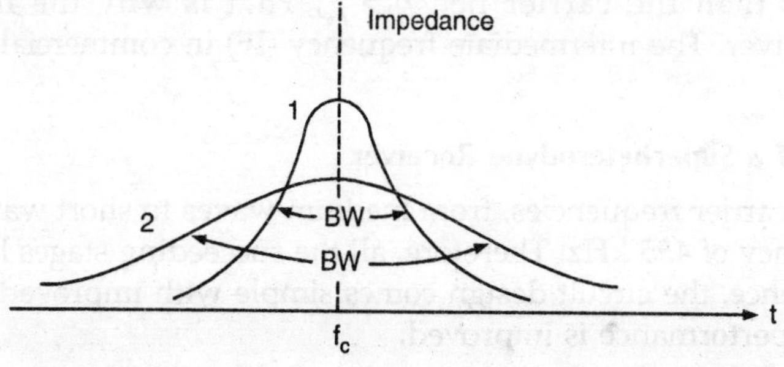

Fig. 2.64 *Resonance Curve of a Tuning Circuit*

The sharpness of the resonance curve depends on Q of the resonant circuit. The higher is the Q, more is the selectivity. In Fig. 2.64 curve 1 is sharper than curve 2, and hence has better selectivity.

2.39.11 Sensitivity

The ability of a receiver to detect the weakest possible signal is known as sensitivity. A receiver with a good sensitivity will provide more output for a similar input signal as against a receiver with poor sensitivity. The sensitivity of a receiver is decided by the gain of its amplifying stage.

2.39.12 Fidelity

The ability of a receiver to reproduce faithfully all frequency components present in the baseband signal is called fidelity. If any component is missed, or attenuated considerably, fidelity suffers and the reproduced signal is distorted. This feature is mainly decided by the audio amplifier which amplifies the baseband signal.

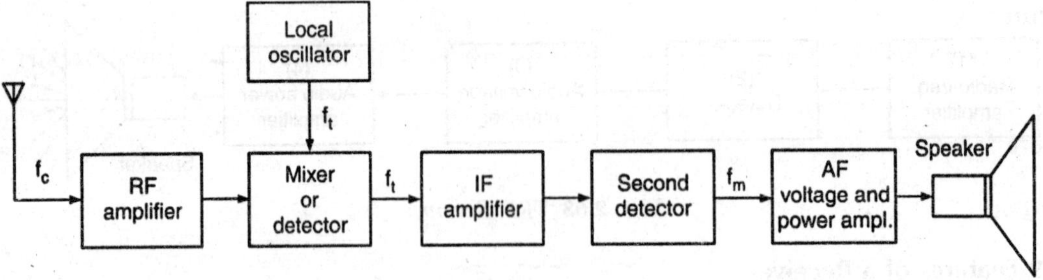

Fig. 2.64 *Superheterodyne Receiver*

2.39.13 Superheterodyne Receiver

The receiver is populárly used. The heterodyning gives a far better performance than the TRF receiver. The block diagram of a superheterodyne receiver is given in Fig. 2.68. Significant feature of the heterodyne receiver is that all incoming radio frequencies are converted into a single intermediate frequency f_i by the heterodyning, mixing, process. The incoming carrier, f_c, and a locally generated signal, f_l, are mixed in mixer, also referred to as first detector. The mixer generates the sum and difference frequencies at the output. The difference frequency, $f_l - f_c$, is selected by a properly tuned circuit. The local oscillator frequency is higher than the carrier i.e., $f_l > f_c$. That is why the receiver is called superheterodyne receiver. The intermediate frequency (IF) in commercial radio receivers is fixed to 445 kHz.

2.39.14 Advantages of a Superheterodyne Receiver

All the incoming carrier frequencies, from medium waves to short waves, are converted into a fixed IF frequency of 455 kHz. Therefore, all the succeeding stages have to operate on a fixed frequency. Hence, the circuit design comes simple with improved performance. Let us now see how the performance is improved.

2.39.15 Sensitivity

The IF signal is amplified in an intermediate frequency amplifier. This is to be designed for a fixed frequency band centered near 455 kHz. Naturally, it is easy to design and develop a high gain amplifier at a fixed carrier frequency. The sensitivity of a superheterodyne receiver is primarily decided by the IF amplifier stage. A high gain IF amplifier provides a better sensitivity than the TRF receiver. In the TRF receiver, the sensitivity is decided by an RF amplifier, which amplifiers the entire band of medium and short waves rather than a fixed carrier frequency. It is difficult to design a RF stage that provides high gain for such a wide range of carrier frequencies, medium waves to short waves. Also, high gain AF amplifiers suffer from instability problem.

2.39.16 Selectivity

The selectivity of a superheterodyne receiver is also decided mainly by the IF amplifier. The tuned circuit associated with the IF amplifier operates at a fixed centre frequency of 455 kHz. Since the bandwidth of the baseband signal is about 10 kHz, the Q required by a tuned circuit is given by

$$Q = \frac{f_c}{\text{bandwidth}} \qquad ...(2.171)$$

where f_c is the centre frequency.

Therefore, $$Q = \frac{455\text{ kHz}}{10\text{ kHz}} = 45.5$$

As the tuning circuit for this fixed and practical value of Q is easy to design, a better selectivity than that of a TRF receiver can be achieved.

The selectivity of a TRF receiver is poor because of the following reasons:

(*i*) The Q factor requirement of the tuned circuits employed in RP amplifier is not fixed. Consider a tuned circuit required to hand a bandwidth of 10 kHz at the lower edge of the broadcasting medium wave, 550 kHz. Q required by the tuned circuit is 55. At the higher edge of the medium wave, 1650 kHz, the Q requirement is 165. Thus, the Q required varies with carrier frequency. The variation in Q is difficult to achieve in practice.

(*ii*) The Q required by the tuned circuits for upper side of the short wave, say 20 MHz, is 2000. It is impossible to achieve such a high value of Q with ordinary circuits.

2.39.17 Detector Design

The detector stage of a superheterodyne receiver operates on a fixed frequency, IF and hence it can be designed to detect the entire baseband without introducing much distortion, as the time constant RC of the detector remains fixed. In a TRF, the detector has to detect signals of various carrier frequencies. The single detector cannot give a good performance for a large range of radio frequencies, as the desired time constant does not remain fixed.

2.39.18 Advantages of TRF Receiver

TRF receivers a cheaper and simpler than superheterodyne receiver. Also, they do not have problems such as image signal rejection or tracking and alignment, that arise in a superheterodyne receiver.

2.40 CONSTITUENT STAGES OF A SUPERHETERODYNE RECEIVER

2.40.1 RF Amplifier

This is a Class C tuned voltage amplifier. The main functions of this stage are:

(*a*) Amplification of the received radio signal to provide better sensitivity and improved signal to noise ratio.

(*b*) Rejection of the unwanted signals and as improved adjacent channel selectivity.

(*c*) Rejection of the image signal.

2.40.2 Image Signal

This is a signal whose frequency is above the local oscillator frequency by the same amount, as the desired signal frequency is below the local oscillator frequency. This is illustrated by the spectrum shown in Fig. 2.66. It is obvious from the figure that

$$f_l - f_c = f_i$$

or

$$f_c = f_l - f_c \quad \text{...(2.172)}$$

The desired signal frequency f_c is below local oscillator frequency f_l by an amount f_i, the IF. By definition, the image signal f_c is above f_l by the same amount f_i i.e., image signal f_c' and f_c is $2f_i$. From Eqns. (2.172) and (2.173)

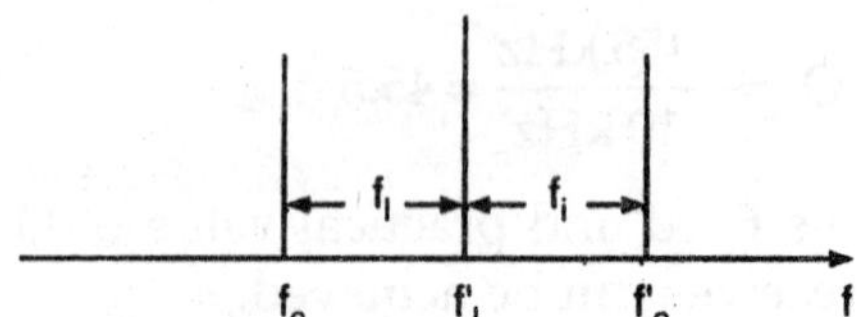

Fig. 2.66 *Image Signal*

$$f_c' = f_l + f_i = f_c + f_i + f_i$$

$$= f_c + 2f_i \quad \text{...(2.173)}$$

Thus, the image signal is $2f_i$ more than the desired carrier signal. If somehow, this image signal is intercepted by an antenna and reaches the mixer, it produces the same IF as produced by the desired signal f_c. The desired signal produces $f_l - l_c = f_i$ whereas the image signal produces $f_c' - f_l = f_i$. The spurious IF signal produces by image signal is also amplified by the IF amplifier and produces interference in the receiver output. Therefore, the image signal should be rejected. The RF amplifier adds one more tuned circuit per stage, and thereby increase the image frequency attenuation.

2.40.3 Mixer Stage

A mixer, Fig. 2.67 is a non-linear device that mixes the incoming signal of frequency f_c with a local oscillator voltage of frequency f_l and generates an output voltage of an intermediate frequency $(f_l - f_c)$ as shown in Fig. 2.66. The non-linear mixer circuit produces the sum and difference frequency components $(f_l \pm f_c)$ alongwith the input frequencies and their harmonics. The desired intermediate frequency $f_i = (f_l - f_c)$ is selected by a tuned circuit known as input intermediate frequency transformer (IFT). The IFT is tuned by adjusting the core of the transformer. The process is known as inductive tuning.

The mixer is also known as first detector. Some mixer circuits use separate devices for mixing and generating local oscillator voltage. Such circuits are referred as separately excited mixers. Another type of mixer circuit, known as self-excited mixer utilises a single devices as mixer and local oscillator. They are also referred to a frequency convertors. The frequency convertors and mixers performs the job of frequency changers.

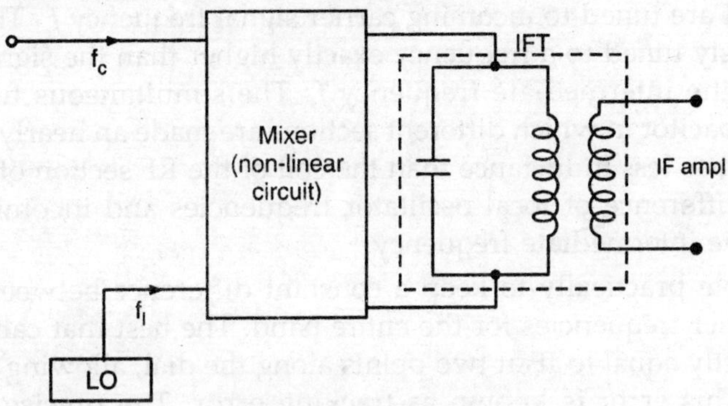

Fig. 2.67 *A Mixer-Circuit*

Figure 2.68 shows a frequency convertor circuit used in broadcast receivers. In this case, the receiver circuit does not contain RF amplifier stage. The incoming signal is applied at the base of a transistor. The local oscillator, LO, voltage develope across the emitter of the same transistor. Tuning capacitors of antenna and local oscillator

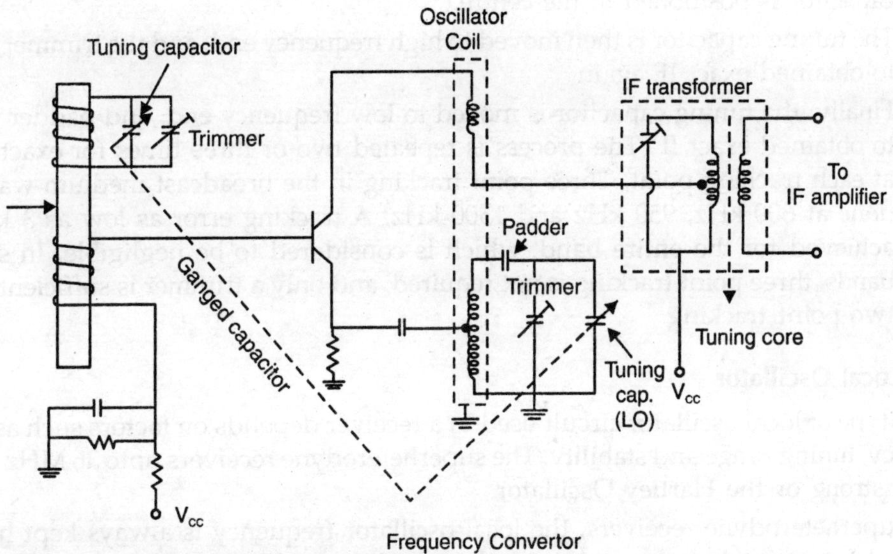

Fig. 2.68 *Frequency Convertor*

circuits are mounted on the same rotating shaft, i.e., they are ganged together to provide single dia tuning. The ganged capacitor is shown by a dotted line in Fig. 2.68.

2.40.4 Tracking and Alignment

For convenient tuning, it is common practice to gang the tuning capacitors of local oscillator, and front-end tuning circuit including the RF amplifier, if present. The ganged capacitor forms a mechanically coupled system for simultaneous tuning of a number of resonant circuits by a single knob. The front-end circuits including RF circuit and mixer input tuned circuit are tuned to incoming carrier signal frequency f_c. The local oscillator has to be simultaneously tuned to a frequency exactly higher than the signal frequency f_c by an amount equal to the intermediate frequency f_i. The simultaneous tuning is obtained by using a ganged capacitor in which different sections are made an nearly identical as possible. The oscillator coil has less inductance than the coil of the RF section of achieve the tracking i.e., to keep the difference of local oscillator frequencies and incoming signal frequency exactly equal to the intermediate frequency.

It is not feasible practically to keep a constant difference between the local oscillator and incoming carrier frequencies for the entire band. The best that can be done is to adjust the difference exactly equal to IF at two points along the dial, allowing some error along the rest of the dial. This error is known as tracking error. The precise alignment of tuned circuits the achieve a zero tracking error at two points along the dial is known as two points tracking. However, it is possible to precisely align the tuned circuits at three points along the dial by including a series capacitance known as padder and a shunt capacitor known as trimmer as shown in Fig. 2.68. The process is called as three point tracking. These process of alignment of the receiver for achieving three point tracking is as follows:

(*a*) Select a proper inductance to achieve exact IF in the middle of the band when tuning capacitor is positioned in the centre.

(*b*) The tuning capacitor is then moved to high frequency end, and the trimmer is adjusted to obtained exact IF again.

(*c*) Finally, the tuning capacitor is moved to low frequency end, and padder is adjusted to obtained exact IF. The process is repeated two or three times for exact alignment at each tracking point. Three point tracking in the broadcast medium wave band is done at 600 kHz, 950 kHz and 1500 kHz. A tracking error as low as 3 kHz can be achieved for the entire band, which is considered to be negligible. In short wave bands, three point tracking is not required, and only a trimmer is sufficient to achieve two point tracking.

2.40.5 Local Oscillator

The type of local oscillator circuit used in a receiver depends on factors such as operating frequency, tuning range and stability. The superheterodyne receivers upto 36 MHz use mostly the Armstrong or the Hartley Oscillator.

In superheterodyne receivers, the local oscillator frequency is always kept higher than the signal frequency by an amount equal to the intermediate frequency, for the following reasons.

(*a*) The maximum to minimum capacitance ratio of the two sections, signal and local oscillator sections, of the ganged capacitor is quite close. For medium wave band, 550 kHz-1650 kHz, the maximum to minimum capacitance ratio required by the signal section of the ganged capacitor is

$$\frac{C_{max}}{C_{min}} = \left(\frac{1650}{550}\right)^2 = 9:1$$

Now, consider the local capacitor section. When local oscillator frequency is kept higher, the maximum to minimum capacitance ratio required is

$$\frac{C_{max}}{C_{min}} = \left(\frac{1650+455}{550+455}\right)^2$$

$$= \left(\frac{2105}{1005}\right)^2 = 4.4:1$$

Which is quite close to that of signal section. The usual ganged capacitors available have a capacitance ratio of 10 : 1, which is well within the limit imposed by the tuning capacitors of both the sections. Therefore, the ganged capacitor with identical sections serves the purpose of tuning the front-end as well as local oscillator circuit.

If the local oscillator frequency is kept lower, the maximum to minimum capacitance ratio required is,

$$\frac{C_{max}}{C_{min}} = \left(\frac{1650-455}{550-455}\right)^2$$

$$= \left(\frac{1195}{95}\right)^2 = 156:1$$

This ratio is beyond the limit imposed by tuning capacitor of signal section and cannot be converted by the oscillator in one sweep. The ganged capacitor with identical sections will not serve the purpose. It is noted that ganged capacitors with identical sections are easy to fabricate.

(*b*) The tracking errors can be reduced to a great extent by keeping constant ratio of local oscillator frequency to signal frequency for the entire band. This ratio varies between $\frac{1005}{550} = 1.83$ to $\frac{2150}{1650} = 1.28$ for medium wave band when local oscillator frequency is kept above the incoming carrier signal frequency. This variation from 1.83 to 1.28 is quite small as compare to the situation when local oscillator operates below the signal frequency, giving a frequency ratio variation between $\frac{550}{95} = 5.79$ to $\frac{1650}{1195} = 1.38$. The variation of ratio from 5.79 to 1.38 is quite large and can result in severe tracking problems.

2.41 IF AMPLIFIER

The intermediate frequency (IF) amplifiers are tuned voltage amplifiers. Most of the receiver gain is provided by this stage. More than one stage of IF amplifier is used to get a good sensitivity. The output of the IF amplifier appears across a tuned transformer circuit. For n stages of IF amplifier, the total number of IF transformers needed is $(n + 1)$.

2.41.1 Choice of Intermediate Frequency

The intermediate frequency in a commercial AM receiver has a fixed value of 455 kHz. This value is chosen as a compromise between two conflicting factors:

(*i*) adjacent channel selectivity and easy tracking for which IF should be low, and

(*ii*) image signal rejection for which IF should be high.

For a proper channel selectivity the intermediate frequency f_i should be low. A lower f_i needs a lower Q, and the proper tank circuit can be easily designed, whereas if f_i is large, it will require a large Q and the tank circuit design will be complex. For example, if $f_i = 455$ kHz and the baseband is 10 kHz, Q value desired by the tank circuit is

$$Q = \frac{455}{10} = 45.5$$

The tuned circuit for this Q can be easily designed. But, if $f_i = 10$ MHz, the Q desired is

$$Q = \frac{10 \times 10^6}{10 \times 10^3} = 1000$$

The design of a tuned circuit with such a high Q is impossible. Hence, for a better selectivity, f_i should be lower. Also, a low value of IF makes the difference between signal and local oscillator frequency small, and as a result tracking becomes easy.

The value of f_i should be large for image signal rejection, from Eqn. 5.7.4 we have

$$f_c' = f_c + 2f_i$$

or

$$\frac{f_c'}{f_c} = 1 + \left(\frac{2f_i}{f_c}\right) \quad \text{...(2.174)}$$

Therefore, if f_i is kept large, the image signal f_c' will be for apart from desired signal f_c and can be easily rejected. In commercial AM receivers, the intermediate frequency is kept fixed at 455 kHz in order to achieve a compromise between the conflicting factors. This value of IF provides good image signal rejection at broadcast band even if no RF amplifier is used, as the ratio $\frac{f_c'}{f_c}$ is large. At short waves, however, the image signal rejection becomes poor due to a phenomenon known as double spotting. In this phenomenon, signals from the same short waves station is picked at two nearby points on the receiver tuning dial. Double spotting is harmful because it can mask a weak station by a nearby strong station at the spurious point on the dial. The spurious point corresponds to an image signal. Double spotting can be avoided by a good front end selectivity to avoid image signal. The RF amplifier add more tuning circuits and improve the front and selectivity of the receiver at short waves.

2.41.2 Second Detector

The linear diode detector is popularly used in radio receivers because of its simple circuit and low cost. The automatic volume control (AVC) bias is obtained from this stage in order to keep the receiver output substantially constant with time for any variations in receivers input voltage. The magnitude of the receiver input voltage varies with time due to fading, or when the receiver is tuned from one station to another having a different signal strength. The AVC eliminate the effect of these variations. Fig. 2.69(*a*) provides a simple AVC bias if diode D in AVC circuit is removed. The AVC circuit samples a fraction of the detector output and converts it to AVC bias voltage. The AVC bias is applied to RF and IF stages to provide them a negative bias. As the input of the receiver signal increase, the AVC bias voltage also increases, and in turn, the negative bias to RF and IF amplifiers is increased, thereby reducing their gain. The output of the receiver is, thus, maintained constant. The simple AVC circuit has a disadvantage in that it is operative even for low magnitude signals at the receiver input. This deteriorates the sensitivity of the receiver. Therefore, it is desirable that AVC should be operative only when the input signal is strong. This arrangement is made by a circuit called delayed AVC, which is obtained by incorporating a forward biased diode D in the AVC filter circuit as shown in Fig. 2.69(*b*). The AVC is operative only if the signal is more than the diode bias voltage. The AVC characteristic can be further improved by amplifying the delayed AVC bias with the help of a dc amplifier. This is known as amplified and delayed AVC. The characteristics of the various types of AVC circuits are shown in Fig. 2.69(*b*). Obviously amplified and delayed AVC is very much near to the idea AVC. An ideal AVC remains in operative for signal strength between certain limits, and provides a constant output for signal strength exceeding this limit.

2.41.3 Audio Amplifier

This stage consists of an RC coupled voltage amplifier followed by a push-pull power amplifier. The fidelity of the receiver is determined by the frequency response characteristic of the stage. The more the bandwidth of the stage, the better the fidelity.

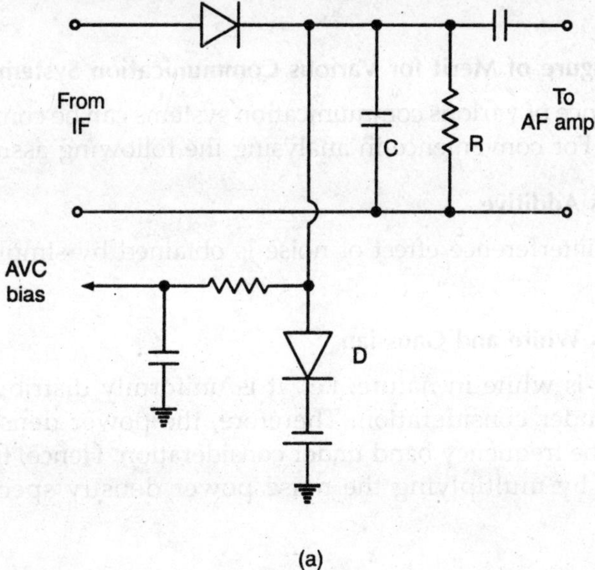

(a)

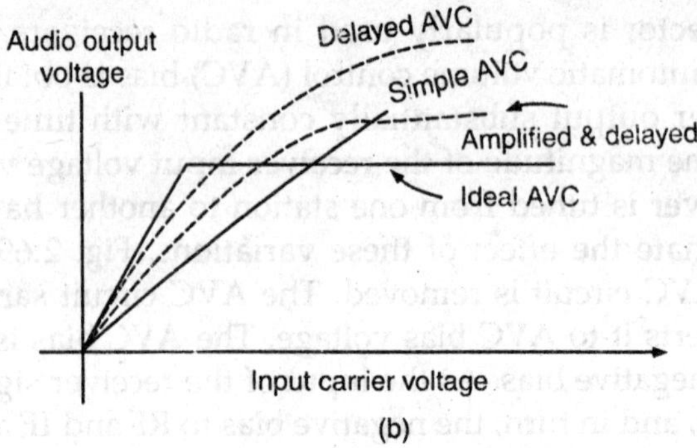

Fig. 2.69 *(a) A Detector Circuit with AVC; (b) AVC Characteristics*

2.41.4 Noise in Amplitude Modulated Systems

In communication system message signal travels from the transmitter to the receiver via a channel, radio or line. This is shown in Fig. 2.70. The channel introduces additive noise in the message and, hence, the message reaching the receiver becomes corrupted. As the receiver detects both noise and message signals, it reproduces a noisy message at the output. The noise characteristics of a modulation system is evaluated by a parameter known as figure of merit denoted by γ. It is defined as the ratio of output signal to noise ratio to input signal to noise ratio of a receiver.

$$\gamma = \frac{S_o/N_o}{S_i/N_i} \quad ...(2.175)$$

where N_i is the input noise power in a frequency range equal to the message bandwidth $2f_m$ including negative frequencies.

A modulation system with higher γ has a better noise performance and adverse effect of noise is less.

2.41.5 Calculation of Figure of Merit for Various Communication Systems

The noise performance of various communication systems can be compared by evaluating their figure of merit γ. For convenience in analysing the following assumptions are made:

2.41.6 Channel Noise is Additive

It is assumed that interference effect of noise is obtained by simple addition of signal $f(t)$ and noise $n(t)$.

2.41.7 Channel Noise is White and Gaussian

Channel noise $n(t)$ is white in nature, i.e., it is uniformly distributed over the entire band of frequencies under consideration. Therefore, the power density spectrum of the noise is uniform over the frequency band under consideration. Hence, the total noise power N is obtained simply by multiplying the noise power density spectrum $\eta/2$ with the bandwidth, i.e.,

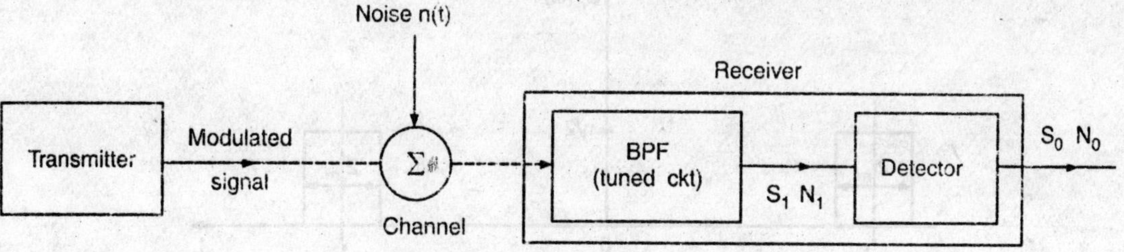

Fig. 2.70 *Channel Noise in Communication System*

$$N = \left(\frac{\eta}{2}\right) \times \text{bandwidth} \qquad ...(2.176)$$

The noise amplitude has a Gaussian distribution.

2.41.8 Bandpass Noise at the Input of Detector

The first stage of every receiver is a tuned circuit that acts as a bandpass filter. This filter allows only the narrowband signal centred about carrier frequency $\pm f_c$ and rejects the other frequencies. Therefore, the noise signal lying outside this band is also rejected, and the bandwidth of the noise at the demodulator input is same as that of the modulated signal. The white noise at the input of the detector has a constant power density spectrum over the passband as shown in Fig. 2.71. The power density is given by

$$S_{ni} = \frac{\eta}{2} \qquad ...(2.177)$$

2.41.9 The Input Noise Power N

For evaluating the figure of merit, Eqn. (2.175) the input noise power N_i is determined for baseband, i.e. $2f_m$ Hz, including negative frequencies. From Eqn. (2.176)

$$N_i = \left(\frac{\eta}{2}\right)(2f_m) = \eta f_m \qquad ...(2.178)$$

Eqn. (2.178) will be taken as reference input noise power for evaluating γ of various AM and FM receivers. However, N_i in Eqn. (2.178) is the true input noise power only for the case of SSB systems.

2.41.10 Noise Calculation for Various AM Systems

The performance of various amplitude modulated systems can be studied by evaluating their individual figure of merit.

2.41.11 DSB-SC

The DSB-SC systems uses synchronous detection at the receiver. A schematic diagram of the transmitter and receiver is shown in Fig. 2.72.

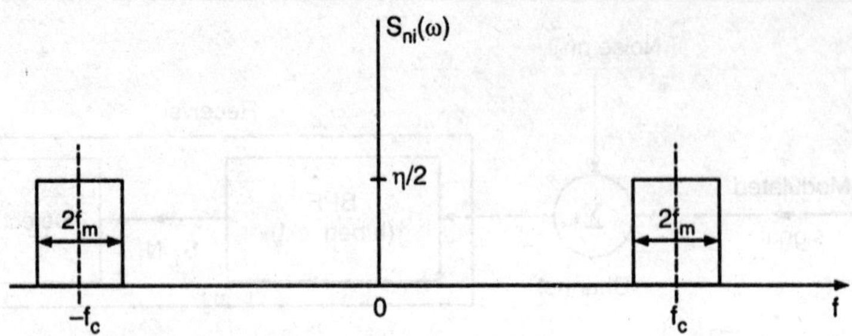

Fig. 2.71 *Power Density Spectrum of the Bandpass White Noise*

2.41.12 Noise Power

The input noise power is given by Eqn. 2.178. Now we will evaluate the noise power N_o at the output of synchronous detector. The noise signal at the input of this detector is a bandpass noise given by,

$$n(t) = n_c(t) \cos \omega_c t - n_s(t) \sin \omega_c t$$

This signal is multiplied by $\cos \omega_c t$ in the synchronous detector. The multiplied signal n_d is given by

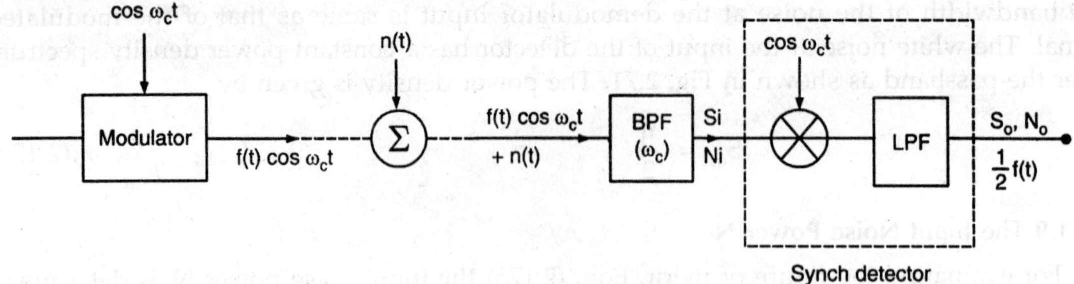

Fig. 2.72 *A DSB-SC System*

$$\begin{aligned}
n_d(t) &= n(t) \cos \omega_c t \\
&= \left[n_c(t) \cos \omega_c t - n_c(t) \sin \omega_c t\right] \cos \omega_c t \\
&= n_c(t) \cos^2 \omega_c t - n_s(t) \cos \omega_c t \sin \omega_c t \\
&= \frac{1}{2} n_c(t)\left[1 + \cos 2\omega_c t\right] - \frac{1}{2} n_s(t) \sin 2\omega_c t \\
&= \frac{1}{2} n_c(t) + \frac{1}{2} n_c(t) \cos 2\omega_c t - \frac{1}{2} n_s(t) \sin 2\omega_c t
\end{aligned}$$

The noise signal $n_d(t)$ is passed through a low pass filter. The terms $\frac{1}{2} n_c \cos 2\omega_c t$ and $\frac{1}{2} n_c(t) \sin 2\omega_c t$ centered near $\pm 2\omega_c$ are filtered out by the low pass filter, and hence, the noise power $n_o(t)$ appearing at the output of the synchronous detector is only $\frac{1}{2} n_c(t)$ i.e.,

$$n_o(t) = \frac{1}{2} n_c(t) \qquad ...(2.179)$$

Since the power density spectrum is proportional to the mean square value of the signal, the power density spectrum of $n_o(t)$ is related to the power density spectrum of $n_c(t)$ as follows:

$$S_{n_o}(\omega) = \frac{1}{4} S_{n_o}(\omega) \qquad ...(2.180)$$

Now for a DSB-SC system,

$$S_{n_c}(\omega) = \eta, \quad |f| \leq f_m \qquad ...(2.181)$$

Substituting this value of $S_{n_c}(\omega)$ in Eqn. (2.180), we get

$$S_{n_o}(\omega) = \frac{\eta}{4}, \quad |f| \leq f_m \qquad ...(2.182)$$

The bandwidth of the baseband signal at the output of the detector is $2f_m$. Hence, the output noise power N_o is given by,

$$N_o = S_{n_o}(\omega) \times (\text{Bandwidth})$$

$$= \frac{\eta}{4} \times 2f_m = \frac{\eta f_m}{2} \qquad ...(2.183)$$

From Eqn. (2.178) we have

$$N_i = \eta f_m$$

Therefore,

$$\frac{N_o}{N_i} = \frac{1}{2} \qquad ...(2.184)$$

2.41.13 Signal Power

The modulated signal at the input of the detector is $f(t) \cos \omega_c t$. Therefore, the input signal power S_i is the MS value of $f(t) \cos \omega_c t$

$$S_i = \overline{[f(t) \cos \omega_c t]^2}$$

$$= \frac{1}{2} \overline{f^2(t)} \qquad ...(2.185)$$

The signal voltage at the output of the detector is $\frac{1}{2} f(t)$. Therefore, the output signal power is the MS value of $\frac{1}{2} f(t)$,

$$S_o = \overline{\left[\frac{1}{2} f(t)\right]^2} = \frac{1}{4} \overline{f^2(t)} \qquad ...(2.186)$$

Comparing with Eqn. (2.185),

$$S_o = \frac{1}{2}S_i \qquad \text{...(2.187)}$$

or

$$\frac{S_o}{S_i} = \frac{1}{2} \qquad \text{...(2.188)}$$

Now figure merit γ is given by

$$\gamma = \frac{S_o/N_o}{S_i/N_i} = \frac{S_o/S_i}{N_o/N_i}$$

Substituting (S_o/S_i) from Eqn. (2.188) and from Eqn. (2.814), we get

$$\gamma = 1 \qquad \text{...(2.189)}$$

Thus the figure of merit for a DSB-SC system is unity. Therefore, the S/N ratios at the input and output of the detector are identical, and there is no improvement in S/N ratio.

EXAMPLE 1

Show that for a DSB-SC system, the power densities of various components of bandpass noise are related as

$$S_{n_c}(\omega) = S_{n_s}(\omega) = 2S_n(\omega) = \eta, \quad |\omega| \leq \omega_m \qquad \text{...(2.190)}$$

Also show that for an SSB-SC system,

$$S_{n_c}(\omega) = S_{n_s}(\omega) = S_n(\omega)$$

$$= \frac{\eta}{2}, \quad |\omega| \leq \omega_m \qquad \text{...(2.191)}$$

The power densities in Eqn. (2.190) and Eqn. (2.191) are zero elsewhere except for $|\omega| \leq \omega_m$.

Solution:

The bandpass noise signal is represented as

$$n(t) = n_c(t) \cos \omega_c t - n_s(t) \sin \omega_c t$$

The power density spectrum of $n(t)$ is related to the power density spectrum of $n_c(t)$ and $n_s(t)$ as given below

$$S_{n_c}(\omega) = S_{n_s}(\omega)$$

$$= \begin{cases} S_n(\omega + \omega_c) + S_n(\omega - \omega_c), & |\omega| \leq \omega_m \\ 0 & \text{elsewhere} \end{cases}$$

Therefore, $S_{n_c}(\omega)$ and $S_{n_s}(\omega)$ can be obtained simply by shifting the spectrum $S_n(\omega)$ by an amount $-\omega_c$ and $+\omega_c$ and then, superposing the two shifted spectra, as shown in Fig. 2.73. It is obvious from the superposed spectrum shown in Fig. 2.73(*d*) that,

$$S_n(\omega + \omega_c) + S_n(\omega - \omega_c) = 2S_n(\omega) = \eta, \quad |\omega| \leq \omega_n$$

Therefore, using the relations defined by Eqn., we get

$$S_{n_c}(\omega) = S_{n_s}(\omega) = \begin{cases} S_n(\omega+\omega_c)+S_n(\omega-\omega_c), & -\omega_m \le \omega \le \omega_m \\ 0 & \text{otherwise} \end{cases}$$

$$S_{n_c}(\omega) = S_{n_s}(\omega) = 2S_n(\omega) = \eta, \text{ for } |\omega| \le \omega_m$$

The power densities are zero elsewhere. This proves Eqn. (2.190).

In a similar way, we can proceed for SSB-SC system. The spectrum $S_n(\omega)$ for a SSB-SC system is shown in Fig. 2.74(*a*) in which the spectrum has exists only for upper sidebands. The spectrum has been shifted by $-\omega_c$ in Fig. 2.74(*b*), and by $+\omega_c$ in Fig. 2.74(*c*). The superposition of the shifted spectra is shown in Fig. 2.74(*d*). It is obvious from the Fig. 2.74(*d*) and Eqn. (2.190) that

$$S_{n_c}(\omega) = S_{n_c}(\omega) = S_n(\omega) = \frac{\eta}{2}, \quad |\omega| \le \omega_m$$

The power densities are zero elsewhere. This proves Eqn. (2.191).

2.41.14 SSB-SC

Like DSB-SC systems, the SSB-SC system too uses synchronous detection at the receiver, and hence, the procedure for evaluating the figure of merit is similar.

2.41.15 Noise Power

The input noise power N_i is given by Eqn. (2.178). The noise power density at the output the synchronous detector is same as given by Eqn. (2.180) for the DSB-SC case.

$$S_{n_o}(\omega) = \frac{1}{4}S_{n_c}(\omega)$$

The power density spectrum $S_{n_o}(\omega)$ for SSB-SC signal is equal to $\eta/2$. Hence

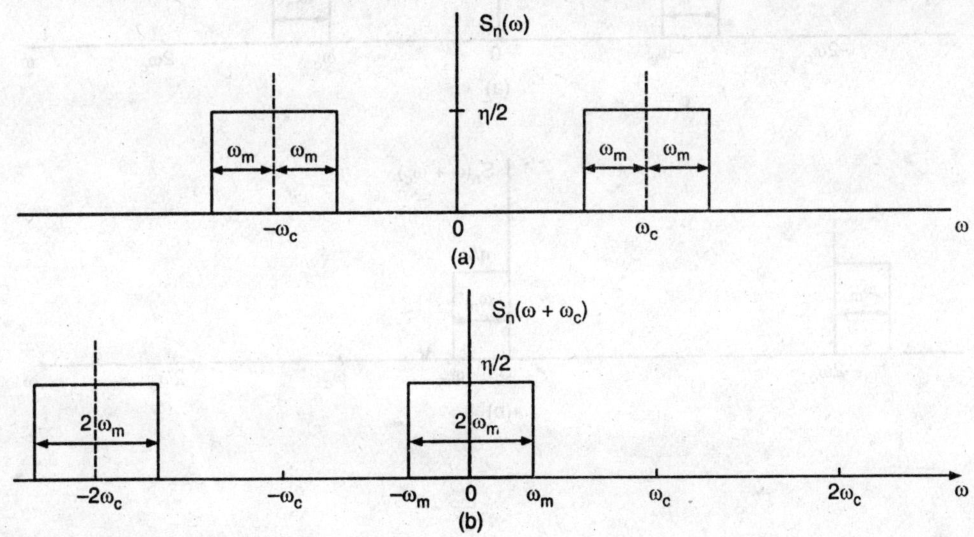

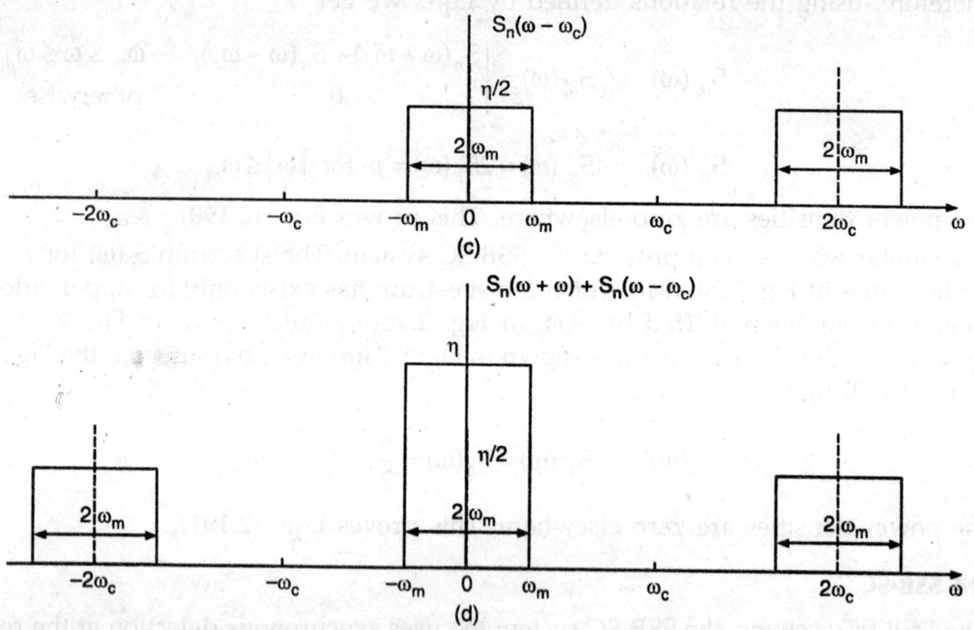

Fig. 2.73 *Noise Power Density in SSB-SC System (a) Spectrum of n(t) (b) Spectrum (a) Shifted by – ω_c (c) Spectrum (a) shifted by + ω_c (d) Super position of (b) and (c)*

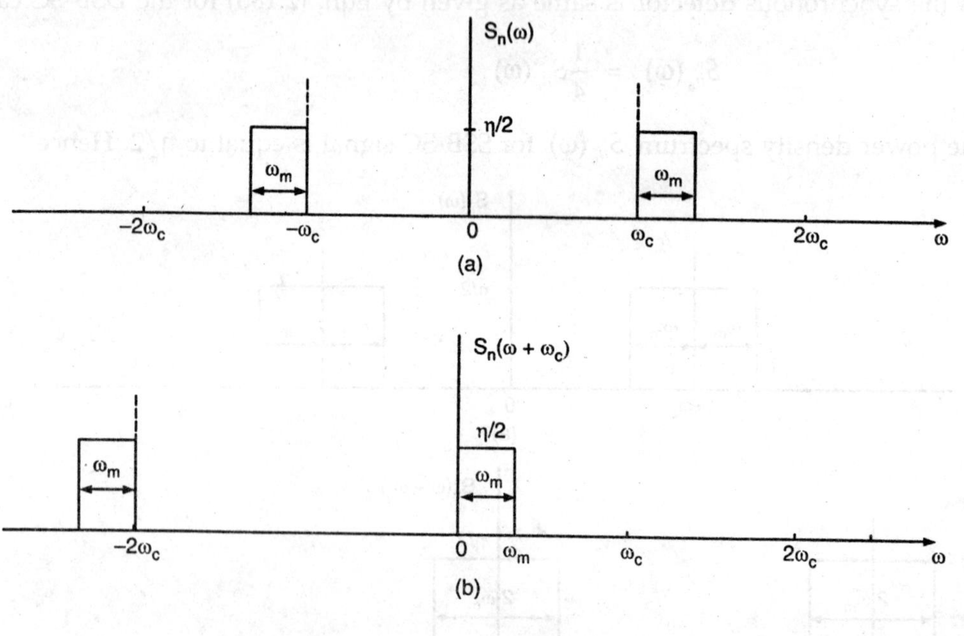

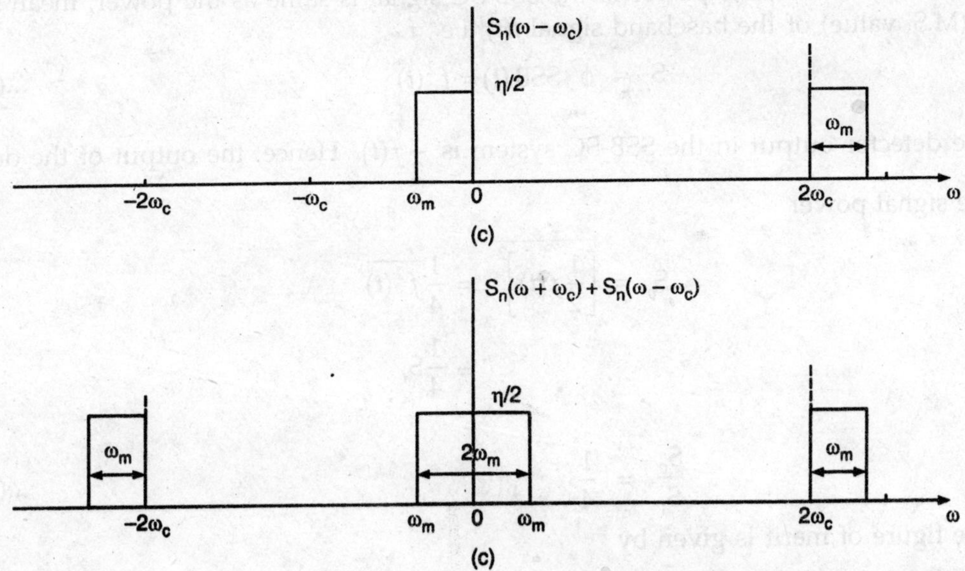

Fig. 2.74 *Noise Power Density in SSB-SC System (a) Spectrum n(t) (b) Spectrum (a) shifted by* $-\omega_c$ *(c) Spectrum of (a) shifted by* $+\omega_c$ *(d) Superposition of (b) and (c)*

$$S_{n_o}(\omega) = \frac{1}{4} \times \frac{\eta}{2} = \frac{\eta}{8} \qquad ...(2.192)$$

The bandwidth of the detected signal at the output of the detector is $2f_m$. It should be kept in mind that both SSB-SC and DSB-SC produce identical spectra at the synchronous detector output. This is obvious from above figures. In the detected output is $\frac{1}{2}f(t)$, with identical spectra. Therefore, the noise power at the output of the detector is

$$N_o = S_{n_o}(\omega) \times (\text{bandwidth})$$

From Eqn. (2.178)

$$N_o = \eta f_m$$

Hence,

$$N_o = \frac{N_i}{4} \qquad ...(2.193)$$

2.41.16 Signal Power

The signal power at the input of the synchronous detector in a SSB-SC receiver can be evaluated by observing the spectrum. The signal power is proportional to the area under the power density spectrum, Parseval's Theorem. The power density spectrum of the signal is proportional to square of the Fourier Transform. It is obvious that the area under the power density spectrum of a SSB-SC signal is same as the area under the power density spectrum F(ω). Hence, the power of the SSB-SC signal is same as the power, mean square value (M.S. value) of the baseband signal $f(t)$, i.e., f

$$S_i = \phi^2 \text{ SSB}(t) = f^2(t) \quad ...(2.194)$$

The detector output in the SSB-SC system is $\frac{1}{2}f(t)$. Hence, the output of the detector has the signal power

$$S_o = \overline{\left[\frac{1}{2}f(t)\right]^2} = \frac{1}{4}\overline{f^2(t)}$$

$$= \frac{1}{4}S_i$$

or

$$\frac{S_o}{S_i} = \frac{1}{4} \quad ...(2.195)$$

The figure of merit is given by

$$\gamma = \frac{S_o/N_o}{S_i/N_i} = \frac{S_o/S_i}{N_o/N_i} \quad ...(2.196)$$

Substituting S_o/S_i from Eqn. (2.195) and N_o/N_i from Eqn. (2.193),

$$\gamma = 1 \quad ...(2.197)$$

This is same as the figure of merit of the DSB-SC system, which means both are identical from noise performance point of the view inspite of the fact that SSB-SC has half bandwidth. This is because, the large bandwidth in the DSB-SC system increases the signal as well as noise power.

2.41.17 Amplitude Modulation System (Envelop Detector)

In amplitude modulation system, a large carrier is accompanied with the two sidebands. If $n_i(t)$ is the additive noise signal, the input to the detector is given by

$$\phi_{AM} = [A + f(t)]\cos\omega_c t + n_1(t) \quad ...(2.198)$$

The input signal S_i is given as

$$S_i = \text{MS value of carrier + MS value of sideband}$$

$$= \frac{1}{2}\left[A^2 + \overline{f^2(t)}\right] \quad ...(2.199)$$

2.41.18 Output Power

The output of the envelope detector, popularly used in AM receivers, will be the envelope of the AM signal $\phi_{AM}(t)$ given in Eqn. (2.198). Therefore, in order to find the detected output, we need to find the envelope of $\phi_{AM}(t)$. Substituting quadrature representation for $n_2(t)$, Eqn. (2.198) becomes

$$\phi_{AM}(t) = [A + f(t)]\cos\omega_c t + n_c(t)\cos\omega_c t - n_s(t)\sin\omega_c t$$

$$= [A + f(t) + n_c(t)]\cos\omega_c t - n_s(t)\sin\omega_c t \qquad ...(2.200)$$

After trigonometric manipulation Eqn. (2.200) can be written as

$$\phi_{AM}(t) = A(t)\cos\{\omega_c t + \psi(t)\} \qquad ...(2.201)$$

where A(t) and $\psi(t)$ respectively are randomly time varying amplitude and phase angle of $\phi_{AM}(t)$. They are given by

$$A(t) = \sqrt{\left[a + f(t) + n_c(t)^2\right] + n_s^2(t)} \qquad ...(2.202)$$

$$\psi(t) = \tan^{-1}\left[\frac{n_s(t)}{A + f(t) + n_c(t)}\right] \qquad ...(2.203)$$

The time varying amplitude A(t) is the envelope of $\phi_{AM}(t)$ and hence, the output of the envelope detector will be the envelope A(t). The envelope A(t) has both the signal and noise components. The noise performance depends on the relative magnitude of the signal and noise. The analysis can be carried out be carried out for two cases, viz.

(*i*) small noise case and

(*ii*) large noise case.

2.41.19 Small-Noise Case

In this case, noise is taken to be much small than the signal,

$$n_i(t) \ll [A + f(t)]$$

A phasor-representation of envelope A(t) is shown in Fig. 2.75. The noise component $n_c(t)$ is shown to be in phase with signal $[A + f(t)]$. whereas $n_s(t)$ is in phase quadrature. Since $n_i(t) \ll [A + f(t)]$, the noise component $n_s(t)$ is also much smaller than $[A + f(t)]$.

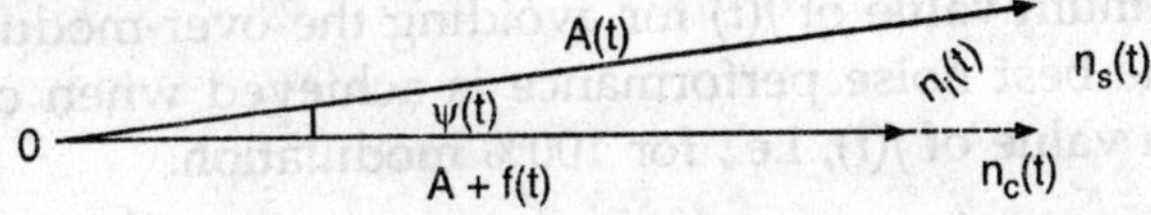

Fig. 2.75 *Phasor Diagram of the Envelope A(t)*

It is evident from the phasor diagram that if $n_s(t)$ is much small, then $\psi(t)$ is also much small and may be taken as zero. When $\psi(t) = 0$, the envelope A(t) becomes,

$$A(t) = A + f(t) + n_c(t) \qquad ...(2.204)$$

Therefore, the output of the envelope detector contains a useful signal $f(t)$ and a noise component $n_c(t)$. The carrier amplitude A carries no useful information. The signal power S_o and the noise power N_o, at the output of the detector, are evaluated as follows:

2.41.20 Output Signal Power S_o

It is the MS value of useful signal $f(t)$, i.e.,

$$S_o = \overline{f^2(t)} \qquad ...(2.205)$$

2.41.21 Output Noise Power

The noise signal $n_o(t)$ at the out put of the detector is $n_c(t)$ with the power density given by

$$S_{n_o}(\omega) = S_{n_c}(\omega) \qquad ...(2.206)$$

As the AM system contains both sidebands, $S_{n_c}(\omega)$ will be identical to that of DSB-SC system, given by Eqn. (2.204) i.e.,

$$S_{n_o}(\omega) = S_{n_c}(\omega) = \eta.$$

The output noise power is given as,

$$N_o = S_{n_o}(\omega)(B\omega) = \eta.(2f_m) = 2\eta f_m \qquad ...(2.207)$$

The figure of merit γ is obtained with the help of Eqns. (2.191), (2.200), (2.205) and (2.207):

$$\gamma = \frac{S_o/N_o}{S_i/N_i} = \frac{\overline{f^2(t)}/(2\eta f_m)}{\frac{1}{2}\left[A^2 \overline{f^2(t)}\right]/\eta f_m}$$

$$= \frac{\overline{f^2(t)}}{A^2 + \overline{f^2(t)}} \qquad ...(2.208)$$

From Eqn. (2.208) it is clear that the noise performance improves with a reduction in carrier amplitude A and it is maximum when A = 0, which is equivalent to the suppressed carrier AM system. Thus, AM system remains always inferior to the suppressed system from the point of view of 'noise performance.' The greatest value of γ that can be achieved will depend on the minimum possible value of carrier amplitude A. The minimum value of A is equal to the maximum value of $f(t)$ for avoiding the over-modulation in the envelope detector. Therefore, the best noise performance is achieved when carrier amplitude A is equal to the maximum value of $f(t)$, i.e., for 100% modulation.

EXAMPLE 2

Single-Tone Modulation

Derive an expression for the figure of merit γ when the modulating signal $f(t)$ is a single-tone sinusoid given by $f(t) = m_a A \cos \omega_m t$, where m_a is the modulation index, and A is the carrier amplitude. Find the value of γ when the depth of modulation is:

(a) 100%;

(b) 50%; and

(c) 30%.

Solution:

$$f(t) = m_a A \cos \omega_m t \qquad ...(2.209)$$

The power content in $f(t)$ is given as

$$\overline{f^2(t)} = \frac{m_a^2 A^2}{2}$$

From Eqn. (2.209), the figure of merit is given as

$$\gamma = \frac{m_a^2 A^2/2}{A + \frac{m_a^2 A^2}{2}}$$

or

$$\gamma = \frac{m_a^2}{2 + m_a^2} \qquad ...(2.210)$$

(*a*) When $m_a = 100\%$, i.e., $m_a = 1$

$$\gamma = \frac{1}{2+1} = \frac{1}{3}$$

$$= 0.33$$

Thus, the AM system must transmit three times more power as compared to the suppressed carrier AM system in order to achieve the same noise performance.

$$m_a = 50\% = \frac{1}{2}$$

$$\gamma = \frac{1/2}{2+\frac{1}{2}} = \frac{1}{5}$$

$$= 0.2$$

Which is less than the one in case (*a*).

(*c*) $m_a = 30\% = 0.3$

$$\gamma = \frac{0.3}{2+0.3} = \frac{3}{23} = 0.13$$

Which is less than even case (*b*).

Thus, the maximum value of γ in envelope detector is 0.33 for 100% modulation.

2.42 SYNCHRONOUS DETECTION OF AM SYSTEM

Even if the synchronous detection is used for AM demodulation, the noise performance is identical to the envelope detector. Moreover, the synchronous detection is both complex and costly; hence, it is never used for AM detection.

2.42.1 Large Noise Case

In this case, $n_i(t) \gg [A + f(t)]$. Therefore, and the performance of the envelope detector is completely different from the low noise case. The envelope of the modulated signal at the input of the detector is given by Eqn. (2.202),

$$A(t) = \sqrt{[A + f(t) + n_c(t)]^2 + n_s(t)^2}$$

$$= \sqrt{[A + f(t)]^2 + 2n_c(t)[A + f(t)] + n_c^2(t) + n_s^2(t)}$$

Since the noise dominates over the signal, the independent signal term $[A + f(t)]^2$ is very small as compared to the other terms. Therefore, $[A + f(t)]^2$ may be ignored, and the expression for A(t) becomes

$$A(t) = \sqrt{n_c^2(t) + n_s^2(t) + 2n_c(t)[A + f(t)]}$$

$$= \sqrt{[n_c^2(t) + n_s^2(t)]\left[1 + \frac{2n_c(t)[A + f(t)]}{n_c^2(t) + n_s^2(t)}\right]}$$

Putting $$R(t) = \sqrt{n_c^2(t) + n_s^2(t)}$$

and $$\theta(t) = \tan^{-1}\left(\frac{n_s(t)}{n_c(t)}\right)$$

We get

$$A(t) = \sqrt{R^2(t)\left[1 + 2\left\{A + f(t).\frac{n_c(t)}{R^2(t)}\right\}\right]}$$

$$= \sqrt{R^2(t)\left[1 + \frac{2\{A + f(t)\}}{R(t)}\cos\theta(t)\right]}$$

$$= R(t)\left[1 + \frac{2\{A + f(t)\}}{R(t)}\cos\theta(t)\right] \quad \text{...(2.211)}$$

Since the noise component $R(t) \gg 2\{A + f(t)\}$, Eqn. (2.211) can be written in approximate form as

$$A(t) = R(t)\left[1 + \frac{1}{2}\frac{2\{A + f(t)\}}{R(t)}\cos\theta(t)\right]$$

$$= R(t) + A\cos\theta(t) + f(t)\cos\theta(t) \quad \text{...(2.212)}$$

It is evident from Eqn. (2.212), that envelope A(t) which appears at the output of the envelope detector has no exclusive term proportional to the baseband modulating signal $f(t)$. The term $f(t)$ is multiplied with a large noise term. cos $\theta(t)$ and hence $f(t)$ cannot be separated from the noise. As the modulating term $f(t)$ is completely mingled with noise, it carries no useful information. The loss of the message $f(t)$ in an envelope detector due to the presence of the large noise is referred to as the *threshold effect*.

2.42.2 Threshold Effect in An Envelope Detector

As we know that when a noise is large as compared to the signal at the input of the envelope detector, the detected output has message signal completely mingled with the noise. It means that if the input signal-to-noise ratio (S_i/N_i) is below a certain level, called threshold level, the noise dominates over the message signal. Threshold is defined as value of the input signal to noise ratio (S_i/N_i) below which the output signal to noise ratio (S_o/N_o) deteriorates much more rapidly than the input signal to noise ratio. The threshold effect starts in an envelope detector when ever the carrier power to noise power ratio approaches unity or less. It is important to stress that the threshold effect is the property of an envelope detector; on the other hand, such an effect is not observed in a coherent detector.

2.42.3 Comparison of Various AM System

We have discovered various forms of amplitude modulation. Now we can divide them into two broad categories accordingly, viz., suppresses carrier AM system; and AM systems with a large carrier amplitude. The suppressed carrier system can be further subdivided into three parts:

(*a*) DSB-SC;
(*b*) SSB-SC; and
(*c*) VSB-SC.

It is worthwhile to compare the various AM systems based on the following points:

2.42.4 Receiver End

For the receiver, an AM is advantageous, because the envelope detector used in such a receiver is very simple and in expensive. Therefore, the AM system is preferably used in public communication systems where a transmitter is associated with a large number of receivers. The receivers cost of important in such cases, as for example in a radio broadcast system, a receiver is needed almost in every house, and can be treated as an essential commodity. Therefore, the cost of the receiver must be low. AM systems with a large carrier are advantageous in this case. The receivers of a suppressed carrier system are complex and costly as they, need additional synchronizing circuits. Therefore, suppressed carrier system is undesirable for public communication systems.

2.42.5 Transmitter End

The suppressed carrier systems need low transmitters as they transmit only sideband power and no carrier power is transmitted. The low power transmitters used in suppressed carrier systems are less expensive than the large power AM transmitters. Suppressed carrier systems are useful in point to communication where we need many transmitters but only few receivers per transmitter. The expensive receivers may be justified in this case for the gain achieved by less expensive transmitters.

2.42.6 Generation of Modulated Signals

AM is easier to generate, even for large power level. On the other hand, balanced modulators that are required to generate suppressed carrier signals are more difficult to design. Further, SSB-SC is more difficult to generate than DSB-SC. VSB-SC is easier to generate than SSB-SC, but at the cost of higher channel-width.

2.42.7 Bandwidth

SSB is most advantageous from bandwidth point of view. This system has a minimum bandwidth, although its generation is most difficult. SSB-SC is used in long range high-frequency communications, especially in audio-communication, where phase-distortion is not significant. Amateur operated radio used SSB-SC systems. VSB-SC system is next to the SSB-SC. It has bandwidth between DSB and SSB and has less problem in its generation. Picture signal transmission in televisions uses vestigial sideband transmission that cuts down the bandwidth of a channel as compared to the DSB-SC system. The DSB system has a maximum bandwidth. It should be recalled that the higher bandwidth of a channel reduces that the number of available channels in a given frequency space.

2.42.8 Selective Fading

The random variation of a signal strength at the receiver due to multipath propagation is called fading. The signals from a transmitter arrive at the receiver by multiple paths, each path having a different length. The path having a different length. The path length varies randomly due to random changes in the ionospheric conditions. Signals arriving at the receiver by different path lengths randomly differ in phase. The strength of a resultant received signal varies randomly with time and this produces a fading effect. The fading is frequency sensitive and hence carrier and sidebands may suffer by different amounts of fading. This is called *selective fading*. This type of fading disturbs the magnitude ratio of carrier and sidebands. In AM, carrier amplitude may be considerably reduced as compared to sidebands due to more severe fading in the former. This effect produces over modulation in AM signal, and when detected by an envelope detector, it will be heavily distorted. Also, the selective fading becomes severe at high frequencies. Therefore, at high frequencies, AM is inferior to suppressed carrier systems. Selective fading produces more distortion in the DSB-SC system as against in the SSB-SC system, because it disturbed the phase relationship of two sidebands of the DSB-SC signal. Since SSB-SC has only one sideband this type of distortion is minimum.

2.42.9 Transmission Efficiency

Efficiency of a suppressed carrier system is 100%, whereas, in an AM system the maximum efficiency is only, 33.3% corresponding to $m_a = 1$.

2.42.10 Noise Performance

SSB-SC and DSB-SC systems have $\gamma = 1$, hence they are equivalent from the noise performance point of view. The AM system, with a large carrier, using the envelope detector has maximum $\gamma = \frac{1}{3}$, and also it suffers from the threshold effect. Therefore, the AM system is always inferior to suppressed carrier systems from the noise performance point of view.

2.42.11 Non-Linear Distortion

Non-linear distortion occurs due to non-linear characteristics of the systems. This type of distortion causes two undesired effects:

(*a*) harmonic generation; and

(*b*) generation of intermodulation frequencies.

In the harmonic distortion, harmonics of the input frequencies are generated. In intermodulation distortion, the sum and difference of input frequencies, and the harmonics are generated. Obviously, non-linear distortions will be more severe in the systems involving signals with more number of frequency terms, because this causes more number of harmonics and intermodulation frequency terms. Therefore, non-linear distortion is expected to be maximum in AM and minimum in SSB-SC systems, whereas in DSB-SC it is in between the AM and SSB-SC. Non-linear distortion not only modifies the quality of a signal, but also it may produce interchannel cross-talk in frequency division multiplexing. This problem is controlled by suitable design of filters.

2.43 FREQUENCY DIVISION MULTIPLEXING (FDM)

We have already discussed that multiplexing is the process of simultaneous transmission of several messages over a common channel without interference. Now let us suppose, that we want to transmit n messages each one band limited to ω_m, i.e., telephone signals from n subscribers. Each message is centered near zero-frequency. The spectra of individual messages are shown in Fig. 2.76(*a*). The bandwidth of each signal depends on the type of message. For example, a telephonic message has a bandwidth of 3 kHz and a broadcast signal has a bandwidth of 5 kHz. If these n signals are transmitted simultaneously over a common channel, line or radio, without multiplexing, they will interfere with each other, and no useful information will be produced at the receiver. However, the can be transmitted without interference if they are multiplexed. In frequency division multiplexing, each baseband signal is translated by analog modulation, amplitude or angle, to different carrier frequencies. The corresponding spectrum of multiplexed signals using AM is shown in Fig. 2.76(*b*). Each carrier is separated from the neighbouring one by at least $2\omega_m$. The multiplexed signal can be transmitted over a common channel without interference.

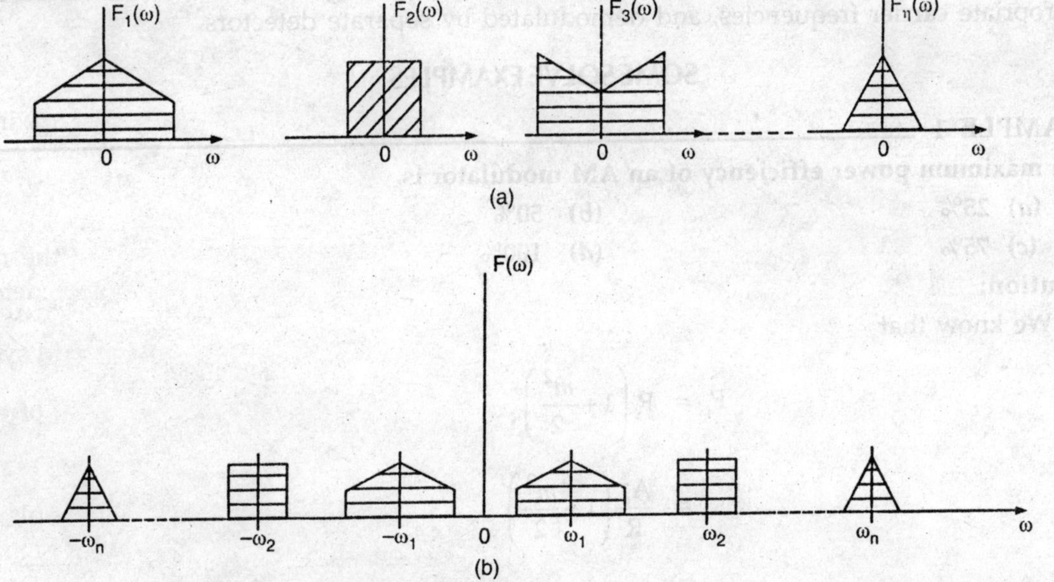

Fig. 2.76 *(a) Spectrum of n-messages (b) Spectrum of Frequency Division Multiplexed Signal*

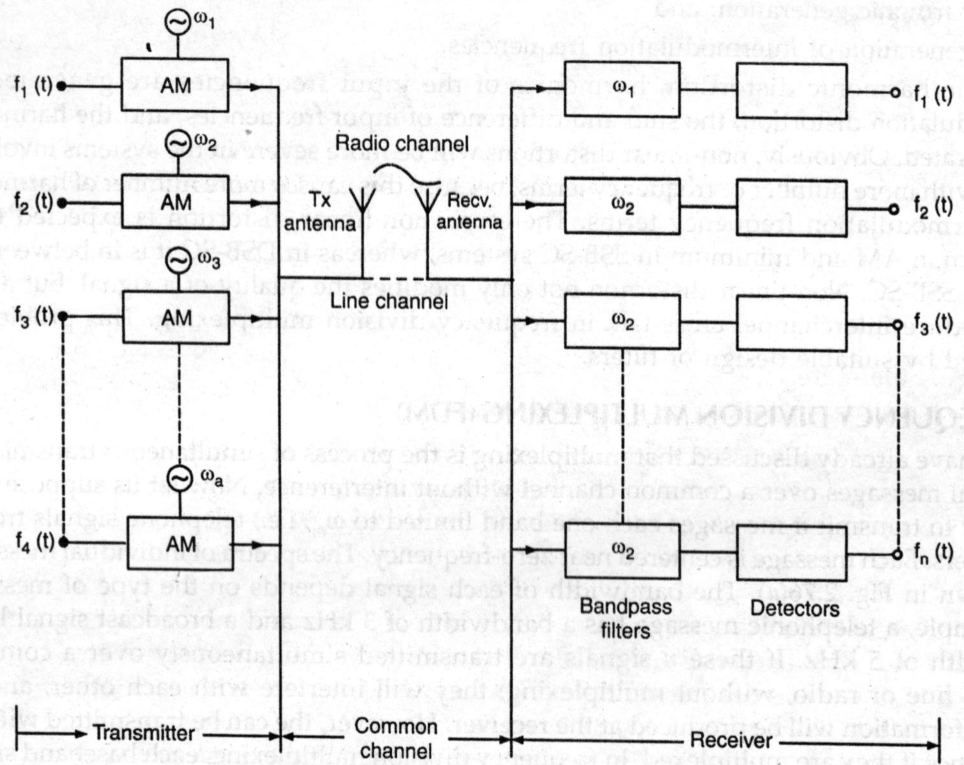

Fig. 2.77 *Schematic Diagram of an FDM System*

Fig. 2.77 shows the schematic diagram of a frequency division multiplexing system. At the transmitter, each message is superimposed on a separate carrier wave using AM. At the receiver, the various carrier frequencies are selected using bandpass filters tuned to appropriate carrier frequencies, and demodulated by separate detectors.

SOME SOLVE EXAMPLES

EXAMPLE 1

The maximum power efficiency of an AM modulator is

(*a*) 25% (*b*) 50%

(*c*) 75% (*d*) 100%

Solution:

We know that

$$P_t = P_c\left(1+\frac{m^2}{2}\right)$$

$$= \frac{A_c^2}{R}\left(1+\frac{m^2}{2}\right)$$

and useful power

$$P_u = \frac{m^2}{2} \cdot \frac{A_c^2}{R}$$

and $$P_c = \frac{A_c^2}{R}$$

Hence, efficiency

$$\eta = \frac{P_u}{P_c} = \frac{m^2}{2}$$

But $m \leq 1$, therefore,

$$\eta_{max} = \frac{1}{2}$$

i.e. 50%.

Hence, option (*b*) is correct.

EXAMPLE 2

A bandwidth of 20 MHz is to be considered for the transmission of AM signals. If the highest audio frequencies used to modulate the carriers are not to exceed 3 kHz, how many station could broadcast within this band simultaneously without interfering with one another.

Solution:

Here, given that

$$\text{Total Bandwidth} = 20 \text{ MHz}$$

$$f_{m_{max}} = 3 \text{ kHz}$$

We know that the maximum bandwidth of each AM station is determined by the maximum frequency of the modulating signal.

$$\begin{aligned} \text{Station BW} &= 2f_{m_{max}} \\ &= 2 \times 3 \times 10^3 \\ &= 6 \times 10^3 \\ &= 6 \text{ kHz} \end{aligned}$$

Thus, the number of stations that can broadcasted simultaneously without interfering with one another will be

$$\frac{20 \times 10^6}{6 \times 10^3} = 3.333 \times 10^3$$

Therefore, Number of stations

$$= 333 \textbf{ Ans.}$$

EXAMPLE 3

Determine the power content of the carrier and each of the sidebands for an AM signal having a per cent modulation of 80% and a total power of 2500 W.

Solution:

Here, given that

$$M = 80\%$$
$$m_a = 0.8$$
$$P_t = 2500 \text{ W}$$

Now, we have to find out

P_c, P_{USB} and P_{LSB}.

We know that the total power of an AM signal is the sum of the power at the carrier frequency and the power contained in the sidebands, i.e.

$$P_t = P_c + P_{USB} + P_{LSB}$$

Using the equation for total power, we write

$$P_t = P_c + \frac{m_a^2 P_c}{4} + \frac{m_a^2 P_c}{4}$$

$$= \frac{m_a^2 P_c}{2}$$

$$P_t = P_c + \frac{m_a^2 P_c}{2}$$

$$2500 = P_c + \frac{(0.8)^2 P_c}{2}$$

$$= P_c + \frac{0.64}{2} P_c$$

$$= 1.32 \, P_c$$

or

$$P_c = \frac{2500}{1.32}$$

$$= 1893.9 \text{ W } \textbf{Ans.}$$

EXAMPLE 4

How many AM broadcast stations can be accommodated in 10 kHz bandwidth if the highest frequency modulating a carrier is 5 kHz?

Solution:

Here, given that

$$\text{Total bandwidth} = 100 \text{ kHz}$$

$$f_{m_{max}} = 5 \text{ kHz}$$

We know that any station being modulated by a 5 kHz signal will produce an upperside frequency 5 kHz below its carrier, thereby requiring a bandwidth of 10 kHz. Thus, we have

Number of stations accommodated

$$= \frac{\text{Total BW}}{\text{BW per station}}$$

$$= \frac{100 \times 10^3}{10 \times 10^3}$$

Hence, number of stations accommodated

= 10 stations **Ans.**

EXAMPLE 5

The total power content of an AM signal is 1000 W. Determine the power being transmitted at the carrier frequency and at each of the sidebands when the per cent modulation is 100%.

Solution:

Here given that

$$P_t = 1000 \text{ W}$$
$$m = 100\% = 1$$

We know that the total power consists of the power at the carrier frequency, at the upper sideband frequency and at the lower sideband frequency. i.e.,

$$P_t = P_c + P_{USB} + P_{LSB}$$

From the equation for total power, we have

$$P_t = P_c + \frac{m_a^2 P_c}{4} + \frac{m_a^2 P_c}{4}$$

$$= P_c + \frac{m_a^2 P_c}{2}$$

Substituting, given that, we get

$$1000 = P_c + \frac{(1.0)^2 P_c}{2}$$

$$= P_c + 0.5P_c$$

or

$$P_c = \frac{1000}{1.5} = 666.67 \text{ W}$$

The leaves 1000 – 666.67 = 333.33 watts to be shared equally between upper and lower sidebands, i.e.,

$$P_{USB} + P_{LSB} = 333.33 \text{ W}$$

But,

$$P_{USB} = P_{LSB}$$

Thus, $2P_{LSB} = 333.33$

or $$P_{LSB} = P_{USB} = \frac{333.33}{2} = 166.66$$ **Ans.**

EXAMPLE 6

The power content of the carrier of an AM wave is 5 kilowatts. Determine the power content of each of the sidebands and the total power transmitter when the carrier is modulated upto 75%.

Solution:

Here, given that,

$$P_c = 5 \text{ kW}$$
$$M = 75\%$$
$$m_a = 0.75$$

We have to find that

P_{USB}, P_{LSB} and P_t.

Since, in an AM wave, the power in each of the sideband is equal, therefore, we have

$$P_{USB} = P_{LSB} = \frac{m_a^2 P_c}{4}$$

$$= \frac{(0.75)^2\,(5000)}{4}$$

$$P_{USB} = P_{LSB} = 703.13 \text{ W}$$

Now, the total power is the sum of the carrier power and the power in the two sidebands. Thus, we have

$$P_t = P_c + P_{USB} + P_{LSB}$$

or $$P_t = 5000 + 703.13 + 703.13$$
$$= 6406.26 \text{ W}$$ **Ans.**

EXAMPLE 7

Which of the following demodulators can be used for demodulating the signal

$$x(t) = 5(1 + 2 \cos 2000\pi t) \cos 2000\pi t$$

(*a*) Envelop demodulator

(*b*) Square-low demodulator

(*c*) Synchronous demodulator

(*d*) None of the above

Solution:

Here, the given signal is

$$x(t) = 5(1 + 2 \cos 2000\pi t) \cos 2000\pi t$$

This is essentially an amplitude modulation, AM, double-sideband signal which can be demodulated by any of the 3 demodulators (*a*), (*b*) or (*c*), though envelope detection using a square law device, such as a diode, is most commonly used.

Hence, answers (*a*), (*b*) and (*c*) are correct.

EXAMPLE 8

Determine the power cent modulation of an amplitude. Modulated wave which has a power content at the carrier of 8 kW and 2 kW in each of its sidebands when the carrier is modulated by a simple audio tone.

Solution:

Here, given that,

$$P_c = 8 \text{ kW}$$

$$P_{USB} = P_{LSB} = 2 \text{ kW}$$

We have to find out M

Knowing the power content of the sidebands and the carrier, the relationship of sideband power can be used to determine the modulation index. Once the modulation index is known, merely multiplying it by 100 provides per cent modulation.

$$P_{USB} = P_{LSB}$$

$$= \frac{m_a^2 P_c}{4}$$

or $$2 \times 10^3 = \frac{m_a^2 (8 \times 10^3)}{4}$$

or $$m_a^2 = \frac{4 \times 2 \times 10^3}{8 \times 10^3} = 1.0$$

or $$m_a = 1.0$$

Also, $$M = m_a \times 100$$

$$= 100\% \textbf{ Ans.}$$

EXAMPLE 9

Find the per cent modulation of an AM wave whose total power content is 2500 W and whose sidebands each contains 400 W.

Solution:

Here, given that,

$$P_t = 2500 \text{ W}$$

$$P_{USB} = P_{LSB} = 400 \text{ W}$$

Now, we have to find out M.

First find the power contained at the carrier frequency. Then use the relationship between sideband power and carrier power to determine the modulation index. Once the modulation index is known the per cent modulation can easily be found merely by multiplying by 100.

The power at the carrier frequency can be determined as follows:

$$P_t = P_c + P_{USB} + P_{LSB}$$

$$2500 = P_c + 400 + 400$$

$$P_c = 2500 - 800 = 1700 \text{ W}$$

$$P_{USB} = P_{LSB} = \frac{m_a^2 P_c}{4}$$

$$400 = \frac{m_a^2(1700)}{4}$$

or $$m_a^2 = \frac{400(4)}{1700} = \frac{1600}{1700} = 0.741$$

or $$m_a = 0.970$$

Thus, $$M = 0.970 \times 100 = 97\% \textbf{ Ans.}$$

EXAMPLE 10

An AM signal in which the carrier is modulated upto 70% contains 1500 W at the carrier frequency. Determine the power content of the upper and lower sidebands for this per cent modulation. Calculate the power at the carrier and the power content of each of the sidebands when the per cent modulation drops to 50%.

Solution:

Here, given that,

$$M = 70\%$$

$$m_a = 0.70$$

$$P_{c\,70} = 1500 \text{ W}$$

We have to find out

$P_{USB\,70}$, $P_{LSB\,70}$, $P_{c\,50}$, $P_{USB\,50}$, $P_{LSB\,50}$

We know that the power content of each of the sidebands is equal to $\frac{m_a^2 P_c}{4}$,

thus, we have

$$P_{USB\,70} = P_{LSB\,70}$$

$$= 183.75 \text{ W}$$

Since in standard AM transmission, carrier power remains the same, regardless of per cent modulation, thus, we have

$$P_c = P_{c\,70} = 1500 \text{ W}$$

$$P_{USB\,50} = P_{LSB\,50}$$

$$= \frac{m_a^2 P_c}{4}$$

$$= \frac{(0.5)^2\,1500}{4}$$

$$P_{c\,50} = 83.75 \text{ W } \textbf{Ans.}$$

EXAMPLE 11

Find the modulation index and the per cent modulation of the signal shown in Fig. 2.78.

Solution:

Here, given that AM signal as shown in Fig. 2.78, we have to find out m_a and M.

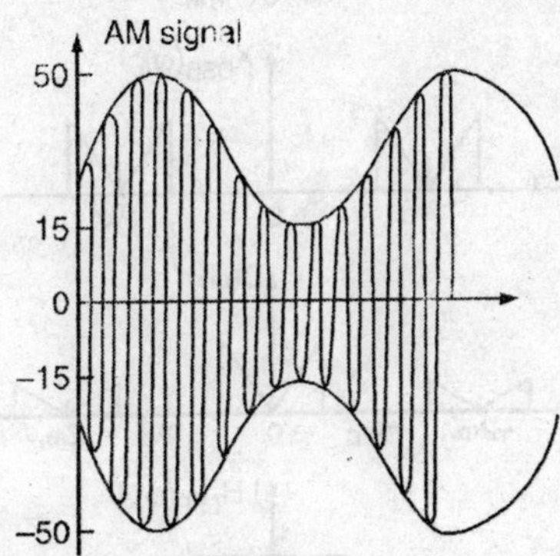

Fig. 2.78

using
$$m_a = \frac{\text{max p.p} - \text{min p.p}}{\text{max p.p} + \text{min p.p}}$$

and values from figure 2.79, we get

$$\text{max p.p} = 2(50) = 100$$
$$\text{min p.p} = 2(15) = 30$$

Thus,
$$m_a = \frac{100 - 30}{100 + 30} = \frac{70}{130}$$
$$= 0.538$$

Hence,
$$M = m_a \times 100 = 53.8\% \textbf{ Ans.}$$

EXAMPLE 12

Verify that the message signal *m*(*t*) is recovered from a modulated DSB signal by first multiplying it by a local sinusoidal carrier and then passing the resultant signal through a low-pass filter LPF, as shown in Fig. 2.79. (*a*) in the time domain and (*b*) in the frequency domain.

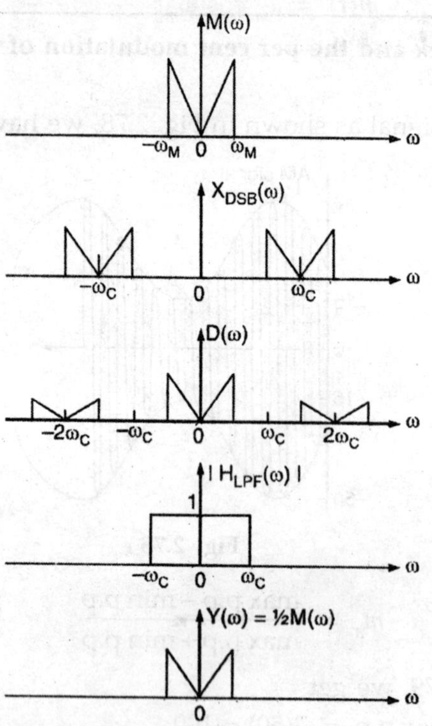

Fig. 2.79 *Demodulation of a DSB signal*

Solution:

(*a*) The output of the multiplier is given by

$$d(t) = x_{DSB}(t) \cos \omega_c t$$
$$= [m(t) \cos \omega_c t] \cos \omega_c t$$

or
$$d(t) = m(t) \cos^2 \omega_c t$$

or
$$d(t) = \frac{1}{2} m(t) + \frac{1}{2} m(t) \cos 2\omega_c t$$

After low-pass filtering of signal $d(t)$, we get

$$y(t) = \frac{1}{2} m(t)$$

Thus, by proper amplification multiplying by 2, we can recover the message signal $m(t)$.

(*b*) Demodulation of signal $x_{DSB}(t)$ in frequency domain has been illustrated in Fig. 2.80.

EXAMPLE 13

Skater the ordinary AM signal for a single-tone modulation with modulation indices of $m_a = 0.5$ and $m_a = 1$.

Solution:

For a single-tone modulation, we have

$$m(t) = a_m \cos \omega_m t$$

Then, the modulation index is

$$m_a = \frac{[\max\{m(t)\}]}{A}$$

$$= \frac{a_m}{A}$$

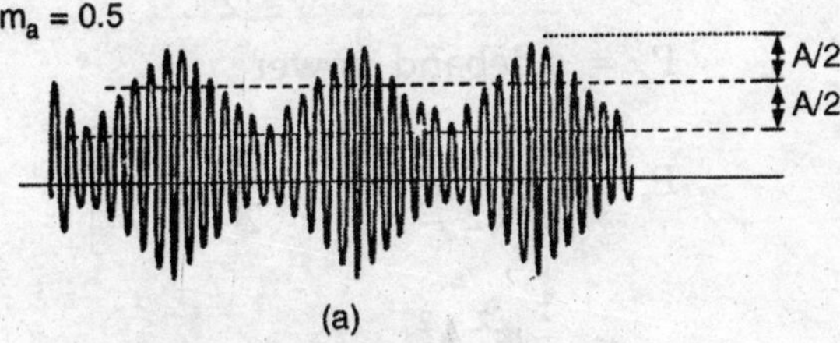

(a)

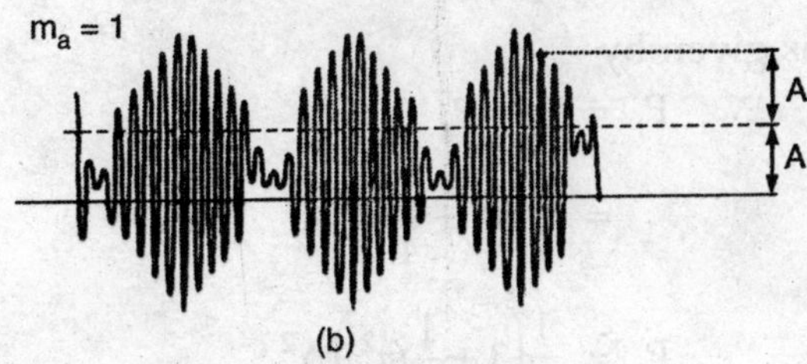

(b)

Fig. 2.80

Hence,

$$m(t) = a_m \cos \omega_m t$$
$$= m_c A \cos \omega_m t$$

Also,

$$x_{AM}(t) = [A + m(t)] \cos \omega_c t$$
$$= A[1 + m_a \cos \omega_m t] \cos \omega_c t$$

Fig. 2.80 (*a*) and 2.80 (*b*) show the ordinary AM sinal corresponding to $m_a = 0.5$ and $m_a = 1$ respectively.

EXAMPLE 14

The efficiency η of ordinary AM is defined as the percentage of the total power carried by the sidebands, that is,

$$\eta = \frac{P_s}{P_t} \times 100\%$$

where P_s is the power carried by the sidebands and P_t is the total power of the AM signal.

(*i*) Find h for $m_a = 0.5$ (50 per cent modulation)

(*ii*) Show that for a single-tone AM, η_{max} is 33.3 per cent at $m_a = 1$

Solution:

A single-tone AM signal can be expressed as

$$x_{AM}(t) = A\cos\omega_c t + A\cos\omega_m t\cos\omega_c t$$

$$= A\cos\omega_c t + \frac{1}{2}m_a A\cos(\omega_c - \omega_m)t + \frac{1}{2}m_a A\cos(\omega_c + \omega_m)t$$

$$P_c = \text{carrier power} = \frac{1}{2}A^2$$

$$P_s = \text{sideband power}$$

$$P_s = \frac{1}{2}\left[\left(\frac{1}{2}m_a A\right)^2 + \left(\frac{1}{2}m_a A\right)^2\right]$$

$$= \frac{1}{4}m_a^2 A^2 \quad \text{...(a)}$$

The total power P_t is given by

$$P_t = P_c + P_s$$

$$= \frac{1}{2}A^2 + \frac{1}{4}m_a^2 A^2$$

$$P_t = \frac{1}{2}\left(1 + \frac{1}{2}m_a^2\right)A^2 \quad \text{...(b)}$$

Thus,

$$\eta = \frac{P_s}{P_t} \times 100\% \quad \text{...(c)}$$

Substituting values of P_s and P_t in eqn. (*c*), we get

or

$$\eta = \frac{\frac{1}{4}m_a^2 A^2}{\left(\frac{1}{2} + \frac{1}{4}m_a^2\right)A^2} \times 100\%$$

$$= \frac{m_a^2}{2 + m_a^2} \times 100\%$$

with the condition that $m_a \le 1$.

(*i*) For $m_a = 0.5$, we have

$$\eta = \frac{(0.5)^2}{2 + (0.5)^2} \times 100\%$$

$$= 11.1\%$$

(*ii*) Since $m_a \le 1$, it can be observed than η occurs at $m_a = 1$ and is given by

$$\eta = \frac{1}{3} \times 100\%$$

$$= 33.3\% \textbf{ Ans.}$$

EXAMPLE 15

Show that the system shown in Fig. 2.81 can be used to demodulate an SSB signal.

Solution:

We know that the upper-side band SSB signal $x_c(t)$ is given by

$$x_c(t) = m(t) \cos w_c t - \hat{m}(t) \sin \omega_c t$$

We also have,

$$\hat{x}_c(t) = m(t) \sin w_c t + \hat{m}(t) \cos \omega_c t$$

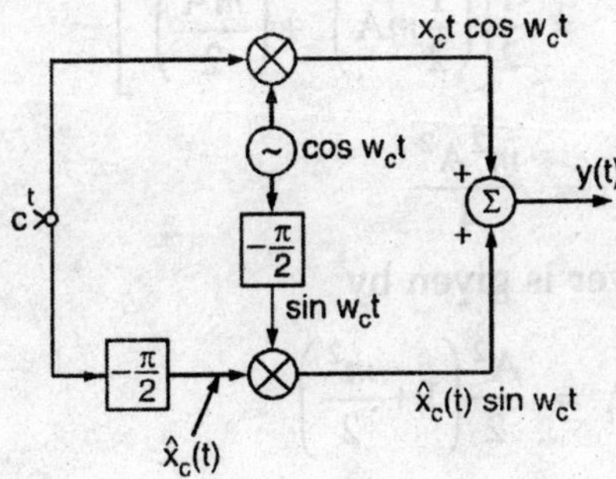

Fig. 2.81 *Phase-shift SSB demodulator for example 3.41*

From Fig. 2.81, we have

$$y(t) = x(t) \cos w_c t + \hat{x}_c(t) \sin \omega_c t$$

or

$$y(t) = m(t)(\cos^2 \omega_c t + \sin^2 \omega_c t) = m(t)$$

In a similar manner, the lower-sideband SSB signal $x_c(t)$ is given by

$$x_c(t) = m(t) \cos \omega_c t + \hat{m}(t) \sin \omega_c t$$

$$\hat{x}_c(t) = m(t) \sin \omega_c t - \hat{m}(t) \cos \omega_c t$$

Again, from Fig. 2.81, we obtain

$$y(t) = x_c(t) \cos \omega_c t + \hat{x}_c(t) \sin \omega_c t$$

$$= m(t)$$

Hence, Fig. 2.81 can be used to demodulate an SSB signal, which is proved.

EXAMPLE 16

A given AM broadcast station transmits a total power of 5 kW when the carrier is modulated by a sinusoidal signal with a modulation index of 0.7071. Compute

(*i*) the carrier power

(*ii*) the transmission efficiency, and

(*iii*) the peak amplitude of the carrier assuming the antenna to be represented by a (50 + *j*0) Ω load.

Solution:

(*i*) An amplitude modulated signal can be expressed as

$$x_{AM}(t) = A\cos\omega_c t + mA\cos\omega_m t\cos\omega_c t$$

or

$$x_{AM}(t) = A\cos\omega_c t + \frac{1}{2}mA\cos(\omega_c - \omega_m)t + \frac{mA}{2}.\cos(\omega_c + \omega_m)t \quad ...(i)$$

Now,

$$P_c = \text{carrier power} = \frac{A^2}{2}$$

$$P_s = \text{sideband power}$$

$$= \frac{1}{2}\left[\left(\frac{1}{2}mA\right)^2 + \left(\frac{mA}{2}\right)^2\right]$$

$$= \frac{m^2A^2}{4} \quad ...(ii)$$

We know that the total power is given by

$$P_t = \frac{A^2}{2}\left(1 + \frac{m^2}{2}\right)$$

$$= P_c\left(1 + \frac{m^2}{2}\right)$$

Given

$$P_t = 50 \times 10^3 \text{ Watt},$$

Then,

$$P_c = \frac{P_t}{1 + \frac{m^2}{2}} = \frac{50 \times 10^3}{1.25}$$

$$= 40 \text{ kW } \textbf{Ans.}$$

(*ii*) Further, we know that the transmission efficiency is given by

$$\eta = \frac{P_s}{P_t} \times 100\%$$

$$= \frac{\frac{1}{4}m^2A^2}{\left(\frac{1}{2} + \frac{m^2}{4}\right)A^2} \times 100\%$$

or
$$\eta = \frac{m^2}{2+m^2}\times 100\%$$

$$= \frac{0.50}{2+0.50}\times 100\%$$

$$= 20\% \textbf{ Ans.}$$

(*iii*) Power delivered to the load will be

$$A_c = \frac{A_c^2}{2R} = 40 \times 10^3$$

$$= \sqrt{2\times 40\times 10^3 \times 50}$$

$$= 2 \text{ kV } \textbf{Ans.}$$

EXAMPLE 17

The spectrum of a message signal *m*(*t*) is shown in Fig. 2.82(*a*). The ensure communication privacy, this signal is applied to a system, known as a scrambler, shown in Fig. 2.82(*b*). Analyze the system and sketch the spectrum of the output *x*(*t*).

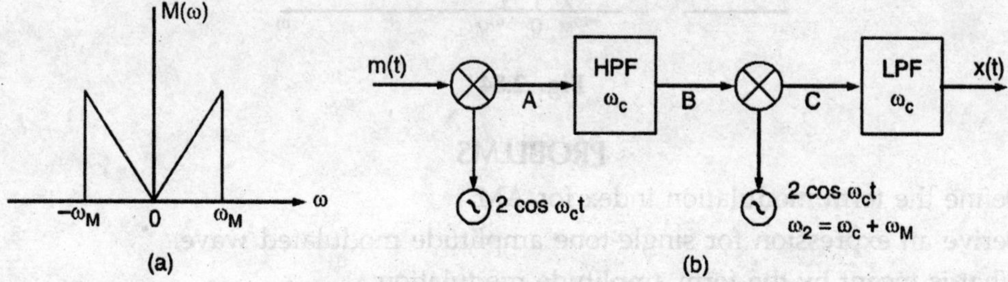

Fig. 2.82

Solution:

The spectrum of the signal at each point is shown in Fig. 2.82. We see that the spectrum of the output *x*(*t*).X(ω), consists of the two reversed lobes of M(ω).

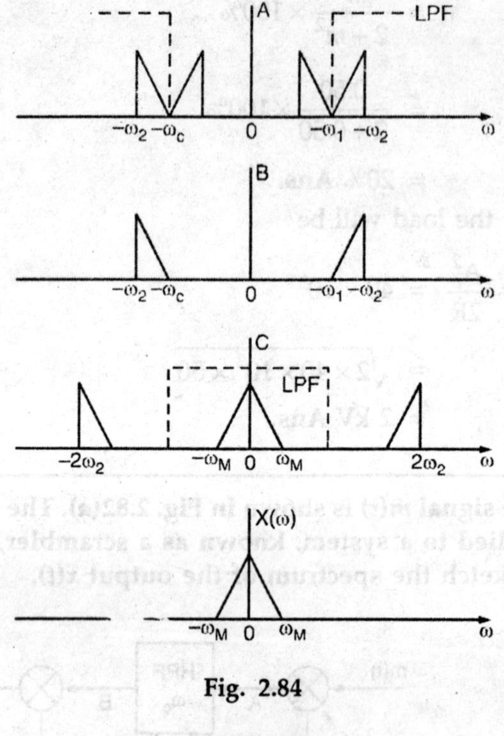

Fig. 2.84

PROBLEMS

1. Define the term modulation index for AM.
2. Derive an expression for single-tone amplitude modulated wave.
3. What is meant by the term amplitude modulation.
4. Draw the single-sided frequency spectrum for a single tone amplitude modulated signal.
5. Derive the current relations for single.
6. Derive the power relations for single-tone amplitude-modulated wave.
7. Explain the low-level and high-level AM modulation methods with help of figures.
8. Derive an expression for the transmission efficiency of AM wave.
9. Explain the square-law diode modulation method for AM generation.

❑❑❑

Angle Modulation 3

3.1 INTRODUCTION

In 1992, J.R. Carson first used the term "frequency modulation" which was in principle totally different from then popular amplitude modulation. He showed that the bandwidth required for frequency modulation is at least twice the highest modulating frequency. But he was wrong when he said "I have proved, mathematically, that this type of modulation inherently distorts without any compensating advantages whatsoever."

It required the genius of E.H. Armstrong to catch Carson off-guarded on this count. A prolific inventor, Armstrong invented regenerative circuits in his college (1912), super heterodyne receiver in 1918 and was holder of 42 patents. On June 9, 1934, he conducted first field test on FM transmission by sending an organ recital, via both AM and FM, from top of the Empire State Building to the home of a trusted old friend on long Island. The FM organ came through loud and clear with a better quality than AM. And he had this to say, "I could never accept findings based almost exclusively on mathematics. It ain't ignorance that causes all the trouble in this world. It's the things people know that ain't so."

Angle Modulation has certain important differences. The spectral components in the modulated waveform depend on the amplitude as well as the frequency of the spectral components in the baseband signal. Furthermore, the modulation system is not linear and superposition does not apply. Such a system results when, in connection with a carrier of constant amplitude, the phase angle is made to respond in some way to a baseband signal. Such a signal has the form

$$V(t) = A\cos\left[\omega_c t + \phi(t)\right]$$

In which A and ω_c are constant but in which the phase angle $\phi(t)$ is a function of the baseband signal. Modulation of this type is called angle modulation for obvious reasons. It is also referred to as phase modulation since $\phi(t)$ is the phase angle of the argument of the cosine function.

3.2 CONCEPT OF ANGLE MODULATION

It may be defined as the process in which the total phase angle of a carrier wave is varied in accordance with the instantaneous value of the modulating or message signal while keeping the amplitude of the carrier constant.

3.2.1 Mathematical Representation

Let us consider that an unmodulated carrier signal is expressed as

$$C(t) = A\cos(\omega_c t + \theta_0) \quad ...(3.1)$$

where

A = amplitude of carrier

ω_c = carrier frequency

θ_0 = some phase angle

Substituting $\omega_c t + \theta_0 = \phi$, we get

$$C(t) = A\cos\phi \quad ...(3.2)$$

where

$$\phi = \omega_c t + \theta_0 \quad ...(3.3)$$

is the total phase angle of the carrier signal.

We may consider the Eqn. (3.2) as the real part of rotating phasor $Ae^{i\phi}$. Let us denote this rotating phasor by C, so that

$$C = Ae^{i\phi} \quad ...(3.4)$$

Also,

$$C(t) = Re(Ae^{i\phi}) = A.Re(e^{i\phi}) = A\cos\phi \quad ...(3.5)$$

This phasor C rotates at a constant angular velocity ω_c. Clearly θ_0 is the phase angle of the unmodulated carrier at $t = 0$.

From Eqn. (3.3) the constant angular velocity ω_c of the phasor C is related to its total phase angle 'ϕ' as

$$\phi = \omega_c t + \theta_0$$

Differentiating both sides of this equation with respect to t, we have

$$\omega_c = \frac{d\phi}{dt} \quad ...(3.6)$$

It may be noted at this point that for unmodulated carrier, the derivative $\frac{d\phi}{dt}$ is constant. However, this derivative $\frac{d\phi}{dt}$ may not be a constant with time, in general i.e., this may vary with time. Therefore, the angular velocity of the phasor C would also vary with time. Hence, this time-dependent angular velocity or angular frequency is known as instantaneous angular velocity or instantaneous angular frequency. This instantaneous angular frequency is denoted by ω_i.

Hence, for this case, Eqn. (3.6) becomes

$$\frac{d\phi}{dt} = \omega_i \quad ...(3.7)$$

where angular frequency ω_i is time-dependent.

From Eqn. (3.7), we get,

$$\phi = \int \omega_i \, dt \qquad ...(3.8)$$

Again writing Eqn. (3.1) of unmodulated carrier as

$$C(t) = A \cos (\omega_c t + \theta_0) = A \cos \phi$$

where ϕ is the total phase angle of the unmodulated carrier expressed as

$$\phi = \omega_c t + \theta_0$$

Now, if this angle ϕ is varied according to the instantaneous value of the message or modulating signal, the carrier signal is then said to be angle modulated.

3.2.2 Types of Angle Modulation

We can vary this phase angle ϕ in two ways and thus there are two types of angle modulation as under:

(*i*) Phase Modulation (PM)

(*ii*) Frequency Modulation (FM)

3.2.2.1 Phase Modulation (PM)

Phase modulation (PM) is that type of angle modulation in which the phase angle ϕ is varied linearly with a baseband or modulating signal $x(t)$ about an unmodulated phase angle $(\omega_c t + \theta_0)$. This means that in Phase Modulation, the instantaneous value of the phase angle is equal to the phase angle of the unmodulated carrier $(\omega_c t + \theta_0)$ plus a time varying component which is proportional to modulating signal $x(t)$.

We know that unmodulated carrier signal is expressed as

$$c(t) = A \cos (\omega_c t + \theta_0) = A \cos \phi$$

where,

$$\phi = \omega_c t + \theta_0$$

Neglecting θ_0, we get, total phase angle of unmodulated carrier is

$$\phi = \omega_c t \qquad ...(3.9)$$

Now, according to Phase Modulation, this phase angle 'ϕ' is varied linearly with a baseband or modulating signal $x(t)$.

Let the instantaneous value of phase angle be denoted by ϕ_i

Therefore,

$$\phi_i = \omega_c t + k_p.xt \qquad ...(3.10)$$

where k_p is the proportionality constant and is known as phase sensitivity of the modulator. This is expressed in radians/volts.

Since, the expression for unmodulated carrier wave is

$$c(t) = A \cos \phi$$

Therefore, the expression for phase modulated wave will be

$$s(t) = A \cos \phi_i \qquad ...(3.11)$$

putting the value of ϕ_i in Eqn. (3.11) from Eqn. (3.10), we get,

$$s(t) = A \cos\left[\omega_c t + k_p.x(t)\right] \qquad ...(3.12)$$

Which is the required mathematical expression for a phase modulated wave.

3.2.2.2 Frequency Modulation

Frequency modulation is that type of angle modulation in which the instantaneous frequency ω_i is varied linearly with a message or baseband signal $x(t)$ about an unmodulated carrier frequency ω_c. This means that the instantaneous value of the angular frequency ω_i will be equal to the carrier frequency ω_c plus a time-varying component proportional to the baseband signal $x(t)$.

We know that, the instantaneous frequency is given by

$$\omega_i = \omega_c + k_f x(t) \quad ...(3.13)$$

where k_f is proportionality constant and is known as the frequency sensitivity of the modulator. This is express in Hz/Volt.

Now, let the expression for unmodulated carrier signal be

$$c(t) = A\cos(\omega_c t + \theta_0) \quad ...(3.14)$$

or

$$c(t) = A\cos\phi \quad ...(3.15)$$

where

$$\phi = \omega_c t + \theta_0 \quad ...(3.16)$$

'ϕ' is the total phase angle of the unmoduıated carrier.

Let ϕ_i be the instantaneous phase angle of the modulated signal.

From Eqn. (3.15), the equation for unmodulated carrier is

$$c(t) = A\cos\phi$$

On frequency modulation amplitude A must remain constant and only angle 'ϕ' will change. Hence, the expression for frequency modulated wave will be

$$s(t) = A\cos\phi_i \quad ...(3.17)$$

where ϕ_i = instantaneous phase angle

From Eqn. (3.16)

$$\phi = \omega_c t + \theta_0$$

On differentiating, we get

$$\frac{d\phi}{dt} = \omega_c$$

or

$$\phi = \int \omega_c \, dt \quad ...(3.18)$$

Based on Eqn. (3.18) we may write the expression for instantaneous phase angle θ_i as

$$\phi_i = \int \omega_i \, dt \quad ...(3.19)$$

where ω_i = instantaneous frequency of frequency modulated wave.

Putting the value of ω_i in Eqn. (3.19) from Eqn. (3.13), we get,

$$\phi_i = \int \left[\omega_c + k_f x(t)\right] dt = \omega_c t + k_f \int x(t)\, dt \quad ...(3.20)$$

Putting the value of ϕ_i in Eqn. (3.17), we get the expression for frequency modulated wave will be

$$s(t) = A\cos\left[\omega_c t + k_f \int x(t)\,dt\right] \quad ...(3.21)$$

Now, if the phase angle of the unmodulated carrier is taken at $t = 0$, then the limit of integration in Eqn. (3.21) will be 0 to t.

In this case, the expression for FM wave will be

$$s(t) = A\cos\left[\omega_c t + k_f \int_0^1 x(t)\,dt\right]$$

Which is the required general expression for FM wave.

3.2.3 Frequency Deviation

We know that the instantaneous frequency of FM wave is given as

$$\omega_i = \omega_c t + k_f x(t) \quad ...(3.22)$$

The instantaneous frequency of FM signal varies with time around the carrier frequency ω_c. This means that the instantaneous frequency of FM signal varies according to the modulating signal.

The maximum change in instantaneous frequency from the average frequency ω_c is called frequency deviation.

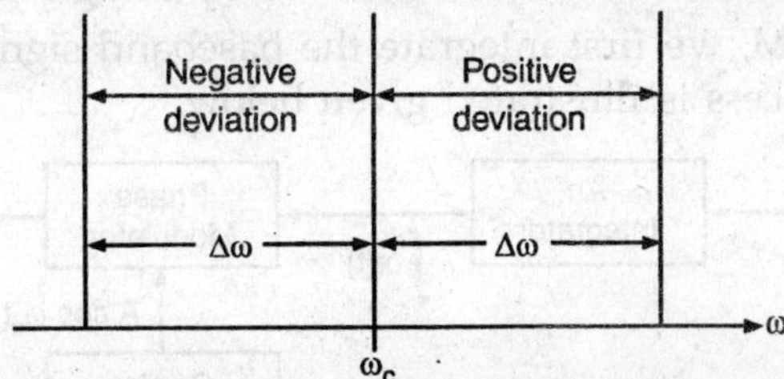

Fig. 3.1 *Illustration of Frequency Deviation*

This maximum change in instantaneous frequency ω_i from the average or carrier frequency ω_c depends upon the magnitude and sign of $k_f x(t)$. This means that the frequency deviation would be either positive or negative depending upon the sign of $k_f x(t)$. However, the amount of frequency deviation in both these cases is given by the maximum magnitude, i.e., $\left|k_f . x(t)\right|_{max}$.

Maximum frequency deviation is generally denoted by $\Delta\omega$.

Therefore,

Frequency deviation $\Delta\omega = \left|k_f . x(t)\right|_{max}$.

3.3 RELATIONSHIP BETWEEN PHASE MODULATION (PM) AND FREQUENCY MODULATION (FM)

We know that an angle modulated wave is given by

$$s(t) = A \cos \phi_i \qquad ...(3.23)$$

where

A = Amplitude

ϕ_i = Instantaneous total phase angle of the angle modulated wave.

The expression for phase modulation (PM) wave is

$$s(t) = A \cos\left[\omega_c t + k_p . x(t)\right] \qquad ...(3.24)$$

Similarly, the expression for frequency modulated (FM) wave is

$$s(t) = A\left[\cos \omega_c t + k_f \int_0^1 x(t)\, dt\right] \qquad ...(3.25)$$

It may be observed from above equations that Phase Modulation (PM) and Frequency Modulation (FM) are closely related to each other because in both the cases there is a variation in the total phase angle. In Phase Modulation (PM), the phase angle varies linearly with baseband signal $x(t)$ whereas in case of Frequency Modulation (FM), the phase angle varies linearly with the integral of baseband signal $x(t)$. This means that FM wave may be obtained by using PM. Conversely, PM wave may be obtained by using FM.

To get FM by using PM, we first integrate the baseband signal and then apply to the phase modulator. This process is illustrated given below.

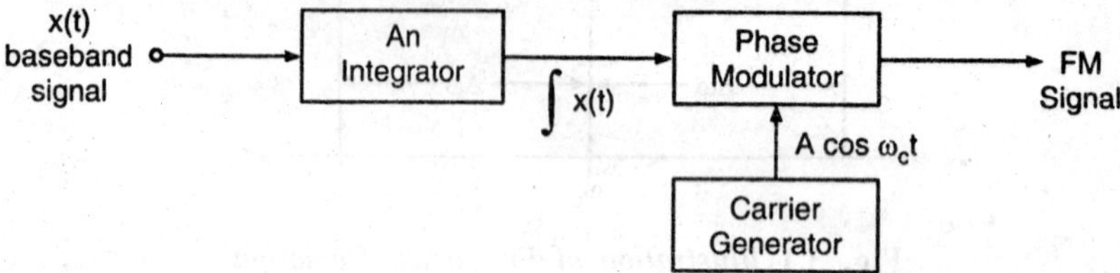

Fig. 3.2 *Generation of FM Using Phase Modulator*

Similarly, PM wave may be generated by using frequency modulator by first differentiating modulating or baseband signal $x(t)$ and then applying to the frequency modulator. This process is illustrated with the help of a block diagram in the Fig. 3.3.

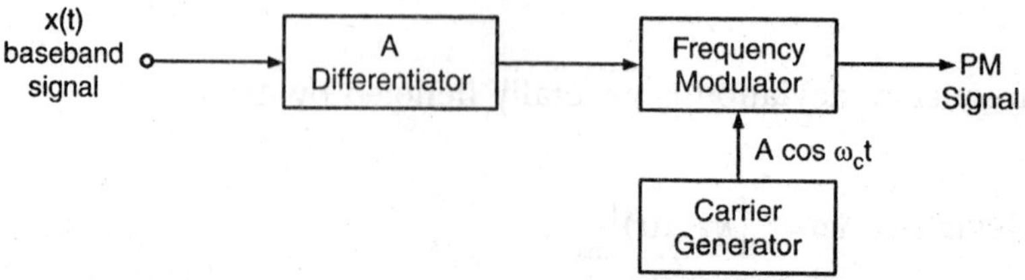

Fig. 3.3 *Generation of PM Using Frequency Modulator*

3.3.1 Single Tone Frequency Modulation

We know that the general expression for FM wave is given as

$$s(t) = A\cos\left[\omega_c t + k_f \int_0^1 x(t)\,dt\right] \quad ...(3.26)$$

From this equation, it is clear that the FM wave $s(t)$ is a nonlinear function of the baseband or modulating signal $x(t)$. This means that frequency modulation is a nonlinear modulation process. Due to this reason unlike amplitude modulation, the spectrum of an FM wave is related in a complex manner with the modulating or baseband signal.

Therefore, to study the spectrum or spectral properties of FM signal, we shall consider the simplest case of FM wave known as single tone frequency modulation.

In general expression of FM wave Eqn. (3.26), we consider a modulating or baseband signal $x(t)$ which may consist of any number of frequency components.

On the other hand, a single tone frequency modulation is that type of frequency modulation (FM) in which the modulating or baseband signal contains a single frequency.

Let us consider a carrier signal as

$$c(t) = A\cos\omega_c t$$

Let the modulating signal be

$$x(t) = V_m \cos\omega_m t$$

where

V_m = maximum amplitude of modulating signal

ω_m = frequency of modulating signal

Then according to the definition of FM, the frequency of carrier signal is varied according to the instantaneous value of the modulating or baseband signal, $x(t)$.

3.4 MATHEMATICAL EXPRESSION FOR SINGLE-TONE FREQUENCY MODULATION

Let the expression for a carrier signal be

$$c(t) = A\cos\omega_c t$$

Let the modulating signal be

$$x(t) = V_m \cos\omega_m t$$

Let the expression for FM wave be

$$s(t) = A\cos\phi_i \quad ...(3.27)$$

where ϕ_i is the instantaneous phase angle of the modulated wave.

We know that the instantaneous frequency of the modulated signal is given by

$$\omega_i = \omega_c + k_f x(t) \quad ...(3.28)$$

Putting the value of $x(t)$, we get,

$$\omega_i = \omega_c + k_f V_m \cos\omega_m t$$

But we know that frequency deviation is given as

$$\Delta\omega = \left|k_f . x(t)\right|_{\max} = k_f \left|x(t)\right|_{\max} = k_f . V_m$$

$$\therefore \quad \omega_i = \omega_c + \Delta\omega.\cos \omega_m t \qquad ...(3.29)$$

The total phase angle ϕ_i of the modulated wave is given as

$$\omega_i = \int \omega_i \, dt$$

Putting the value of ω_i from Eqn. (3.29), we get,

$$\phi_i = \int [\omega_c + \Delta\omega \cos \omega_m t] dt = \omega_c t + \frac{\Delta\omega}{\omega_m} \sin \omega_m t \qquad ...(3.30)$$

But, modulation index m_f is given as

$$m_f = \frac{\text{Frequency deviation}}{\text{Modulating Frequency}} = \frac{\Delta\omega}{\omega_m}$$

Therefore, putting the value of m_f in Eqn. (3.30), we obtain,

$$\phi_i = \omega_c t + m_f \sin \omega_m t$$

Substituting this value of ϕ_i in Eqn. (3.27), we get the expression for single-tone FM wave

$$s(t) = \text{A} \cos \phi_i = \text{A} \cos \left[\omega_c t + m_f \sin \omega_m t \right] \qquad ...(3.31)$$

This is the required mathematical expression for single tone FM wave.

3.5 PHASOR REPRESENTATION OF ANGLE MODULATION

The expression for FM wave is

$$s(t)_{\text{FM}} = \text{A} \cos \left[\omega_c t + k_f \int_0^t x(t) \, dt \right] \qquad ...(3.32)$$

Similarly, the expression for PM wave is

$$s(t)_{\text{PM}} = \text{A} \cos \left[\omega_c t + k_p . x(t) \right] \qquad ...(3.33)$$

We may consider the sign a 1st $s(t)_{\text{FM}}$ and $s(t)_{\text{PM}}$ as real parts of phasors C_{FM} and C_{PM} respectively expressed as

$$\text{C}_{\text{FM}} = \text{A}e^{j\left[\left(\omega_c t + k_f \int x(t) dt\right)\right]}$$

and

$$\text{C}_{\text{PM}} = \text{A}e^{j\left[\left(\omega_c t + k_p . x(t)\right)\right]}$$

Assuming

$$\int x(t) \, dt = y(t)$$

So that,

$$\text{C}_{\text{FM}} = \text{A}e^{j\left[\left(\omega_c t + k_p . y(t)\right)\right]} \qquad ...(3.34)$$

3.6 TYPES OF FREQUENCY MODULATION (FM)

As a matter of fact, if the frequency deviation is high, the bandwidth will be large. Similarly, if the frequency deviation is low, the bandwidth will be small. This is discussed previously in frequency deviation topic already that the bandwidth of an FM signal depends upon the frequency deviation.

$$\Delta\omega = k_f x(t).$$

Since deviation is given by

$$\Delta\omega = k_f x(t),$$

it means that for any given $x(t)$, the frequency deviation and therefore the bandwidth will depend upon frequency sensitivity k_f. Hence, when k_f is quite small then the bandwidth will be narrow and when k_f is large then the bandwidth will be large.

Thus depending upon the value of frequency sensitivity k_f FM may be divided as under

(*i*) Narrowband FM.

In this case, k_f is small and hence the bandwidth of FM is narrow.

(*ii*) Wideband FM.

In this case, k_f is large and hence the FM signal has a wide bandwidth.

3.6.1 Narrowband Frequency Modulation

We know that the general expression for FM wave in the phasor form is given by

$$C_{FM} = Ae^{j[\omega_c t + k_f \cdot y(t)]} \quad ...(3.35)$$

However, for a narrowband FM, we have $k_f \times y(t) \ll 1$ for all the values of t. Thus

$$e^{jk_f \cdot y(t)} = 1 + jk_f y(t)$$

Hence, the FM phasor expression, now, becomes,

$$C_{FM} = Ae^{j\omega_c t} \cdot e^{jk_f \cdot y(t)} = A\left[1 + jk_f \cdot y(t)\right] e^{j\omega_c t}$$

Also, since the FM signal is the real part of its phasor representation, therefore, we write

$$s(t)_{FM} = R_e\left[C_{FM}\right] = A\cos\omega_c t - Ak_f\, y(t)\sin\omega_c t \quad ...(3.36)$$

In the same way, on the other hand, the expression for narrowband PM is given as

$$s(t)_{PM} = A\cos\omega_c t - Ak_p\, x(t)\sin\omega_c t \quad ...(3.37)$$

Thus, the expressions for narrowband FM and PM are quite similar to the expression for AM signal with a little modification. Therefore, the bandwidth of a narrowband FM is almost same as that of the AM signal.

Now, we can also know that the signals $y(t)$ and $x(t)$ have the same bandwidth.

Since,

$$y(t) = \int_0^1 x(t)\,dt$$

$$\therefore \qquad F[y(t)] = F\left[\int_0^1 x(t)\,dt\right] = \frac{1}{j\omega}.X(\omega)$$

or,

$$Y(\omega) = \frac{1}{j.\omega}.X(\omega) \qquad ...(3.38)$$

3.6.2 Conclusion

Hence, from above expression it is clear that if $X(\omega)$ is band limited to ω_m, then $Y(\omega)$ is also band limited to ω_m. Both FM and PM wave expressions have carrier and sideband terms which are identical to the AM expression except with differing phase relations between carrier and sideband terms.

3.6.3 Generation of Narrowband FM

We know that the expressions for narrowband FM and narrowband PM are given by

$$s(t)_{FM} = A\cos\omega_c t - Ak_f.y(t)\sin\omega_c t \qquad ...(3.39)$$

and

$$s(t)_{PM} = A\cos\omega_c t - Ak_p.x(t)\sin\omega_c t \qquad ...(3.40)$$

The Eqns. (3.39) and (3.40) provide an idea of generate narrowband FM and narrowband PM respectively. This means that the sideband terms are generated by a balanced modulator, as in a DSB-SC system, and then the carrier term is added to sideband terms.

The given Fig. 3.4 shows the method to generate narrowband FM and narrowband PM. It may be observed that the block diagram satisfy the corresponding equation for narrowband FM and narrowband PM.

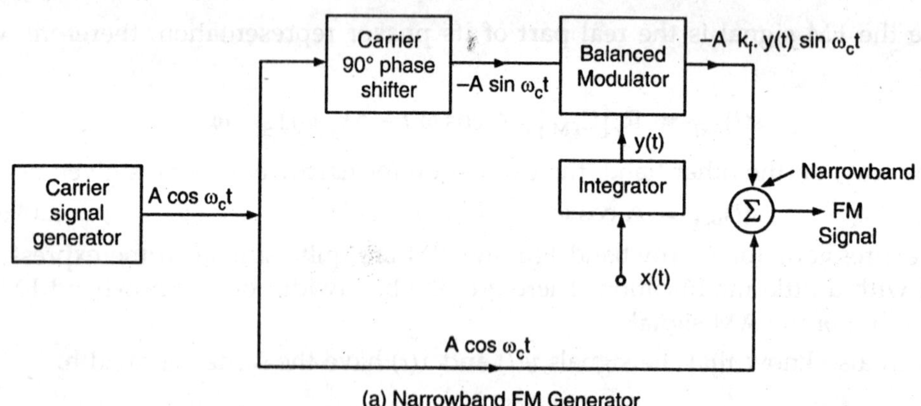

(a) Narrowband FM Generator

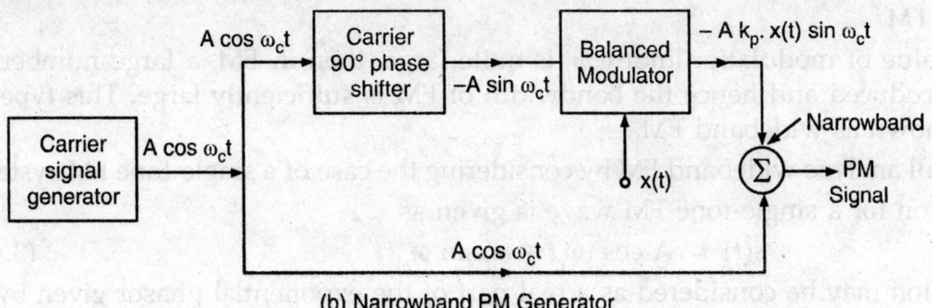

Fig. 3.4 *Generation schemes for narrowband FM and PM*

3.6.4 Expression for Single-Tone Narrowband FM

We know that the general expression for narrowband FM is given as

$$s(t)_{FM} = A\cos\omega_c t - Ak_f.y(t)\sin\omega_c t \qquad ...(3.41)$$

here,

$$y(t) = \int x(t)\,dt$$

where $x(t)$ is any modulating signal and A cos $\omega_c t$ is a carrier signal.

For a single-tone narrowband FM, we take modulating signal as considering of a single frequency

i.e.,

$$x(t) = V_m \cos\omega_m t \qquad ...(3.42)$$

(here $x(t)$ is taken as a voltage signal)

where ω_m is the single modulating frequency.

Thus,

$$y(t) = \int x(t)\,dt$$

$$y(t) = \int V_m \cos\omega_m t = \frac{V_m}{\omega_m}\sin\omega_m t$$

Putting this value of $y(t)$ in the Eqn. (3.41), we get,

$$s(t)_{FM} = A\cos\omega_c t - Ak_f.\frac{V_m}{\omega_m}\sin\omega_m t.\sin\omega_c t$$

But,

also, we know that,

$$\frac{k_f.V_m}{\omega_m} = m_f = \text{modulation index.}$$

Hence,

$$s(t)_{FM} = A\cos\omega_c t - Am_f\sin\omega_m t\sin\omega_c t \qquad ...(3.43)$$

Which is the required expression for a single-tone narrowband FM.

3.6.5 Wideband FM

When the value of modulation index m_f is quite large, then in FM, a large number of sidebands are produced and hence the bandwidth of FM is sufficiently large. This type of FM system is known as wideband FM.

Now, we shall analyze wideband FM by considering the case of a single-tone FM system. The expression for a single-tone FM wave is given as

$$s(t) = A \cos (\omega_c t + m_f \sin \omega_m t) \quad \text{...(3.44)}$$

This expression may be considered as a real part of the exponential phasor given by

$$C_{FM}(t) = A e^{j(\omega_c t + m_f \sin \omega_m t)}$$

$$C_{FM}(t) = A e^{j\omega_c t} e^{jmf \sin \omega_m t} \quad \text{...(3.45)}$$

In above expression, the second exponential is a periodic function of period $\frac{1}{f_m}$ and may be expanded in the form of a complex Fourier series as under:

$$e^{jm_f \sin \omega_m t} = \sum_{n=-\infty}^{\infty} C_n e^{jn\omega_m t} \quad \text{for } -\frac{1}{2f_m} \le t \le \frac{1}{2f_m} \quad \text{...(3.46)}$$

Hence, the coefficient C_n is given by

$$C_n = f_m \int_{-\pi/\omega_m}^{\pi/\omega_m} e^{j(m_f \sin \omega_m t)} e^{-jn\omega_m t} \, dt$$

Substituting $x = \omega_m t$, we have,

$$C_n = \frac{1}{2\pi} \int_{-\pi}^{\pi} e^{j(m_f \sin x - nx)} \, dx \quad \text{...(3.47)}$$

In above expression, the integral on the right hand side is the nth order Bessel function of the first kind and argument m_f. This function is represented by $J_n(m_f)$.

Thus,

$$C_n = J_n(m_f) \quad \text{...(3.48)}$$

Putting the value of C_n in Eqn. (3.46), we get,

$$e^{jm_f \sin \omega_m t} = \sum_{n=-\infty}^{\infty} J_n(m_f) e^{jn\omega_m t}$$

Substituting above expression in Eqn. (3.45), we have,

$$C_{FM}(t) = A e^{j\omega_c t} \sum_{n=-\infty}^{\infty} J_n(m_f) e^{jn\omega_m t}$$

$$C_{FM}(t) = A \sum_{n=-\infty}^{\infty} J_n(m_f) e^{j(\omega_c + n\omega_m)t} \quad \text{...(3.49)}$$

In above expression, the real part of the right hand side provides the expression for FM signal.

i.e.,

$$s(t) = A \sum_{n=-\infty}^{\infty} J_n(m_f)\cos(\omega_c + n\omega_m t) \quad ...(3.50)$$

Thus, we have converted original single tone FM expression into modified form which consists of Bessel function.

Now, the Bessel function $J_n(m_f)$ may be expanded in a power series given by

$$J_n(m_f) = \sum_{m=0}^{\infty} \frac{(-1)^m \left(\frac{1}{2}m_f\right)^{n+2m}}{\angle m \angle n+m} \quad ...(3.51)$$

The above expression i.e., the Bessel function $J_n(m_f)$ with respect to m_f has been plotted in Fig. 3.5 for various positive integer values of n.

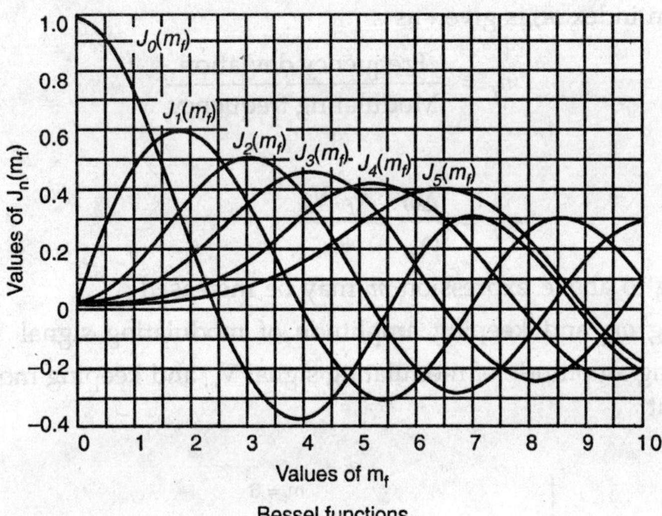

Fig. 3.5 *Bessel Functions*

Few important properties of the Bessel function may be summarized as follows:

(i) $J_n(m_f) = J_{-n}(m_f)$ for even n

$J_n(m_f) = -J_{-n}(m_f)$ for odd n

(ii) For a small value of m_f, we have

$J_0(m_f) = 1$

$J_1(m_f) = \frac{m_f}{2}$

$J_n(m_f) = 0$ for $N > 1$

(iii) $$\sum_{n=-\infty}^{\infty} J_n^2(m_f) = 1$$

Making use of first property, the Eqn. (3.51) may be expressed as

$$s(t) = AJ_0(m_f)\cos\omega_c t + AJ_1(m_f)[\cos(\omega_c+\omega_m)t - \cos(\omega_c-\omega_m)t] + AJ_2(m_f)[\cos(\omega_c+2\omega_m)t + \cos(\omega_c-2\omega_m)t] + AJ_3(m_f)[\cos(\omega_c+3\omega_m)t - \cos(\omega_c-3\omega_m)t] + \ldots \quad \ldots(3.52)$$

3.7 EFFECT OF VARIATION OF MODULATION INDEX M_F ON THE SPECTRUM OF FM SIGNAL

The number of sidebands produced in FM increases with increase in the value of modulation index m_f

The modulation index m_f is given as

$$m_f = \frac{\text{Frequency deviation}}{\text{Modulating frequency}}$$

or,

$$m_f = \frac{\Delta\omega}{\omega_m} = \frac{k_f . V_m}{\omega_m} \quad \ldots(3.53)$$

Thus according to above expression m_f may be increased

(i) by reducing ω_m and keeping amplitude of modulating signal V_m constant.

(ii) by increasing amplitude of modulating signal V_m and keeping modulating frequency ω_m constant.

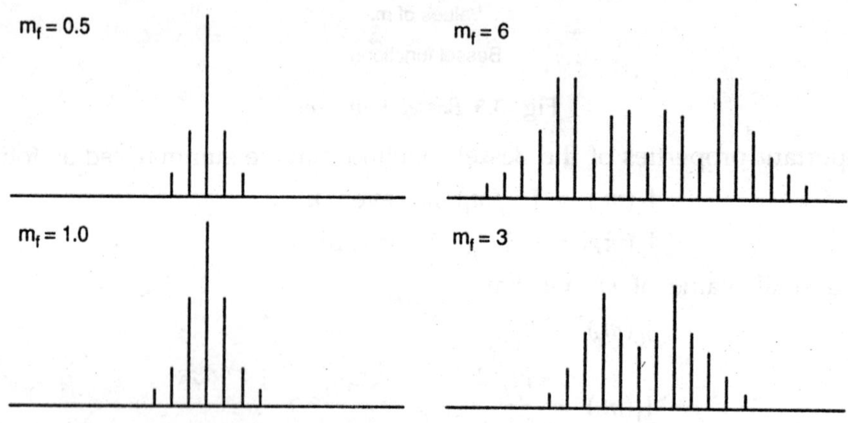

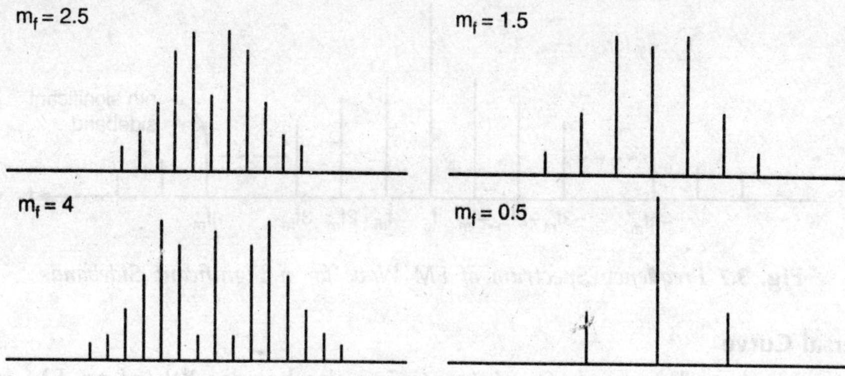

Fig. 3.6

Infact, it is possible to evaluate the size of the carrier and each sideband for each specific value of the modulation index. When this is done, the frequency spectrum of the FM wave for that particular value of modulation index m_f may be plotted. This has been done in Fig. 3.6. Which shows these frequency spectrum first for increasing deviation and f_m constant and then for decreasing modulating frequency f_m and frequency deviation Δf constant. Also, we may observed that as modulation depth increases, so does the bandwidth and also that reduction in modulating frequency f_m increases the number of sidebands, though not necessarily the bandwidth. Another point which may be observed is that although the number of sideband components is theoretically infinite, but in practice a lot of the higher sidebands have insignificant relative amplitudes and this is why they are not shown in the spectrum.

3.8 TRANSMISSION BANDWIDTH OF FM SIGNAL

The bandwidth of FM signal depends upon the value of modulation index m_f. With the increase of modulation index m_f, more and more number of sidebands acquire significant amplitudes and thus bandwidth is increased. For a particular value of modulation index m_f, we can note the number of significant sidebands from further steps.

Basically the effective bandwidth is the separation between the two extreme significant side frequencies on either side of the carrier.

Fourier analysis reveals that the number of side frequencies which contain a significant amount of power and thus the effective bandwidth of the FM signal is dependent on the modulation index of the modulated wave.

For practical purpose, a side frequency is considered to be significant if it's amplitude is at least one per cent of the unmodulated carrier amplitude.

Let us consider that in the spectrum of FM wave the number of significant sidebands is n. Since the upper sidebands are separated by ω_m, therefore, they form a frequency band of $n\omega_m$.

A similar frequency band of $n\omega_m$ is formed by the lower sidebands.

Thus, for n sidebands the bandwidth of FM wave is given by

$$BW = 2n\omega_m \text{ radians/sec} = 2nf_m \text{ Hz.} \quad ...(3.54)$$

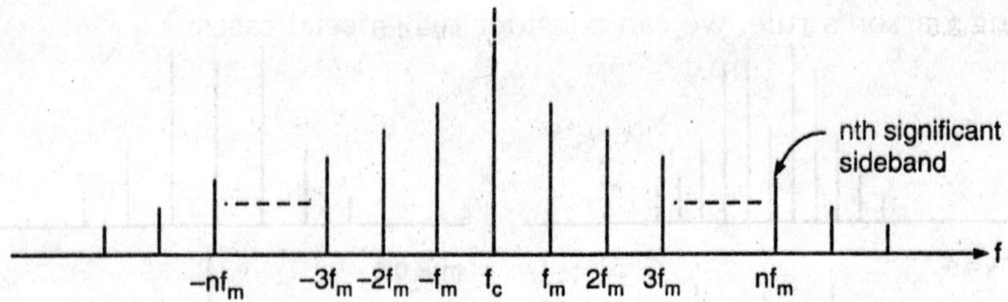

Fig. 3.7 *Frequency Spectrum of FM Wave for n Significant Sidebands*

3.8.1 Universal Curve

Schwartz developed a graph for determining the bandwidth of an FM signal if the modulation index is known. The graph has been shown below. Schwartz uses as his criterion the rule of thumb that any frequency component with a signal strength (voltage) less than 1% of that of the unmodulated carrier will be considered too small to be significant. This graph is also called universal curve which shows the variation of the Bandwidth B normalised with respect to Δf against m_f.

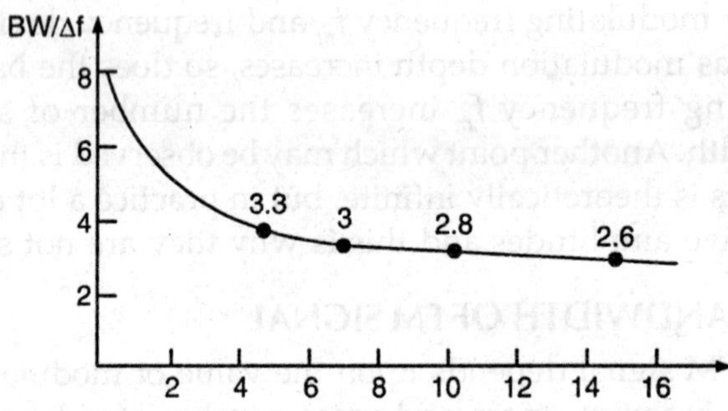

Fig. 3.8 *Schwartz's Universal Curve*

3.8.2 Carson's Rule

Carson's rule provides a thumb formula to calculate the bandwidth of a single-tone wideband FM. According to this rule, the FM bandwidth is given as twice the sum of the frequency deviation and the highest modulating frequency. However, it must be remembered that this role is just an approximation.

Mathematically,

$$BW = 2(\Delta\omega + \omega_m)$$

but

$$m_f = \frac{\Delta\omega}{\omega_m}$$

Therefore,

$$BW = 2(m_f.\omega_m + \omega_m) = 2(1 + m_f)\,\omega_m \qquad ...(3.55)$$

Regarding Carson's rule, we can consider two special cases:

(i) $\because$ $$BW = 2(1 + m_f)\,\omega_m$$

In $\Delta\omega \ll \omega_m$ i.e., $m_f \ll 1$ as in the case for narrowband FM

then

$$BW = 2\omega_m \quad \text{(this case is equivalent to AM)} \qquad ...(3.56)$$

(ii) $\because$ $$BW = 2(1 + m_f)\,\omega_m$$

In $\Delta\omega \gg \omega_m$, i.e., $m_f \gg 1$ as in the case for wideband FM

then

$$BW = 2(m_f)\,\omega_m$$

But

$$m_f \omega_m = \Delta\omega$$

Thus

$$BW = 2\Delta\omega \qquad ...(3.57)$$

(for large values of m_f, this bandwidth relation has a very small error and hence can be assumed to be true bandwidth for all practical purposes.)

3.9 FM BANDWIDTH FOR AN ARBITRARY MODULATING SIGNAL X(F)

Let us consider that the modulating signal $x(t)$ is an arbitrary signal i.e., it may consists of large number of frequency components. In this case to calculate the bandwidth, we first find a parameter known as deviation ratio D defined as

$$D = \frac{\text{Peak frequency deviation corresponding to the maximum possible amplitude of } x(t)}{\text{Maximum frequency component present in the modulating signal } x(t)} \qquad ...(3.58)$$

3.10 NARROWBAND FM VERSUS WIDEBAND FM

Fig. 3.9 shows that at high values of m_f the curve tends towards a horizontal asymptote and at low values of m_f it tends toward the vertical. Detailed mathematical study shows that the bandwidth of an FM signal for which m_f is less than $\pi/2$ is dependent mainly upon the frequency of the modulating signal and is quite independent of frequency deviation. Further analysis shows that the bandwidth of an FM signal for which m_f is less than $\pi/2$ is equal to twice the modulating frequency.

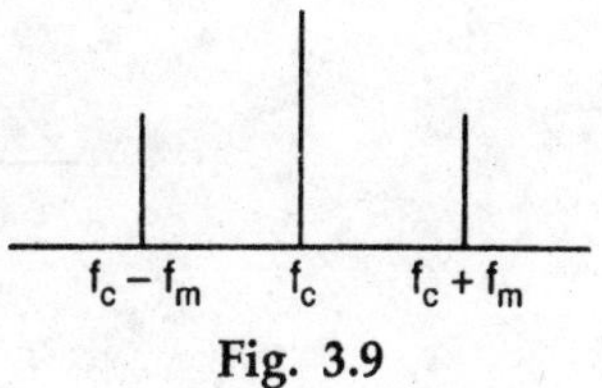

Fig. 3.9

$$\text{Bandwidth} = 2f_m \quad \text{for} \quad m_f < \frac{\pi}{2}$$

Just as with AM, and unlike the situation in which $m_f > \frac{\pi}{2}$, two side frequencies show up for each modulating frequency, one above and one below the frequency of the carrier, each spaced f_m away from the carrier frequency. Because of the limited bandwidth of FM signals with $m_f < \frac{\pi}{2}$, such modulations are known as narrowband FM, and FM signals with $m_f > \frac{\pi}{2}$ are known as wideband FM.

Though the spectrum form an AM signal and a narrowband FM signal appear to be the same, a Fourier analysis shows that the magnitude and phase relationships for AM and FM are quite different. In the given figure, it shows the frequency spectrum of a narrowband FM signal.

Many of the advantages obtained with wideband FM, such as noise reduction, are not available with narrowband FM. Why, then, would one want to use narrowband FM rather than AM? One reason is that with narrowband FM the power content at the carrier frequency decreases as the modulation increases so that we have describe situation of putting the power where the information is.

3.11 MULTIPLE FREQUENCY MODULATION

We already discussed a specific case of a single frequency-modulating signal. Now, let us extend these results to the case of multiple frequencies. First only two frequencies will be considered. It can then be generalized to any number of frequencies.

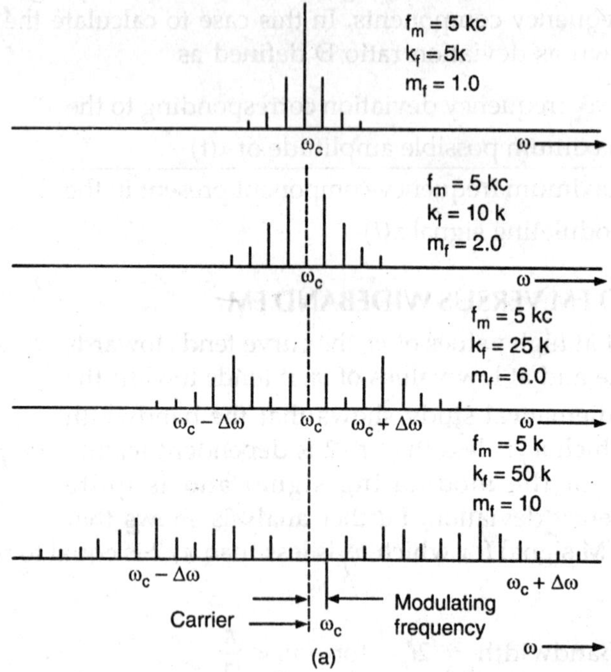

(a)

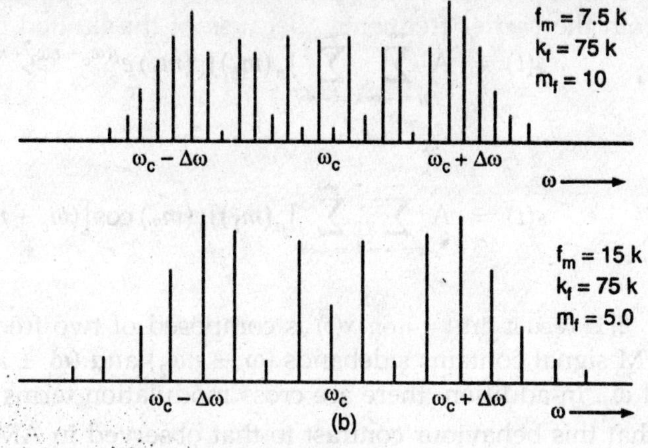

Fig. 3.10

Let us consider

$$x(t) = a_1 \cos \omega_1 t + a_2 \cos \omega_1 t$$

$$\omega_i = \omega_c + k_f x(t) = \omega_c + k_f (a_1 \cos \omega_1 t + a_2 \cos \omega_1 t)$$

The maximum frequency deviation will be

$$\Delta\omega = (a_1 + a_2)\, k_f$$

and

$$\phi = \int \omega_i \, dt = \omega_c t + \frac{a_1 k_f}{\omega_2} \sin \omega_1 t + \frac{a_2 k_f}{\omega_2} \sin \omega_2 t$$

or

$$\phi = \omega_c t + m_1 \sin \omega_1 t + m_2 \sin w_2 t$$

where,

$$m_1 = \frac{a_1 k_f}{\omega_1} \quad \text{and} \quad m_2 = \frac{a_2 k_f}{\omega_2}$$

Also,

$$\hat{s}(t) = \mathrm{A} e^{j\theta(t)} = \mathrm{A} e^{j(\omega_c t + m_1 \sin \omega_1 t + m_2 \sin \omega_2 t)}$$

or

$$\hat{s}(t) = \mathrm{A} e^{j\omega_c t} (e^{jm_1 \sin \omega_1 t})(e^{jm_2 \sin \omega_2 t}) \qquad \text{...(3.59)}$$

The exponentials in the brackets are obviously periodic function with periods $2\pi/\omega_1$ and $2\pi/\omega_2$, respectively. These exponentials can be represented by Fourier series using Bessel functions.

Hence, we have

$$s(t) = \mathrm{A} e^{j\omega_c t} \left[\sum_{n=-\infty}^{\infty} \mathrm{J}_n(m_1)\, e^{jn\omega_1 t} \right] \left[\sum_{n=-\infty}^{\infty} \mathrm{J}_n(m_2)\, e^{jn\omega_2 t} \right]$$

or

$$s(t) = A \sum_{n=-\infty}^{\infty} \sum_{k=-\infty}^{\infty} J_n(m_1) J_k(m_2) e^{j(\omega_c + n\omega_1 + k\omega_2)t} \quad ...(3.60)$$

and

$$s(t) = A \sum_{n=-\infty}^{\infty} \sum_{k=-\infty}^{\infty} J_n(m_1) J_k(m_2) \cos\left[(\omega_c + n\omega_1 + k\omega_2)t\right]$$

...(3.61)

It is evident from this result that when $x(t)$ is composed of two frequencies ω_1 and ω_2, the spectrum of the FM signal contains sidebands $(\omega_c \pm n\omega_1)$ and $(\omega_c \pm n\omega_2)$, corresponding to frequencies ω_1 and ω_2. In addition, there are cross modulation terms $(\omega_c \pm n\omega_1 \pm k\omega_2)$.

It may be noted that this behaviour contrast to that observed in AM. In AM, each new frequencies added to the modulating signal gives rise to its own side bands only. There are no cross modulation terms. For this reason, AM is called linear modulation, whereas FM is a non-linear type of modulation.

3.12 SQUARE WAVE MODULATION

Let us consider one more special case of FM modulation where the modulating signal $x(t)$ is a square wave as shown in Fig. 3.11(*a*). The method discussed here is very general and is applicable to any periodic modulating signal. For a square wave, the instantaneous frequency of the FM carrier is given by,

$$\omega_i = \omega_c + k_f x(t)$$

and the phase ϕ is given by

$$\phi = \int \omega_i \, dt = \omega_c t + k_f \int x(t) \, dt$$

or

$$\phi = \omega_c t + \psi(t)$$

where,

$$\psi(t) = k_f \int x(t) \, dt$$

This is a triangle periodic function with period T.

It may be noted that the maximum carrier frequency deviation $\Delta\omega$ in this case is k_f since $|x(t)|_{max} = 1$.

Hence,

$$k_f = \Delta\omega$$

and

$$y(t) = \begin{cases} (\Delta\omega)t & \text{for } -T/4 < t < T/4 \\ (\Delta\omega)[T/2] & \text{for } T/4 < t < 3T/4 \end{cases} \quad ...(3.62)$$

and

$$y(t) = y(t \pm nT)$$

Also

$$\hat{s}(t) = Ae^{j\theta(t)} = Ae^{j\psi(t)}\, e^{j\omega_c t}$$

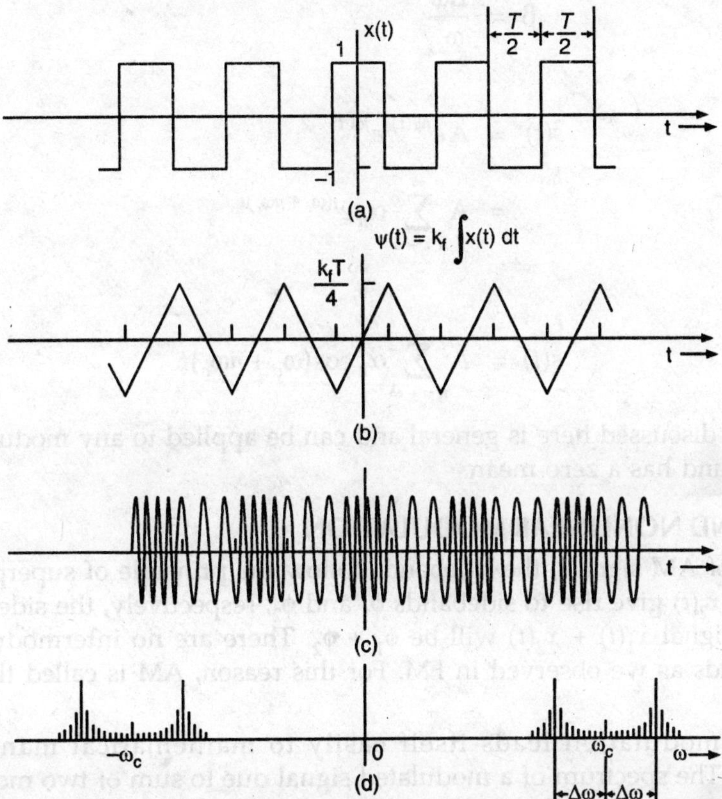

Fig. 3.11

The function $e^{j\psi(t)}$ is itself a periodic function of period T, and it can be expressed by Fourier series as

$$e^{j\psi(t)} = \sum_{n=-\infty}^{\infty} \alpha_n\, e^{jn\omega_s t}$$

$$\omega_s = \frac{2\pi}{T}$$

where,

$$\alpha_n = \frac{1}{T}\int_{-T/4}^{3T/4} e^{j\psi(t)}\, e^{-jn\omega_s t}\, dt$$

Substituting of equation in the above equation and the subsequent integration will yield

$$\alpha_n = \frac{1}{2}\left\{S_a\left[\frac{\pi}{2}(m_f - n)\right] + (-1)^n S_a\left[\frac{\pi}{2}(m_f + n)\right]\right\} \qquad \text{...(3.63)}$$

where,

$$\beta = \frac{\Delta\omega}{\omega_s}$$

Hence,

$$\hat{s}(t) = A e^{j\psi(t)} e^{j\omega_c t}$$

$$= A \sum_{n=-\infty}^{\infty} \alpha_n e^{j(\omega_c + n\omega_c)t}$$

and

$$s(t) = A \sum_{n=-\infty}^{\infty} \alpha_n \cos(\omega_c + n\omega_c)t \qquad \text{...(3.64)}$$

The method discussed here is general and can be applied to any modulating signal $x(t)$ that is periodic and has a zero mean.

3.13 LINEAR AND NONLINEAR MODULATION

In the case of AM signals, the sidebands follow the principle of superposition. Thus, if signals $x(t)$ and $x_2(t)$ give rise to sidebands ϕ_1 and ϕ_2, respectively, the sideband created by the composite signal $x_1(t) + x_2(t)$ will be $\phi_1 + \phi_2$. There are no intermodulation or cross-product sidebands as we observed in FM. For this reason, AM is called the linear type of modulation.

The linear modulation leads itself easily to mathematical manipulations and generalizations. The spectrum of a modulated signal due to sum of two modulating signals can be found by calculating the spectrum due to each individual signal and then adding them together. This proves very useful in noise calculations in communication systems. For linear modulation systems, the effect of additive noise over the channel can be calculated by assuming the signal to be zero. This is not true of a nonlinear modulation where cross-modulation terms arise. For these reasons, we are interested in approximating a nonlinear modulation by a linear model. This is analogous to the case of system analysis where one can approximate a nonlinear system by a linear one over a limited range of signal amplitudes. Now, let us show that FM closely approximates a linear behaviour for a small modulation index.

3.13.1 Linearization of FM signal for Small Modulation Index

For a small modulating index, FM closely exhibits linear behaviour. Let us consider again the case of modulating signal $x(t)$ containing two frequencies ω_1 and ω_2. The FM signal in this case is given as

$$\hat{s}(t) = A e^{j\omega_c t}(e^{jm_1 \sin \omega_1 t})(e^{jm_2 \sin \omega_2 t})$$

If m_1 and $m_2 < 1$, then for $x(t)$, we have

$$x(t) = a_1 \cos \omega_1 t + a_2 \cos \omega_c t$$

$$\hat{s}(t) \equiv A(1 + jm_1 \sin \omega_1 t)(1 + jm_2 \sin \omega_2 t)\, e^{j\omega_c t}$$

$$\equiv A(1 + jm_1 \sin \omega_1 t + jm_2 \sin \omega_2 t)\, e^{j\omega_c t}$$

Note that if

$$x_1(t) = a_1 \cos \omega_1 t \quad \text{and} \quad m_1 \ll 1$$

then

$$\hat{s}(t) = A e^{jm_1 \sin \omega_1 t} e^{j\omega_c t}$$

or

$$\hat{s}(t) = A\left[1 + jm_1 \sin \omega_1 t\right] e^{j\omega_c t}$$

If

$$x_2(t) = a_2 \cos \omega_2 t \quad \text{and} \quad m_2 \ll 1$$

then

$$\hat{s}(t) = A(1 + jm_2 \sin \omega_2 t)\, e^{j\omega_c t}$$

It is easy to see that under he conditions $m_1, m_2 \ll 1$, the sidebands due to modulating signal $x_1(t) + x_2(t)$ are the sum of sidebands due to individually. Hence, FM can be assumed to be a linear modulation for a small modulation index. The cross-modulation terms under this assumption can be ignored.

3.14 PHASE MODULATION: AN ANALYTIC VIEW

We have already discussed earlier as, phase modulation may be analysed in a similar way at the FM signal. The PM has some distinct features as compared to FM. The main distinct feature of phase-modulation is that the deviation in the carrier frequency ω_c in PM is linearly proportional to the baseband or modulating frequency ω_m. However in FM, the deviation is independent of the baseband or modulating frequency.

Let us find an expression for frequency deviation for a PM wave.

We know that the total phase angle of a PM wave is expressed as

$$\phi_i = \omega_c t + k_p . x(t) \qquad ...(3.65)$$

For a single-tone modulating signal, we have

$$x(t) = V_m \cos \omega_m t \qquad ...(3.66)$$

Thus,

$$\phi_i = \omega_c t + k_p . V_m \cos \omega_m t \qquad ...(3.67)$$

In above expression, the maximum departure in the phase is $k_p . V_m$. This is called as phase deviation denoted by θ_d.

Hence phase deviation, $\theta_d = k_p . V_m$ ...(3.68)

Now, the expression for the PM wave becomes

$$s(t)_{PM} = A \cos\phi_i = A\cos\left[\omega_c t + \theta_d \cos\omega_m t\right] \quad ...(3.69)$$

Also, the instantaneous frequency related to ϕ_i is expressed as

$$\omega_i = \frac{d\phi_i}{dt} = \frac{d}{dt}\left[\omega_c t + k_p V_m \cos\omega_m t\right]$$

or,

$$\omega_i = \omega_c - k_p V_m \omega_m \sin\omega_m t \quad ...(3.70)$$

Thus, the maximum departure in the frequency from ω_c is $k_p.V_m.\omega_m$.

Hence, for PM, the frequency deviation is given by

$$\Delta\omega_{PM} = k_p.V_m.\omega_m \quad ...(3.71)$$

Which depends upon the modulating frequency ω_m. On the other hand, the frequency deviation for FM is

$$\Delta\omega_{FM} = k_f.V_m \quad ...(3.72)$$

Thus, we conclude that for an equal bandwidth in FM and PM, we have,

$$k_f = k_p.\omega_m \quad ...(3.73)$$

3.14.1 Translation Bandwidth of PM

We may estimate the transmission bandwidth for PM using Carson's rule as

$$BW_{PM} \cong 2(\Delta\omega) \cong 2k_p.V_m.\omega_m \quad ...(3.74)$$

Hence, the bandwidth of the PM wave varies fastly with the variation in the modulating frequency ω_m. On the other hand, the FM bandwidth varies slowly with modulating frequency ω_m.

For PM, the modulation index m_p will be same as the deviation θ_d and will be expressed as

$$m_p = k_p.V_m = \theta_d \quad ...(3.75)$$

3.15 COMPARISON OF ANGLE MODULATION WAVES AND AMPLITUDE MODULATED WAVE

The given Fig. 3.12 shows a single-tone modulating signal, a carrier signal, amplitude-modulated (AM) wave and angle-modulated waves. On comparison, we note the following differences between two techniques.

(*i*) The envelope of FM wave or PM wave is constant and is equal to the unmodulated carrier amplitude.

On the otherhand, the envelope of AM waves is dependent on the modulating signal $x(t)$.

(*ii*) The zero crossings of a FM wave or a PM wave no longer exhibit a perfect regularity in their spacing like AM wave. Thus this makes the instantaneous frequency of the angle modulated wave depend upon time.

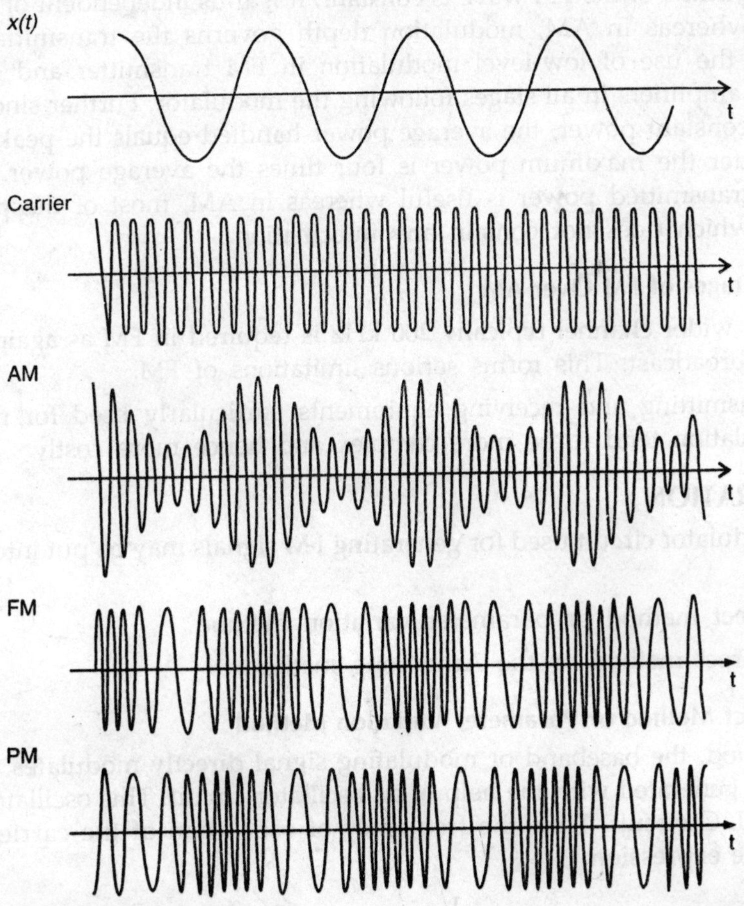

Fig. 3.12 *AM, FM, and PM waveforms.*

3.16 COMPARISON OF FREQUENCY MODULATION AND AMPLITUDE MODULATION

3.16.1 Advantages of FM over AM

(*i*) FM receivers may be fitted with amplitude limiters to remove the amplitude variations caused by noise.

This makes FM reception a good deal more immune to noise than AM reception.

(*ii*) It is possible to reduce noise still further by increasing the frequency deviation. This is a feature which AM does not have because it is not possible to exceed 100 per cent modulation without causing severe distortion.

(*iii*) Standard Frequency Allocations provide a guard band between commercial FM stations. Due to this, there is less adjacent channel interference than in AM.

(*iv*) FM broadcasts operate in upper VHS and UHF frequency ranges at which there happens to be less noise than in the MF and HF ranges occupied by AM broadcasts.

(*v*) The amplitude of the FM wave is constant. It is thus independent of the modulation depth, whereas in AM, modulation depth governs the transmitted power. This permits the use of low-level modulation in FM transmitter and use of efficient Class C amplifiers in all stages following the modulator. Further since all amplifiers handle constant power, the average power handled equals the peak power. In AM transmitter the maximum power is four times the average power. Finally in FM, all the transmitted power is useful whereas in AM, most of the power is carrier power which does not contain any information.

3.16.2 Disadvantages of FM Over AM

(*i*) A much wider channel typically 200 kHz is required in FM as against only 10 kHz in AM broadcast. This forms serious limitations of FM.

(*ii*) FM transmitting and receiving equipments particularly used for modulation and demodulation tend to be more complex and hence more costly.

3.17 FM GENERATION

The FM modulator circuit used for generating FM signals may be put into two categories as under.

(*i*) The direct method or parameter variation method.

(*ii*) the indirect method or the Armstrong method.

3.17.1 The Direct Method or Parameter Variation Method

In this method, the baseband or modulating signal directly modulates the carrier. The carrier signal is generated with the help of an oscillator circuit. This oscillator circuit uses a parallel tuned L-C circuit. Thus the frequency of oscillation of the carrier generation is governed by the expression.

$$\omega_c = \frac{1}{\sqrt{LC}}$$

Now, we can make the carrier frequency ω_c, to vary in accordance with the baseband or modulating signal $x(t)$ if L or C is varied according to $x(t)$. An oscillator circuit whose frequency is controlled by a modulating voltage is called voltage controlled oscillator (VCO). The frequency of VCO is varied according to the modulating signal simply by putting a shunt voltage variable capacitor with its tuned circuit. This voltage variable capacitor is called varactor or varicap. This type of property is exhibit by reverse biased semiconductor diodes. Also the capacitance of bipolar transistor (BJT) and field-effect transistor (FET) is varied by the Miller-effect. This miller capacitance may be utilized for frequency modulation.

In addition to this, the electron tubes may provide variable reactance which is proportional to modulating or baseband signal. This type of tubes are called reactance tubes and may be used for FM generation.

The inductance L of the tuned circuit may also be varied in accordance with the baseband or modulating signal $x(t)$. The FM circuit using such inductors is called saturable reactor modulator. Frequency modulation can also be achieved from voltage controlled devices such as PIN diode. Klystron oscillators and multivibrators.

3.17.2 Varactor Diode Method For FM Generation

The varactor diode is a semiconductor diode whose junction capacitance changes with d.c. bias voltage. This varactor diode is connected in shunt with the tuned circuit of the carrier oscillator.

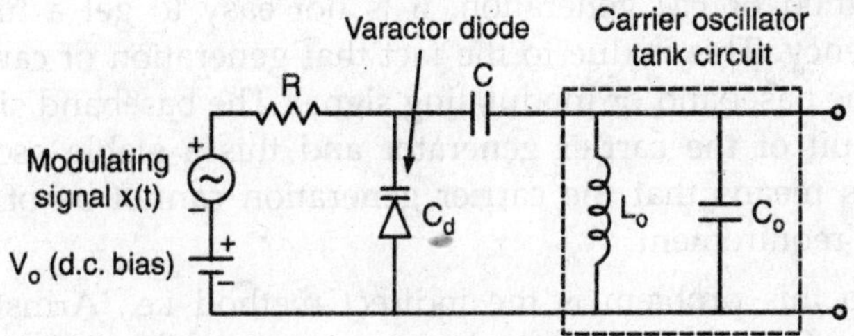

Fig. 3.13 *Varactor diode method of FM generation*

In varactor diode FM generation arrangement, the capacitor C is made much smaller than the varactor diode capacitance C_d so that the radio frequency (RF) voltage from oscillator across the diode is small as compared to reverse bias d.c. voltage across the varactor diode. In addition to this, the reactance of the capacitor C at the highest modulating frequency is made large enough compared to resistor R so that the shunting of the baseband or modulating signal through the tuned circuit may be checked.

3.17.2.1 Mathematical Analysis

The capacitance C_d of the Varactor diode is expressed as

$$C_d = \frac{k}{\sqrt{V_D}} = k(V_D)^{-1/2} \qquad ...(3.76)$$

Here, V_D is the total instantaneous voltage across the varactor diode and is given by

$$V_D = V_o + x(t) \qquad ...(3.77)$$

Also, k is a constant of proportionality.

The oscillation frequency is given as

$$\omega_c = \frac{1}{\sqrt{LC}} \qquad ...(3.78)$$

Now, the total capacitance of the oscillator tank circuit will be $C_o + C_d$ and thus the instantaneous frequency of oscillation ω_i is expressed as

$$\omega_i = \frac{1}{\sqrt{L_o(C_o + C_d)}} \qquad ...(3.79)$$

In above equation, substituting the value of C_d from equation, we have

$$\omega_i = \frac{1}{\sqrt{L_o(C_o + kV_D)^{-1/2}}}$$

We conclude that the instantaneous frequency ω_i of FM signal depends upon V_D which in turn depends upon the value of modulating signal $x(t)$. Thus, the instantaneous oscillator frequency ω_i also depends upon the baseband or modulating signal $x(t)$ and hence frequency modulation is generated.

3.18 DRAWBACKS OF DIRECT METHOD FOR FM GENERATION

(*i*) In direct method of FM generation, it is not easy to get a high order stability in carrier frequency. This is due to the fact that generation of carrier signal is directly affected by the baseband or modulating signal. The baseband signal directly controls the tank circuit of the carrier generator and this a stable oscillator circuit cannot be used. This means that the carrier generation cannot be of high stability which is necessary requirement.

A solution to this problem is the indirect method i.e., Armstrong method of FM generation. In fact, in this method, the carrier oscillator does not respond to the modulating signal directly rather the generation of the carrier is made isolated from other parts of the circuit. Thus stable crystal oscillators may be used to generate carrier signal.

(*ii*) The non-linearity of the varactor diode produces a frequency variation due to harmonics of the modulating or baseband signal and therefore the FM signal is distorted. We will have to take the proper cares to keep this type of distortion minimum.

Despite all above drawbacks, the direct method is utilized for high power FM generation in several applications.

3.19 THE INDIRECT OR ARMSTRONG METHOD OF FM GENERATION

In this method, we can get very high frequency stability since in this case the crystal oscillator may be used as a carrier frequency generator. The working principle of Armstrong method is to generate a narrowband FM (NBFM) indirectly by utilizing the phase-modulation technique and then changing this narrowband FM into a wideband FM (shown in below figure).

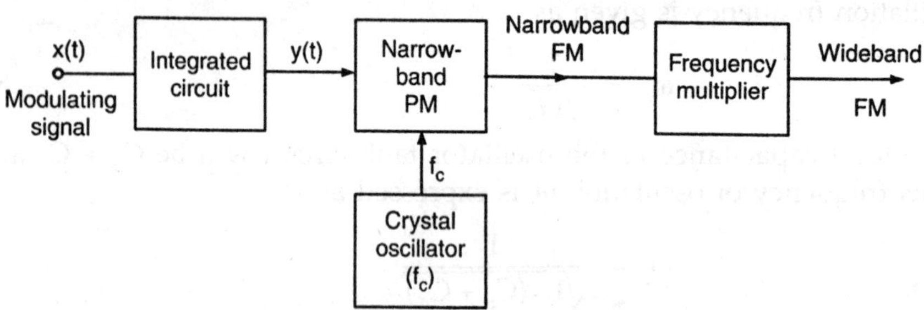

Fig. 3.14 *Armstrong Method for FM Generation*

Since in narrowband FM the modulation index is small, therefore the distortion is very low in narrowband FM. Here we prefer phase modulation technique circuit apart from multiplying the carrier frequency also increases the frequency deviation and hence the narrowband FM is converted into wideband FM.

3.20 PRACTICAL ARMSTRONG METHOD FOR FM GENERATION

In this, we will discuss a practical method of generating FM signal.

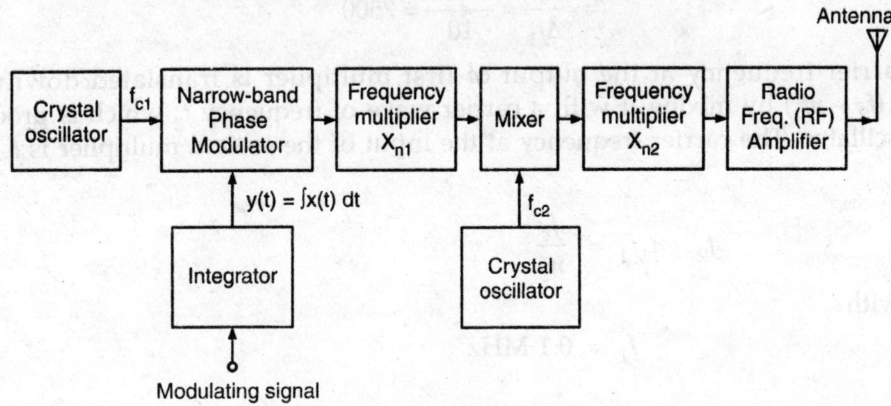

Fig. 3.15

We know that in commercial use we require to transmit audio signal consisting frequencies in the range of 50 Hz to 15 kHz and the value of Δf = 75 kHz. Let us assume that the final carrier frequency of the FM required is f_c = 100 MHz.

Let us begin with narrowband FM having a carrier frequency f_{c1} = 100 kHz generated by a crystal oscillator. To check the harmonic distortion produced by narrowband phase modulator, the modulation index m_f to a maximum value of 0.3 radians.

Let us assume m_{f1} = 0.2 radian. The lowest modulation frequency of 100 Hz produces a frequency deviation of

$$\Delta f_1 = 0.2 \times 50 = 10 \text{ Hz}$$

at the narrowband phase modulator output whereas the largest modulating frequency 15 kHz produces a frequency deviation of

$$\Delta f_2 = 0.2 \times 15 = 3 \text{ kHz} \quad \text{...(3.80)}$$

Thus, the lowest modulating frequency is of immediate concern. We take the value of Δf_1 = 10 Hz, so that at the highest modulating frequency, m_f becomes even less.

To produce a frequency deviation of Δf = 75 kHz at the output, a frequency multiplication is required. As an example Δf_1 = 10 Hz and the required deviation is Δf = 75 kHz. We thus require a total frequency multiplication by a factor

$$n = \frac{75000}{10} = 7500 \quad \text{...(3.81)}$$

However, a straight forward frequency multiplication equal to this value leads to a very high value of carrier frequency than the required 100 MHz.

To get the required frequency-deviation and carrier frequency we use a two-stage frequency multiplier. This arrangement makes use of two multipliers and a mixer. With the help of a mixer, the carrier frequency is translated suitably without altering frequency-deviation Δf. The final stage multiplier provides the required carrier frequency and deviation.

Let us assume that n_1 and n_2 are the frequency multiplication factors for the two multipliers, so that

$$n = n_1.n_2 \quad ...(3.82)$$

$$= \frac{\Delta f}{\Delta f_1} = \frac{75000}{10} = 7500 \quad ...(3.83)$$

The carrier frequency at the output of first multiplier is translated downwards to frequency $(f_2 - nf_1)$ by mixing it with a carrier wave of frequency f_2 which is produced by another oscillator. The carrier frequency at the input of the second multiplier is f_c/n_2.

Hence,

$$f_2 - n_1 f_1 = \frac{f_c}{n_2}$$

Thus with

$$f_1 = 0.1 \text{ MHz}$$

and,

$$f_2 = 8.5 \text{ MHz},$$

we have

$$8.5 - 0.1n_1 = \frac{100}{n_2}$$

Using above equation, we have

$$n_1 = 100, \text{ and}$$
$$n_2 = 75$$

3.21 FM DEMODULATORS OR DETECTORS

The process of getting a modulating or baseband signal from a frequency modulated (FM) signal is called demodulation or detection. The electronics circuits which perform the demodulation process are known as FM demodulators or detectors. The FM demodulator or detector performs the extraction of modulating signal in two steps as follows:

(*i*) It converts the frequency modulated (FM) signal into a corresponding amplitude modulated (AM) signal with the help of frequency dependent circuits, i.e., the circuits whose output voltage depends upon the input frequency. These circuits are generally known as Frequency Discriminators.

(*ii*) The original modulating or baseband signal is recovered from this AM signal with the help of the linear diode detector or envelope detector.

A simple R-L circuit may be used as a discriminator, but this circuit has a very poor sensitivity as compared to a tuned L-C circuit. Therefore, L-C circuits are generally used as frequency discriminators.

3.22 TYPES OF FM DEMODULATORS

FM demodulators are of following types:

3.22.1 Slope Detector

The principle of operation of slope detectors depends upon the slope of the frequency response characteristics of a frequency selective network. The two main FM detectors, which use detuned resonant circuits come under this category, are as under:

(*i*) Single-tune detector circuit or simple slope-detector.

(*ii*) Stagger-tuned detector circuit or balanced slope detector.

3.22.2 Phase Difference Detectors

The following two circuits come under this category:

(*i*) Foster-Seeley Detector

(*ii*) Ratio-Detector

3.22.3 Slope Detector

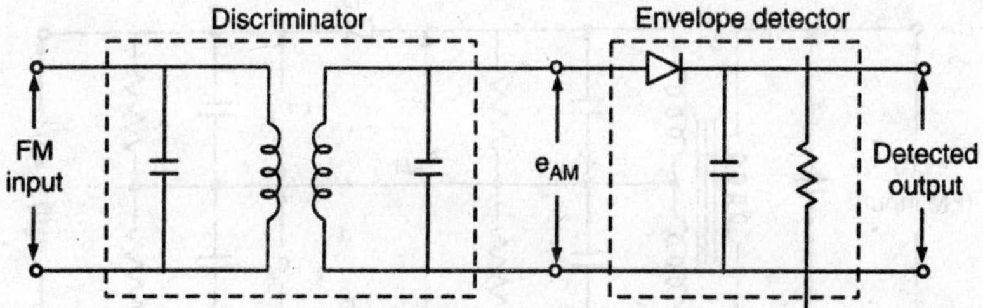

Fig. 3.16 *Circuit Diagram of a Slope Detector*

The circuit consists of a tuned circuit which is slightly detuned from the carrier frequency ω_c. In other words, the circuit uses two different frequencies. First one is tuned to the incoming FM carrier frequency ω_c whereas the second one is tuned to a frequency slightly different from the carrier frequency ω_c. Therefore, this portion of the circuit which contains two tuned circuits tuned to different frequencies, is called discriminator.

This circuit converts the FM signal into an AM signal as shown in the slope detector characteristic curve. The another portion of the circuit is envelope detector. The AM signal from the output of the discriminator is applied at the input of envelope detector.

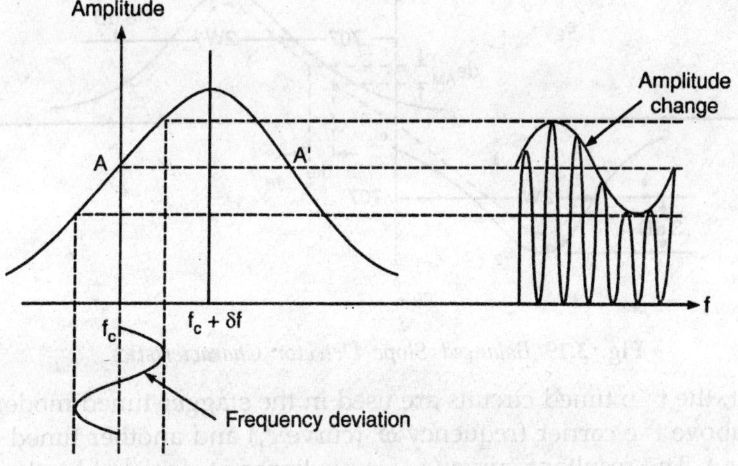

Fig. 3.17 *Slope Detector Characteristic Curve*

At the output of envelope detector, the original modulating or baseband signal is obtained. Although this circuit is simple and inexpensive, it has following drawbacks.

(*i*) The circuits non-linear characteristic produces a harmonic distortion. The non-linearity is obvious from the fact that the slope is not the same at each point of the characteristics.

(*ii*) The circuit does not eliminate the amplitude variations and the output is sensitive to any amplitude variations in the input FM signal which is obviously not a desirable feature. A good discriminator circuit must respond only to frequency variations and not to amplitude variations.

3.22.4 Balanced Slope Detector

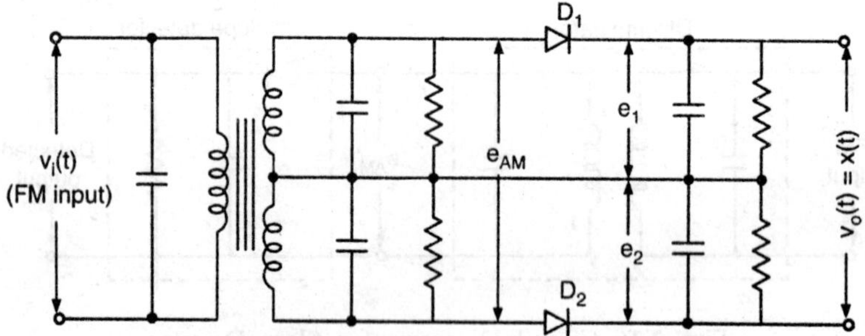

Fig. 3.18 *Circuit Diagram of a Balanced Slope Detector*

The balanced slope detector is an important over the simple slope detector. This circuit of balanced slope detector consists of two L-C circuits. The frequency response for the circuit has been shown below:

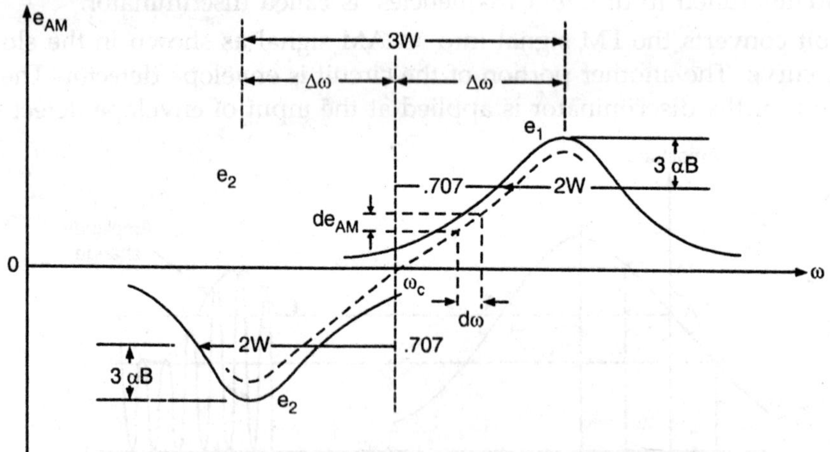

Fig. 3.19 *Balanced Slope Detector Characteristics*

In this circuit, the two tuned circuits are used in the stagger-tuned mode, i.e., one tuned circuit is tuned above the carrier frequency ω_c (curve e_1) and another tuned circuit is tuned below ω_c (curve e_2). The resultant curve $(e_1 + e_2)$ is linear as depicted by the dotted line.

The slope $de_{AM}/d\omega$ is identical at all the points on the linear portion of the characteristics curve. Hence, no harmonic distortion is caused when the operation is restricted to the linear region.

3.22.5 Disadvantages of the Balanced Slope Detector

The following are the disadvantages of balanced slope detector:

(*i*) The tuned circuit output is not purely bandlimited and thus the low pass RC filter of the envelope detector introduces distortion.

(*ii*) The discriminator characteristics depends critically upon the amount of detuning of the resonant circuits.

(*iii*) The linear characteristics is limited to a small frequency deviation $\Delta\omega$. The frequency deviation $\Delta\omega$, for which the resultant characteristics is linear, depends upon the 3 dB bandwidth of each tuned circuit which is equal to 2 W. The linearity is limited to a range of 3 W.

Thus the deviation $\Delta\omega$, for which the output is given as

$$2(\Delta\omega) = 3\text{ W}$$

or,

$$\Delta\omega = 1.5\text{ W}$$

if the frequency deviation of an output FM wave is more than the one expressed by above equation, the distortion will occur due to the non-linearity of the characteristics. Hence, the operation of this detector is limited to small deviation only.

However, in spite of all these drawbacks, this circuit can be designed to keep the distortions within tolerable limits and provide a satisfactory operation.

3.23 PHASE DIFFERENCE DISCRIMINATORS:

In the phase difference discriminators we discuss Foster-Seeley (Center-Tuned) discriminator and Ratio-Discriminator.

3.23.1 Foster-Seeley Discriminator

This type of discriminator is very widely used. The circuit arrangement is shown given below. The circuit comprises of an inductively coupled double-tuned circuit in which both primary and secondary are tuned to same frequency. The centre of the secondary is connected to the top of the primary (collector end) through a capacitor C. This capacitor C performs the following functions.

- it blocks the dc from primary to secondary.
- it couples the signal frequency from primary to center-tapping of the secondary.

The primary voltage V_3 (signal voltage) thus appears across the inductance L. Almost, entire V_3 appears across L, except a small drop across C. By proper choice of C and L the drop across C may be kept negligible.

The centre tapping of the secondary has an equal and opposite voltage across each half winding. Therefore, V_1 and V_2 are equal in magnitude but opposite in phase.

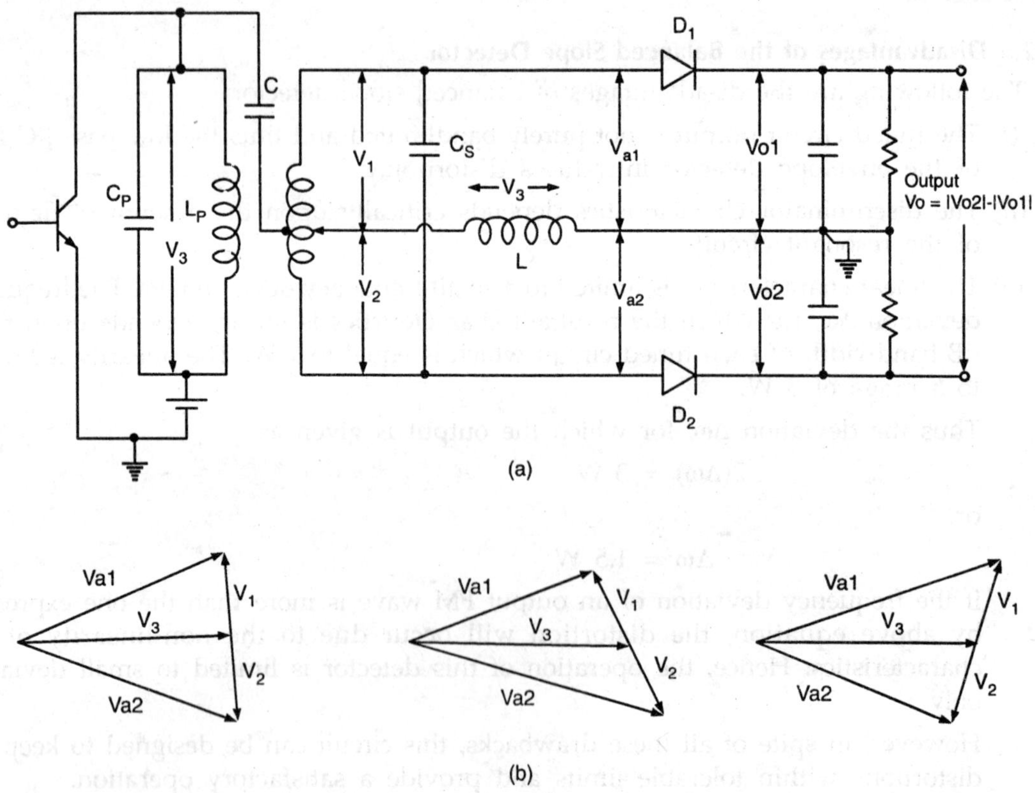

Fig. 3.20 *(a) Circuit Diagram of a Foster-Seeley Discriminator (b) Phasor Diagram*

The ratio frequency voltages V_{a1} and V_{a2} applied to the diodes D_1 and D_2 are given as:

$$V_{a1} = V_3 + V_1$$

$$V_{a2} = V_3 - V_2$$

Voltage V_{a1} and V_{a2} depend on the phasor relations between V_1, V_2 and V_3. The phasor diagrams V_1 and V_2 are always equal and are in phase opposition. However the phase position of V_1 and V_2 relative to V_3 will depend on the tuned secondary at the resonance, or off the resonance, as discussed as follows:

3.23.2 At resonance

When an input voltage has a frequency equal to the resonant frequency f_{if} of the tuned secondary, V_3 is inphase quadrature (90 out phase) with V_1 and V_2. This is shown in first phasor diagram. The resultant voltages V_{a1} and V_{a2} are equal in magnitude.

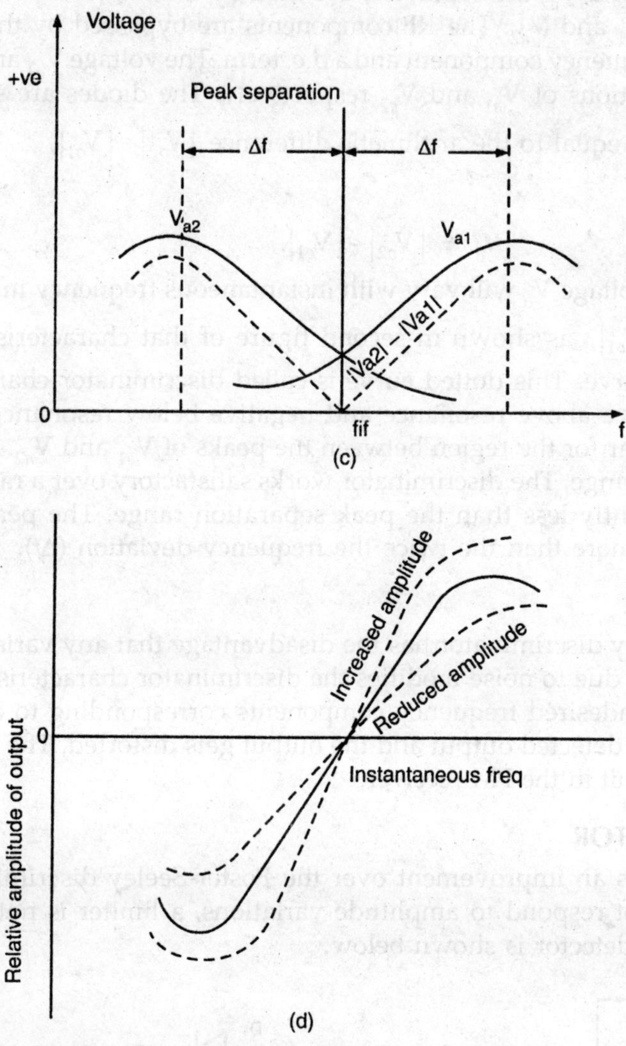

Fig. 3.21 Discriminator Characteristics

3.23.3 Off resonance

When the input voltage is off the resonant frequency f_{if} of the tuned secondary, the phase position of V_1 and V_2 relative to V_3 will be different from 90°. Let Q_s be the quality factor of the tuned secondary. When an input signal frequency is above the resonant frequency f_{if} by a amount $(f_{if}/2Q_s)$ the phase difference between V_3 and V_1 is 45°, as shown in the second phasor diagram. Since V_2 is in phase opposition of V_1, the phase difference between V_3 and V_2 is 135°. The phasor diagram reveals that V_{a1} is reduced whereas V_{a2} is increased. The situation is reversed when the input voltage has a frequency below f_{if} which is evident from the third phasor diagram. Thus, the amplitude of the voltages V_{a1} and V_{a2} will vary with the instantaneous frequency f in the manner shown in discriminator characteristics.

The RF voltage V_{a1} and V_{a2} are separately rectified by the diodes D_1 and D_2 respectively to produce voltages V_{a1} and V_{a2}. The RF components are bypassed by the capacitors leaving only modulating frequency component and a d.c. term. The voltage V_{a1} and V_{a2} then represent the amplitude variations of V_{a1} and V_{a2} respectively. The diodes are so arranged that the output voltage V_o is equal to the arithmetic difference $|V_{o2}| - |V_{o1}|$,

i.e.,

$$V_o = |V_{o2}| - |V_{o1}|.$$

Therefore, the voltage V_o will vary with instantaneous frequency in accordance with the difference $|V_{o2}| - |V_{o1}|$, as shown in second figure of that characteristics curve, and it is shown by dotted curve. This dotted curve is called discriminator characteristics. It is zero at resonance, positive above resonance and negative below resonance. The discriminator characteristic is linear for the region between the peaks of V_{a1} and V_{a2}. This range is known as peak-separation range. The discriminator works satisfactory over a range of instantaneous frequency only slightly less than the peak separation range. The peak separation range, therefore, must be more than the twice the frequency deviation (Δf).

3.23.4 Disadvantage

The Foster-Seeley discriminator has the disadvantage that any variation in amplitude of the input FM signal due to noise modifies the discriminator characteristics as shown by the dotted lines. The undesired frequency components corresponding to amplitude variations are produced in the detected output and the output gets distorted. The distortion is reduced using a limited circuit in the FM receiver.

3.24 RATIO DETECTOR

Ratio detector is an improvement over the Foster-Seeley discriminator and is widely used. As it does not respond to amplitude variations, a limiter is not needed. The circuit diagram of a ratio detector is shown below.

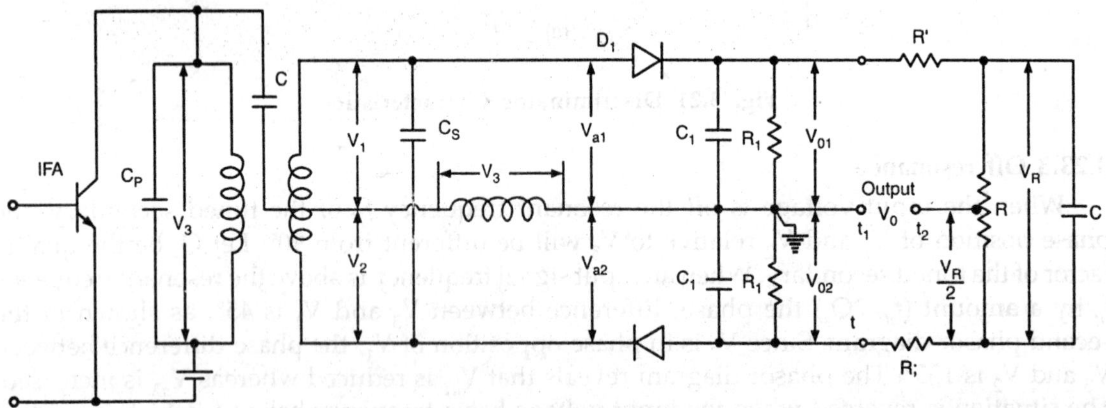

Fig. 3.22 *Ratio Detector*

The circuit is similar to the circuit of Foster-Seeley discriminator, except the following:

(*i*) The polarity of diode D_2 has been reversed.

(*ii*) The output V_o is taken from the centre tap of a resistor R that shunts the load impedance of the two diodes. The output voltage varies with the input signal frequency (FM) exactly in the same way as it does in the Foster-Seeley discriminator, but its magnitude is reduced to half.

This can be shown as follows:

For the time being, we will ignore the two resistances (both denoted by R) and the capacitor C. The voltage V_{o1} and V_{o2} have the same magnitude as in the case of a Foster-Seeley discriminator, but V_{o2} is now reversed in polarity. Therefore the voltage V_R is now sum of V_{o1} and V_{o2} unlike in the Foster-Seeley case, where it was the difference of V_{o2} and V_{o1}. Thus,

$$V_R = |V_{o1}| + |V_{o2}|.$$

The output voltage V_o is taken across the terminal t_1 and t_2. From the circuit diagram, we have,

$$V_o = V_{t1,t2} = |V_{t,t1}| - |V_{t,t2}|$$

$$= |V_{o2}| - \left|\frac{V_R}{2}\right|$$

But,

$$V_R = |V_{o1}| + |V_{o2}|$$

$$\therefore \quad V_o = V_{o2} - \frac{V_{o1} + V_{o2}}{2}$$

$$V_o = \frac{V_{o2} - V_{o1}}{2}$$

This output is only half of the output given by Foster-Seeley discriminator. Thus the ratio detector has exactly same behaviour except that its output is reduced.

3.24.1 Merits of the Ratio Detector

The circuit has an advantage over the Foster-Seeley discriminator that it does not respond to the amplitude variations simultaneously present in the FM input signal. The effect of amplitude fluctuations (AM) is suppressed by shunting the capacitor C with R and introducing the two resistances R_1 of a suitable value in the circuit. The reactance of C at the lowest modulation frequency must be small as compared to R_2. Then the capacitance C bypasses the undesired amplitude fluctuations but not the amplitude variations that are generated due to FM.

3.24.2 Demerits of the Ratio Detector

The ratio detector may not tolerate the long-period variation in signal strength. This requires an AGC signal. The dc voltage across C may be used as an AGC control voltage.

3.25 FM RECEIVER

The function of the FM receiver is to intercept the FM signal incoming from an FM transmitter, and then recover the original modulating signal. Given below figure shows the block diagram of an FM broadcast receiver which is similar to an AM superheterodyne receiver.

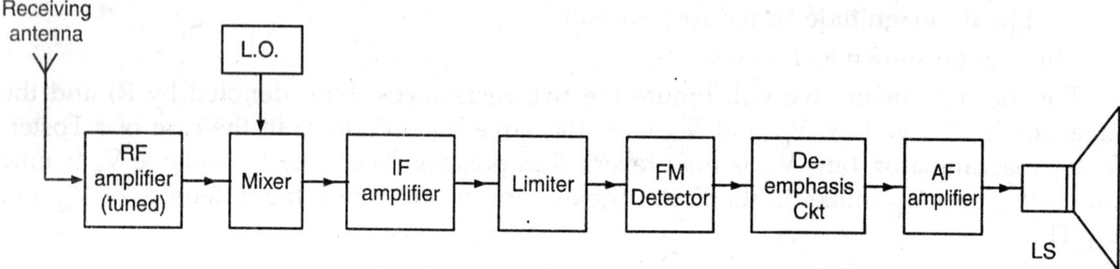

Fig. 3.23 *Block Diagram of FM Receiver*

The stages of the block diagram are described as follows:

3.26 RF AMPLIFIER

This stage amplifies the received radio signal. FM uses radio frequency ranging from 40 MHz to 1 GHz for various application such as FM broadcasting, television sound transmission, police radio, military system etc.

The bandwidth needed for an RF amplifier may also be large (150 kHz). The RF amplifier also reject the image signal, as in AM receivers.

3.27 FREQUENCY MIXER AND LOCAL OSCILLATOR

Separate active devices are used for the mixer and local oscillator as the frequency involved is in VHF and UHF range. The IF frequency in an FM receiver is much higher than AM. The FM broadcast receives use 10.7 MHz as intermediate frequency.

3.28 IF AMPLIFIER

This stage amplifies the intermediate frequency signals. It comprises of multistage double-tuned or stagger-tuned amplifiers to provide a high gain and a high overall bandwidth (150 kHz). This stage is responsible for sensitivity and selectivity of the receiver.

3.29 LIMITER

The limiter keeps the IF amplifier output voltage constant to a pre-determined value and removes all amplitude fluctuations due to noise, or any other interference. This is essential because the FM detector needs a constant amplitude FM voltage at its input for a satisfactory operation. The limiter may be ignored if the ratio detector has been used. Simple diodes and amplifying devices are used in making a limiter circuit. The leaky type bias provided at the gate by R_G-C_G combination gives the limiting action.

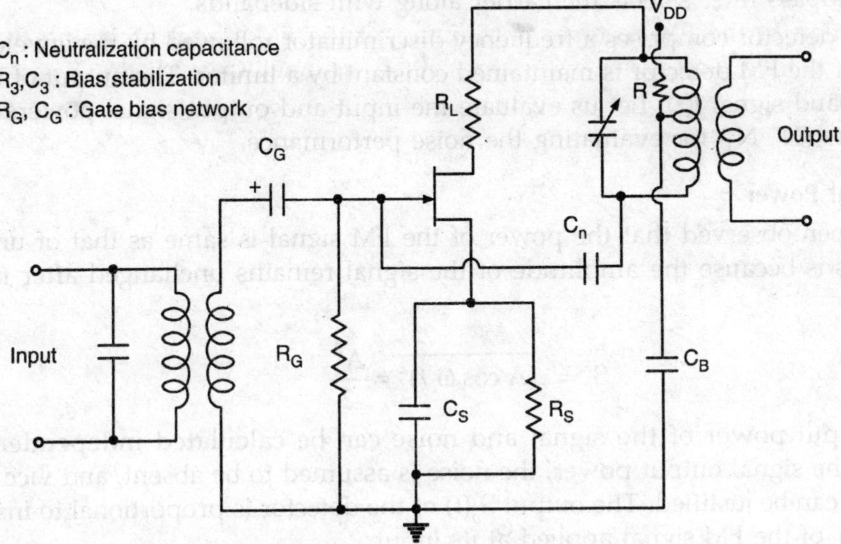

Fig. 3.24 *A Limiter Circuit*

As the input voltage increases, the bias at CS increases in such a way that the gain of the amplifier is reduced and the output voltage remains constant irrespective of variations in input voltage amplitude. Thus, the limiting action is achieved.

3.30 FM DETECTOR

It recovers the modulating signal from the IF signal. Various types of FM detector circuit have already been discussed. De-emphasis circuit does the inverse job of the pre-emphasis circuit. The high modulating frequency terms boosted by pre-emphasis are brought back to original amplitude level by de-emphasis circuit.

3.31 AF AMPLIFIER AND SPEAKER

This stage amplifies the audio frequency (AF) modulating signal recovered by the FM detector. The amplifier has wider bandwidth than that of AM receiver. The loud speaker converts the electrical signal into sound signal.

3.32 NOISE IN ANGLE MODULATED SYSTEMS FREQUENCY MODULATION

The noise performance of amplitude modulated systems as well as the noise performance of angle modulated signal for both frequency and phase can be discuss with the help of the block diagram of FM system shown in Fig. 3.25.

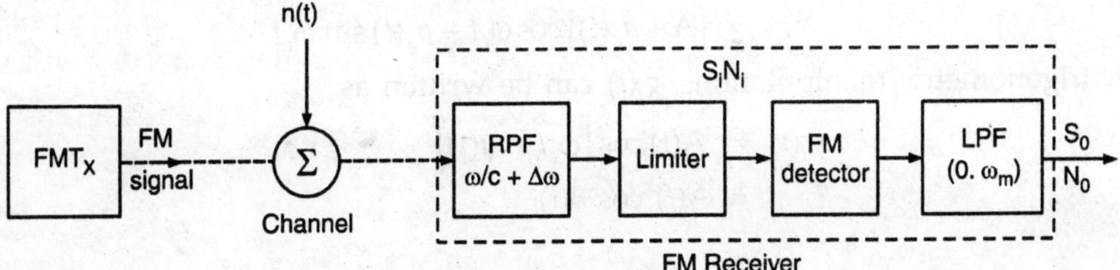

Fig. 3.25 *FM System*

The bandpass filter selects the carrier along with sidebands.

The FM detector comprises a frequency discriminator followed by the envelop detector. The input of the FM detector is maintained constant by a limiter. The output of the receiver is the baseband signal $f(t)$. Let us evaluate the input and output signal powers (S_i, S_o) and noise powers (N_i, N_o) for evaluating the noise performance.

3.32.1 Signal Power

It has been observed that the power of the FM signal is same as that of unmodulated carrier. This is because the amplitude of the signal remains unchanged after modulation. Therefore

$$S_i = \overline{(A \cos \omega_c t)^2} = \frac{A^2}{2} \quad ...(3.84)$$

The output power of the signal and noise can be calculated independently i.e. for evaluating the signal output power, the noise is assumed to be absent, and vice versa. This assumption can be justified. The output $S_o(t)$ of the detector is proportional to instantaneous frequency ω_i of the FM signal applied at its input,

$$S_o(t) \alpha \omega = K\omega_i \quad ...(3.85)$$

where K is the constant of proportionality. The angular frequency depends on the modulating signal $f(t)$ and is given as

$$\omega_i = \omega_c + K_f f(t)$$

Hence,

$$S_o(t) = K\omega_c + KK_f f(t)$$

in which the useful message signal is $KK_f f(t)$. The signal power due to this message term is

$$S_o = K^2 K_f^2 \overline{f^2(t)} \quad ...(3.86)$$

3.32.2 Noise Power

The noise power is evaluated for the baseband width i.e. bandwidth of the message signal $f(t)$ which is $2\omega_m$. The input noise power is same as given

$$N_i = \eta f_m \quad ...(3.87)$$

The output noise power is computed assuming the message signal term $f(t)$ to be zero. Therefore, input $g_i(t)$ in the detector will have a carrier term plus noise term.

$$g_i(t) = A \cos \omega_c(t) + n_i(t)$$

Substituting quadrature components of $n_i(t)$ we get

$$g_i(t) = A \cos \omega_c t + n_c(t) \cos \omega_c t - n_s(t) \sin \omega_c t$$

$$= [A + n_c(t)] \cos \omega_c t - n_s(t) \sin \omega_c t$$

After trigonometric manipulations, $g_i(t)$ can be written as

$$g_i(t) = A(t) \cos\{\omega_c t + \psi(t)\} \quad ...(3.88)$$

$$= A(t) \cos \theta(t) \quad ...(3.89)$$

where

$$A(t) = \sqrt{[A + n_c(t)]^2 + n_s^2(t)} \quad ...(3.90)$$

$$\psi(t) = \tan^{-1}\left\{\frac{n_s(t)}{A + n_c(t)}\right\} \quad ...(3.91)$$

Obviously,

$$\theta(t) = \omega_c(t) + \psi(t) \quad ...(3.92)$$

Since, the output of the FM detector depends on the instantaneous frequency ω_i which is $\frac{d}{dt}\{\theta(t)\}$, hence $A(t)$ is not of our interest, we are interested only in $\theta(t)$ and $\psi(t)$ which are responsible for the FM detector output. Again the analysis, depends on the fact whether the noise signal is small or large in comparison to the carrier amplitude. Now consider two cases separately, viz.

(*i*) $A \gg n_c(t)$ and $n_s(t)$ (small noise case)

(*ii*) $A \ll n_c(t)$ and $n_s(t)$ (large noise case)

3.32.3 Small Noise Case

In this case $A \gg n_c(t)$ and $n_s(t)$ and hence Eqn. (3.91) becomes

$$\psi(t) \tan \simeq \tan^{-1}\left[\frac{n_s(t)}{A}\right] \quad ...(3.93)$$

It is well known that for a small angle ϕ, $\tan \phi = \phi$, hence,

$$\psi(t) \simeq \frac{n_s(t)}{A} \quad ...(3.94)$$

The corresponding value of $\theta(t)$ is given as, from Eqn. (3.92)

$$\theta(t) = \omega_c(t) + \frac{n_s(t)}{A} \quad ...(3.95)$$

The instantaneous frequency is given as

$$\omega_i = \frac{d}{dt}[\theta(t)] = \frac{d}{dt}\left[\omega_c(t) + \frac{n_s(t)}{A}\right]$$

$$= \omega_c + \frac{\dot{n}_s(t)}{A} \quad ...(3.96)$$

where

$$\dot{n}_s(t) = \frac{d}{dt}\{n_s(t)\} \quad ...(3.97)$$

The detected output $e_d(t)$ is proportional to ω, hence

$$e_d(t) = K\left[\omega_c + \frac{\dot{n}_s(t)}{A}\right]$$

$$= K\omega_c + \frac{K\dot{n}_s(t)}{A}$$

The noise part of this detected output is given as

$$n_o(t) = \frac{K\dot{n}_s(t)}{A} \quad ...(3.98)$$

The power density spectrum of $n_o(t)$ is given by

$$S_{n_o}(\omega) = \frac{K^2}{A^2} S_{\dot{n}_s}(\omega) \quad ...(3.99)$$

where $S_{\dot{n}_s}$ is the power density spectrum of $\dot{n}_s(t)$. Assume that $n_s(t)$ has the power density spectrum $S_{n_s}(\omega)$, then its derivatives $\dot{n}_s(t)$ will have its power density spectrum given by

$$S_{\dot{n}_s}(\omega) = \omega^2 S_{n_s}(\omega) \quad ...(3.100)$$

By putting this value in Eqn. (3.100) we get

$$S_{n_o}(\omega) = \frac{K^2\omega^2}{A^2} S_{n_s}(\omega) \quad ...(3.101)$$

The detected output is bandlimited to ω_o and hence Eqn. (3.101) holds only for $|\omega| \le \omega_m$, and is zero elsewhere. The value of $S_{n_s}(\omega)$ for a while noise is similar to the DSB-SC case and is given as follows

$$S_{n_s}(\omega) = \eta, \quad |\omega| \le \omega_m$$

Substituting this value in Eqn. (3.101), we get

$$S_{n_o}(\omega) = \begin{cases} \frac{K^2\omega^2\eta}{A^2}, & |\omega| \le \omega_m \\ 0 & \text{elsewhere} \end{cases} \quad ...(3.102)$$

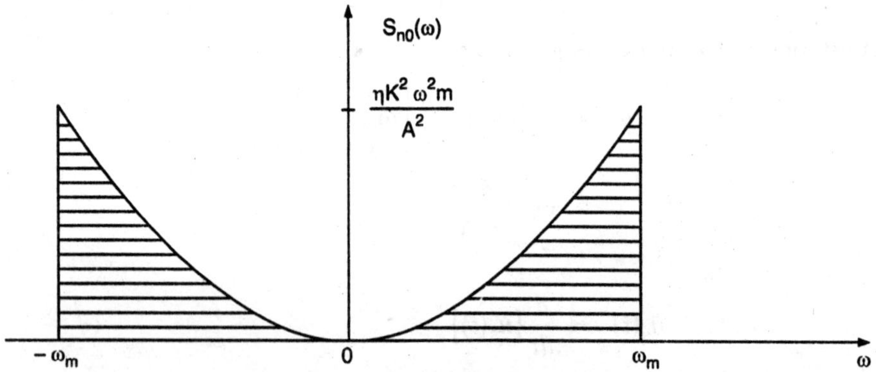

Fig. 3.26 *Noise Power Density Spectrum at the Discriminator Output*

The spectrum $S_{n_o}(\omega)$ is shown in Fig. 3.26. It is evident that the spectrum is a parabolic function of frequency (ω).

The output noise power N_o is given by

$$N_o = \frac{1}{\pi}\int_0^{\omega_m} S_{n_o}(\omega)\, d\omega = \frac{K^2\eta}{\pi A^2}\int_0^{\omega_m} \omega^2\, d\omega$$

$$= \frac{\eta K^2 \omega_m^2}{3\pi A^2} \qquad \text{...(3.103)}$$

3.32.4 Noise-quieting Effect

It is revealed from Eqn. (6.103) that the output noise power of the FM detector varies inversely with carrier power ($A^2/2$). This decrease in the noise power with an increase in carrier power is known as noise-quieting effect of the carrier.

The output signal to noise ratio (S_o/N_o) is obtained by Eqns. (3.86) and (3.103)

$$\frac{S_o}{N_o} = \frac{3\pi K_f^2 \overline{f^2(t)} A^2}{\eta\omega_m^3} \qquad \text{...(3.104)}$$

3.32.5 Figure of Merit γ

The figure of merit γ for the FM system may be obtained by using Eqns. (3.84), (3.87) and (3.104)

$$\gamma_{FM} = \frac{(S_o/N_o)}{(S_i/N_i)} = \frac{3K_f^2 \overline{f^2(t)}\,(2\pi f_m)}{\omega_m^3}$$

$$= \frac{3K_f^2 \overline{f^2(t)}}{\omega_m^3} \qquad \text{...(3.105)}$$

3.32.6 Single-tone Modulator

EXAMPLE

A single-tone modulating signal $E_m \cos \omega_m t$, frequency modulates a carrier $A \cos \omega_c t$. Show that

(*a*) The detector output signal to noise ratio is proportional to the square of bandwidth of the M signal.

(*b*) The figure of merit γ is given by

$$\gamma_{FM} = \left(\frac{3}{2}\right) m_f^2 \qquad \text{...(3.106)}$$

Solution:

(*a*)
$$f(t) = E_m \cos \omega_m t$$
$$K_f f(t) = K_f E_m \cos \omega_m t$$

The mean square value of both the side yields

$$K_f^2 \overline{f^2(t)} = K_f^2 \frac{E_m^2}{2} \quad ...(3.107)$$

Frof above equation

$$K_f^2 E_m^2 = (\Delta\omega)^2 \quad ...(3.108)$$

and hence

$$K_f^2 \overline{f^2(t)} = \frac{(\Delta\omega)^2}{2} \quad ...(3.109)$$

Substituting this value in Eqn. (3.104), we get,

$$\frac{S_o}{N_o} = \frac{3\pi(\Delta\omega)^2 A^2}{2\eta\omega_n^3}$$

$$= 3\pi\left(\frac{\Delta\omega}{\omega_m}\right)^2 \frac{(A^2/2)}{2\pi f_m \eta} \quad ...(3.110)$$

This shows that (S_o/N_o) is proportional to $(\Delta\omega)^2$. Now the bandwidth of FM is twice the deviation $(\Delta\omega)$, and hence, the (S_o/N_o) ratio is proportional to the square of bandwidth of the FM signal.

(*b*) By definition $\frac{\Delta\omega}{\omega_m} = m_f$, hence Eqn. (3.110) becomes

$$\frac{S_o}{N_o} = \frac{3}{2} m_f^2 \frac{(A^2/2)}{\eta f_m}$$

$$= \frac{3}{2} m_f^2 \left(\frac{S_i}{N_i}\right)$$

Therefore,

$$\gamma_{FM} = \frac{(S_o/N_o)}{(S_i/N_i)} = \frac{3}{2} m_f^2 \quad ...(3.111)$$

3.32.7 Noise Performance of FM as Compared to AM

It is worthwhile to compare the noise performance of FM with that of AM. The Comparison may be conveniently made by considering the single-tone modulation. From Eqn. (3.111)

$$\gamma_{FM} = \frac{3}{2}m_f^2$$

For AM with a large carrier, we have

$$\gamma_{AM} = \frac{m_a^2}{2+m_a^2}$$

The best noise performance is obtained for $m_a = 1$,

$$\gamma_{AM} = \frac{1}{3}$$

Therefore

$$\frac{\gamma_{FM}}{\gamma_{FM}} = \frac{1}{2}m_f^2 \qquad \text{...(3.112)}$$

The performance of FM and AM will be same if the right-hand side unity,

$$\frac{1}{2}m_f^2 = 1$$

$$m_f = \frac{\sqrt{2}}{3} = 0.5 \qquad \text{...(3.113)}$$

This shows that FM has no improvement over AM when $m_f \geq 0.5$. Thus $m_f = 0.5$ may be considered as a transition point between the wideband and narrowband FM. It should be recalled that narrowband FM and AM are identical in bandwidth. This implies that the noise improvement depends on the bandwidth. Since narrowband FM and AM have the bandwidth, there is no improvement in the noise performance of narrowband FM over AM. The improvement in noise performance is at the expense of the bandwidth as,

Bandwidth of AM signal,

$$W_{FM} = 2(\Delta\omega)$$

$$= 2m_f\omega_m$$

and

Bandwidth of AM signal,

$$W_{AM} = 2\omega_m$$

Hence

$$\frac{W_{FM}}{W_{AM}} = m_f$$

Substituting this value of m_f in Eqn. (3.112), we get

$$\frac{\gamma_{FM}}{\gamma_{AM}} = \frac{1}{2}\left(\frac{W_{FM}}{W_{AM}}\right)^2 \qquad \text{...(3.114)}$$

This reveals that if a bandwidth ratio is increased by a factor 2, the $\left(\frac{\gamma_{FM}}{\gamma_{AM}}\right)$ increases by a factor 4 i.e. 6 dB, which verifies the fact that noise performance is improved by increasing the bandwidth. This exchange of bandwidth and noise performance studied in detail by Hertley and Shannon, is feasible in FM and not in AM, because AM has a fixed bandwidth.

3.32.8 Phasor Representation of FM Noise

A qualitative study of FM noise can be done by its phasor diagram representation. Assume that the message signal is absent i.e. $f(t) = 0$ and only $n_i(t)$ is associating the carrier at the input of the detector. Then the input of the detector is given as

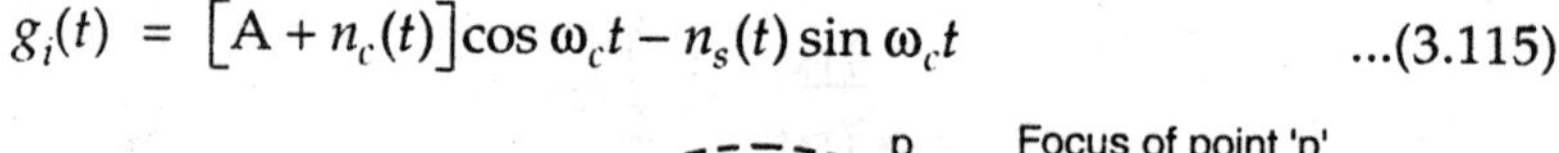

$$g_i(t) = \left[A + n_c(t)\right]\cos\omega_c t - n_s(t)\sin\omega_c t \qquad ...(3.115)$$

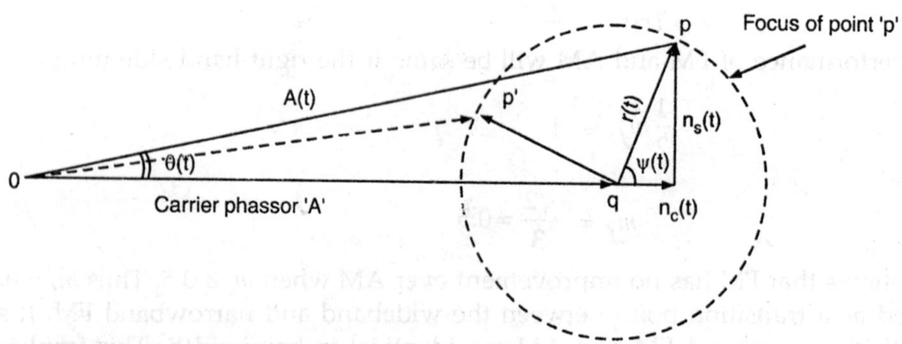

Fig. 3.27 *Phasor Diagram of FM Noise*

Obviously, the term $n_c(t)$ is in phase with carrier term A and the term $n_s(t)$, in phase quadrature with $n_c(t)$. The phasor diagram is shown in Fig. 3.27. The coordinate system is assumed to be rotating counter clockwise with a frequency ω_c. The resultant of quadrature components of the noise is $r(t)$. Now from Eqns. (3.88) and (3.89).

$$\begin{aligned} g_i(t) &= A(t)\cos\left[\omega_c t + \psi(t)\right] \\ &= A(t)\cos\theta(t) \end{aligned} \qquad ...(3.116)$$

where A(t) is the resultant of the carrier vector A and the noise vector $r(t)$.$\psi(t)$ is the angle that A(t) makes with the carrier vector A. Noise parameters $r(t)$ and $\psi(t)$ vary randomly with time, accordingly the noise vector $r(t)$ rotates around the point q. The rotation path is locus of points p and is random in nature. For convenience if we assume $r(t)$ constant, the path will be circular as shown by dotted curve. As this point p rotates around the point q, the resultant vector A(t) and its phase $\theta(t)$ also varies. One such case is shown by point p' for which both A(t) and $\theta(t)$ is less as compared to point p. For a low-noise case, $r(t) \ll A$ and hence, locus of p is restricted near the point q. But as the noise signal $r(t)$ increases, the locus researches near the origin and even may sweep the origin 'O' if noise vector $r(t)$ exceeds the carrier amplitude A which is large noise case.

A interfering noise vector $r(t)$ modulates the carrier both in amplitude and in angle. The amplitude variation in A(t) does not produce any adverse effect in the FM receiver, as it is not responded to by the FM receiver due to the features of limiter, or ratio detector. However, the angle fluctuation produces a phase deviation of ± $\psi(t)$ about the carrier phasor. This phasor is responded by FM receiver. As the noise signal $r(t)$ increases in comparison to carrier, the phase deviation $\psi(t)$ also increases. This increases the distortion produced by noise. This type of distortion may be caused by any type of interfering signal accompany the FM signal.

3.32.9 Threshold Effect in FM

When the carrier to noise ratio is slightly less than unity, the frequency of spike generation is small, and each spice produces individual clicking sound in the receiver. But, when the carrier to noise ratio is further decreased so that the ratio is moderately less than unity, the spikes are generated rapidly and the clicks merge into a sputtering sound. This phenomenon is known as *threshold effect* in FM. The minimum carrier to noise ratio for which the FM noise improvement is not deteriorated significantly as compared to the small-noise case is defined as *threshold*. If the carrier to noise is less than the threshold value, the sputter effect causes distortion in the FM receiver and the desired message signal cannot be resolved from the noise. The threshold effect is more severe in FM than AM.

3.33 GENERATION OF SPIKES AND TRIPLET PULSES

A positive or negative impulse (spike) is generated at the discriminator output under large noise condition; whereas a pulse-triplet (combination of three pulses) is generated under low-noise condition. A pulse-triplet contributes very small noise energy at the detector output as compared to a spike.

3.33.1 Pulse Triplet Generation

The phasor diagram representation for low noise case, and a triplet path is shown in Fig. 3.28. The end point of the resultant vector $r(t)$ rotates near the point q, and as the

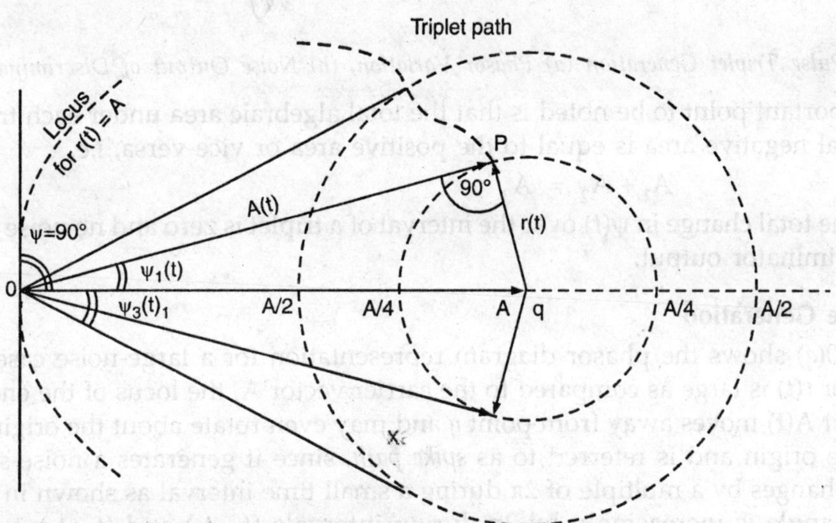

Fig. 3.28 *Phase Deviation in Low Noise Case*

noise increase, the end point moves away from the point q. The locus of the end point of A(t) may reach near the origin 'O', but never encircle the origin. This locus is known as *triplet path,* because this type of path causes a triplet pulse at the FM discriminator output. When the noise vector $r(t)$ is slightly less than the carrier vector A, $\psi(t)$ varies from π to $-\pi$ or vice-versa as shown in Fig. 3.29(a). The waveform $\psi(t)$ from a pulse doublet for each phase reversal over a small interval, e.g., t_1 to t_2, and t_3 to t_4. The output of the discriminator is proportional to the instantaneous frequency ω_i (= $d\psi/dt$) and is obtained by differentiating the waveform of Fig. 3.29(a). This differentiated waveform ($d\psi/dt$) generates a pulse triplet for each pulse doublet, as shown in Fig. 3.29(b).

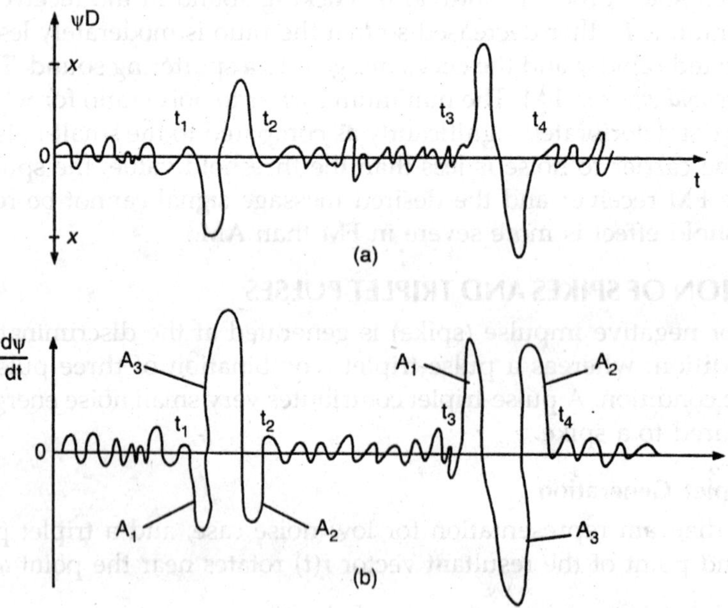

Fig. 3.29 *Pulse Triplet Generation (a) Phasor Variation, (b) Noise Output of Discriminator (triplet)*

The important point to be noted is that the total algebraic area under each triplet is zero because total negative area is equal to the positive area or vice-versa, i.e.

$$A_1 + A_2 = A_3$$

Thus, the total change in $\psi(t)$ over the interval of a triplet is zero and no noise is produced at the discriminator output.

3.33.2 Spike Generation

Fig. 3.30(a) shows the phasor diagram representation for a large-noise case. When the noise phasor $r(t)$ is large as compared to the carrier vector A, the locus of the end point P of the resultant A(t) moves away from point q and may even rotate about the origin. The locus incircles the origin and is referred to as *spike path,* since it generates a noise-spike. Phase angle $\psi(t)$ changes by a multiple of 2π during a small time interval as shown in Fig. 3.30(b). The phase angle is increasingly by 2π during intervals (t_1, t_2) and (t_3, t_4); whereas it is decreasing by 2π during the interval (t_5, t_6).

Note that phase angle $\psi(t)$ changes in any one direction over an interval (either increasing or decreasing), and not positive to negative or vice versa as in the low-noise case. Hence, an unidirectional impulse appears over an interval at the FM discriminator output which is obvious from the waveform shown in Fig. 3.30(*c*). These noise impulses shown in Fig. 3.30(*c*) are called *noise-spikes*. When $A(t)$ rotates counterclockwise, the phase $\psi(t)$ changes by -2π and generates a positive spike; and if $A(t)$ rotates clockwise the phase $\psi(t)$ changes by -2π generating a positive spike.

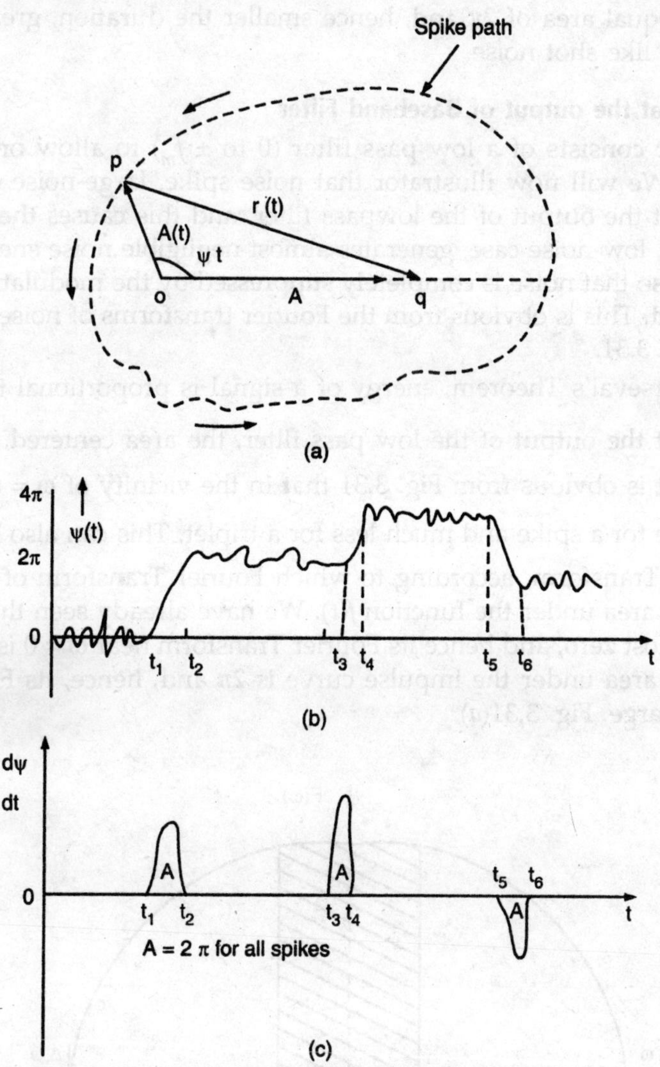

Fig. 3.30 *Spike Generation (a) Phasor Diagram; (b) Phase Angle Variation; (c) Noise-output of Discriminator*

The area under each spike in Fig. 3.30(c) is 2π. This can be seen by integrating the first impulse occurring for the interval (t_1, t_2).

$$A = \int_{t_1}^{t_2} \frac{d\psi}{dt} dt$$

$$= [\psi]_1^2 = 2\pi \quad ...(3.117)$$

Each spike has equal area of 2π and, hence smaller the duration, greater is its height. These spikes behave like shot noise.

3.33.3 Noise Energy at the output of Baseband Filter

The FM detector consists of a low pass filter (0 to $\pm f_m$) to allow only the baseband modulating signal. We will now illustrator that noise spike, large-noise case, generates a large noise energy at the output of the lowpass filter, and this causes the threshold effect. But the phase-triplet, low-noise case, generates almost negligible noise energy at the output of a baseband filter; so that noise is completely suppressed by the modulating signal and no threshold is observed. This is obvious from the Fourier transforms of noise-spike and noise-triplet shown in Fig. 3.31.

According to Parseval's Theorem, energy of a signal is proportional to the area under its $[f(\omega)]^2$ curve. At the output of the low pass filter, the area centered in the range 0 to $\pm\,\omega_m$ is of interest. It is obvious from Fig. 3.31 that in the vicinity of $\omega = 0$, the area under $|F(\omega)|$ is much more for a spike and much less for a triplet. This can also be proved from a property of Fourier Transform, according to which Fourier Transform of a function $f(t)$ at $\omega = 0$ is equal to the area under the function $f(t)$. We have already seen that the area under a triplet curve is almost zero, and hence its Fourier Transform near $\omega = 0$ is negligibly small, Fig. 3.31(b). But the area under the impulse curve is 2π and, hence, its Fourier Transform near $\omega = 0$ is quite large, Fig. 3.31(a).

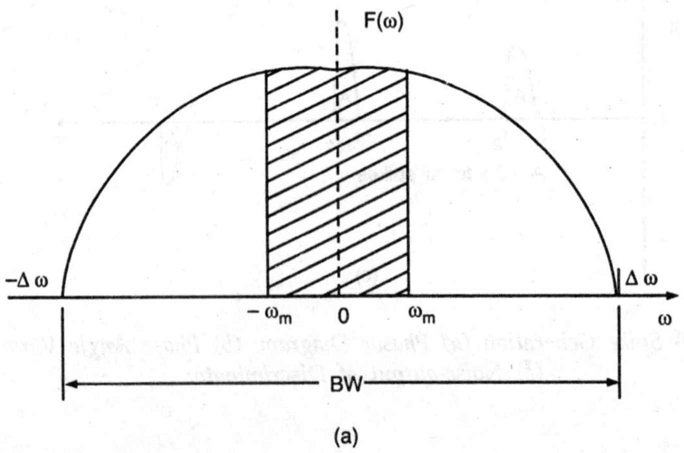

(a)

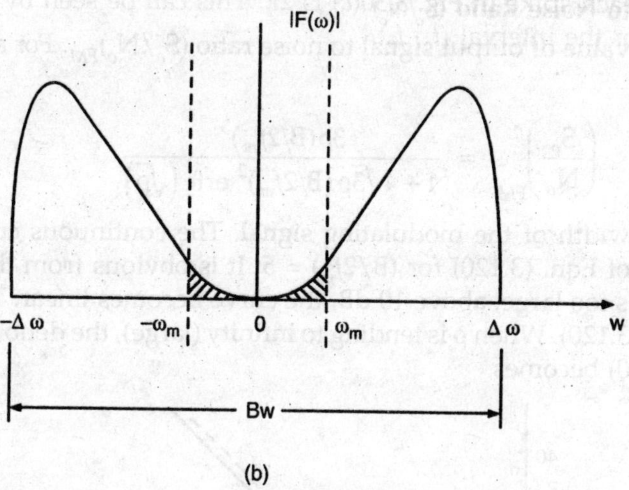

Fig. 3.31 *Fourier Transform of Noise at the Output of the Discriminator (a) Noise Spike Case; (b) Noise Triple Case*

Thus, we arrive at the conclusion that a spike produces large energy at the output of an FM detector causing threshold, whereas, a triplet produces negligible energy, avoiding threshold effect.

3.33.4 Spike Duration

Since the spectrum in Fig. 3.31(*a*) extends over a 0 to ± Δf the spike duration will be given by the reciprocal of (Δf), i.e. ($1/\Delta f$).

3.33.5 Frequency of Spike Occurrence (Unmodulated Carrier)

If a carrier frequency is in the centre of a band the noise spectral components are symmetrically placed on both sides of carrier frequency, and the average number of positive spikes per second f_+ is same as the average number of negative going spikes per second f_-. For an ideal IF filter of bandwidth B, centered at a carrier frequency f_c, the frequency of spike generation, f is given by

$$f = f_+ = f_-$$

$$= \frac{B}{2\sqrt{3}} \operatorname{erfc}\left(\sqrt{\rho}\right) \quad \text{...(3.118)}$$

where ρ is the carrier power to noise power ratio given by

$$\rho = \frac{A^2}{2B\eta} \quad \text{...(3.119)}$$

and erfc denotes the complementary error function. $\eta/2$ is the noise spectral density. Eqn. (3.119) reveals that the frequency of spike occurrence is proportional to the bandwidth B.

3.33.6 Output Signal to Noise Ratio (S_o/N_o)

The approximate value of output signal to noise ratio $(S_o/N_o)_{FM}$. For a large noise case is given as

$$\left(\frac{S_o}{N_o}\right)_{FM} = \frac{3\rho(B/2f_m)^3}{1+4\sqrt{3}\rho\,(B/2f_m)^2\ \text{erfc}\left(\sqrt{\rho}\right)} \qquad \text{...(3.120)}$$

where f_m is the bandwidth of the modulating signal. The continuous curve *'a'* in Fig. 3.32 shows a typical plot of Eqn. (3.120) for $(B/2f_m) = 5$. It is obvious from the curve that when carrier to noise ratio is too large, above 10 dB, the curve becomes linear. The linearity can be explained from Eqn. (3.120). When ρ is tending to infinity (large), the denominator approaches unity, and Eqn. (3.120) becomes

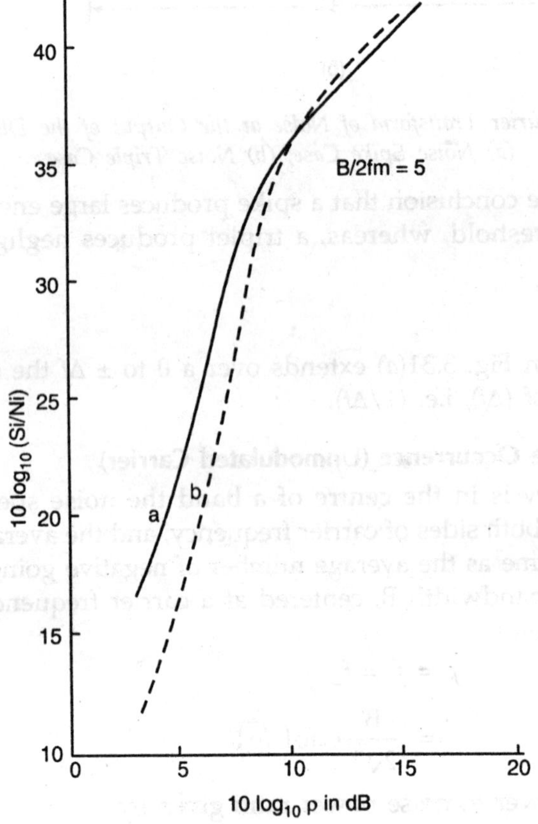

Fig. 3.32

$$\left(\frac{S_o}{N_o}\right)_{FM} = 3\rho\left(\frac{B}{2f_m}\right)^3 \qquad \text{...(3.121)}$$

Accordingly, (S_o/N_o) has a linear variation with ρ and threshold effect does not occur. When ρ is less than 10 dB, i.e. carrier to noise ratio is less than 10, non-linearity in the curve *'a'* is observed.

3.33.7 Effect of Modulation

Curve *'b'* in Fig. 3.32 shows the effect of the single-tone modulation. In this case, non-linear variation is observed for ρ less than 11 dB which is very close to experimentally observed threshold value of 13 dB. Below this threshold value of ρ, (S_o/N_o) falls more rapidly than the unmodulated curve, and the *threshold effect is more severe*. But above this threshold value of ρ, the noise performance is better than the unmodulated case.

In practice, the threshold effect occurs when input carrier to noise ratio is below 13 dB (i.e. 20). If ρ is kept above 20, the average number of spikes generated per second is very small and the threshold effect may be avoided. Hence, for avoiding the FM threshold, $\rho \geq 20$. From Eqn. 3.119, we have

$$\frac{A^2}{2\eta B} \geq 20$$

or

$$\frac{A^2}{2} \geq 20\eta B \qquad ...(3.122)$$

Hence, the FM threshold effect may be avoided by keeping the carrier power $\left(\frac{A^2}{2}\right)$ greater than $20\eta B$.

When a carrier is modulated by a single-tone modulating frequency, and $\rho < 20$, then the baseband signal appearing at the discriminator output is accompanied by noise-spikes. The rate of spike generation is increases after modulation. Negative spikes appear most commonly in positive extremity, whereas, positive spikes appear in negative extremity as shown in Fig. 3.33.

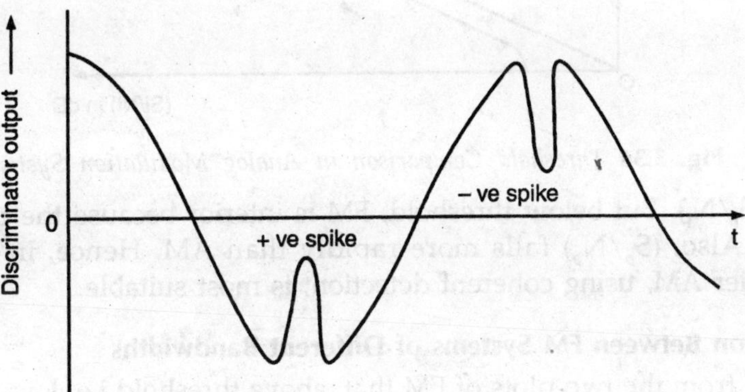

Fig. 3.33 *Spikes in Single-tone Modulating Signal*

3.34 COMPARISON OF THRESHOLD IN ANALOG MODULATED SYSTEMS

It is worth comparing the noise performance and threshold in various analog modulation systems. The comparison has been shown in Fig. 3.34. The figure shows the plots (S_o/N_o) vs (S_i/N_i) based on the expressions of the figure of merit γ derived earlier in this chapter for different modulation systems. It should be kept in mind that the curved showing threshold at lower (S_i/N_i) are better because in this case the threshold can be avoided with comparatively weaker signals.

The following compulsions can be drawn from the plots shown in Fig. 3.34.

3.34.1 Comparison between AM-Systems

The SSB-SC and the AM-SC systems have no threshold effect, which is obvious from the complete linearity of the curve; whereas, AM, envelop detector, exhibits a threshold effect for low (S_i/N_i). Hence noise performance of the suppressed AM system is better than AM using envelop detector. Since for the same (S_i/N_i), the former provides a higher (S_o/N_o). Thus, suppressed carrier AM systems using coherent detection is superior to AM that uses envelope detection from the noise performance point of view; SSB-SC having the lowest bandwidth is optimum.

3.34.2 Comparison Between AM and FM

FM is superior to AM above threshold since the former provides more (S_o/N_o) for

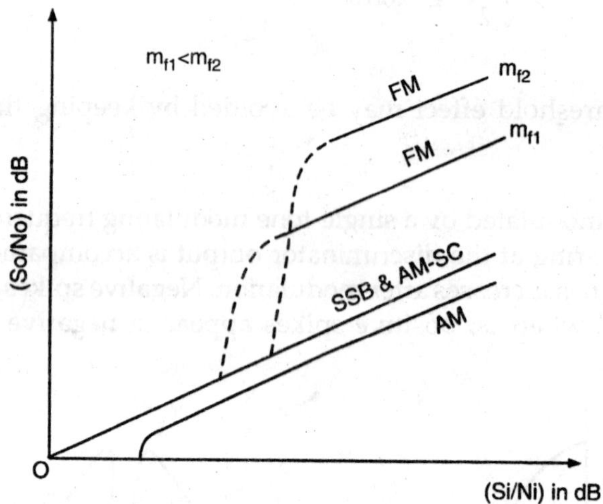

Fig. 3.34 *Threshold Comparison in Analog Modulation System*

an equivalent (S_i/N_i). But below threshold, FM is interior because the threshold occurs at higher (S_i/N_i). Also, (S_o/N_o) falls more rapidly than AM. Hence, in large-noise cases, suppressed carrier AM, using coherent detection, is most suitable.

3.34.3 Comparison Between FM Systems of Different Bandwidths

It is obvious from the two plots of FM that, above threshold i.e. low-noise, system with a larger modulation index m_{f2} (large bandwidth) is superior than the one with a smaller modulation index m_{f1} (lower bandwidth). However, below threshold, the system with a larger bandwidth exhibits faster decay in (S_o/N_o) and threshold occurs at higher (S_i/N_i). Hence it is inferior to the lower bandwidth FM.

From the above discussion we come at the ultimate conclusion that, for large-noise systems, below threshold, coherent detection is most suitable. On the contrary, for small-noise systems, above threshold, the wideband FM is most appropriate. Narrowband FM is equivalent to AM.

3.35 NOISE IN PHASE MODULATION

The phase modulation is equivalent to FM, except that the modulating signal $f(t)$ is first differentiated, and then it frequency modulates the carrier. If we use an FM detector to demodulate a phase-modulated signal, the modulated signal will be a derivative of $f(t)$, i.e., $\frac{df}{dt}$. The original modulating signal can be obtained by integrating the output of the FM demodulator. Fig. 3.35 shows the schematic diagram of phase modulation, and demodulation using a frequency modulator.

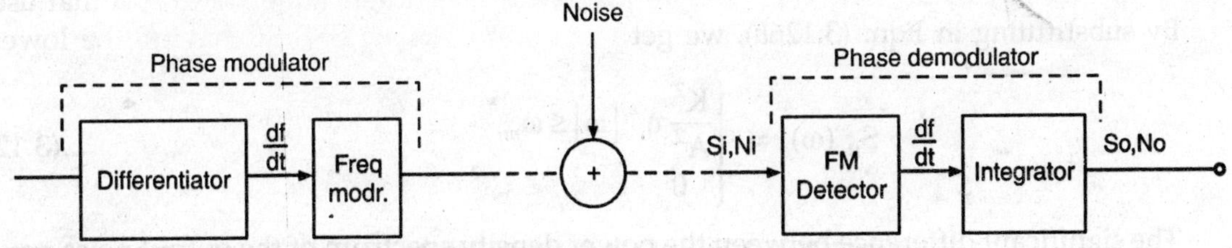

Fig. 3.35 *Phase Modulation and Demodulation*

The differentiate circuit preceding the frequency modulator is equivalent to a pre-emphasis circuit, whereas, the integrator is equivalent to a de-emphasis circuit. These circuits are used to improve the noise performance in FM receiver. Thus, a frequency modulation system equipped with the fertility of pre-emphasis and de-emphasis is equivalent to a phase-modulation system. Obviously, FM has better noise performance as compared to phase modulation. This can be seen by evaluating the figure of merit for phase modulation system.

3.35.1 Figure of Merit (γ_{PM})

The figure of merit γ_{PM} for phase modulation can be derived by adopting the procedure used for frequency modulation. The expressions for the input signal and noise power is identical to the FM case, and is given as Eqn. 3.84 and 3.87.

$$S_i = \frac{A^2}{2} \quad \text{and} \quad N_i = \eta f_m \qquad \text{...(3.123)}$$

The signal power at the FM detector output is proportional of $\frac{df}{dt}$ and, after integrator circuit, the output becomes $f(t)$. Hence, the output signal power is obtained by using Eqn. (3.86).

$$S_o = K^2 K_p^2 \overline{f^2(t)} \qquad \text{...(3.124)}$$

Similarly, the noise signal at the output of FM detector is similar to Eqn. (3.98) and is given as

$$n_0(t) = \frac{K\dot{n}_s(t)}{A}$$

After integration (assumed ideal), the noise power at the output of the PM detector is

$$n_0(t) = \frac{Kn_s(t)}{A} \qquad ...(3.125)$$

The power density spectrum of $n_0(t)$ is given as

$$S_{n_o}(\omega) = \frac{K^2}{A^2} S_{n_s}(\omega) \qquad ...(3.126)$$

$S_{n_s}(\omega)$ for a white noise is given as

$$S_{n_s}(\omega) = \eta, \quad |\omega| \leq \omega_m$$

By substituting in Eqn. (3.126*b*), we get

$$S_{n_o}(\omega) = \begin{cases} \frac{K^2}{A^2}\eta & |\omega| \leq \omega_m \\ 0 \end{cases} \qquad ...(3.127)$$

The significant difference between the power density spectrum of the output noise power in FM and PM may be seen by comparing Eqns. (3.122) and (3.102). In FM the power density has a quadratic relation with frequency, whereas in PM, it is independent of frequency as shown in Fig. 3.36.

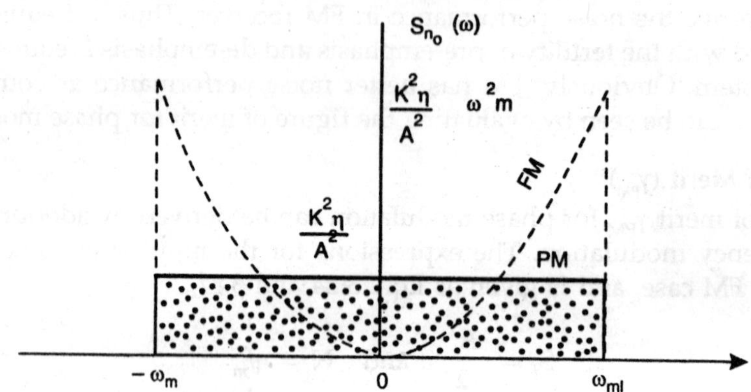

Fig. 3.36

The output noise power N_o is given by

$$N_o = \frac{1}{\pi}\int_0^{\omega_m} S_{n_o}(\omega)\, d\omega$$

$$= \frac{1}{\pi}\int_0^{\omega_m} \frac{K^2}{A^2}\eta\, d\omega$$

$$= \frac{2K^2\eta f_m}{A^2} \qquad ...(3.128)$$

From Eqns. (3.124) and (3.128), the output signal to noise ratio is given by

$$\left(\frac{S_o}{N_o}\right)_{PM} = \frac{A^2K_p^2}{2\eta f_m}\overline{f^2(t)} \qquad ...(3.129)$$

Using Eqns. (3.123) and (3.129) we get the figure of merit γ_{PM} as given below

$$\gamma_{PM} = K_p^2\,\overline{f^2(t)} \qquad ...(3.130)$$

EXAMPLE

A single-tone modulating signal $f(t) = E_m \cos \omega_m t$ phase modulates a carrier $A \cos \omega_c t$. Show that figure of merit γ_{PM} is given by

$$\gamma_{PM} = \frac{1}{2}m_f^2 \qquad ...(3.131)$$

Solution:

$$f(t) = E_m \cos \omega_m t$$

$$K_p f(t) = K_p E_m \cos \omega_m t$$

M.S. value of both sides are given as

$$K_p^2\,\overline{f^2(t)} = K_p^2\frac{E_m^2}{2} \qquad ...(3.132)$$

From Eqn. (3.85) the phase deviation

$$K_p E_m = \frac{\Delta\omega}{\omega_m} = m_f$$

Hence,

$$K_p^2\,\overline{f^2(t)} = \frac{m_f^2}{2} \qquad ...(3.133)$$

Substituting Eqn. (3.133) in Eqn. (3.130), we get

$$\gamma_{PM} = \frac{1}{2}m_f^2$$

Thus, the nature of noise figure dependence on the modulation index (i.e. bandwidth) is identical to FM system.

3.35.2 Comparison of Noise Performance in Phase Modulation and Frequency Modulation

The noise performance can be compared by matching the figure of merits γ_{FM} and γ_{PM}. For convenience, let us compare the figure of merit for the single-tone case. Thus from Eqns. (3.106) and (3.131), we get

$$\frac{\gamma_{FM}}{\gamma_{PM}} = 3 \qquad ...(3.134)$$

Thus, FM has a somewhat better overall noise performance for the audio band. This is also clear from Eqn. (3.101) in which the noise power area over the band $(0, \pm\, \omega_m)$ in the vicinity of $\omega = 0$ is greater in PM than FM. However, as the frequency increases, the noise power in FM dominates over PM. A comparison of output noise power N_o for FM and PM can be made with the help of Eqns. (3.103) and (3.128). Thus

$$\frac{(N_o)_{FM}}{(N_o)_{PM}} = \frac{\omega_m^2}{3} \qquad ...(3.135)$$

The ratio of the output noise powers Eqn. (3.135) increases as square of the modulating frequency ω_m. But the ratio of figure of merits is independent of ω_m Eqn. (3.136). This is because the ratio of output signal powers also increases with the square function of ω_m. This can be seen from Eqns. (3.86) and (3.124).

$$\frac{(S_o)_{FM}}{(S_o)_{PM}} = \frac{K^2K_f^2\,\overline{f^2(t)}}{K^2K_p^2\,\overline{f^2(t)}} = \frac{K_f^2}{K_p^2}$$

For an equivalent bandwidth of FM and PM, $K_f = \omega_m K_p$, and

$$\frac{(S_o)_{FM}}{(S_o)_{PM}} = \omega_m^2 \qquad ...(3.136)$$

Thus the ratio of output signal powers increases three times faster than the ratio of noise powers. This is the reason why figure of merit in FM is better than in PM. However, in multichannel FDM system, PM offers some advantage over FM. In this system, a number of voice-channels are multiplexed using SSB-SC modulation. The sub-carriers are separated by 4 kHz. The multiplexed signal forms a composite baseband signal that is used to angle-modulate a carrier signal.

3.35.3 Comparison Between AM and FM

Let us see the comparison between amplitude modulation and frequency modulation or angle modulation from various aspects:

3.35.3.1 Noise Performance

We have seen that in the low-noise case, no threshold wideband FM has a better noise performance than AM. The greater the bandwidth, better is the noise performance. Obviously, the narrowband FM has a noise performance equivalent to AM. On the contrary, in the large-noise case, AM is better than FM because the squattering sound due to the threshold effect in FM appears at a higher (S_i/N_i) ratio than in AM.

3.35.3.2 Common Channel Interference

FM is better than AM, because of capture effect. The interfering weak signals received from the neighbouring channels are almost completely suppressed in FM, but produce distortion in AM.

3.35.3.3 Externally Generated Noise Pulses

Let us consider a situation when the FM signal is weaker than the noise pulses produced by external sources, such as ignition systems, lighting, galactic noise, etc. Such noise pulses rarely occur, but have extremely high amplitude for a small duration. The rise time of the receiver is normally larger than the pulse duration because of the finite bandwidth of the receiver. At the output of the receiver, the pulse width will increase. The pulse width becomes comparable with reciprocal of receiver bandwidth. Increase in the duration of pulse causes reduction in its peak magnitude by the same factor, since the area under the pulse remains unchanged. The FM receiver responds slightly to noise pulses because of its wideband RF and IF circuits. Also, the FM receiver has a limiter circuit which limits the amplitude of noise pulses so that it may not exceed the signal amplitude at the detector output. However, if the FM receiver is slightly mistuned, its ability to suppress noise pulses is highly reduced whereas, in AM such exact tuning is not essential.

3.35.3.4 Channel Bandwidth

The wideband FM has a larger bandwidth as compared to AM because the former produces a larger number of sidebands. In a typical broadcast system, each channel bandwidth in AM is 15 kHz, whereas, in FM, it is 150 kHz. Therefore, FM has a disadvantage over AM. In a given frequency space, FM accommodates less number of channels as compared to AM. Hence, in a given frequency space, more number of transmitters can be accommodated using AM.

3.35.3.5 Operating Carrier Frequency

FM utilizes higher carrier frequency, above 30 MHz, because of its higher bandwidth. If FM is used at the lower carrier frequencies (HF and MF band), the number of available of available channels will be too less.

3.35.3.6 Transmission Efficiency

We know that FM has more efficiency than AM. However, the efficiency of the suppressed carrier AM is better than FM, in the sense that the later can have 100% efficiency only for some specific value of the modulation index m_f, whereas former always has 100% efficiency. Thus, FM has better efficiency than AM, but better than AM-SC.

3.36 THRESHOLD IMPROVEMENT IN DISCRIMINATORS

The threshold effect is more serious in FM as compared to AM, because in the former, the signal to noise ratio (S_i/N_i) at the input of a detector, at which threshold effect starts, is higher. We have already seen that lower the threshold level (S_i/N_i), better is the system, because threshold can be avoided at a comparatively lower (S_i/N_i) ratio, and a small signal is needed to avoid threshold for an equivalent noise power. Hence, it is desirable to lower the threshold level in the FM receivers. The process of lowering the threshold level is known as *threshold improvement*, or *threshold reduction*. Now we will discuss the following two methods used for the improvement of the threshold.

1. Pre-emphasis and De-emphasis Circuits and
2. FMFB (Frequency and De-emphasis

3.36.1 Pre-emphasis and De-emphasis circuits

In the pre-emphasis and de-emphasis method, simple R-C networks are used to improve the threshold. It has been observed in Fig. 3.26 that the noise power density spectrum increases as a parabolic function with frequency. The result is that signal to noise ratio (S_o/N_o) at the output of the detector becomes very low towards the higher edge of the message-band and may cause threshold effect, in spite of the fact that (S_o/N_o) is large enough at lower edge of the message-band. The threshold effect may be avoided by improving (S_o/N_o) at the higher edge of the message band. This is done by a simple R-C network known as pre-emphasis circuit, Fig. 3.37 which boosts the signal amplitude of higher frequencies in the message-band before they modulate the carrier. The boosted signal at transmitter increases the (S_o/N_o) ratio at the detector output.

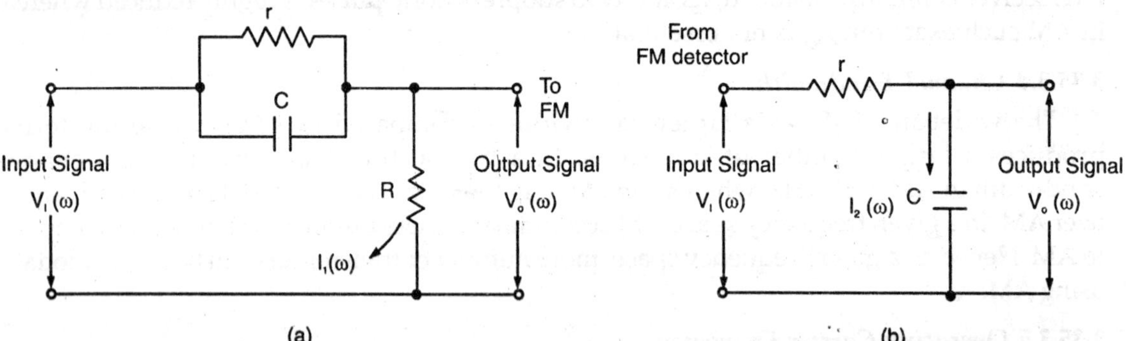

Fig. 3.37 *(a) Pre-emphasis Circuit, (b) De-emphasis Circuit*

Thus, (S_o/N_o) ratio becomes large enough to improve the threshold level over the entire message-band. However, the high frequency components of the message signal reproduced by the detector are at a raised amplitude level, and thereof, amplitude distribution of the base band is disturbed. An inverse action is needed at the discriminator output to bring back the original level of high frequency components, and restore the amplitude distribution of message-band. This is done by another R-C network known as de-emphasis circuit Fig. 3.37. The transfer function, $H_d(\omega)$ of the de-emphasis, and $H_p(\omega)$ of pre-emphasis circuits, have inverse relationship, so that their product is constant for the entire message band, i.e.

$$H_d(\omega) = \frac{K}{H_p(\omega)}$$

or

$$H_d(\omega)\, H_p(\omega) = K \qquad \text{...(3.137)}$$

The pre-emphasis circuit at the transmitter is a high pass network which behaves like a differentiation. Similarly the de-emphasis circuit at the receiver is a low-pass network which behaves like an integration. Thus, an FM system equipped with the facility of pre-emphasis and de-emphasis has a block diagram just like a phase modulation system shown in Fig. 3.35. A typical pre-emphasis and de-emphasis circuits are shown in Figs. 3.37(*a*) and 3.37(*b*) respectively.

Assuming $R \ll r$ in the pre-emphasis circuit, its transfer function is given by

$$H_p(\omega) = \frac{R}{r}\left(1 + j\frac{\omega}{\omega_1}\right) \qquad \text{...(3.138)}$$

where

$$\omega_1 = \frac{1}{rc} \qquad \text{...(3.139)}$$

Transfer function $H_p(\omega)$ is plotted in Fig. 3.38. It is obvious that the frequency components between break frequencies ω_1 and ω_2 have been boosted. The second break frequency ω_2 is given by

$$\omega_2 = \frac{1}{RC} \quad \text{...(3.140)}$$

The rate of increase in amplitude is 6 dB/octave. For the FM broadcast purpose, the lower break frequency f_1 is about 2.1 kHz and the higher break frequency f_2 is chosen to be much higher than the highest frequency term in the message band, so that f_2 lies outside the baseband spectral range. For audio range, f_2 may be taken as 30 kHz. The corresponding de-emphasis circuit has an inverse characteristic. The transfer function $H_d(\omega)$ of the de-emphasis circuit is given by

$$H_d(\omega) = \frac{1}{1 + j\left(\frac{\omega}{\omega_1}\right)} \quad \text{...(3.141)}$$

The transfer function is plotted in Fig. 3.38. The amplitude of frequency components above ω_1 are reduced at a rate of – 6 dB octave and thus does the inverse action of pre-emphasis circuit to restore the amplitude distribution of the message-band.

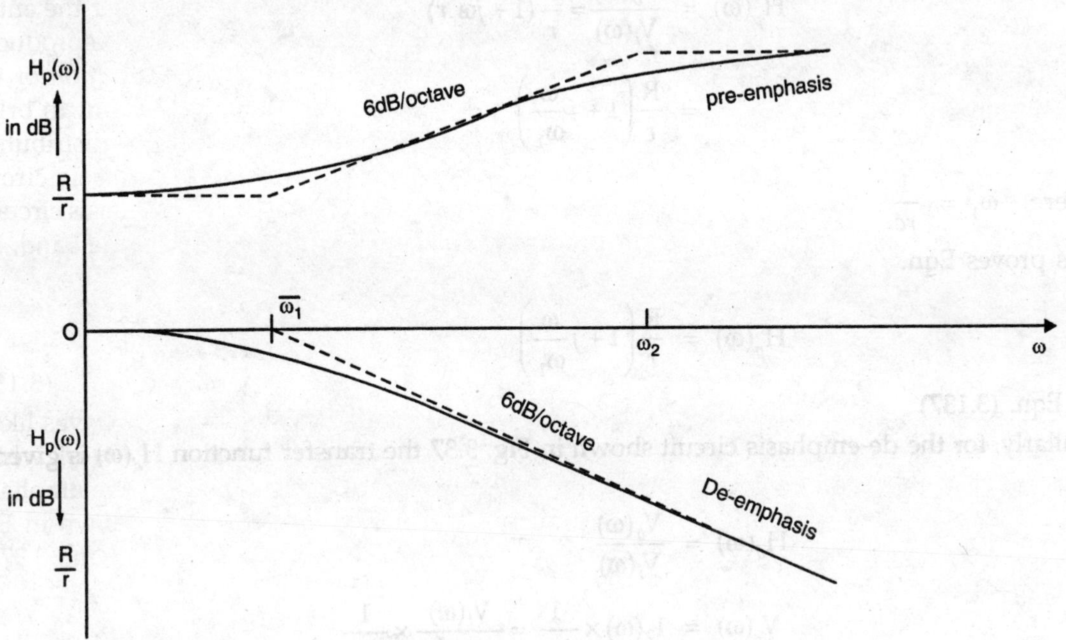

Fig. 3.38 *Transfer Function of Pre-emphasis and De-emphasis Circuits*

EXAMPLE

Prove that the transfer function of the pre-emphasis and de-emphasis circuits shown in Fig. 3.37 is given by Eqns. (3.138) and (3.141) respectively, and hence show that their product is constant.

Solution:

The transfer function $H_p(\omega)$ of pre-emphasis circuit shown in Fig. 3.37(*a*) is given as

$$H_p(\omega) = \frac{V_o(\omega)}{V_i(\omega)}$$

Assuming $R \ll r$, the current $I_1(\omega)$ is given by

$$I_1(\omega) = V_i(\omega)\left(\frac{1}{r} + j\omega_c\right)$$

Hence,

$$V_o(\omega) = R\left(\frac{1}{r} + j\omega_c\right)V_i(\omega)$$

The transfer function is,

$$H_p(\omega) = \frac{V_o(\omega)}{V_i(\omega)} = \frac{R}{r}(1 + j\omega_c r)$$

$$= \frac{R}{r}\left(1 + j\frac{\omega}{\omega_1}\right)$$

where, $\omega_1 = \frac{1}{rc}$

This proves Eqn.

$$H_p(\omega) = \frac{R}{r}\left(1 + j\frac{\omega}{\omega_1}\right)$$

i.e. Eqn. (3.137)

Similarly, for the de-emphasis circuit shown in Fig. 3.37 the transfer function $H_d(\omega)$ is given by

$$H_d(\omega) = \frac{V_o(\omega)}{V_i(\omega)}$$

$$V_o(\omega) = I_2(\omega) \times \frac{1}{j\omega_c} = \frac{V_i(\omega)}{r + \frac{1}{j\omega_c}} \times \frac{1}{j\omega_c}$$

Hence,

$$H_d(\omega) = \frac{V_o(\omega)}{V_i(\omega)} = \frac{1}{1 + j\left(\frac{\omega}{\omega_1}\right)}$$

where $\omega_1 = \dfrac{1}{rc}$

this proves Eqn.

$$H_d(\omega) = \frac{1}{1+j\left(\dfrac{\omega}{\omega_1}\right)}$$

i.e. Eqn. 3.141

Therefore the product of transfer function is given by

$$H_p(\omega)\, H_d(\omega) = \left(\frac{r}{R}\right)$$

Which is constant with frequency.

3.36.2 Improvement Factor IR

The pre-emphasis circuit does not increase the transmitted signal power because it is passive network. The improvement in the signal to noise ratio at the detector output is because of the reduction in the noise power due to the de-emphasis circuit. The de-emphasis circuit reduce the high frequency components of the noise, and cuts down the overall noise because of its low pass filtering characteristic. The improvement in output signal to noise ratio, in effect, improves the threshold level. The extent of the improvement in signal to noise ratio depends upon the noise reduction by the de-emphasis circuit. The improvement factor (IR) is defined as

$$IR = \frac{\text{Mean output noise power without de-emphasis } (N_o)}{\text{Mean output noise power with de-emphasis } (N_o)_d}$$

Since input of the de-emphasis circuit is the same as the output of the FM detector, the noise power density at the input of the de-emphasis circuit will be the same as given by Eqn. (3.102). Hence, the power density at the output of the de-emphasis circuit will be given by,

$$S_{od}(\omega) = \frac{K^2\eta\omega^2}{A^2}|H_d(\omega)|^2$$

Substituting $|H_d(\omega)|$ from Eqn. (3.141), we get

$$S_{od}(\omega) = \frac{K^2\eta\omega^2}{A^2}\left|\frac{1}{1+j(\omega/\omega_1)}\right|$$

$$= \frac{K^2\eta\omega}{A^2}\cdot\frac{\omega_1^2}{\omega^2+\omega_1^2}$$

The noise power $(N_o)_d$ is obtained by integrating $S_{od}(\omega)$ over the frequency range 0 to ω_m, i.e.

$$(N_o)_d = \frac{1}{\pi}\int_0^{\omega_m} S_{od}(\omega)\, d\omega$$

$$= \frac{K^2\eta\omega_1^2}{\pi A^2}\int_0^{\omega_m} \frac{\omega_1^2}{\omega^2+\omega_1^2}\, d\omega$$

$$= \frac{K^2\omega_1^2\eta}{\pi A^2}\left[\frac{\omega_m}{\omega_1} - \tan^{-1}\left(\frac{\omega_m}{\omega_1}\right)\right] \quad ...(3.142)$$

The noise improvement factor is given by

$$IR = \frac{N_o}{(N_o)_d} = \frac{(\omega_m/\omega_1)^3}{3\left[(\omega_m/\omega_1) - \tan^{-1}(\omega_m/\omega_1)\right]} \quad ...(3.143)$$

Thus, the improvement factor is independent of the carrier power A, and hence, the carrier power have plays no role in noise improvement. Although it is obvious that IR increases with an increase in ratio (ω_m/ω_1), noise improvement can be achieved by increasing ω_m and by decreasing ω_1. In commercial broadcast systems, ω_1 = 2.1 kHz. If ω_m is taken to be 5 kHz, the improvement ratio is about 22, or 13 dB. This shows a considerable reduction in noise due to de-emphasis circuit. The pre-emphasis and de-emphasis method is quite satisfactory for the broadcast systems, with an additional advantage being that the circuit is simple and inexpensive.

3.36.3 Frequency Modulation with Feedback (FMFB)

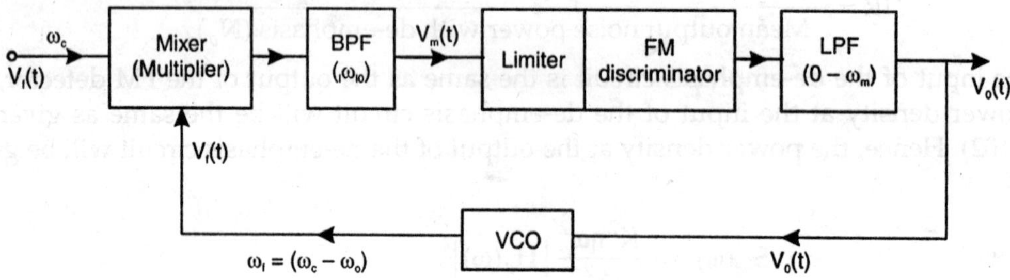

Fig. 3.39 *Frequency Modulation with Feedback (FMFB)*

In this method, a feedback path is used for the threshold improvement, as shown in the block diagram of Fig. 3.39. The feedback path comprises of a voltage controlled oscillator (VCO). The mixer circuit is a voltage multiplier circuit which generates a signal at its output which is a multiplication of the two inputs of the mixer, i.e.,

$$\text{Mixer output} = v_1(t)\, v_f(t) \quad ...(3.144)$$

where, v_i is the FM signal mixed with noise, and $v_f(t)$ is the feedback signal coming from the FCO output. The FM signal $v_i(t)$ is given by

$$v_i(t) = A(t)\cos\left[\omega_c t + \psi_n(t) + \psi_s(t)\right] \quad ...(3.145)$$

where, A(t) is the envelope of the carrier plus noise signal; $\psi_n(t)$, and $\psi_s(t)$, respectively are the phase deviation produced in the FM wave by the noise and message signals respectively.

3.37 VOLTAGE CONTROLLED OSCILLATOR (VCO)

Voltage controlled oscillator is an electronic since wave oscillator. Its frequency is controlled by an externally applied voltage. Thus, VCO is a kind of frequency modulator. VCO is available in the form of IC chip. A VCO can be adjusted to generate a sine wave of frequency ω_c in the absence of any external control voltage. Now, when an external control voltage is applied, the frequency of VCO departs from ω_c. The amount of frequency change per unit input voltage is defined as the sensitivity of the VCO, denoted by G_o, i.e.

$$G_o = \frac{d\omega_i}{dv} \text{ rad/v} \qquad \text{...(3.146)}$$

where $d\omega_i$ is the change in the instantaneous frequency for a small change dv in the external input voltage. Therefore, the frequency change for an external control voltage $v_o(t)$ will be $G_o v_o(t)$, and the instantaneous frequency of VCO output becomes

$$\omega_i = \omega_c + G_o v_o(t) \qquad \text{...(3.147)}$$

The corresponding phase angle of the VCO output voltage is

$$\phi(t) = \int \omega_i \, dt$$

$$= \omega_c(t) + G_o \int v_o(t)\, dt \qquad \text{...(3.148)}$$

Hence, the VCO output voltage is given as

$$V_{VCO} = C \cos\left[\omega_o t + G_o \int_{-\alpha}^{t} v_o(t)\, dt\right] \qquad \text{...(3.149)}$$

where C is amplitude of VCO output voltage.

3.38 OUTPUT OF THE DISCRIMINATOR WITH AND WITHOUT FEEDBACK

We will now see that a demodulation process can be achieved with feedback (FMFB). The demodulated output voltage is similar to an open-loop, without feedback, discriminator output, except that the output of FMFB is reduced. In the absence of a control voltage, the VCO is offset from ω_c by an amount ω_o, i.e., the VCO is adjusted to have a frequency $(\omega_c - \omega_o)$. The output of the VCO then is given as by Eqn. (3.149)

$$v_f(t) = C \cos\left[(\omega_c - \omega_p)t + G_o \int_{-\alpha}^{t} v_o(t)\, dt\right] \qquad \text{...(3.150)}$$

The mixer output produces the sum and difference frequencies of its input signals $v_i(t)$ and $v_f(t)$. The bandpass filter, tuned to ω_o, picks up only the difference frequency, and rejects the sum frequency. Thus, the output of the bandpass filter is given by

$$v_m(t) = D \cos\left[\omega_o(t) + \psi_s(t) + \psi_n(t) - G_o \int_{-\alpha}^{t} v_o(t)\, dt\right] \qquad \text{...(3.151)}$$

The amplitude D has been arbitrary chosen because it has no significance in FM demodulation. The output $v_o(t)$ of FMFB is given by

$$v_o(t) = K\left[\frac{d\psi_s(t)}{dt} + \frac{d\psi_n(t)}{dt} - G_o v_o(t)\right] \quad ...(3.152)$$

where K is a constant of proportionality. Solution of Eqn. (3.152) is given as

$$v_o(t) = \frac{K}{1+KG_o}\frac{d}{dt}(\psi_s + \psi_n) \quad ...(3.153)$$

The output of conventional discriminator without feedback is proportional to a change in the instantaneous frequency of the input signal $v_i(t)$ given by Eqn. (3.146). Thus in an open-loop discriminator

$$v_o(t) = K\frac{d}{dt}[\psi_s(t) + \psi_n(t)] \quad ...(3.154)$$

This output is similar to the output voltage of FMFB given by Eqn. (3.153), except that the amplitude of the FMFB output is reduced by a factor $(1 + KG_o)$. Thus, the demodulation process is identical in both the cases. Since the amplitude reduction is identical for the noise and message signals, S_o/N_o ratio remains same in both cases.

3.39 BANDWIDTH OF FMFB

The bandwidth B_p of the FM signal at the input of the FMFB discriminator is less, as compared to the bandwidth B_c of conventional open-loop FM discriminator, i.e.,

$$B_p < B_c \quad ...(3.155)$$

This can be verified as follows. In FMFB the input of the discriminator is received from a bandpass filter tuned to ω_o and having a bandwidth B_p. The output of the bandpass filter is given by Eqn. (3.151), which may be rewritten as by using Eqn. (3.153).

$$v_m(t) = D\cos\left[\omega_o t + \frac{1}{1+KG_o}(\psi_s + \psi_n)\right] \quad ...(3.156)$$

While comparing it with the conventional FM discriminator input as given by Eqn. (3.146). We find that the frequency deviation $(\Delta\omega)_p$ in FMFB is reduced by a factor $(1 + KG_o)$ as signal the deviation $(\Delta\omega)$ of the conventional discriminator, i.e.,

$$(\Delta\omega)_p = \frac{1}{1+KG_o}(\Delta\omega) \quad ...(3.157)$$

Thus, the frequency deviation in the FMFB is reduced by a factor $(1 + KG_o)$. By Carson's rule, the bandwidth of FM signal is determined by its frequency deviation. Therefore, the bandwidth B_p of the FMFB is less than that of the conventional FM case. For example, CCITT recommends a frequency deviation of 75 kHz. If we take a modulating signal of the bandwidths 15 kHz, the bandwidth B_c, conventional FM, is given as

$$B_c = 2(\Delta f + f_m)$$
$$= 2(75 = 5) = 180 \text{ kHz}$$

whereas, the bandwidth B_p for FMFB is given as,

$$B_p = 2\left(\frac{\Delta f}{1+KG_o} + f_m\right)$$

Assuming $KG_o = 9$, we get

$$B_p = 2\left(\frac{75}{10} + 15\right) = 45\,\text{kHz}$$

Thus, the FMFB bandwidth B_p is reduced by a factor of 4, i.e.,

$$B_p = \frac{1}{4}B_c$$

EXAMPLE

Prove that

$$B_p = \frac{\left[m_f/(1+KG_o)\right]+1}{m_f+1}B_c \qquad \text{...(3.158)}$$

Solution: By Carson's rule, the bandwidth of an FM signal at the input of a conventional discriminator is,

$$B_c = 2(\Delta\omega + \omega_m)$$

$$= 2\omega_m\left(\frac{\Delta\omega}{\omega_m} + 1\right)$$

$$= 2\omega_m(m_f + 1) \text{ radians}$$

$$= 2f_m(m_f + 1) \text{ Hz} \qquad \text{...(3.159)}$$

Similarly, for the case of FMFB,

$$B_p = 2\left(\frac{\Delta\omega}{1+KG_o} + \omega_m\right)$$

$$= 2\omega_m\left(\frac{m_f}{1+KG_o} + 1\right) \text{ radians}$$

$$= 2\left(\frac{m_f}{1+KG_o} + 1\right)f_m \text{ Hz} \qquad \text{...(3.160)}$$

Comparing this with Eqn. (3.159), we find the relation given in Eqn. (3.158).

3.40 THRESHOLD EXTENSION IN FMFB

In law noise case, the signal to noise ratio at the output of the FMFB remains same as the case of conventional discriminator. However, in the large noise case, the threshold level in FMFB is extended due to a reduction in the frequency of spike generation. This is because the frequency of spike occurrence is proportional to the bandwidth of the FM signal at the input of a discriminator, Eqn. (3.118). The bandwidth for the FMFB case, B_p, is less than the bandwidth B_c for the case of conventional discriminator. Hence, the frequency of spike generation is less in FMFB. Thereby due to the reduction in bandwidth, the threshold is extended. With reduction in the bandwidth B_p by a factor of 4, a threshold improvement of 6 dB may be obtained. In practice, FMFB can be designed to provide threshold improvement of 5.7 dB. This is significant improvement for the low power FM systems.

3.41 PHASE LOCKED LOOP FM DEMODULATOR

Phase locked loop, PLL, is a negative feedback system. In PLL, an error signal generated is equal to the phase difference between the input and feedback signal. It can demodulate FM signals and provide threshold improvement properties that are similar to those of FMFB detectors. PLL circuits are available in the form of expensive IC chips and can be used as a self-contained FM demodulator with an improved threshold. The PLL system has two main components as shown in Fig. 3.40. (*i*) Phase Comparator, and (*ii*) VCO. The comparator comprises of a voltage multiplier and a low pass filter (LPF).

The filter is designed in such a way that it selects only the difference-frequency components, and rejects the sum frequency components produced by the multiplier. The VCO, as discussed earlier, is a part of the FMFB. The PLL system is similar to the FMFB system as Fig. 3.39 except with the following difference.

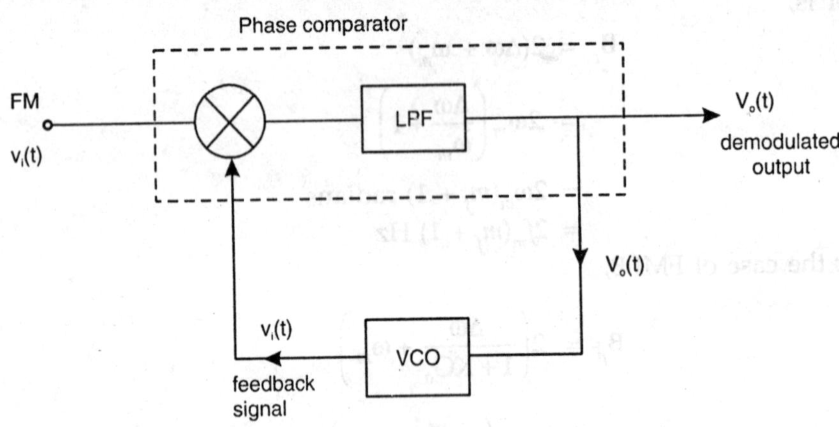

Fig. 3.40 *Phase Locked Loop (First Order)*

(*i*) PLL has on BPF and discriminator; thus PLL is self-contained demodulator, rather than just a method for threshold improvement of the conventional demodulator like FMFB, or de-emphasis.

(*ii*) When a control signal $v_0(t)$ is zero, VCO in PLL is adjusted to generate the same frequency as the unmodulated carrier; as against as offset from carrier, as in FMFB.

3.41.1 Initial Equilibrium

Initially, the modulating signal $f(t)$ is zero. Let us assume that initially when the control voltage $v_0(t)$ is zero. VCO has been adjusted to satisfy the following two conditions:

(*i*) The frequency of VCO has been set precisely at an unmodulated carrier frequency ω_c.

(*ii*) The VCO output $v_f(t)$ is in phase quadrature with the unmodulated carrier wave.

This establishes an initial equilibrium state in which the comparator output is zero when $v_o(t)$ is zero.

3.41.2 FM Demodulation

Let us assume that the FM signal at the PLL input is given by

$$v_i(t) = A\sin\left[\omega_c t + K_f \int_{-\alpha}^{t} f(t)\,dt\right]$$

$$= A\sin\left[\omega_c t + \phi_1(t)\right] \qquad ...(3.161)$$

where $\phi_1(t)$ is the angle deviation produced by the modulating signal $f(t)$, we have

$$\phi_1(t) = K_f \int_{-\alpha}^{t} f(t)\,dt \qquad ...(3.162)$$

The FM signal produces an output $v_o(t)$ which, in turn, will increase the frequency of VCO. Let the VCO output, which acts as feedback, be given as by using Eqn. (3.149).

$$v_f(t) = B\cos\left[\omega_c t + G_o \int_{-\alpha}^{t} v_o(t)\,dt\right]$$

$$= B\cos\left[\omega_c t + \phi_2(t)\right]$$

where

$$\phi_2(t) = G_o \int_{-\alpha}^{t} v_o(t)\,dt \qquad ...(3.163)$$

The FM signal $v_i(t)$ and the VCO output $v_f(t)$ are applied at the input of phase comparator, which produces an output consisting of only the difference frequencies. It can be seen that the phase comparator output is given by

$$v_o(t) = ABG_m \sin\left[\phi_1(t) - \phi_2(t)\right]$$

$$= ABG_m \sin\left[\psi_c(t)\right] \qquad ...(3.164)$$

where G_m is the gain of the multiplier, and $\psi_c(t)$ is the phase error defined by

$$\psi_c(t) = \psi_1(t) - \psi_2(t) \qquad ...(3.165)$$

The initial equilibrium is disturbed when $f(t)$ modulates the carrier and $v_o(t)$ is generated. An increase in the frequency of the FM carrier generated a positive $v_o(t)$ at PLL output, which in turn increases the frequency of VCO. The equilibrium is reestablished when the VCO frequency is increased to an extent that it becomes equal to the frequency of the input signal, but no longer in phase quadrature. In this case, the error signal becomes zero, i.e.,

$$\psi_c(t) = 0 \qquad ...(3.166)$$

or

$$\phi_1(t) = \phi_2(t) \qquad ...(3.167)$$

The PLL is said to be phase locked when the phase error $\psi_c(t)$ is equal to zero. Substituting ϕ_2 from Eqn. (3.163), we get

$$\phi_1(t) = G_o \int_{-\alpha}^{t} v_o(t)\, dt \qquad ...(3.168)$$

Differentiating both the sides,

$$\frac{d\phi_1(t)}{dt} = \frac{d}{dt}\left[G_o \int_{-\alpha}^{t} v_o(t)\, dt\right]$$

$$= G_o v_o(t) \qquad ...(3.169)$$

The time derivative of $\phi_1(t)$ is equal to the frequency change, denoted by ω, in the FM signal, i.e.,

$$\frac{d\phi_1(t)}{dt} = \omega$$

and hence,

$$\omega = G_o v_o(t)$$

or

$$v_o(t) = \frac{\omega}{G_o} \qquad ...(3.170)$$

The frequency change ω in the FM wave is proportional to the modulating signal $f(t)$, i.e.,

$$\omega = K_f f(t).$$

Therefore Eqn. (3.170) becomes

$$v_o(t) = \left(\frac{K_f}{G_o}\right) f(t) \qquad ...(3.171)$$

Thus, the output $v_o(t)$ of the PLL is proportional to the modulating signal $f(t)$, and frequency demodulation is achieved.

3.41.3 Phase Error $\psi_e(t)$

The phase error $\psi_e(t)$ is time dependent. The output of PLL $v_o(t)$ varies sinusoidally with ψ_e. Fig. 3.41(*a*) shows the sinusoidal variation of $v_o(t)$ with $\psi_e(t)$. However, if we convert the input sinusoidal signal to a square wave before application to the multiplier, the output $v_o(t)$ will have a piecewise linear variation as shown in Fig. 3.41(*b*). The linear variation in the region $-\pi/2$ to $+\pi/2$ is represented by the line PQ, which may be defined as

$$v_o(t) = \psi_e \tan\theta$$

where θ is the slope of the line PQ.

Since

$$\tan\theta = \frac{ABG_m}{\pi/2}$$

Hence,

$$v_o(t) = \frac{ABG_m}{\pi/2}\psi_e, \quad -\frac{\pi}{2} \le \psi_e \le \frac{\pi}{2} \qquad \text{...(3.172)}$$

Similarly, the relation between $v_o(t)$ and ψ_e can be obtained for other regions of ψ_e.

3.41.4 First Order PLL

The block diagram shown in Fig. 3.40 is a first order PLL because it can be represented mathematically by a first order differential equation. This will be illustrated in the discussion to follow. For convenience, we will assume the piecewise-linear phase variation shown in Fig. 3.41(*b*). However, the result will be equally valid for a sinusoidal variation also. Differentiating Eqn. (3.167), we get

$$\frac{d\psi_e}{dt} = \frac{d\phi_1}{dt} - \frac{d\phi_2}{dt} \qquad \text{...(3.173)}$$

For convenience, if the argument (*t*) is ignored then from Eqn. (3.163),

$$\frac{d\phi_2}{dt} = G_o v_o$$

hence

$$\frac{d\psi_e}{dt} + G_o v_o = \frac{d\phi_1}{dt} \qquad \text{...(3.174)}$$

By putting the value of v_o from Eqn. (3.172), we get

$$\frac{d\psi_e}{dt} + \frac{\psi_e}{\tau} = \frac{d\phi_1}{dt}, \quad |\psi_e| \le \frac{\pi}{2} \qquad \text{...(3.175)}$$

where

$$\tau = \frac{\pi}{2ABG_m G_o} \qquad \text{...(3.176)}$$

Alternatively, we can put the value of ψ_e into Eqn. (3.174) from Eqn. (3.172). We now get a relation between ϕ_1 and $v_o(t)$ as follows,

$$\frac{dv_o}{dt} + \frac{v_o}{\tau} = \frac{1}{G_o\tau}\frac{d\phi_1}{dt}, \quad |\psi_e| \le \frac{\pi}{2} \qquad \text{...(3.177)}$$

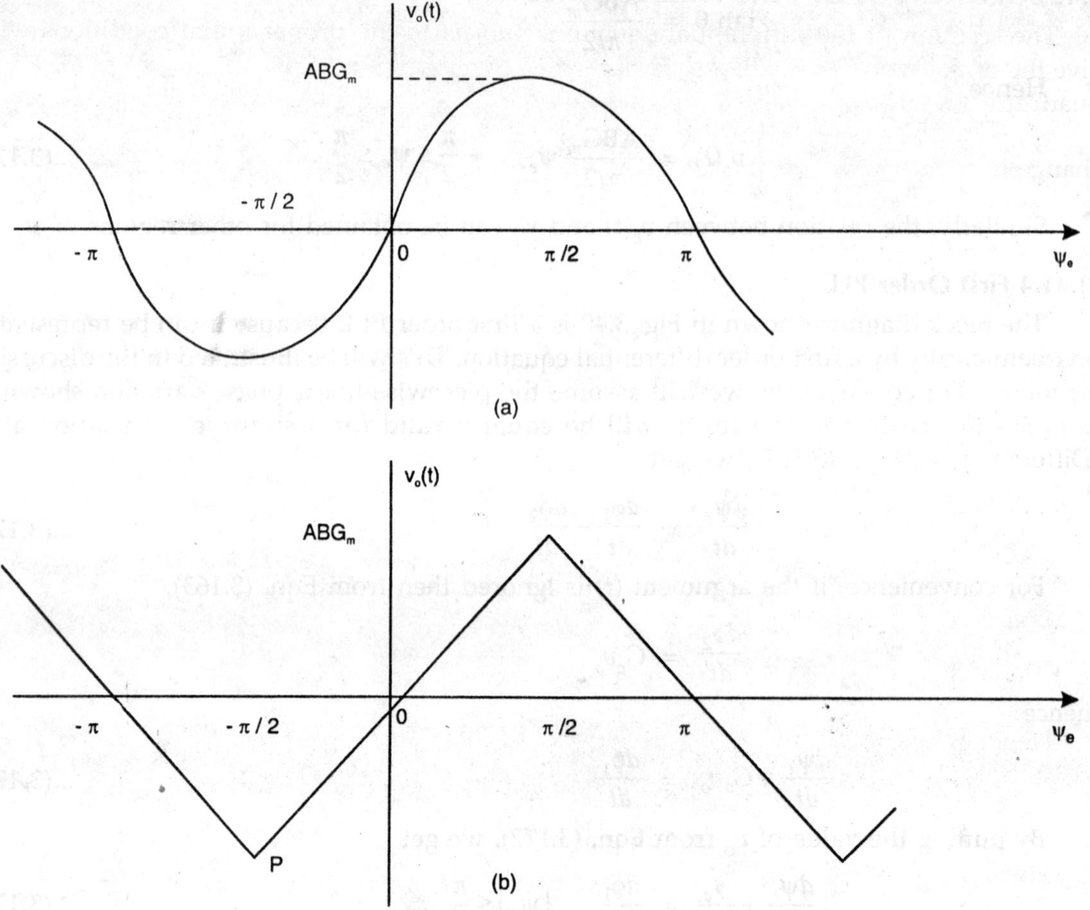

Fig. 3.41 *(a) Sinusoidal Variation (b) Piecewise-linear Variation*

Eqns. (3.175) and (3.177) are different equations of the first order. For sinusoidal characteristic, the PLL differential equation will be

$$\frac{d\psi_e}{dt} + \frac{\sin\psi_e}{\tau_s} = \frac{d\phi_i}{dt} \quad \text{...(3.178)}$$

where

$$\tau_s = \frac{1}{ABG_oG_m}$$

For a small ψ_e, Eqn. (3.178) becomes identical to the piecewise linear case given by Eqn. (3.175), as $\sin\psi_e = \psi_e$. However the time τ is modified due to difference in slopes at the origin in Figs. 3.41(*a*) and 3.41(*b*).

3.42 SOLUTION OF DIFFERENTIAL EQUATION

The solution of the differential equations subject to the proper initial conditions will give the performance of the PLL. Initially, when the input and feedback signals are in phase quadrature and have identical frequencies, the phase error $\psi_e = 0$. Then, for an abrupt change in carrier frequency by an amount ω, $\left(\omega = \frac{d\phi_1}{dt}\right)$ the solution of Eqn. (3.175)

$$\psi_e = \omega\tau\,(1 - e^{-t/\tau}), \quad -\frac{\pi}{2} \le \psi_e \le \frac{\pi}{2}$$

Thus, ψ_e grows with the time constant τ and reaches a new steady state value (when $t \to \alpha$) given by

$$\psi_e = \omega\tau, \quad -\frac{\pi}{2} \le \psi_e \le \frac{\pi}{2} \qquad \text{...(3.179)}$$

or

$$\frac{\psi_e}{\tau} = \omega$$

From Eqn. (3.172) and (3.176), we get

$$\frac{\dot{\psi}_e}{\tau} = v_o(t)\,G_o$$

Hence,

$$v_o(t)G_o = \omega$$

or

$$v_o(t) = \frac{\omega}{G_o} \qquad \text{...(3.180)}$$

Thus, the solution gives a relationship defined by Eqn. (3.170).

3.42.1 Operating Range of PLL

Eqn. (3.179) is valid only for the linear region PQ in Fig. 3.41(*b*), i.e., for $|\psi_e| \le \frac{\pi}{2}$. The frequency change ω of the input carrier corresponding to $|\psi_e| \le \frac{\pi}{2}$, is defined as operating range R of the PLL. In Eqn. (3.179), putting $|\psi_e| \le \frac{\pi}{2}$ and ω = R, we get

$$R = \frac{\pi}{2\tau} \qquad \text{...(3.181)}$$

It can be seen hat for a sinusoidal variation of ψ_e, as Fig. 3.41(*a*), he range is given by

$$R = \frac{1}{\tau_s} \qquad \text{...(3.182)}$$

Thus the range behaviour of PLL depends on the time constant τ or τ_s, which in turn, depends on G_o, G_m, and the amplitudes A, B of the PLL input signals.

The larger the range R, the greater is the speed of the PLL response to frequency change ω. The range may be increased by making τ smaller. In the limit $\tau \to 0$, ψ_e remains close to the origin of Fig. 3.41, and PLL behaves like a fast-acting loop. The fast-acting loop keeps the phase deviation $\phi_1(t)$ close to the phase of VCO and error ψ_e is negligibly small. Thus, the range R determines the speed of PLL.

3.42.2 Bandwidth

The PLL behaves like a low pass filter. This can be verified by taking the Laplace transform, denoted by L, of the Eqn. (3.175).

Let us consider

$$L[\psi_e(t)] \equiv \psi_e(s).$$

and

$$L\left[\frac{d\phi_1}{dt}\right] = L[\omega(t)]$$

$$= \Omega(s)$$

Then from Eqn. (3.175), we have

$$\psi_e(s) = \frac{\Omega(s)}{s + \frac{1}{\tau}} \qquad ...(3.183)$$

The transfer function H(s) of the PLL is given by

$$H(s) = \frac{\psi_e(s)}{\Omega(s)} = \frac{1}{s + \frac{1}{\tau}} \qquad ...(3.184)$$

The transfer function represents a single pole in s-plane, accordingly PLL is termed as a first order PLL. Putting $s = j\omega$ in Eqn. (3.184), we get

$$H(\omega) = \frac{\gamma}{1 + j\omega\tau} \qquad ...(3.185)$$

Which represents the transfer function of a low pass filter with a 3 dB cutoff frequency ω_l given by

$$\omega_l = \frac{1}{\tau} \qquad ...(3.186)$$

Therefore, PLL will produce distortion at its output if the modulating frequency ω_m is greater than ω_l, just like a low pass RC filter. The distortion may be reduced by keeping ω_l much higher than the expected highest modulating term ω_m. In other words, we can say that the bandwidth of PLL is given by

$$(BW)_{PLL} = \frac{1}{\tau} \qquad ...(3.187)$$

3.42.3 Instability in PLL

The first order PLL has been represented by Eqn. (3.175) for a linear variation of ψ_e with a positive slope θ, Fig. 3.41(*b*). If θ is negative, Eqn. (3.175) becomes

$$\frac{d\psi_e}{dt} - \frac{\psi_e}{\tau} = \frac{d\phi_1}{dt}$$

By solving the above equation, it can be seen that the PLL equilibrium may be unstable.

3.42.4 Limitation of the First Order PLL

Both bandwidth and range are controlled by a single parameter τ in first order PLL. Therefore, we cannot vary the range and bandwidth independently. If τ is adjusted to change the range, then the bandwidth also changes, and vice versa. This is an undesirable feature, and a remedy to this is the second order PLL.

3.42.5 The Second Order PLL

The second order PLL is obtained by deliberately introducing another filter of transfer function $H_p(s)$ in the first order PLL as shown in Fig. 3.42. The filter $H_p(s)$ has distinct features as compared to the filter incorporated in the phase comparator. While the former will have an influence on the PLL performance, the latter has no such influence. For convenience in the discussion, the angle modulated signals encountered in the Fig. 3.42 are referred to by their phase angles in *s*-domain. The VCO has been characterized by a transform G_o/s to represent the integration of Eqn. (3.163). From Eqn. (3.165), we have

$$\psi_e(s) = \phi_1(s) - \frac{G_m G_o \psi_e(s) H_p(s)}{s} \qquad \text{...(3.188)}$$

Assume a proportional plus integral filter $H_p(s)$ characterized by

$$H_p(s) = 1 + \frac{b}{s} \qquad \text{...(3.189)}$$

where *b* is a constant, substituting $H_p(s)$ in Eqn. (3.188) and solving for $\psi_e(s)$, we get

$$\psi_e(s) = \frac{s\,\Omega(s)}{s^2 + \frac{s}{\tau} + \frac{b}{\tau}}; \quad \psi \le \frac{\pi}{2} \qquad \text{...(3.190)}$$

where $$s\,\phi_1(s) = L\left[\frac{d\phi_1(t)}{dt}\right] = L(\omega) = \Omega(s)$$

and $$\tau = \frac{1}{G_m G_o}$$

The transform of the output voltage $v_o(t)$ is given by

$$v_o(s) = G_m \psi_e(s)\, H_p(s) \qquad \text{...(3.191)}$$

From Eqns. (3.189), (3.190) and (3.191), we have

$$V_o(s) = \frac{(s+b)\, G_m \Omega(s)}{s^2 + (s/\tau) + (b/\tau)} \qquad \text{...(3.192)}$$

Eqns. (3.190) and (3.192) represent the second order differential equations having two poles in *s*-planes. Therefore, Fig. 3.42 represents a second order PLL.

The steady state response of the second order PLL may be obtained by using final value theorem. According to this theorem, a steady state value ($t \to \alpha$) of a signal $f(t)$ may be obtained from its Laplace Transform F(*s*) in the limit $s \to 0$, i.e.,

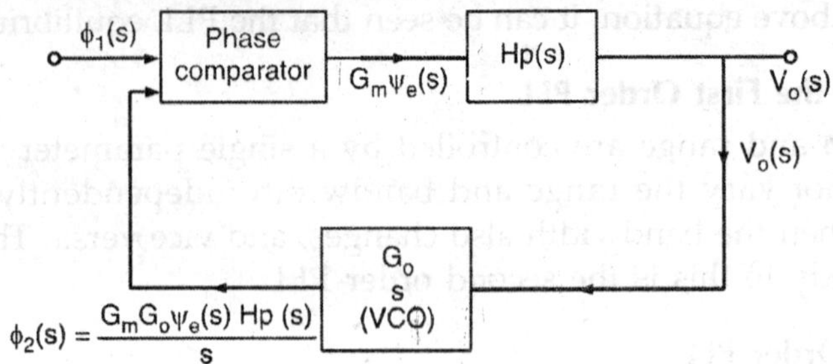

Fig. 3.42 *PLL (Second Order)*

$$\lim_{t \to \alpha} f(t) = \lim_{s \to \alpha} [sF(s)] \quad ...(3.193)$$

Applying the final value theorem to Eqns. (3.190) and (3.192), the steady state values for ψ_e and $v_o(t)$ can be determined.

From the basic definition of the Laplace transform, we have

$$\Omega(s) = L(\omega) = \int_0^{\alpha} \omega e^{-st}\, dt = \frac{\omega}{s}$$

Putting this value of $\Omega(s)$ in Eqn. (3.190), and applying final value theorem, we get

$$\psi_e(\alpha) = \lim_{s \to 0} [s\psi_e(s)] = 0 \quad ...(3.194)$$

Similarly, from Eqn. (3.192),

$$v_o(\alpha) = \lim_{s \to 0} [sV_o(s)]$$

$$= G_m \omega \tau = \frac{\omega}{G_o} \quad ...(3.195)$$

It is observed that the output steady state response v_o is identical to the first order PLL, and hence, provides an equivalent demodulation of the FM signals. But, unlike the first order PLL, the steady state phase error ψ_e in the second order PLL is zero. Therefore, the second order PLL settles back to its starting point after any transient displacement from its initial equilibrium.

Further Eqn. (3.192) relating to the output $v_o(s)$ and frequency change $\Omega(s)$ represents the transfer function of a low pass filter. Therefore, by a proper choice of cutoff frequency, the PLL bandwidth may be adjusted to get the proper response for the entire baseband. On the other hand, Eqn. (3.192) relating to the phase error $\psi_e(s)$, and $\Omega(s)$ represents a bandpass filter, hence, the operating point of the phase comparator does not respond to the modulating signal. The operating point always remains close to the initial equilibrium point.

3.42.6 (S/N) ratio at the Output of a PLL

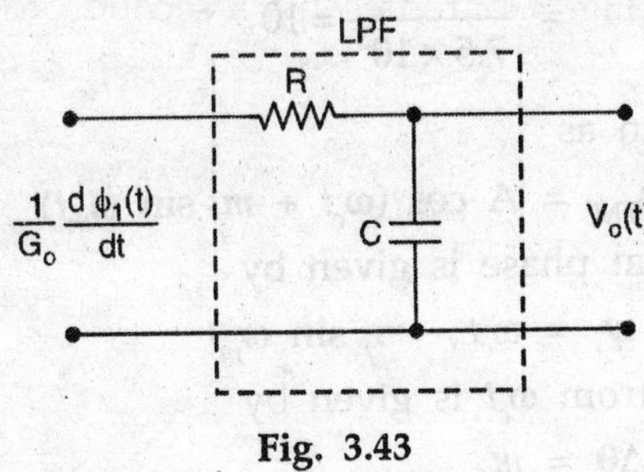

Fig. 3.43

Let us consider that an FM signal $A\cos\left[\omega_c t+\phi_1(t)\right]$ is applied to the input of PLL. The demodulated signal is given by Eqn. (3.170)

$$v_o(t) = \frac{\omega}{G_o} = \frac{1}{G_o}\left(\frac{d\phi_1}{dt}\right)$$

Thus, the demodulated output signal of PLL is identical to the conventional discriminator output given as $v_o(t) = K\frac{d\phi_1}{dt}$ except that the constant K is replaced by $(1/G_o)$. However, the low pass filtering property of PLL provides better performance under large noise conditions. The low pass filtering property of PLL is identical with an R-C network, Fig. 3.43, having a time constant RC, equal to the τ of PLL. The low pass filter behaves like a baseband filter of a conventional discriminator, provided its cutoff frequency is adjusted to pass the entire modulation band. Above Threshold, i.e., low noise case, the spikes are rare and the performance of PLL is identical with a conventions discriminator. But below threshold, i.e., large noise case, the performance of PLL is better because the number of spikes generated are less which results a reduction in threshold. Typically PLL improves the threshold level by 3 dB for $m_f = 12$. The spikes generated are reduced because most of the spikes generate a doublet at the PLL output provided the range R is restricted.

SOME SOLVE EXAMPLES

EXAMPLE 1

A single-tone modulating signal cos ($15\pi\ 10^3 t$) frequency modulate a carrier of 10 MHz and produces a frequency deviation of 75 kHz. Find (*i*) the modulation index and (*ii*) phase deviation produced in the FM wave (*iii*) If another modulating signal produces a modulation index of 100 while maintaining the same deviation, find the frequency and amplitude of the modulating signal, assuming K_f = 15 kHz per volt.

Solution:

(*i*) The FM modulation index is given by

$$m_f = \frac{\Delta\omega}{\omega_m} = \frac{\Delta f}{f_m}$$

$$= \frac{75 \times 10^3}{7.5 \times 10^3} = 10$$

(*ii*) The FM wave is given as

$$\varphi_{FM} = A \cos(\omega_c t + m_f \sin \omega_m t)$$

The instantaneous total phase is given by

$$\psi_i = \omega_c t - m_f \sin \omega_m t$$

The phase deviation from $\omega_c t$ is given by

$$\Delta\theta = m_f$$

Therefore, the phase deviation is given as,

$$\Delta\theta = 10 \text{ rod}$$

(*iii*)

$$\Delta\theta = \frac{\Delta f}{f_m} = m_f$$

$m_f = 100$, $\Delta f = 75$ kHz

$$f_m = \frac{\Delta f}{m_f} = \frac{75 \times 10^3}{100}$$

or

$$f_m = 750 \text{ Hz.}$$

The amplitude of the modulating signal E_m is given by

$$E_m = \frac{\Delta f}{K_f} = \frac{75}{15}$$

$$= 5 \text{ volt.}$$

EXAMPLE 2

A 100 MHz carrier wave has a peak voltage of 5 volts. The carrier is frequency modulated by a sinusoidal modulating signal or waveform of frequency 2 kHz such that the frequency deviation Δf is 75 kHz. The modulated waveform passes through zero and is increasing at $t = 0$. Determine the expression for the modulated carrier waveform.

Solution:

We know that frequency modulated carrier waveform passes through zero and is increasing at $t = 0$, therefore, the FM signal must be sine wave signal. Thus

$$s(t) = A \sin\left[2\pi f_c t + m_f \sin(2\pi f_m t)\right]$$

where

$$m_f = \text{modulation index of FM wave}$$

$$= \frac{\Delta f}{f_m}$$

Here given

$$f_c = \text{carrier wave frequency} = 100 \times 10^6 = 10^8 \text{ Hz}$$

$$\Delta f = \text{frequency deviation} = 75 \text{ kHz} = 75 \times 10^3 \text{ Hz}$$

$$f_m = \text{modulating frequency} = 2 \text{ kHz} = 2 \times 10^3 \text{ Hz}$$

$$A = \text{peak voltage of carrier wave} = 5 \text{ Volt.}$$

Now
$$m_f = \frac{\Delta f}{f_m} = \frac{75 \times 10^3}{2 \times 10^3} = 37.5$$

Substituting these all the values in above equation, we have

$$s(t) = 5 \sin\left[2\pi \times 10^8 t + 37.5 \sin(2\pi \times 2 \times 10^3 t)\right]$$

$$= 5 \sin\left[2\pi \times 10^8 t + 37.5 \sin(4\pi \times 10^3 t)\right]$$

EXAMPLE 3

(*i*) Show with a phasor diagram that the signal *s*(*t*) expressed by

$$s(t) = \cos(2\pi \times 10^6 t) + 0.02 \cos\left[2\pi(10^6 + 10^3)t\right]$$

represents a carrier wave which is modulated both in amplitude and frequency.

(*ii*) Represent *s*(*t*) in the form

$$s(t) = \left[1 + m \cos(2\pi \times 10^6 t)\right] \cos\left[2\pi \times 10^6 t + m_f \sin(2\pi \times 10^3 t)\right]$$

Find the value of *m* and m_f. Also, write an expression for instantaneous frequency as a function of time *t*. Verify that both amplitude and frequency vary approximately sinusoidally with frequency 1 kHz.

Solution:

(*i*) The phasor diagram of the signal

$$s(t) = \cos(2\pi \times 10^6 t) + 0.02 \cos\left[2\pi(10^6 + 10^3)t\right]$$

has been shown in the figure as

Observation:

It may be easily observed from this figure that the resultant amplitude differs from the carrier amplitude A and the resultant is not in phase with the carrier. Hence, the given signal is modulated both amplitude and phase.

(*ii*) The given signal $s(t)$ can be expressed as

$$s(t) = \cos(2\pi\times 10^6 t) + 0.02\cos\left[2\pi(10^6+10^3)t\right]$$

$$= \cos(2\pi\times 10^6 t) + 0.02\cos(2\pi\times 10^6 t)\cos(2\pi\times 10^3 t) - 0.02\sin(2\pi\times 10^6 t)\sin(2\pi\times 10^3 t)$$

or
$$s(t) = \left[1+0.02\cos(2\pi\times 10^3 t)\right]\cos(2\pi\times 10^6 t) - 0.02\sin(2\pi\times 10^6 t)\sin(2\pi\times 10^3 t)$$

$$s(t) = \left\{\left[1+0.02\cos(2\pi\times 10^3 t)\right]^2 + \left[0.02\sin(2\pi\times 10^3 t)\right]^2\right\}^{1/2}$$

$$\times\cos\left\{2\pi\times 10^6 t + \tan^{-1}\frac{0.02\sin(2\pi\times 10^3 t)}{1+0.02\cos(2\pi\times 10^3 t)}\right\}$$

$$s(t) = \left[1+0.02\cos(2\pi\times 10^3 t)\right]\cos\left[(2\pi\times 10^6 t)+0.02\sin(2\pi\times 10^3 t)\right]$$

$$s(t) = \left\{\left[1+m\cos(2\pi\times 10^3 t)\right]\cos\left[2\pi\times 10^6 t + m_f\sin(2\pi\times 10^3 t)\right]\right\}$$

Hence, modulation index of AM part, m= 0.02 and modulation index of FM part, m_f = 0.02. The instantaneous frequency is found as

$$f_i = \frac{1}{2\pi}\frac{d}{dt}[\theta_i(t)]$$

$$= \frac{1}{2\pi}\frac{d}{dt}\left[2\pi\times 10^6 t + 0.02\sin(2\pi\times 10^3 t)\right]$$

or
$$f_i = 10^6 = 20\cos(2\pi\times 10^3 t)$$

It can be observed that the amplitude and instantaneous frequency of modulated signal vary with $\cos(2\pi\times 10^3 t)$.

EXAMPLE 4

The maximum deviation allowed in FM broadcast system is 75 kHz. If the modulating signal is a single-tone sinusoidal of 10 kHz, find the bandwidth of the FM signal. What will be the change in the bandwidth, if modulating frequency is doubled? Determine the bandwidth when modulating signals amplitude is also doubled.

Solution:

We have here

$$\Delta f = 75 \text{ kHz}$$
$$f_m = 10 \text{ kHz}$$

According to the Carson's Rule, the FM bandwidth is given

$$\begin{aligned} BM &= 2(\Delta\omega + \omega_m) \\ &= 2(\Delta\omega) + 2(\omega_m) \\ &= 2(\Delta f) + 2f_m \end{aligned}$$

Because $\Delta f = \Delta\omega = 75$ kHz

and $f_m = \omega_m = 10$ kHz

Hence $BM = 75 \times 2 + 10 \times 2 = 170$ kHz

When modulating frequency is doubled, the bandwidth changes slightly,

$$\begin{aligned} BW &= 75 \times 2 + 2 \times 20 \\ &= 190 \text{ kHz} \end{aligned}$$

When the modulating signal amplitude is doubled, the frequency deviation becomes 2 × 75 = 150 kHz.

Now, the bandwidth becomes

$$\begin{aligned} BW &= 2 \times 150 + 2 \times 20 \\ &= 340 \text{ kHz} \end{aligned}$$

Thus, the bandwidth is almost doubled.

EXAMPLE 5

A modulating signal 5 cos 2π 15 × 10³*t*, angle modulates a carrier A cos $\omega_c t$.

(*a*) Find the modulation index and the bandwidth for (*i*) the FM system, (*ii*) the PM system

(*b*) Determine the change in the bandwidth and the modulation index for both FM and PM, if f_m is reduced to 5 kHz.

Assume $K_p = K_f = 15$ kHz/Volt.

Solution:

(*a*) Here $E_m = 5$ V, $f_m = 15$ kHz

(*i*) FM System

Frequency deviation

$$\begin{aligned} \Delta f &= \Delta\omega = K_f E_m \\ &= 15 \times 10^3 \times 5 \\ &= 75 \text{ kHz} \end{aligned}$$

Therefore,

$$m_f = \omega_f = \frac{75}{15} = 5$$

$$\begin{aligned} \text{Bandwidth} &= 2(\Delta f + f_m) \\ &= 2(75 + 15) = 2 \times 90 \\ &= 180 \text{ Hz.} \end{aligned}$$

(*ii*) PM System

Frequency deviation,

$$(\Delta f) = K_p E_m f_m$$
$$= 15 \times 10^3 \times 5 \times 15 \times 10^3 = 1125 \text{ MHz}$$

The corresponding bandwidth

$$BW = 2(\Delta f + f_m) = 2\Delta f$$
$$= 2(1125 \text{ MHz})$$
$$= 2250 \text{ MHz.}$$

The bandwidth is quite large as compared to FM. This is because the modulation index in PM is quite large.

$$m_p = K_p E_m = 5 \times 15 \times 10^3 = 75{,}000$$

This tremendous value of the modulation index m_p produce a large number of sidebands and this is the reason for a large bandwidth.

(*b*) Now $f_m = \omega_m = 5 \text{ kHz}$

For FM the deviation (Δf) is independent of f_m, and remains 75 kHz. Hence

$$m_f = \frac{75}{5} = 15$$

$$\text{The bandwidth} = 2(\Delta f + f_m)$$
$$= 2(75 + 5)$$
$$= 160 \text{ kHz.}$$

Thus, in FM the modulation index changes considerably with a change in the modulating frequency, but the bandwidth only slightly. For PM: deviation (Δf) is dependent on f_m and is given as

$$\Delta f = K_p E_m f_m = 15 \times 10^3 \times 5 \times 5 \times 10^3$$
$$= 375 \text{ MHz}$$

Hence,

$$BW = 2(\Delta f + f_m)$$
$$\approx 2(\Delta f)$$
$$= 750 \text{ kHz.}$$

Thus, the bandwidth has changes considerably as compared to case (*ii*).

The modulation index is independent of f_m,

$$m_p = K_p E_m$$
$$= 15 \times 10^3 \times 3$$
$$= 75 \times 10^3$$
$$= 75 \text{ kHz}$$

Which is same as in case (*ii*).

Thus, in PM, the band.

EXAMPLE 6

A semiconductor junction diode is used to modulate the frequency of an oscillator. The junction capacitance is the total tuning capacitance of the oscillator tank circuit. When a d.c. bias voltage of 15 Volts is applied t the diode, the oscillator frequency generated is 5 MHz. If a single-tone modulating voltage 4 sin 12560*t* modulates the carrier: find

(*a*) the percentage second harmonic distortion; and

(*b*) the frequency modulation index.

Solution:

The polarizing voltage V_o = 15 Volt.

The oscillator frequency before modulation is $15\text{ MHz} \simeq 10\pi\, 10^6$ rad/s

Now from the equation

or $$(L_o K)^{1/2} = \frac{V_o^{1/4}}{\omega_c} = \frac{15^{1/4}}{10\pi\, 10^6}$$

The modulating frequency ω_m = 12560 rad/s and its amplitude V_m is 4 V.

(*a*) The percentage second harmonic distortion is given by equation

$$\frac{3V_m}{8V_o} \times 100 = \frac{3 \times 4}{8 \times 15} \times 100$$

$$= 10\%$$

i.e. percentage distortion = 10%.

(*b*) The modulation index is given by equation

$$m_f = \frac{\Delta\omega}{\omega_m} = \frac{V_m}{4\omega_m \left[L_o K V_o^{3/2}\right]^{1/2}}$$

$$= \frac{V_m}{4\omega_m (L_o K)^{1/2} V_o^{3/4}}$$

or $$m_f = \frac{4 \times 10\pi \times 10^6}{4 \times 12560 \times 15^{1/4} \times 15^{3/4}}$$

$$= \frac{\pi 10^6}{1256 \times 15} = 166.66.$$

PROBLEMS

1. Define the following terms for FM waves:

   (*a*) Carrier Swing (*b*) Frequency deviation

   (*c*) Per cent modulation
2. Describe the difference between narrow-band FM and wideband FM.
3. Derive an expression for a single-tone frequency-modulated wave.
4. Explain the term instantaneous frequency in case of angle-modulation.
5. Describe the salient features of wideband FM system.
6. What is meant by the term 'significant sidebands' in frequency modulated wave?
7. Explain the Armstrong method for the generation of wideband FM.
8. What are the principal merits and limitations of FM?
9. Draw the circuit diagram of varactor diode modulator and explain its working.
10. State the typical applications of narrowband FM system.

❑❑❑

Signal Analysis 4

4.1 INTRODUCTION

The most fundamental function in electrical communications is the sinusoidal signal because the response of a sine wave to a linear time invariant system is also sinusoidal, i.e. if a sinusoidal signal is applied at the input of a linear system, the output waveform is also sinusoidal with the same frequency as the input, except for a constant transmission delay and gain or attenuation.

Also the sinusoidal analysis of electrical networks is even more simple and convenient.

The terms signal and function are interchangeably used as there is no difference between the two. A function is defined as a set of rules which substitute one number for another. A "signal" is a function of a time; and therefore, "signal" may be defined as a set of rules to assign a number $f(t)$ for every number t, a time domain.

However, communication systems involve waveforms that are complex in nature, and it is desirable to resolve them in terms of sinusoidal functions. Signal analysis is a tool for achieving this aim. Signal analysis may be defined as the method of resolving a function in terms of an infinite series. The method suggested by J.B.J. Fourier for signal analysis provides the origin of the theory of Fourier series and Fourier transform.

4.2 CLASSIFICATION OF SIGNAL

A signal may be function of time, temperature, position, pressure, distance etc. Some signals in our daily life are music, speech, picture and video signals. Systematically, we can define a signal as "A function of one or more independent variables which contains some information is called a signal." In electrical sense, the signal can be voltage or current. The voltage or current is the function of time as independent variable.

In daily life, we come across several electric signals such as radio, TV signal, computer signal etc. Signals may be classified as:

(*a*) Causal and noncausal signal

(*b*) Real and complex signal

(*c*) Analog and digital signal

(*d*) Periodic and aperiodic signal
(*e*) Even and odd signal
(*f*) Energy and power signal
(*g*) Deterministic and random signals
(*h*) Continuous time and discrete time signals
(*i*) Singularity function

4.2.1 Causal and Noncausal Signal

A causal signal $v(t)$ is the one for which $v(t) = 0$ for $t < 0$ while for noncausal signal it is not. Practical real-life signals are causal in nature. The causal nature of the signal poses an interesting constraint on realized systems which we'll discuss with frequency description of a signal by Fourier method.

4.2.2 Real and Complex Signal

As the name suggests real signals have only real components while complex signal has both real and imaginary components. Complex signal representation has an important place in communication as it conveys both amplitude and phase, i.e. delay information. A complex signal in exponential form can be written as follows

$$v(t) = A\, e^{j\omega t}$$

where

$$j = \sqrt{(-1)}$$

$$v(t) = \text{energy of a signal}$$

and means square power S can be written as

$$S = \lim_{T \to \alpha} \frac{1}{T} \int_{-T/2}^{+T/2} |v(t)|^2\, dt$$

The magnitude information of above is given by $|v(t)| = A$ and phase information is given by $< v(t) = \omega t$. Signals described in Example 1 and 2 are real signals.

EXAMPLE 1

Calculate energy of the signal

$$u(t) = 2e^{-3t}, \quad t \geq 0 \text{ and}$$

$$v(t) = 0, \quad \text{elsewhere}$$

Solution:

Substituting $v(t) = 2e^{-3t}$ in equation of energy of a signal $v(t)$ as

$$E = \int_{-\alpha}^{+\alpha} |v(t)|^2\, dt$$

and considering $v(t) = 0$ for $t < 0$, we get

$$E = \int_{0}^{+\alpha} 4e^{-6t}\, dt + 4\int_{0}^{+\alpha} e^{-6t}\, dt$$

$$= 4e^{-6t}/(-6)\Big|_{0}^{+\alpha}$$

$$= -\left(\frac{2}{3}\right)\left[e^{-6,\alpha} - e^{-6,0}\right]$$

$$= -\left(\frac{2}{3}\right)[0-1] = \frac{2}{3}$$

EXAMPLE 2

Calculate power of the signal $v(t) = 2 \sin 0.5\pi t$.

Solution:

Since $v(t)$ is periodic, i.e. respective, averaging over $-\alpha$ to α and $-T_0/2$ to will give same result, where T_0 is the time period of $v(t)$.

Here, $$T_0 = \frac{2\pi}{(0.5\pi)} = 4$$

Now from equation

$$S = \frac{1}{4}\int_{-2}^{+2} 4 \sin^2 (0.5\pi t)\, dt$$

$$= \int_{-2}^{+2} \sin^2 (0.5\pi t)\, dt$$

$$= 2\int \sin^2 (0.5\pi t)\, dt$$

$$= 2\int_{0}^{+2} 0.5[1 - \cos (\pi t)]\, dt$$

$$= 2 \times 0.5\left[t\Big|_{0}^{+2} \frac{-\cos (\pi t)}{\pi}\Big|_{0}^{+2}\right]$$

$$= [2-0] - \cos (2\pi) - \frac{\cos (0.\pi)}{\pi}$$

$$= 2 - 0 = 2$$

Following above steps, we can find for any $v(t) = A \sin \omega t$, power $S = \frac{A^2}{2}$ and RMS $= \sqrt{S} = \frac{A}{\sqrt{2}}$.

4.2.3 Analog and Digital Signal

Analog signals are continuous in nature while digital signals are generated from discrete events or are sampled version of analog signal, with amplitude axis quantized. Analog signals are represented as $v(t)$ where t is continuous in time while digitals are represented by $v(n)$ where n is sampling time index. If sampling is done at regular interval of time gap T then

$$v(n) = v(t)\big|_{t=nT}, \quad n = \text{integer}$$

The above assumes that there are infinite numbers of quantization levels for amplitude axis. In reality, thus infinite but quantization error can be reduced by increasing number of quantization levels. Nyquiwst Theory states that to representation analog signal by its sampled version, the sampling frequency should be greater than or equal to twice the maximum frequency content present in the analog signal.

4.2.4 Periodic and Aperiodic Signals

We are familiar with periodic signals in the form of sine, cosine functions. A periodic signals is that type of signal which has a definite pattern and repeats over and over with a repetition period of T. In other words, a signal is called periodic if it exhibits periodically as follow:

$$v(t + T) = v(t), \quad -\mu < t < \mu$$

T is the period of the signal. The smallest value of period T which satisfies above equation is called the fundamental period T_0. Fig. 4.1 shows a periodic signal.

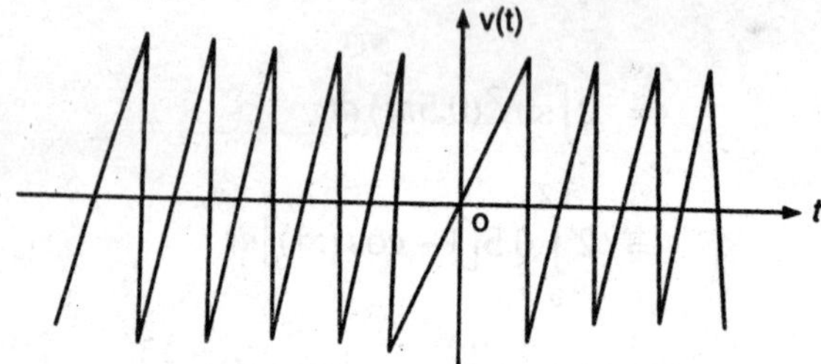

Fig. 4.1 *A Periodic Signal*

For periodic signals the above relationships are not obeyed. A signal is said to be Aperiodic if it does not repeat. Sometimes aperiodic signals are said to have a period equal to infinity. Fig. 4.2 shows an aperiodic signal. This is a decaying exponential pulse given by the equation:

$$v(t) = e^{-at}$$

Now,

$$v(t + T_0) = e^{-a(t+T_0)}$$

$$= e^{-a(t+\alpha)}$$

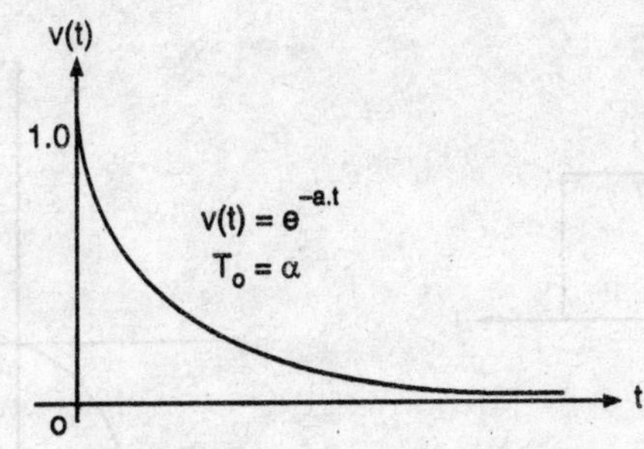

Fig. 4.2 *An Aperiodic Signal*

or $$(vt + T_0) = e^{-at} . e^{-\alpha}$$
$$= e^{-at} . 0$$
$$v(t + T_0) = 0$$

Which is not equal to $v(t)$.

This shows that the signal having period $T_0 = \alpha$ is nothing but aperiodic signal.

EXAMPLE 1

Find out whether the following signals are periodic or aperiodic:

(*i*) $v(t)$ = sin 15pt

(*ii*) $v(t) = \sin \sqrt{2}\pi t$

Solution:

(*i*) $v(t) = \sin 15\pi t$ is a periodic signals.

The fundamental period T is given as

$$T = \frac{2\pi}{\omega} = \frac{2\pi}{15\pi} = \frac{2}{15}$$

or $$T = 0.133 \text{ seconds}$$

(*ii*) $v(t) = \sin \sqrt{2}\pi t$ is a periodic signal.

The fundamental period T is given as

$$T = \frac{2\pi}{\omega} = \frac{2\pi}{\sqrt{2}\pi}$$

or $$T = \frac{2}{\sqrt{2}} = \sqrt{2} = 1.41 \text{ seconds.}$$

4.2.5 Even and Odd Signals

An even signal is symmetric about $t = 0$ axis such that $v(t) = v(-t)$ and an odd signal shows anti-symmetry for the same where $v(t) = -v(-t)$. The signal $v(t) = A \cos \omega t$ is an even signal while $v(t) = A \sin \omega t$ is an odd signal. Any arbitrary signal $v(t)$ can be written as sum of even and odd parts as follows

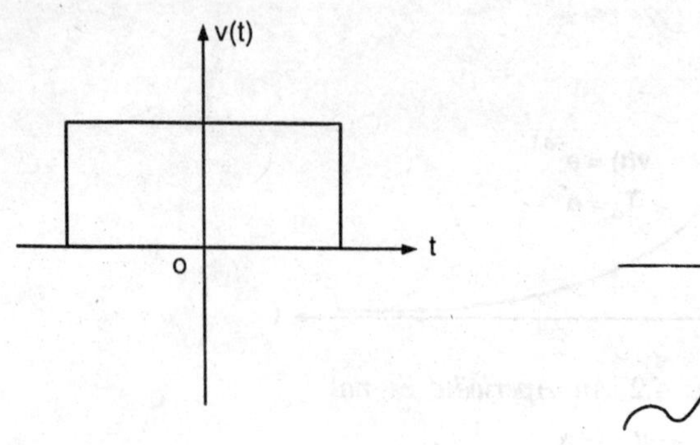

Fig. 4.3 *Even Signal*

Fig. 4.4 *Odd Signals*

$$v(t) = v_e(t) + v(t)$$

where $$v_e(t) = \frac{[v(t) + v(-t)]}{2}$$

and $$v_0(t) = \frac{[v(t) - v(-t)]}{2}$$

(Subscript *e* and 0 stand for even and odd respectively).

4.2.6 Energy and Power Signal

A signal with finite energy is known as energy signal while a signal with finite but greater than zero power is known as power signal. However, there are some signals which can neither be classified as energy signals nor power signals.

Hence, $v(t)$ is an energy signal, if

$0 < E < \alpha$ and $P = 0$

where E is the energy and P is the power of the signal $v(t)$.

And hence, $v(t)$ is a power signal, if

$0 < P < \alpha$ and $E = \alpha$

However, if the signal does not satisfy any of the above two conditions, then it is neither an energy signal nor a power signal.

4.2.7 Deterministic and Random Signals

Those signals which can be completely specified in time are known as Deterministic signals. The pattern of this type of signals is regular and can be characterized mathematically. In addition to this, the nature and amplitude of such a signal at any time can be predicted.

Let us see few examples of deterministic signals:

(*a*) $$v(t) = Bt$$

This is a ramp signal whose amplitude increases linearly with time and the slop is B.

(*b*) $$v(t) = A \sin \omega t$$

For this signal, the amplitude varies sinusoidally with time and its maximum amplitude is A.

A random signal is one whose occurrence is always random in nature. The pattern of such a signal is quite irregular. Random signals are also called non-deterministic signals.

A typical example of a random signal is thermal noise, generated in an electric circuit. Such a noise signal has probabilistic behaviour.

4.2.8 Continuous-time and Discrete-time Signals

That signal in which time t is continuous variable the signal $v(t)$ is a continuous-time signal. Thus a continuous-time signal is defined continuously in the time-domain.

If time t is a discrete variable then $v(t)$ is a discrete-time signal. Since a discrete-time signal is defined as discrete-time, it is often identified as a sequence of numbers and is denoted by $v[n]$, where n = integer.

Fig. 4.1 shows the graph of continuous-time and discrete-time signal.

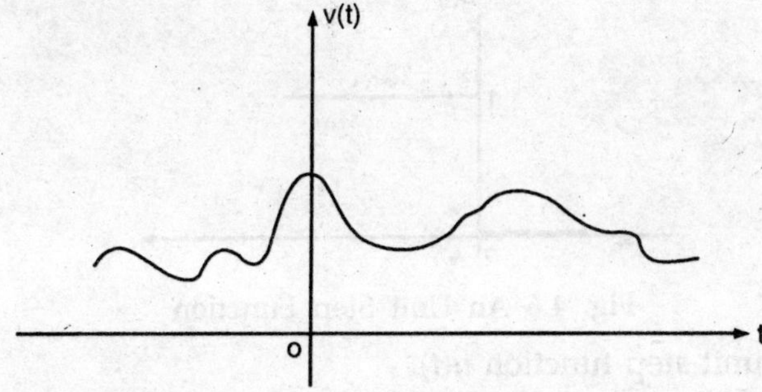

Fig. 4.5 *(a) A Continuous-time Signal*

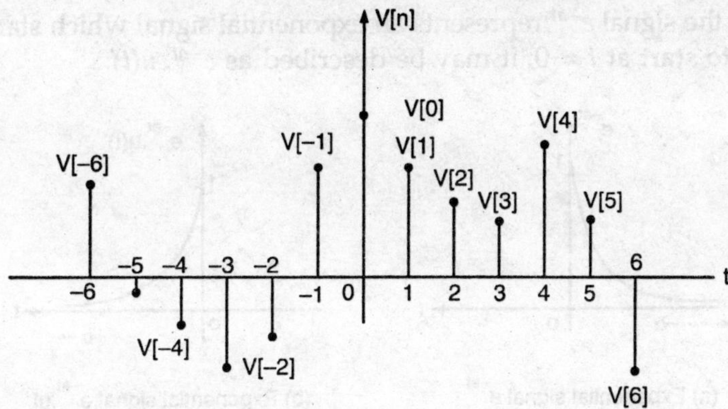

Fig. 4.5 *(b) A Discrete-time Signal*

4.2.9 Singularity Functions

There are several singularity functions which play vital role in the study of communication system. These singularity functions serve as basic building blocks for the construction of more complex signals. Infact, these singularity functions may be used to model a large number of physical signals which occur in nature. These singularity functions may be given as:

1. Unit Step Function,
2. Unit Impulse Function
3. Ramp Function.

Unit Step Function

The unit step function is that type of singularity function which exists only for positive side and is zero for negative side. Also the unit step function is discontinuous at $t = 0$. The unit-step function is denoted by $u(t)$ and may be expressed as

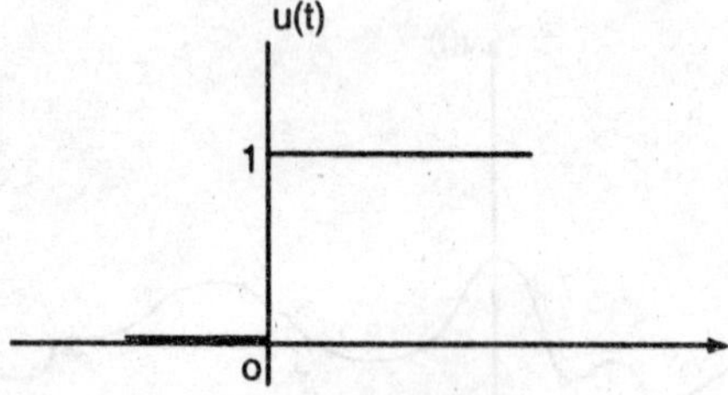

Fig. 4.6 An Unit Step Function

Fig. 4.6 shows the unit step function $u(t)$.

The usefulness of the unit step function lies in the fact that if we want a signal to start at $t = 0$, so that it may have a value of zero for $t < 0$, we only need to multiply the given with unit step function $u(t)$.

Let us suppose, the signal e^{-at} represents an exponential signal which starts at $t = -\alpha$. But if want this signal to start at $t = 0$, it may be described as $e^{-at}.\ u(t)$.

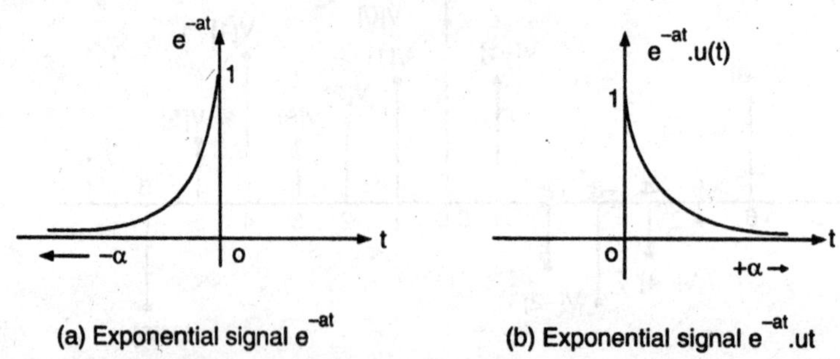

Fig. 4.7 *(a) Exponential Signal* e^{-at} *(b) Exponential Signal* $e^{-at}.u(t)$

Fig. shows the signals e^{-at} and $e^{-at}.\ u(t)$.

Unit Impulse Function

The unit impulse function is the most widely used elementary function used in the analysis of communication system.

An unit-impulse function is denoted as $\delta(t)$ and may be expressed as

$$\delta(t) = 0 \quad t \neq 0$$

and

$$\int_{-\alpha}^{+\alpha} \delta(t)\, dt = 1$$

The unit impulse function is also known as Dirac-Delta function.

Figure shows the graphical representation of an unit impulse function.

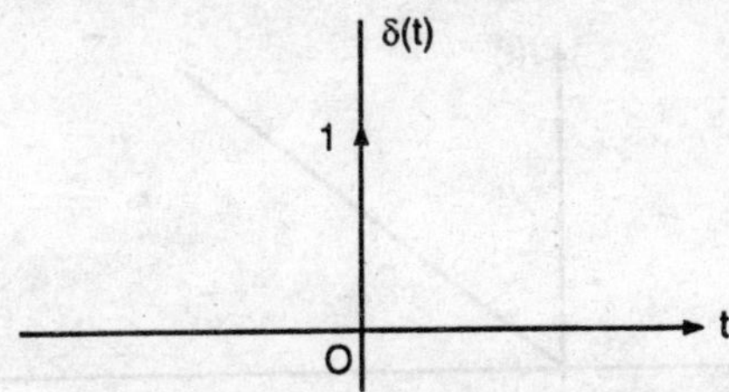

Fig. 4.8 *The Unit Impulse Function*

If unit impulse function is assured in the form of a pulse, then following points may be observed about an unit impulse function:

(*i*) the width of the pulse is zero. This means that the pulse exists only at $t = 0$.

(*ii*) the height of the pulse-goes to infinity.

(*iii*) the area under the pulse-curve is always unity.

(*iv*) the upward pointing arrow indicates the actual value of infinity.

(*v*) the height of arrow indicates the total area under the impulse. As in Fig. , the height of arrow is 1, therefore, the total area under the impulse is unity.

The total area under the pulse is also called the strength of the impulse.

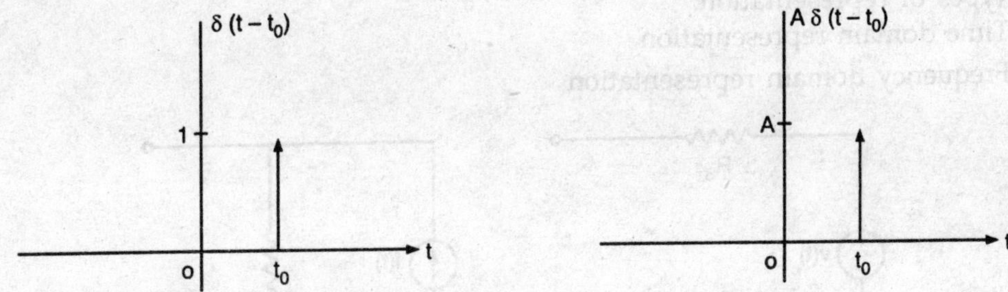

Fig. 4.9 *Unit Impulse Function Centred at $t = t_0$*

Fig. 4.10 *Impulse Function Centred at $t = t_0$ and having area A*

Fig. shows the unit impulse function centered at $t = t_0$. It is denoted by $\delta(t - t_0)$. Fig. shows the impulse function centred at $t = t_0$ and having area equal to A.

Importance

The unit impulse function is generally used to represent point mass, point charge, point square and other physical entities used in science and engineering.

Unit Ramp Function

A continuous time unit ramp function is that type of function which starts at $t = 0$ and increases linearly with time t. It is denoted by $r(t)$.

Mathematically,

$$r(t) = \begin{cases} 0 & \text{for } t < 0 \\ t & \text{for } t < 0 \end{cases}$$

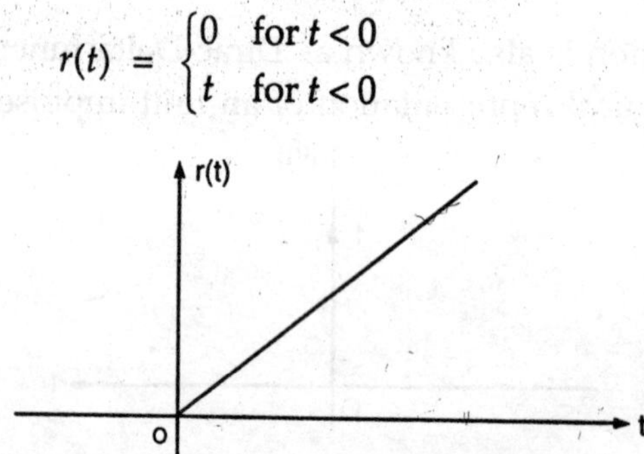

Fig. 4.11 *Continuous time unit ramp function*

Fig. 4.11 shows the graphical representation of an unit ramp function.

4.3 SIGNALS REPRESENTATION

Signals contain information about a variety of things and activities present in the physical world. Electrical signal may be represented in two equivalent forms i.e., as a voltage signal or a current signal. This means that an electrical signal may be represented either in the form of voltage source or in the form of a current source. These two forms are known as Thevenin's form and Norton's form of signal representation respectively.

Thus, an electrical signal either a voltage signal or a current signal may further be given by two types of representation:

(*i*) Time domain representation

(*ii*) Frequency domain representation

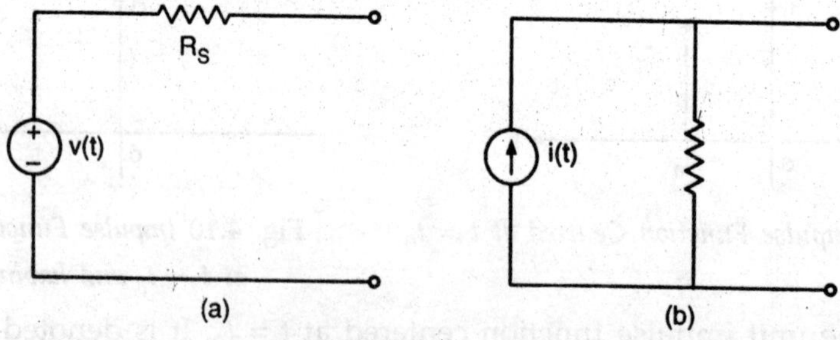

Fig. 4.12 *Representation of Signals (a) Thevenin's Form (b) Norton's Form*

4.3.1 Time-domain Representation

In time-domain representation, a signal is a time-varying quantity.

4.3.2 Frequency-domain representation

In frequency-domain, a signal is represented by its frequency spectrum and for this Fourier Series and Fourier Transform are used.

Fourier series is used to get frequency spectrum of a time-domain signal, when the signal is a periodic function of time. With the help of Fourier series, a given periodic function of time may be expressed as the sum of an infinite number of sinusoids whose frequency are harmonically related. Fig. 4.13 represents as arbitrary time-domain signal.

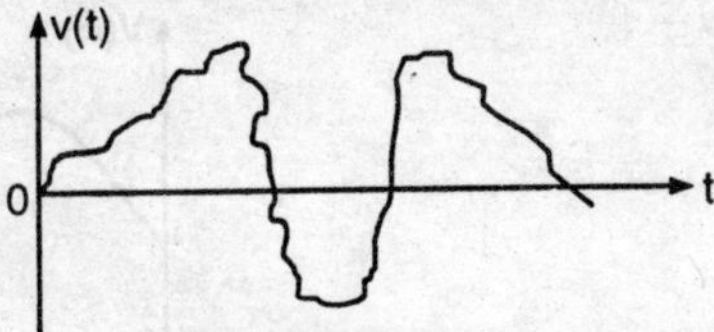

Fig. 4.13 *An Arbitrary Time-domain Signal*

Fig. 4.13 represents the time-domain square wave.

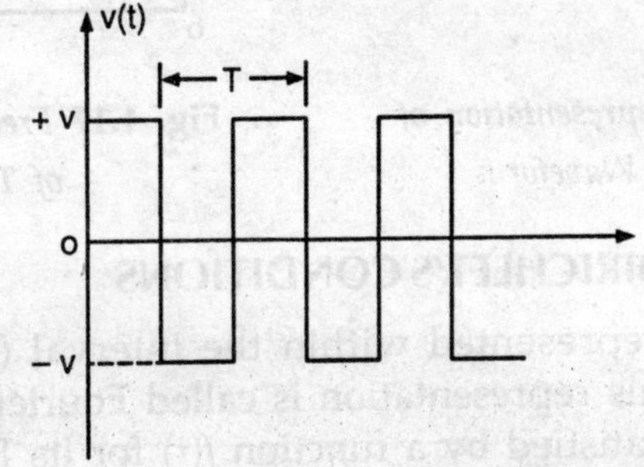

Fig. 4.14 *A Symmetrical Square-Wave Signal (Time-domain Representation)*

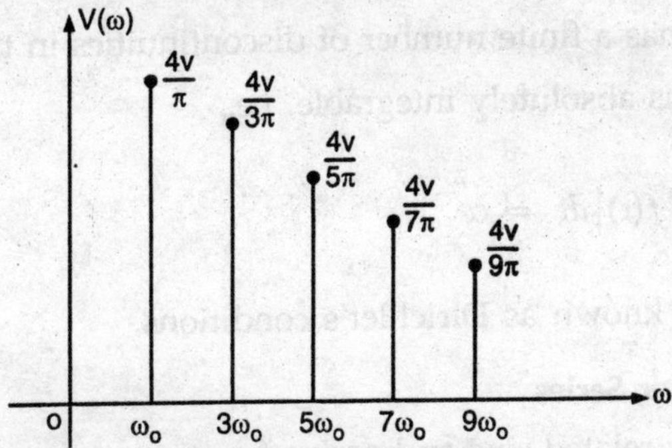

Fig. 4.15 *Frequency Spectrum of Square Wave Signal (Frequency-domain Representation)*

When Fourier Series is applied, we get

$$v(t) = \frac{4v}{\pi}\left(\sin \omega_0 t + \frac{1}{3}\sin \omega_0 t + \frac{1}{5}\sin 5\omega_0 t + ...\right)$$

From above representation, it is clear that the frequency spectrum of a periodic signal is discrete.

Fourier transform is used to get frequency spectrum of a time-domain signal. When signal waveform is an arbitrary function of time or signal waveform is a nonperiodic function of time. Also the Fourier transform provides a continuous frequency spectrum of an arbitrary time domain signal waveform.

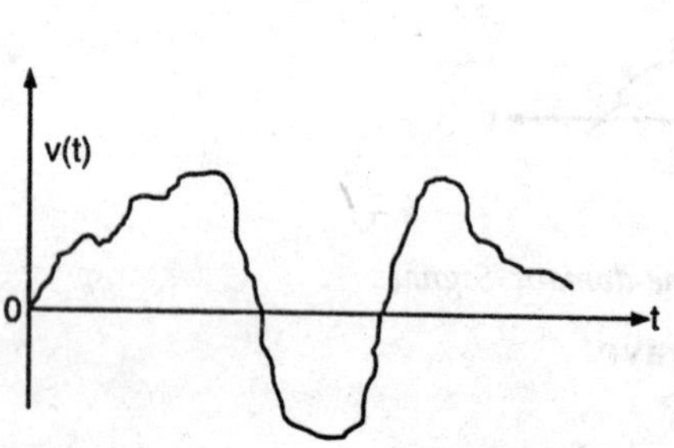

Fig. 4.16 *Time-domain representation of an Arbitrary Signal Waveform*

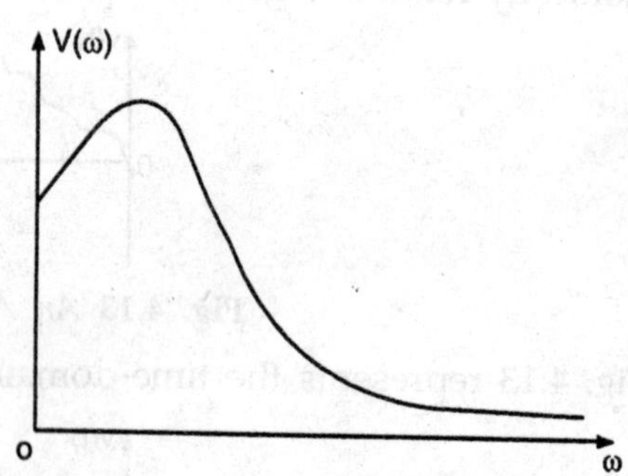

Fig. 4.17 *Frequency-domain Representation of Time-domain Signal*

4.4 FOURIER SERIES AND DIRICHLET'S CONDITIONS

A function $f(t)$ can be represented within the interval (t_1, t_2) by a complete set of orthogonal functions, and this representation is called Fourier Series. There are sufficient conditions that need to be satisfied by a function $f(t)$ for its Fourier series representation within the interval (t_1, t_2). These are as follows:

1. The function $f(t)$ is a single-valued function of the variable t within the interval (t_1, t_2).
2. The function $f(t)$ has a finite number of discontinuities in the interval (t_1, t_2).
3. The function $f(t)$ is absolutely integrable, i.e.,

$$\int_{t_1}^{t_2} |f(t)|\, dt = \alpha$$

These conditions are known as Dirichlet's conditions.

4.4.1 Trigonometric Fourier Series

A set of harmonically related sine and cosine functions, i.e., cos $n\,\omega_0 t$ and sin $n\,\omega_0 t$ (n = 0, 1, 2, ...) forms a complete orthogonal set over the interval $t_0 \le t \le t_0 + \frac{2\pi}{\omega_0}$.

Therefore, a function $f(t)$ can be represented by a Fourier Series comprising the following sine and cosine functions:

$$f(t) = a_0 + a_1 \cos \omega_0 t + a_2 \cos 2\omega_0 t + \ldots + a_n \cos n\omega_0 t + \ldots + b_1 \sin \omega_0 t + b_2 \cos 2\omega_0 t + \ldots + b_n \cos n\omega_0 t + \ldots$$

$$= a_0 + \sum_{n=1}^{\infty} (a_n \cos n\omega_0 t + b_n \sin n\omega_0 t), \quad (t_0 \le t \le t_0 + \mathrm{T}) \qquad \ldots(4.1)$$

where $\mathrm{T} = \dfrac{2\pi}{\omega_0}$

Eqn. (4.1) is the trigonometric Fourier series representation of $f(t)$ over an interval (t_0 + T). The constant a_0 corresponds to $n = 0$, and is given by

$$a_0 = \frac{1}{\mathrm{T}} \int_{t_0}^{t_0+\mathrm{T}} f(t)\, dt \qquad \ldots(4.2)$$

It is evident that a_0 is the average value, i.e., the d.c. component of $f(t)$ over the interval (t_0, t_0 + T). Also that constant $b_0 = 0$, as $\sin n\omega_0 t$ is zero for $n = 0$.

The other constant corresponding to a_n and b_n are given by

$$a_n = \frac{2}{\mathrm{T}} \int_{t_0}^{t_0+\mathrm{T}} f(t) \cos n\omega_0 t\, dt \qquad \ldots(4.3)$$

$$b_n = \frac{2}{\mathrm{T}} \int_{t_0}^{t_0+\mathrm{T}} f(t) \sin n\omega_0 t\, dt \qquad \ldots(4.4)$$

4.5 POLAR FOURIER SERIES REPRESENTATION

The polar Fourier series is the modified form of the trigonometric Fourier series. The polar Fourier series may be derived from trigonometric Fourier series as follow:

Trigonometric Fourier series is given by

$$v(t) = a_0 + \sum_{n=1}^{\alpha} (a_n \cos n\omega_0 t + b_n \sin n\omega_0 t)$$

or

$$v(t) = a_0 + \sum_{n=1}^{\alpha} \sqrt{a_n^2 + b_n^2} \left[\frac{a_n}{\sqrt{a_n^2 + b_n^2}} \cos n\omega_0 t + \frac{b_n}{\sqrt{a_n^2 + b_n^2}} \sin n\omega_0 t \right]$$

Putting

$$\frac{a_n}{\sqrt{a_n^2 + b_n^2}} = \cos \phi_n \text{ and } \frac{b_n}{\sqrt{a_n^2 + b_n^2}} = \sin \phi_n$$

So that
$$\tan \phi_n = \frac{b_n}{a_n}$$

or
$$\phi_n = \tan^{-1}\left(\frac{b_n}{a_n}\right)$$

$$v(t) = a_0 + \sum_{n=1}^{\alpha} \sqrt{a_n^2 + b_n^2}\left[\cos n\omega_0 t \cos \phi_n + \sin n\omega_0 t \sin \phi_n\right]$$

or

$$v(t) = a_0 + \sum_{n=1}^{\alpha} \sqrt{a_n^2 + b_n^2}\,.\,\cos(n\omega_0 t + \phi_n)$$

Now, let us suppose
$$D_0 = a_0$$

and
$$D_n = \sqrt{a_n^2 + b_n^2}$$

So that
$$v(t) = D_0 + \sum_{n=1}^{\infty} D_n \cos(n\omega_0 t + \phi_n) \quad \text{...(4.5)}$$

$$\phi_n = \tan^{-1}\left(\frac{b_n}{a_n}\right) \quad \text{...(4.6)}$$

The coefficient D_n are called spectral amplitudes. This means D_n is the amplitude of the spectral component $D_n \cos(n\omega_0 t - \phi_n)$ having frequency $n\omega_0$. Similarly, D_n is also called an amplitude spectrum of function $v(t)$. ϕ_n specifies the phase information of the spectral component $n\omega_0$. And then, ϕ_n is also called as phase spectrum of the function $v(t)$.

4.6 COMPLEX FOURIER EXPONENTIALLY SERIES

The exponential form of Fourier series is simpler and more compact and hence this is most widely used in signal analysis.

We will derive an expression for complex Fourier Exponential Series with the help of trigonometric Fourier Series.

The trigonometric Fourier series is given as

$$v(t) = a_0 + \sum_{n=1}^{\infty} a_n \cos n\omega_0 t + b_n \sin n\omega_0 t \quad \text{...(4.7)}$$

We know, Euler's identity
$$e^{j\theta} = \cos\theta + j\sin\theta$$

and
$$e^{-j\theta} = \cos\theta - j\sin\theta$$

Adding
$$2\cos\theta = e^{j\theta} + e^{-j\theta}$$

or $$\cos\theta = \frac{e^{j\theta} + e^{-j\theta}}{2j}$$

Subtracting $$2j\sin\theta = e^{j\theta} + e^{-j\theta}$$

or $$\sin\theta = \frac{e^{j\theta} - e^{-j\theta}}{2j}$$

Hence,

$$\cos n\omega_0 t = \frac{e^{jn\omega_0 t} + e^{-jn\omega_0 t}}{2}$$

and $$\sin n\omega_0 t = \frac{e^{jn\omega_0 t} - e^{-jn\omega_0 t}}{2}$$

Substituting the values of cos $n\omega_0$t and sin $n\omega_0$t in Eqn. (4.7)

$$v(t) = a_0 + \sum_{n=1}^{\alpha} a_n\left[\frac{-e^{jn\omega_0 t} + e^{-jn\omega_0 t}}{2}\right] + b_n\left[\frac{e^{jn\omega_0 t} - e^{-jn\omega_0 t}}{2}\right]$$

or $$v(t) = a_0 + \sum_{n=1}^{\alpha}\left[\frac{(a_n - jb_n)\,e^{jn\omega_0 t}}{2} + \frac{(a_n + jb_n)\,e^{-jn\omega_0 t}}{2}\right] \quad ...(4.8)$$

Let us suppose

$$C_n = \frac{1}{2}(a_n - jb_n)$$

$$C_{-n} = \frac{1}{2}(a_n + jb_n)$$

$$C_0 = a_0$$

where C_{-n} is the complex conjugate of C_n.

Putting these values in Eqn. (4.8)

$$v(t) = C_0 + \sum_{n=1}^{\alpha} C_n . e^{jn\omega_0 t} + \sum_{n=1}^{\alpha} C_{-n} . e^{-jn\omega_0 t}$$

or $$v(t) = C_0 + \sum_{n=1}^{\alpha} C_n . e^{jn\omega_0 t} + \sum_{n=-\alpha}^{-1} C_n . e^{jn\omega_0 t}$$

or $$v(t) = \sum_{n=-\alpha}^{\alpha} C_n . e^{jn\omega_0 t}$$

C_0 has been included under the $\sum$ sign by using the values of $n = 0$.

Now, substituting expression for coefficients a_n and b_n in the expression

$$C_n = \frac{1}{2}(a_n - jb_n)$$

or
$$C_n = \frac{1}{T}\int_{-\pi/2}^{\pi/2} v(t)\left[\cos n\omega_0 t - j\sin n\omega_0 t\right]dt$$

or
$$C_n = \frac{1}{T}\int_{-\pi/2}^{\pi/2} v(t)\, e^{-jn\omega_0 t}\, dt \qquad ...(4.9)$$

Similarly
$$C_{-n} = \frac{1}{T}\int_{-\pi/2}^{\pi/2} v(t)\, e^{jn\omega_0 t}\, dt \qquad ...(4.10)$$

4.6.1 Concept of Negative Frequency

The exponential form of Fourier series represents an arbitrary function $f(t)$ over the interval $(t_0, t_0 + T)$ by a linear combination of complex exponential functions of frequencies, $0, \pm\omega_0, \pm 2\omega_0, \pm 3\omega_0 ....$ The presence of negative frequency is simply because the mathematical model of a signal requires the use of negative frequency. In fact, a real function $f(t)$ cannot be represented by the complex exponential function $e^{jn\omega_0 t}$ without associating them with their opposite rotating counterparts $e^{-jn\omega_0 t}$. The signal $e^{-jn\omega_0 t}$ of frequency $-n\omega_0 t$ (negative frequency) represents the counterpart of the signal $e^{jn\omega_0 t}$ (positive frequency), and when combined they yield a real function of time. Indeed, the complex-valued basic functions, (viz., $e^{jn\omega_0 t}$ and $e^{-jn\omega_0 t}$) having no physical meaning separately, but yield real functions ($\cos n\omega_0 t$ or $\sin n\omega_0 t$). When combined together i.e., negative frequency signals are not physical signals, rather they are only a mathematical concept required to give a compact mathematical description of a real signal by a combination of complex exponentials of positive and negative frequencies.

4.7 DEVIATION OF TRIGONOMETRIC FOURIER SERIES FROM COMPLEX EXPONENTIAL SERIES

The proceeding discussions show that both the series are not different. Indeed, one series can be derived from the other. Let us consider the following exponential series:

$$f(t) = F_0 + F_1 e^{j\omega_0 t} + F_2 e^{j2\omega_0 t} + ... + F_n e^{jn\omega_0 t} + ... + F_{-1} e^{-j\omega_0 t}$$
$$+ F_{-2} e^{-j2\omega_0 t} + F_{-n} e^{-jn\omega_0 t} + ...$$

Replacing the exponential functions by their trigonometric identities, the series becomes

$$f(t) = F_0 + F_1(\cos\omega_0 t + j\sin\omega_0 t) + ... + F_n(\cos n\omega_0 t$$
$$+ j\sin n\omega_0 t) + ... + F_{-1}(\cos\omega_0 t - j\sin n\omega_0 t) + ...$$
$$+ F_{-n}(\cos n\omega_0 t + j\sin n\omega_0 t) + ...$$

Rearranging the right-hand side, we have

$$f(t) = F_0 + (F_1 + F_{-1}) \cos \omega_0 t + \ldots + (F_n + F_{-n}) \cos n\omega_0 t + \ldots$$
$$+ j(F_1 + F_{-1}) \sin \omega_0 t + \ldots + j(F_n - F_{-n}) \sin n\omega_0 t + \ldots$$

Which is a trigonometric series representation of $f(t)$ derived from exponential series. Comparing this derived trigonometric series with actual form of trigonometric series Eqn. 4.5. The following identities between coefficients of exponential and trigonometric series can be achieved

$$a_0 = F_0 \quad \ldots(4.11)$$

$$a_n = F_n + F_{-n} \quad \ldots(4.12)$$

$$b_n = j(F_n - F_{-n}) \quad \ldots(4.13)$$

Similarly, an exponential series can be derived from the trigonometric series by replacing the sine and cosine functions with their exponential equivalents. The Eqns. (4.11), (4.12) and (4.13) may be written as

$$F_0 = a_0 \quad \ldots(4.14)$$

$$F_n = \frac{1}{2}(a_n - jb_n) \quad \ldots(4.15)$$

$$F_{-n} = \frac{1}{2}(a_n + jb_n) \quad \ldots(4.16)$$

Which represent the coefficients of exponential series in terms of coefficients of trigonometric series. The trigonometric series has no negative frequencies as against the exponential series which contains negative frequencies.

4.7.1 Fourier Series Representation of a Periodic Function Over the Entire Interval ($-\alpha < t < \alpha$)

As far as, we have considered the Fourier series representation of an arbitrary function over a finite interval between to and $(t_0 + T)$. The function and its Fourier series may not be equal outside this interval. A periodic function which repeats after very T seconds is expected to have the same Fourier series for the entire interval $(-\alpha, \alpha)$. A periodic function has an identical Fourier series for the entire interval $(-\alpha, \alpha)$ as for the interval $(t_0, t_0 + T)$ since the same function repeats after every T seconds. Thus for a periodic function $f(t)$ we have,

$$f(t) = \sum_{n=-\infty}^{\alpha} F_n e^{jn\omega_0 t}, \quad (-\alpha < t < \alpha) \quad \ldots(4.17)$$

$$F_n = \frac{1}{4} \int_{t_0}^{t_0+T} f(t) e^{-jn\omega_0 t}\, dt, \quad \omega_0 = \frac{2\pi}{T} \quad \ldots(4.18)$$

The choice of t_0 is immaterial for a periodic function; however, it is significant for the series expansion of a function in a finite interval, and such a series expansion is therefore not unique, whereas, the Fourier series expansion of a periodic function is unique, irrespective of the location of to in the waveform.

4.8 PROPERTIES OF CONTINUOUS-TIME FOURIER SERIES (CTFS)

Let us see the properties of Continuous-time Fourier Series (CTFS). The continuous-time Fourier series representation consists of a number of properties which may be used to develop conceptual fundamentals and in reducing the complexity of Fourier series of several signals.

4.8.1 Linearity

The complex exponential Continuous-time Fourier Series (CTFS) is expressed as

$$x(t) = \sum_{n=-\alpha}^{\alpha} F_n . e^{jn\omega_0 t}$$

where $x(t)$ is any Continuous-time periodic signal and F_n is Fourier series coefficient. Thus $x(t)$ has coefficients F_n.

For simplicity, let us denote these coefficients as F_{xn}. This means that we can represent the signal $x(t)$ and its coefficients F_{xn} as a continuous-time Fourier series pair i.e.,

$$x(t) \xleftrightarrow{\text{CTFS}} F_{xn}$$

Thus the linearity property may be stated as

If $$x(t) \xleftrightarrow{\text{CTFS}} F_{xn}$$

and $$y(t) \xleftrightarrow{\text{CTFS}} F_{yn}$$

Then $$z(t) = ax(t) + by(t) \xleftrightarrow{\text{CTFS}} F_{zn} = aF_{xn} + bF_{yn}$$

4.8.2 Time Shifting Property

The time shifting property states that if

$$x(t) \xleftrightarrow{\text{CTFS}} F_{xn}$$

Then

$$x(t - t_0) \xleftrightarrow{\text{CTFS}} e^{-jn\omega_0 t_0} . F_{xn}$$

Thus, magnitude of Fourier series coefficients remain unchanged when a periodic signal is shifted in time.

4.8.3 Frequency Shifting Property

Frequency shifting property states that if

$$x(t) \xleftrightarrow{\text{CTFS}} F_{xn}$$

Then

$$e^{jn_0\omega_0 t} x(t) \xleftrightarrow{\text{CTFS}} F_{x(n - n_0)}$$

Thus, from above, it is clear that shifting the frequency components by n_0 is equivalent to multiplying $x(t)$ by $e^{jn_0\omega_0 t}$.

4.8.4 Time Scaling Property

The scaling property states that if $x(t)$ is a periodic signal with time T_0 then $x(t)$ will be a periodic signal with period T_0/α, where α is a positive real number. Thus, with the application of time-scaling property, the period of a signal changes.

Now, it may be observed that since the time-scaling property applies to each of the harmonic components of a signal $x(t)$, therefore the Fourier series coefficients for each of these harmonic components remain the same.

Mathematically we can see if

$$x(t) \xleftrightarrow{\text{CTFS}} F_{xn}$$

then

$$x(at) \xleftrightarrow{\text{CTFS}} F_{xn}$$

Thus, after time-scaling, the Fourier series coefficients are same. However, the spacing between the frequency component changes from $\frac{1}{T_0}$ to $\frac{a}{T_0}$.

4.8.5 Time-Differentiation

Time-differentiation property states that if

$$x(t) \xleftrightarrow{\text{CTFS}} C_{xn}$$

then

$$\frac{d}{dx}x(t) \xleftrightarrow{\text{CTFS}} j.\frac{2\pi n}{T_0}C_n$$

Thus, differentiating a signal $x(t)$ is equivalent to multiplying its coefficients by $\frac{j2\pi n}{T_0}$.

4.8.6 Time-Convolution

The time-convolution property states that if

$$x(t) \xleftrightarrow{\text{CTFS}} C_{xn}$$

and

$$y(t) \xleftrightarrow{\text{CTFS}} C_{yn}$$

then

$$z(t) = x(t) \otimes y(t) \xleftrightarrow{\text{CTFS}} C_{zn}$$

$$= T_0.C_{xn}C_{yn}$$

Thus, convolution of two periodic signals results in multiplication of their coefficients and period T_0.

4.8.7 Multiplication Property

The multiplication property states that if

$$x(t) \xleftrightarrow{\text{CTFS}} C_{xn}$$

and
$$y(t) \xleftrightarrow{\text{CTFS}} C_{yn}$$

then
$$z(t) = x(t).y(t) \xleftrightarrow{\text{CTFS}} C_{zn}$$
$$= T_0 . C_{xn} \otimes C_{yn}$$

Thus, it is clearly that the multiplication of the two signals in time-domain results in convolution of their coefficients.

4.8.8 Symmetry Properties

Symmetry properties states that if $x(t)$ is real then

$$C_n^* = C_{-n}$$

If $x(t)$ is imaginary then

$$C_n^* = -C_{-n}$$

4.8.9 The Fourier Transform

So far, the discussion was confined to the use of Fourier series in the analysis of the following cases of wave forms:

(*i*) an arbitrary wave function form of a finite interval.

(*ii*) a periodic waveform over an entire interval (– α, + α).

However, it is desirable to analyse waveform, periodic or not, over an entire interval (– α, + α) because a vast majority of interesting signals extend for all time (– α, + α) and are nonperiodic in nature.

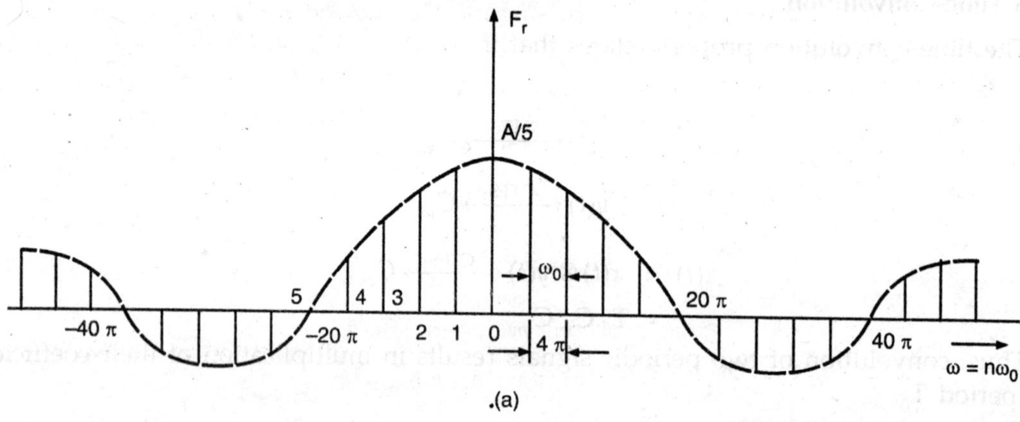

.(a)

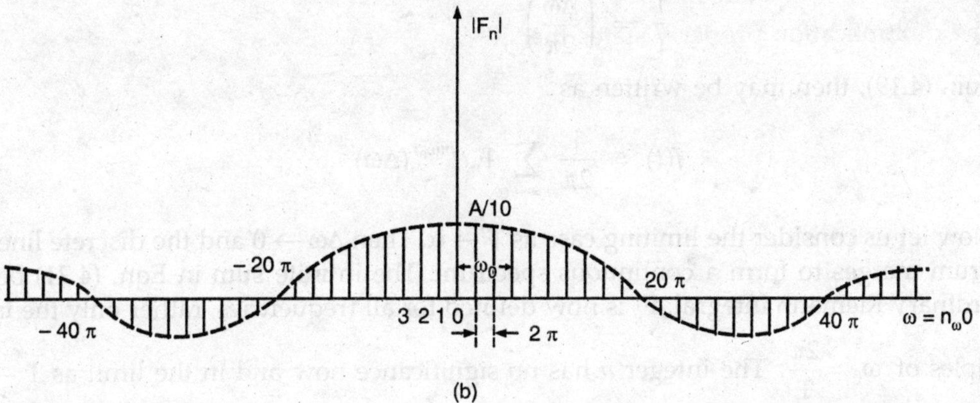

Fig. 4.18 *Spectrum of a Gate Function for T = 1/2 s (b) Spectrum of a Gate function for T = 1 s*

4.9 ANALYSIS OF A NON-PERIODIC FUNCTION OVER ENTIRE INTERVAL

A nonperiodic signal can be viewed as a limiting case of a periodic signal; where the period of the signal approaches infinity. This approach will be use to develop the frequency domain representation of a nonperiodic signal over an entire interval.

Consider a periodic function $f(t)$ with period T. The Fourier series representation of this function is

$$f(t) = \frac{1}{T}\sum_{n=-\alpha}^{\alpha} F_n e^{jn\omega_0 t}$$

where

$$\omega_0 = \frac{2\pi}{T} \qquad \text{...(4.19)}$$

$$F_n = \int_{-T/2}^{T/2} f(t)e^{-jn\omega_0 t}\, dt \qquad \text{...(4.20)}$$

A typical discrete spectrum plot would look give Fig. 4.19.

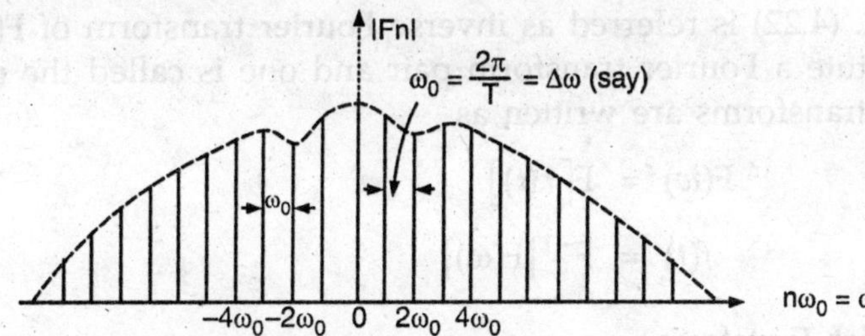

Fig. 4.19 *A Typical Discrete Spectrum*

The spacing between successive harmonics is

$$w_0 = \frac{2\pi}{T} = \Delta\omega$$

or
$$\frac{1}{T} = \left(\frac{\Delta\omega}{2\pi}\right)$$

Eqn. (4.19), then may be written as

$$f(t) = \frac{1}{2\pi} \sum_{n=-\alpha} F_n e^{jn\omega_0 t} (\Delta\omega) \qquad ...(4.21)$$

Now let us consider the limiting case as $T \rightarrow \alpha$. Then $\Delta\omega \rightarrow 0$ and the discrete line in the spectrum merges to form a continuous spectrum. The infinite sum in Eqn. (4.21) becomes the ordinary Riemann integral. F_n is now defined for all frequencies, rather only the integral multiples of $\omega_0 = \frac{2\pi}{T}$. The integer n has no significance now and in the limit as $T \rightarrow \alpha$; F_n becomes a continuous function $F(\omega)$ given as,

$$F(\omega) = \lim_{T \rightarrow \alpha} F_n$$

$$= \int_{-\alpha}^{\alpha} f(t) e^{-j\omega t} dt$$

The Fourier series given in Eqn. 4.21 for a periodic signal now becomes the Fourier integral representation for a non-periodic function over entire interval, given by

$$f(t) = \frac{1}{2\pi} \int_{-\alpha}^{\alpha} F(w) e^{j\omega t} d\omega \qquad ...(4.22)$$

$$F(\omega) = \int_{-\alpha}^{\alpha} f(t) e^{-j\omega t} dt \qquad ...(4.23)$$

In general, $F(\omega)$ is a complex function of ω and may be written as

$$F(\omega) = |F(\omega)| e^{j\theta(\omega)} \qquad ...(4.24)$$

$F(w)$ represented by Eqn. (4.23) is called Fourier transform of $f(t)$. Similarly $f(t)$ represented by Eqn. (4.22) is referred as inverse Fourier transform of $F(\omega)$. The functions $f(t)$ and $F(\omega)$ constitute a Fourier transform pair and one is called the mate of the other. Symbolically those transforms are written as

$$F(w) = F[f(t)]$$

and
$$f(t) = F^{-1}[F(\omega)]$$

4.9.1 Merits of Fourier Transform

A transform is a set of rules substituting one function for another. A function $f(t)$ can have a variety of transforms. Fourier transform is most useful for analysing signals involved in communication systems. Some of the main advantages of this transform are:

(*a*) The original time functions can be uniquely recovered from it.

(*b*) It has a property analogous to common logarithm that helps in evaluation convolution integrals.

(*c*) Although Laplace transformers is extensively used for solving the transient problems of electrical systems, Fourier transform is much more useful in communication system because here the phase and amplitude characteristics are readily known. Laplace transform, on the other hand, is more useful in electrical systems, where network transfer function as a ratio of polynomials in $s(s = 6 + j\omega)$ is readily specified and the analysis is based on poles and zeros.

4.9.2 Limitations of Fourier Transform

The Fourier transform ($s = j\omega$) is closely related with Laplace transform. In Fourier transform, damping factor $\sigma = 0$, and hence the Fourier transform may not converge for many time-functions. In other words, there are many time functions for which Fourier transform does not exist. Such functions are not absolutely integrals i.e., their Fourier integral does not converge in the limit $T \to \alpha$. The Laplace transform for such function may exist because damping factor σ makes the Laplace integral to converge.

4.9.3 Existence of the Fourier Transform

For a function $f(t)$ to be Fourier transformable, it is sufficient that $f(t)$ satisfies Dirichlet's conditions given below:

(*a*) The function $f(t)$ is a single valued with a finite number of maxima and minima; and a finite number of discontinuities in any finite time interval.

(*b*) The function $f(t)$ is absolutely integral, i.e.,

$$\int_{-\alpha}^{\alpha} |f(t)|\, dt < \alpha$$

4.9.4 Continuous Spectrum

The spectrum of a non-periodic function is continuous. Let us consider the amplitude spectrum plot shown in Fig. 13.1. This is a discrete line spectrum corresponding to a periodic function with period T. Successive harmonics are spaced by $(\Delta\omega) = \omega_0 = \frac{2\pi}{T}$. The spectrum can be extended to a corresponding non-periodic function by considering the limiting case as $T \to \alpha$. Then, $\Delta\omega \to 0$; the discrete lines in the spectrum merge and we get a continuous frequency spectrum as shown in Fig. . The amplitude of a single frequency component becomes infinitesimal. However, the amplitude for a range of frequency does exist, and is proportional to $F(\omega)$. The Fourier transform $F(\omega)$ is also known as the spectral density function because $F(\omega).d(\omega)$ gives the amplitude of frequency components lying in the range '$d(\omega)$'.

4.9.5 Fourier Transform Involving Impulse Function

The sufficient condition required for the Fourier transform of a function to exist is that the function $f(t)$ should be obsolutely integrated i.e.

$$\int_{-\alpha}^{\alpha} f(t)\,dt < \alpha$$

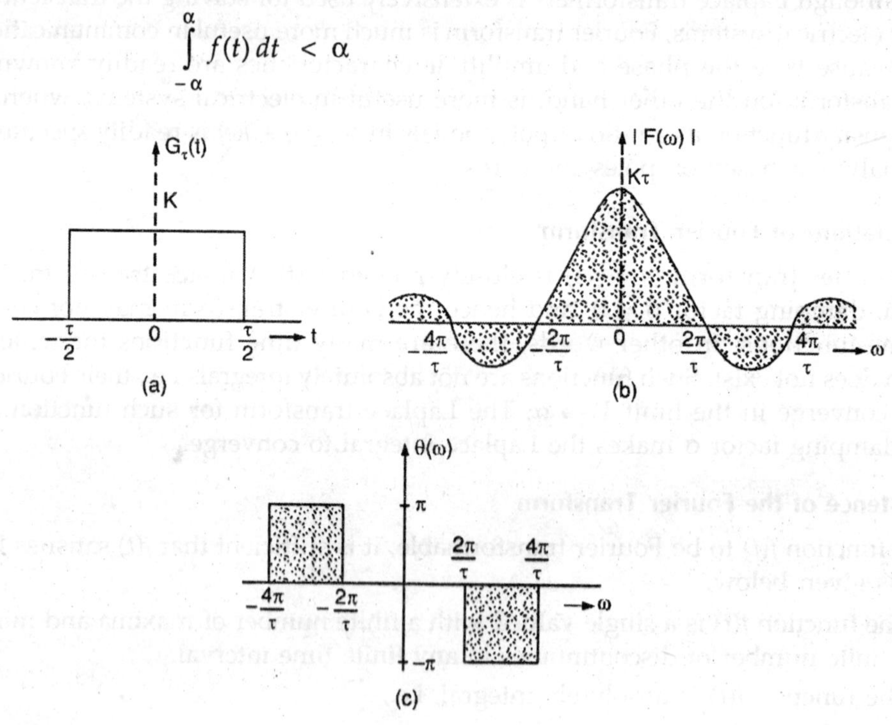

Fig. 4.20

But this is not the necessary condition for existance of the Fourier transform. A number of useful signals in communication systems are not absolutely integrable, i.e. abc signa, but their Fourier transform exists in the limiting condition. The Fourier transform for these signals are obtained by using the concept of impulse function or Delta function denoted by $\delta(t)$. The impulse function or Delta falls under the category of singularity function.

4.9.6 Singularity Function

Singularity functions are discontinuous functions, or have discontinuous derivatives. These functions may have continuous derivatives only upto finite order.

Let us now examine the significance of singularity functions in signal analysis. When a unit step voltage is applied to a capacitor as shown in Fig. 4.2(*a*), the current $i(t)$ flowing through the circuit is

$$i(t) = \frac{du(t)}{dt}$$

The idealized unit function, shown in Fig. 4.21(*b*) is unity for $t > 0$, but discontinuous at $t = 0$. Hence, the current $i(t)$ is zero everywhere except at $t = 0$, where the derivative of the step function is undefined.

The unit impulse function, which is mathematically defined as derivative of the unit step function, is not a true function in the mathematically sense where a function should be defined at every value of t. The unit impulse fuinction, however, can be defined as a 'limiting case of derivatives of unidealized step functions.' The concept is made clear by the following example.

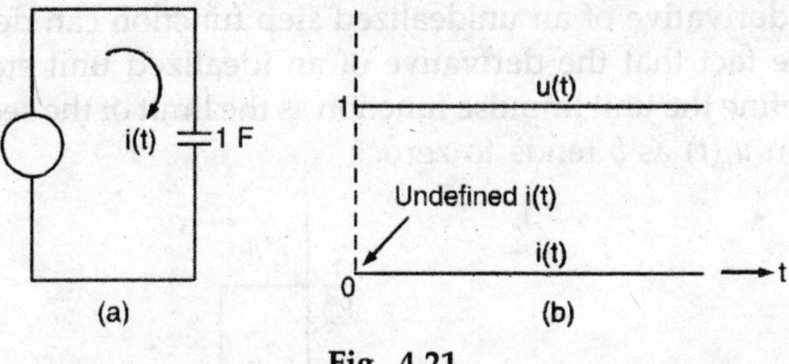

Fig. 4.21

Consider an unidealized voltage function $u_b(t)$ shown in Fig. 4.22(a). This voltage function becomes a unit step function if the limit b goes to zero, i.e.

$$\lim_{b \to 0} u_b(t) = u(t)$$

When this voltage $u_b(t)$ is applied to a capacitor, the current

$$i = \frac{du_b(t)}{dt}$$

will be a rectangular pulse of height $\frac{1}{b}$, and width b as shown in Fig. 4.22(b).

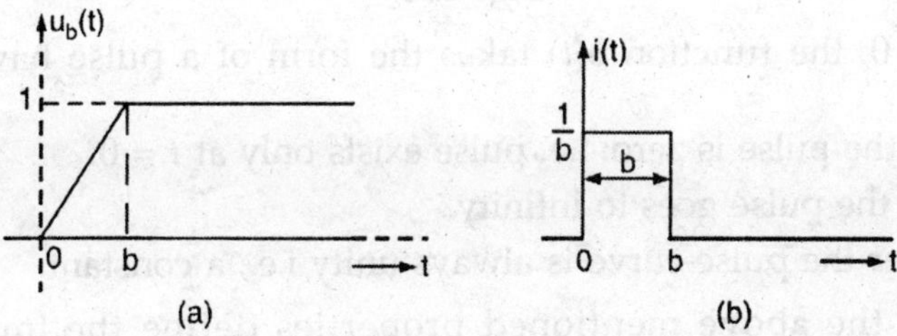

Fig. 4.22

Let us now analyse same situations for various decreasing values of b. For $b = b_1$, b_2, b_3 and b_4, the function $u_b(t)$ and the resulting respective currents i_1, i_2, i_3 and i_4 are shown in Fig. 4.23. The current pulse corresponding to $b = b_1$ has the width b_1 and height $\frac{1}{b}$, with the resulting area of unity. The current pulse corresponding to $b = b_2$ has reduced in width, but increased in height to result in the same area of unity. Some is the case for $b = b_3$ and $b = b_4$. Thus as the value of b reduces. The height of the pulse sequences increases, and the area under the current pulse curve is always unity, i.e. a constant.

In the limit when b goes to zero, the width of the pulse becomes zero, and the height of the pulse approaches infinity, maintaining the area under the pulse-curve unity. In the limit

b goes to zero. $u_b(t)$ becomes a unit step function $u(t)$, and the corresponding current pulse becomes an impulse function $\delta(t)$, given by

$$\delta(t) = \lim_{b \to 0} \frac{d}{dt} \{u_b(t)\}$$

It is clear that the derivative of an unidealized step function can defined a unit impulse function, inspite of the fact that the derivative of an idealized unit step function does not exist. Hence, we can define the unit impulse function as the limit of the sequence of derivatives of unidealized function $u_b(t)$ as b tends to zero.

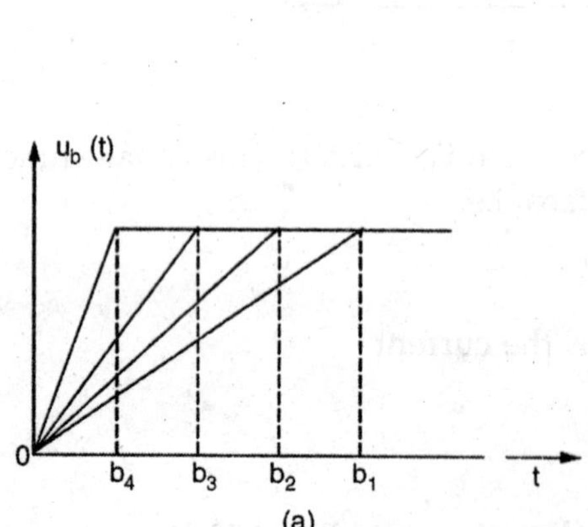

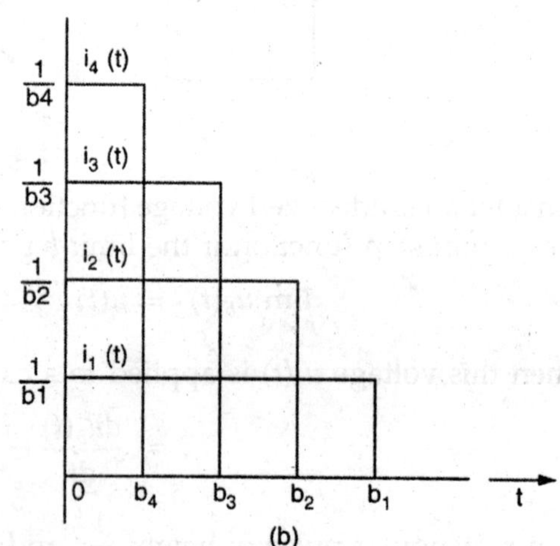

Fig. 4.23

In the limit $b \to 0$, the function $\delta(t)$ takes the form of a pulse having the following properties:

(*a*) The width of the pulse is zero: i.e. pulse exists only at $t = 0$.

(*b*) The height of the pulse goes to infinity.

(*c*) The area under the pulse-curve is always unity i.e., a constant.

Mathematically, the above mentioned properties define the impulse function as follows:

$$\delta(t) = 0, t = 0 \quad \text{...(4.25}a\text{)}$$

$$\int_{-\alpha}^{\alpha} \delta(t)\, dt = 1 \quad \text{...(4.25}b\text{)}$$

The delta function may also be defined as a "limiting case of sequences of regular functions". A number of regular functions defining impulse function by their sequences under the limiting case, are shown in Fig. 4.24.

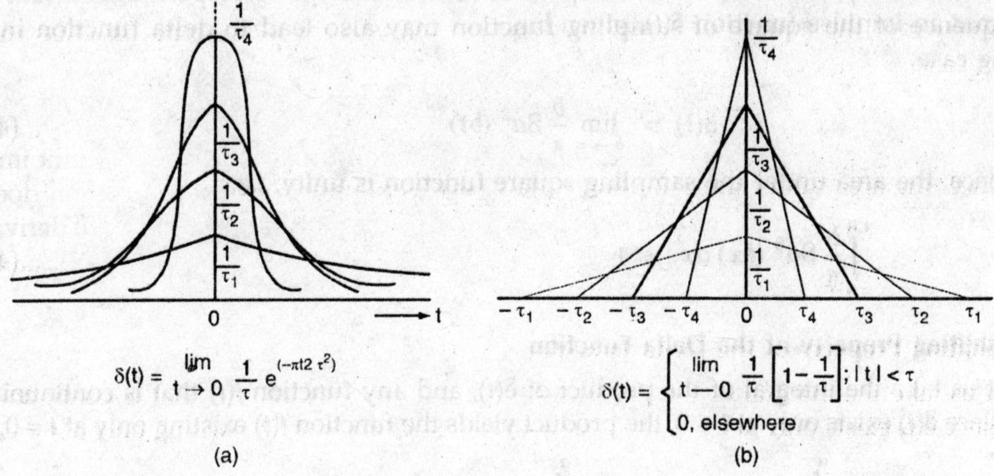

Fig. 4.24 (*a*) Gaussian Pulse Sequence; (*b*) Triangular Pulse Sequence

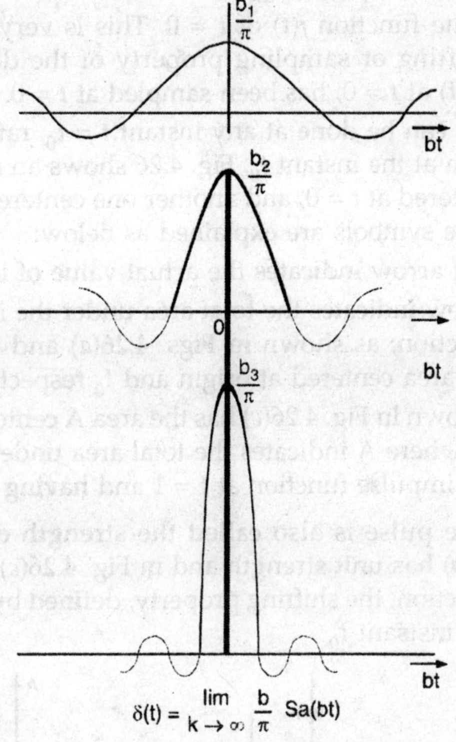

Fig. 4.25 Sampling Function Sequence

The definition of the impulse function is given along with the corresponding figure, sequence of sampling functions are shown in Fig. 4.25. In the limit, area under the curves is unity.

4.9.7 Sampling Square Function

Sequence of the square of sampling function may also lead to delta function in the limiting case,

$$\delta(t) = \lim_{b \to \propto} \frac{b}{\pi} \text{Sa}^2 (bt) \qquad ...(4.26)$$

Hence, the area under the sampling square function is unity, i.e.

$$\int_{-\alpha}^{+\alpha} \frac{b}{\pi} \text{Sa}^2 (bx)\, dx = 1 \qquad ...(4.27)$$

4.9.8 Shifting Property of the Delta Function

Let us take the integral of the product of $\delta(t)$, and any function $f(t)$ that is continuous at $t = 0$. Since $\delta(t)$ exists only at $t = 0$, the product yields the function $f(t)$ existing only at $t = 0$, i.e.,

$$\int_{-\alpha}^{\alpha} f(t)\, \delta(t)\, dx = f(0) \int_{-\alpha}^{\alpha} \delta(t)\, dt$$

$$= f(0).1 = f(0) \qquad ...(4.28)$$

where; $f(0)$ is the value of the function $f(t)$ of $t = 0$. This is very significant result, and is appropriately known as shifting or sampling property of the delta function because the impulse shifts the value of $f(t)$ at $t = 0$; has been sampled at $t = 0$.

The sampling or shifting can be done at any instant $t = t_0$, rather than only at $t = 0$, by defining the impulse function at the instant t_0. Fig. 4.26 shows an impulse function centered at $t = 0$, and another one centered at $t = 0$, and another one centered at $t = t_0$, denoted by $\delta(t)$ and $\delta(t - t_0)$ respectively. The symbols are explained as below:

(*a*) The upward pointing arrow indicates the actual value of infinity.

(*b*) The height of the arrow indicates the total area under the impulse. This area is unity for unit impulse function; as shown in Figs. **4.26**(*a*) and (*b*). These represent delta functions of the unit area centered at origin and t_0 respectively.

(*c*) The delta function shown in Fig. 4.26(*c*) has the area A centered at $t = t_0$. Thus impulse written as $A\delta(t - t_0)$ where A indicates the total **area under** the pulse. Similarly, Fig. 4.26(*d*) represents an impulse function at $t = 1$ and **having** unity area.

The total area under the pulse is also called the **strength** of the impulse. Thus, the impulse in Fig. 4.26(*a*) and (*b*) has unit strength and in Fig. **4.26**(*c*) has a strength of A, with this definition of impulse function, the shifting property, defined by Fig. 4.28 can be extended for shifting to any arbitrary instsant t_0

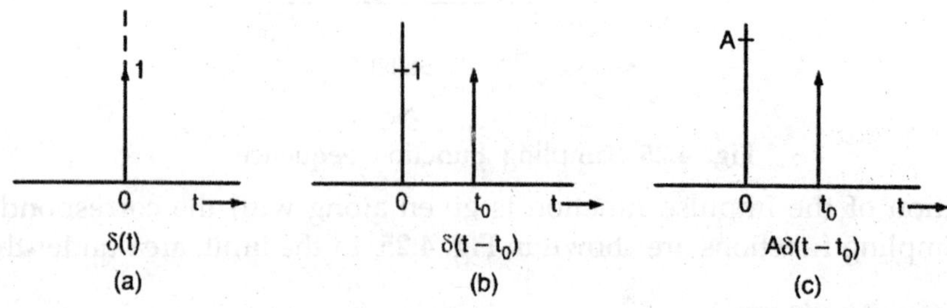

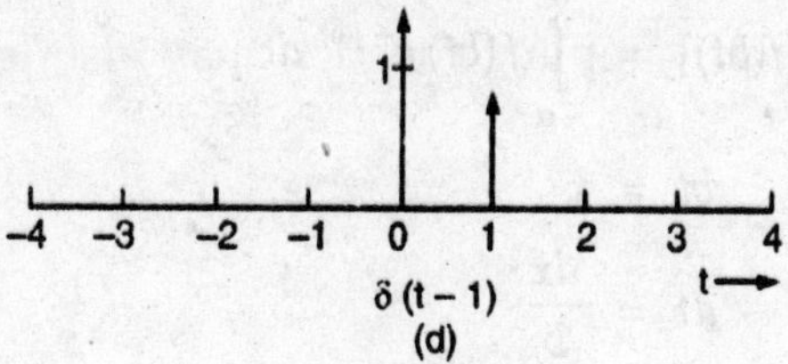

Fig. 4.26

$$\int_{-\alpha}^{\alpha} f(t)\,\delta(t-t_0)\,dt = f(t_0) \qquad ...(4.29)$$

which states that the multiplication of a continuous function $f(t)$ with an impulse function $\delta(t-t_0)$ yields the sampled value of $f(t)$ at $t = t_0$, i.e. $f(t_0)$.

Impulse function is frequently used to represent point mass, point charge, point source and other similar physical entities used in science and engineering. The step function and the impulse function with all their higher derivatives are known as singular function.

4.10 PROPERTIES OF FOURIER TRANSFORM

Fourier transform has many important properties. Apart from giving simple solutionsof complicated Fourier transform, these properties also help in finding the effect of various time domain operations on frequency domain.

4.10.1 Lineary Property

$$f_1(t) \longleftrightarrow F_1(\omega)$$
$$f_2(t) \longleftrightarrow F_2(\omega)$$
$$f_3(t) \longleftrightarrow F_3(\omega)$$
$$\cdots$$
$$\cdots$$
$$f_n(t) \longleftrightarrow F_n(\omega)$$

Then,

$$a_1f_1(t) + a_2f_2(t) + a_3f_3(t) \ldots + a_nf_n(t) + a_1f_1(\omega) + a_2f_2(\omega) + a_3f_3(\omega) + \ldots + a_nF_n(\omega) \qquad ...(4.30)$$

where $a_1, a_2, a_3, \ldots a_n$ are arbitrary constant.

This property is proved easily by linearity property of integrals used in defining Fourier transform.

4.10.2 Time-Scalling Property

Let $\qquad f(t) \longleftrightarrow F_1(\omega)$

then $\qquad f(bt) \longleftrightarrow \dfrac{1}{|b|}F\left(\dfrac{\omega}{b}\right) \qquad ...(4.31)$

where b is a real constant.

Proof: $$\mathrm{F}[f(bt)] = \int_{-\alpha}^{\alpha} f(bt)\, e^{-j\omega t}\, dt$$

Set $$bt = x$$

so that $$dt = \frac{dx}{b}$$

Case 1: When $b > 0$, b becomes a positive constant

$$\mathrm{F}\{f(bt)\} = \frac{1}{b}\int_{-\alpha}^{\alpha} f(x)\, e^{j\left(\frac{\omega}{b}\right)x}\, dx$$

$$= \frac{1}{b}\mathrm{F}\left(\frac{\omega}{b}\right)$$

Case 2: When $b < 0$, b becomes a negative constant.

As in Case 1

$$b = -b$$

$$f(bt) \longleftrightarrow \frac{1}{|b|}\mathrm{F}\left(\frac{\omega}{b}\right)$$

$$\mathrm{F}[f(bt)] = -\frac{1}{b}\mathrm{F}\left(\frac{\omega}{b}\right)$$

Combined, the two cases are expressed as,

$$f(bt) \longleftrightarrow \frac{1}{|b|}\mathrm{F}\left(\frac{\omega}{b}\right)$$

Significance: The function $f(bt)$ represents compression of $f(t)$ in time domain, i.e., the period of the function is reduced by a factor b. The function

$$\mathrm{F}\left(\frac{\omega}{b}\right)$$

represent expansion of $\mathrm{F}(\omega)$ by the same factor. Thus, the scaling property states that "compression of a function in time domain by a factor b is equivalent to expansion of its frequency domain by the same factor b or vice-versa."

Compression in the time domain means reducing the period of the wave. Since frequency is reciprocal of the period, the range of frequency components present in the waveform will increase, i.e., the spectrum is expanded. Let us take the example of a function sin $\omega_0 t$, a signal of frequency ω_0. When compressed in time domain by a factor 2, it becomes sin $2\omega_0 t$. The frequency now is $2\omega_0$, which is double as compared to the previous case. Thus, compression of a sine wave in the time domain, reducing the period, is equivalent to expression in the frequency domain, increasing the frequency.

In the special case when

$$b = -1,$$

the property yields,

$$f(-t) \longleftrightarrow F(-\omega) \qquad ...(4.32)$$

The theorem is also known as "scaling uncertainty principle" as it implies that both the signal and its Fourier transform cannot be of short duration. A signal of shorter duration yields a wider spectrum.

4.10.3 Duality or Symmetry Property

Let us consider

$$f(t) \longleftrightarrow F(\omega)$$

then $\quad f(t) \longleftrightarrow 2\pi f(-\omega)$

or $\quad F(t) \longleftrightarrow f(-f)$

For example, for the Gate function in Fig. 4.27, the Fourier transform is a sampling function. According to symmetry property, a sampling function has a Fourier transform which is a Gate function as in Fig. 4.27.

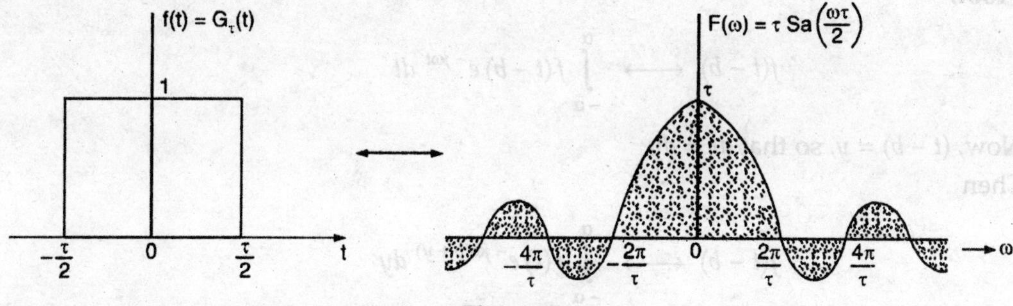

Fig. 4.27

Proof: The inverse Fourier transform of F(ω) is given by

$$f(t) = \frac{1}{2\pi}\int_{-\alpha}^{\alpha} F(\omega)\, e^{j\omega t}\, d\omega$$

Therefore,

$$2\pi f(-t) = \int_{-\alpha}^{\alpha} F(\omega)\, e^{-j\omega t}\, d\omega$$

Since, ω is a dummy variable, we can interchange the variable t and ω, i.e.,

$$2\pi f(-\omega) = \int_{-\alpha}^{\alpha} F(t)\, e^{-j\omega t}\, dt$$

$$= F[F(t)]$$

Thus, the Fourier transform of the time function $F(t)$ is

$$2\pi f(-\omega),$$

$$F(t) \longleftrightarrow 2\pi f(-\omega)$$

For even function

$$f(-\omega) = f(\omega)$$

and hence

Thus is called a perfect symmetry.

4.10.4 Time Shifting Product

Let us see

$$f(t) \longleftrightarrow F(\omega)$$

Then

$$f(t-b) \longleftrightarrow F(\omega)\, e^{-j\omega b} \qquad ...(4.33)$$

Statement: This theorem states that a shift in the time domain by an amount b is equivalent to multiplication by $e^{-j\omega b}$ in the frequency domain, i.e., magnitude F(ω) spectrum remains unchanged, but phase spectrum is changed by $-\omega b$. A shift in time domain does not change the magnitude of a frequency component. A shift of b at a frequency ω is equivalent to a phase shift of ωb.

Proof:

$$f(t-b) \longleftrightarrow \int_{-\alpha}^{\alpha} f(t-b)\, e^{-j\omega t}\, dt$$

Now, $(t-b) = y$, so that $dt = dy$

Then

$$f(t-b) \longleftrightarrow \int_{-\alpha}^{\alpha} f(t)\, e^{-j\omega(b+y)}\, dy$$

$$= \int_{-\alpha}^{\alpha} f(y)\, e^{-j\omega y} e^{-j\omega b}\, dy$$

Since y is a dummy variable

$$f(t-b) \longleftrightarrow F(\omega)\, e^{-j\omega b}$$

4.10.5 Frequency Shifting Property

If

$$f(t) \longleftrightarrow F(\omega)$$

Then

$$e^{j\omega_c t} f(t) \longleftrightarrow F(\omega - \omega_c) \qquad ...(4.34)$$

Proof: The Fourier transform F(ω) is given as

$$F(\omega) = \int_{-\alpha}^{\alpha} f(t)\, e^{-j\omega y}\, dt$$

Therefore,

$$f(t)\,e^{-j\omega_c t} \longleftrightarrow \int_{-\alpha}^{\alpha} e^{j\omega_c t} f(t)\, f(t)\, e^{-j\omega t}\, dt$$

$$= \int_{\alpha}^{\alpha} f(t)\, e^{-j(\omega-\omega_c)t}\, dt$$

or

$$\mathrm{F}\left[f(t)\,e^{j\omega_c t}\right] = \mathrm{F}(\omega - \omega_c)$$

Significance: According to this property, multiplication of a function $f(t)$ by $e^{j\omega_c t}$ is equivalent to shifting its Fourier transform F(ω) in the positive direction by an amount ω_c, i.e., spectrum F(ω) is translated by an amount ω_c. Therefore, this theorem is also known as **Frequency translation theorem**.

Translation of a spectrum is a very important phenomenon in communication system. This translation helps in simultaneous transmission of messages using frequency division multiplexing and is achieved by a process known as modulation. This process can be performed by multiplying the known signal $f(t)$ by a sinusoidal signal. This is because a sinusoidal function can be expressed as a sum of exponentials as follows:

$$f(t)\cos\omega_c t = \frac{1}{2}\left[f(t)\,e^{j\omega_c t} + f(t)\,e^{-j\omega_c t}\right]$$

Therefore,

$$\mathrm{F}[f(t)\cos\omega_c t] = \frac{1}{2}[\mathrm{F}(\omega+\omega_c)+\mathrm{F}(\omega-\omega_c)] \qquad \text{...(4.35}a\text{)}$$

Similary

$$\mathrm{F}[f(t)\sin\omega_c t] = \frac{1}{2}[\mathrm{F}(\omega+\omega_c)-\mathrm{F}(\omega-\omega_c)] \qquad \text{...(4.35}b\text{)}$$

The multiplication of a time function $f(t)$ with a sinusoidal function translates the whole spectrum F(ω) to $\pm\,\omega_c$. The same result can be obtained by the process of modulation. This result, also known as modulation theorem, states that the multiplication of a time function $f(t)$ with a sinusoidal function of frequency ω_c translates the spectrum F(ω) by an amount ω_c. It is noted that $e^{j\omega_c t}$ can also provide frequency translation, but it is not a real signal, whereas, sinusoidal function is a real signal. Hence, sinusoidal function is used in practical modulation systems.

4.10.6 Area Under the Curve

(*i*) The area under a function $f(t)$ is equal to the value of its Fourier transform F(ω) at ω = 0, i.e., if

$$f(t) \longleftrightarrow \mathrm{F}(\omega)$$

then

$$\int_{-\alpha}^{\alpha} f(t)\, dt = \frac{1}{2\pi}\mathrm{F}(0) \qquad \text{...(4.36}a\text{)}$$

This can be proved by putting ω = 0 in the expression of F(ω).

(*ii*) The area under the Fourier transform F(ω) of a function $f(t)$ is equal to the value of the function $f(t)$ at t = 0, i.e.,

$$f(0) = \int_{-\alpha}^{\alpha} F(\omega)\, d\omega \qquad \text{...(4.36}b\text{)}$$

The result can be proved by taking t = 0 in the expression of inverse Fourier transform of F(ω).

4.10.7 Time Differentiation and Integration

(*i*) Differentiation of a function $f(t)$ in the time domain is equivalent to multiplication of its Fourier transform by a factor $j\omega$,

$$\frac{df(t)}{dt} \longleftrightarrow j\omega F(\omega) \qquad \text{...(4.37}a\text{)}$$

Proof: The result can be proved by taking the first derivative of both the sides of the expression for inverse Fourier transform,

$$f(t) = \frac{1}{2\pi}\int_{-\alpha}^{\alpha} F(w)\, e^{j\omega t}\, d\omega$$

$$\frac{df(t)}{dt} = \frac{1}{2\pi}\frac{d}{dt}\left[\int_{-\alpha}^{\alpha} F(\omega)\, e^{j\omega t}\, d\omega\right]$$

Changing the order of differentiation and integration, we get

$$\frac{df(t)}{dt} = \frac{1}{2\pi}\int_{-\alpha}^{+\alpha} \frac{d}{dt}\left\{F(\omega)\, e^{+j\omega t}\right\} d\omega$$

$$= \frac{1}{2\pi}\int_{-\alpha}^{\alpha} j\omega F(\omega)\, e^{j\omega t}\, d\omega$$

$$= F^{-1}[j\omega F(\omega)]$$

Hence,

$$F\left[\frac{df(t)}{dt}\right] = j\omega F(\omega)$$

or

$$\frac{df(t)}{dt} \longleftrightarrow j\omega F(\omega)$$

The theorem is logical because Function $f(t)$ is expressed as the continuous sum of exponential function $e^{j\omega t}$. Hence, the differentiation of $f(t)$ yields $j\omega$ in its Fourier transform since time derivative of $e^{j\omega t}$ is $j\omega e^{j\omega t}$. In general,

$$\frac{d^n f(t)}{dt^n} = (j\omega)^n \, F(\omega) \qquad \text{...(4.37b)}$$

(*ii*) Integration of a function $f(t)$ in time domain is equivalent to division of its Fourier transform by the factor $j\omega$, an integration of $e^{j\omega t}$ yields $\frac{1}{j\omega} e^{j\omega t}$. The theorem is expressed as,

$$\int_{-\alpha}^{\alpha} f(\tau)\, d\tau \longleftrightarrow \frac{1}{j\omega} F(\omega) \qquad \text{...(4.37c)}$$

Provided

$$F(0) = 0$$

The theorem can be proved with an approach similar to the time differentiation theorem.

4.10.8 Conjugate Functions

If

$$f(t) \longleftrightarrow F(\omega)$$

then

$$f^*(t) \longleftrightarrow F^*(-\omega) \qquad \text{...(4.38)}$$

Where, the asterisk denotes the complex conjugate operation. The theorem can be proved by taking the complex conjugates of both sides of the equation for inverse Fourier transform of $F(\omega)$ and then by replacing f with $-f$.

4.10.9 Frequency Differentiation, Dual of Time Differentiation

If

$$f(t) \longleftrightarrow F(t)$$

then

$$-j\,t\,f(t) \longleftrightarrow \frac{dF}{d\omega} \qquad \text{...(4.39)}$$

The theorem can be proved by taking frequency derivative of $F(\omega)$, and then changing the order of integration and differentiation.

PROBLEMS

1. Starting from fundamental, find the exponential Fourier series of the Figure given and derive corresponding trignometric series.

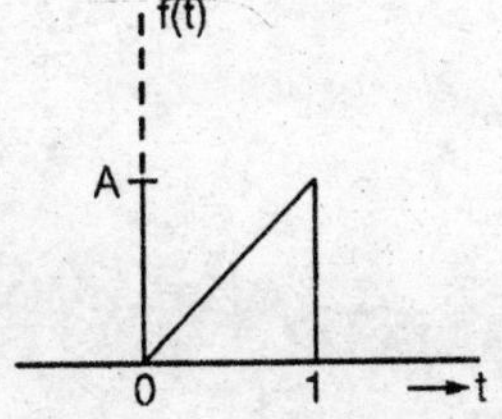

Fig. 4.28 A Triangular Pulse

2. Find the Fourier series of a periodic impulse function with period T and strength A. Draw the spectrum.
3. Prove the following properties of Fourier transform
   (*a*) Conjugate function
   (*b*) Frequency differentiation theorem
   (*c*) Time integration theorem
4. Derive an exponential series from a trignometric series by replacing the sine and cosine functions with their exponential equivalents, and show that the co-efficients are related as follows:

$$F_o = a_o$$

$$F_n = \frac{1}{2}(a_n - jb_n)$$

$$F_{-n} = \frac{1}{2}(a_n + jb_n)$$

5. Find the Fourier transform of the function $f(t)$ given below
   (*a*) $f(t) = t\, e^{-at}u(t)$
   (*b*) $f(t) = |t|$
6. Find the Fourier transform of a Gaussian pulse given by

$$f(t) = e^{\left(\frac{-t^2}{2\sigma^2}\right)}$$

7. Show that set of function $\cos n\,\omega_o t$ and $\sin n\,\omega_o t$, $n = 0, 1, 2, \ldots$, form a complete orthogonal set over the internal $t_0 \le t \le t_0 + T$ for any choice of $t_0 . \left(T = \frac{2\pi}{\omega_0}\right)$.
8. Show that a set of functions $e^{jn\omega_o t}$ $(n = 0, \pm 1, \pm 2, \pm 3, \ldots)$ is a closed orthogonal set over the interval $(t_0, t_0 + T)$.

❑❑❑

Noise

5.1 INTRODUCTION

Undesired electrical signals which are introduced with a message signal during the transmission, or processing of the latter, are called **Noise**. Noise, thus is an unwanted signal that corrupts a desired message signal. In general, noise may be predictable, or unpredictable (random), in nature. The predictable noise can be estimated, and eliminated, by proper engineering design. Some examples of such noise are: power supply hum, spurious oscillation in feedback amplifiers, ignition radiation pickup, radiation pickup generated by electrical appliances, and fluorescent lighting. The predictable noise, generally is man-made and can be reduced or eliminated.

Unpredictable noise varies randomly with time and, as such we have no control over this noise. Identification of the message signal at the receiver depends upon the amount of noise accompanied by the message during the process of communication. In the absence of noise, identification of message signal at the receiver is perfect. The presence of noise complicate the systems of communication. The amount of noise power present in the received signal decides the minimum power level of the desired message signal at the transmitter. The term noise is normally used to refer the unpredictable, or random noise.

5.2 SOURCES OF NOISE

There are various sources of random noise. They are broadly classified as:

(*a*) External Noise, and

(*b*) Internal Noise.

5.2.1 External Noise

External noise may be defined as that type of noise which is generated external to a communication system i.e., whose sources are external to the communication system.

External noise can not be analysed quantitatively. In addition to this, for a given geographical point or location, external noise can be controlled. Thus, to reduce the effect of external noise, the only way is to shift the communication system to other place or location which has comparatively smaller external noise. Thus due to this reason, the satellite earth stations are generally located in noise-free valleys.

External noise may be classified as under:

(i) Atmospheric Noise

(ii) Extraterrestrial Noise

(iii) Industrial Noise.

5.2.2 Atmospheric Noise

Atmospheric Noise, which is also called static, is produced by lightning discharges in thunderstorms and other natural electrical disturbances which occur in the atmosphere. These electric impulse are random in nature. Thus this energy is spread over the complete frequency spectrum used for radio communication. Atmospheric noise contains spurious radio signals which are distributed over a wide frequency range. In fact, these spurious radio waves constituting the noise, propagate over the earth in the manner as the required radio waves of same frequency. Thus, due to this reason, at any given receiving point, the receiving antenna picks up not only the required signal but also the static from all the thunderstorms.

It has been observed that the field strength of atmospheric noise varies approximately inversely with the frequency. This concludes that large atmospheric noise is produced in low and medium frequency bands where as very small noise is produced in the VHF and UHF bands.

Thus Atmospheric noise becomes less severe at frequencies above about 30 MHz.

5.2.3 Extraterrestrial Noise

There are several types of extraterrestrial noise or space noise depending upon their sources. Extraterrestrial noise may be divided into following two sub-groups as under:

(i) Solar Noise

(ii) Cosmic Noise.

Solar Noise

Solar noise is the electrical noise eliminating from the sun. Under steady condition, there is a regular radiation of noise from the sun. This radiation of noise from the sun is due to the fact that sun is a big body at an extremely high temperature and it radiates electric energy in the form of noise over a very wide frequency spectrum including also the frequency spectrum which is occupied by radio communication.

However, the condition of the Sun varies and follows an eleven years cycle. Hence at the peak of this eleven years cycle, the electrical disturbances are caused. This form of noise from electrical disturbance is an addition to the quiet noise from the Sun. Although this additional noise is produced from a small portion of the Sun's surface, it can still be larger than the quiet solar noise. The solar cycle repeats this type of electrical disturbances nearly every eleven years.

Cosmic Noise

Distant stars can also be considered Suns. These distant stars have high temperatures and therefore radiate noise in the same manner as the Sun. The noise received from these distant stars is thermal noise and is distributed almost uniformly over the entire sky. The noise is also received from the centre of our own galaxy, for other distant galaxies and from other virtual point sources such as quasars and pulsars. This type of galactic noise is quite intense but since it comes from very distant sources, the angle subtended by the earth is quite small. Therefore, the strength of galactic noise received on the earth gets diminished.

As a conclusion, we can say that the space noise is quite, significant at frequencies in the 1.43 GHz. Also, in the frequency range of 20 to 12 MHz, the space noise becomes the strongest noise component next only to the industrial noise. But, below 20 MHz, the space noise does not penetrates much through the ionosphere to reach the earth.

5.2.4 Industrial Noise

The industrial noise or man-made noise is that type of noise which is produced by such sources as automobiles and aircraft ignition, electrical motors, switch gears and leakage from his voltage transmission lines and several other heavy electrical equipments. Such type of noise is produced by the arc discharge taking place operation of all these machines or equipments. Industrial or man-made noise is quite intensive in industrial areas, dense populated urban areas. Industrial noise in such areas much stronger than all other sources of noise in the frequency range extending from about 1 MHz to 600 MHz.

Since industrial or man-made noise is highly variable and hence can be analyzed only statistically.

5.2.5 Internal Noise

Internal noise is that type of noise which is generated internally or within the communication system or receiver. Internal noise may be treated quantitatively and can also be reduced or minimized by proper system design. Since internal noise is randomly distributed over the entire frequency spectrum, the noise present in a given bandwidth B is the same at any frequency in the frequency spectrum. Hence, this random noise power is proportional to the bandwidth over which it is measured. Internal noise may be classified as under:

(*i*) Shot Noise
(*ii*) Partition Noise
(*iii*) Low Frequency or Flicker Noise
(*iv*) High Frequency or Transit-time Noise
(*v*) Thermal Noise

Shot Noise

Shot noise arises in active devices due to random behaviour of charge carries. In electron tubes, shot noise is generated due to the random emission of electron from cathodes, whereas in semiconductor devices shot noise is generated due to the random diffusion of minority carries or simply random generation and recombination of electron-hole pairs.

In fact, the current in electron devices (i.e. tubes or solid state device) flows in the form of discrete pulses, every time a charge carrier moves from one point to the other (i.e. cathode to plate). Hence although the current appears to be continuous, it is still a discrete phenomena. Fig. 5.1 shows this nature of current variation with time.

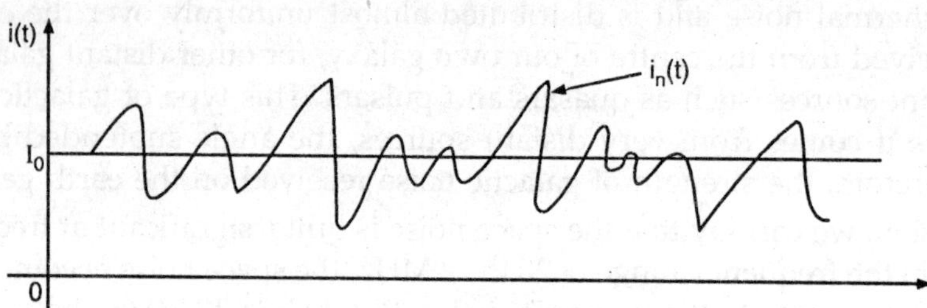

Fig. 5.1

The current fluctuates about a mean value I_0. The current $i_n(t)$ wiggles around the mean value. This current is called as shot noise current. The wiggling nature of this current is not visualized by normal instruments and normally it is assumed that the current is a constant equal to I_0. This wiggling nature of the current may be observed in a fast sweep oscilloscope.

Therefore, the total current $i(t)$ can be expressed as

$$i(t) = I_0 + i_n(t) \qquad ...(5.1)$$

Here I_0 is the mean (constant) and $i_n(t)$ is the fluctuating (shot-noise) current.

Power Density that shot noise current is given by the expression

$$i(t) = I_0 + i_n(t) \qquad ...(5.2)$$

The time-varying component $i_n(t)$ of the current $i(t)$ is always random in nature and hence it cannot be expressed as a function of time. This means that it is an indeterministic function. However, this indeterministic random function $i_n(t)$ may be specified by its power density spectrum.

In fact, the number of electrons, which contribute to the random current $i_n(t)$ is large.

Assuming that the electrons do not interact with each other during their movement or emission, this process (event) can be treated as statistically independent. Now, according to central limit theorem, such a process has a Gaussian distribution. Therefore, the shot-noise is Gaussian distributed with a zero mean value.

We can treat the total diode current as the sum of the current pulses and each pulse is formed by the transit of an electron from the cathode, to the anode. For all practical purposes, it may be observed that the power density spectrum of the statistically independent non-interacting random noise current $i_n(t)$ is expressed as

$$S_i(\omega) = qI_0 \qquad ...(5.3)$$

Here q is the electronic charge and I_0 is the mean value of the current in amperes.

It may be noted that the power density spectrum in Eqn. (5.3) is frequency independent. This type of frequency independence is only up to a frequency range determined by the transit time of an electron to reach from anode to the cathode. Beyond this frequency range, power density varies with frequency as shown in Fig. 5.2.

The transit time of an electron in a diode depends upon anode voltage V and may be expressed as

$$\tau = 3.36 \times \frac{d}{\sqrt{V}} \mu \text{ sec.}$$

Here *'d'* is the spacing between anode and the cathode.

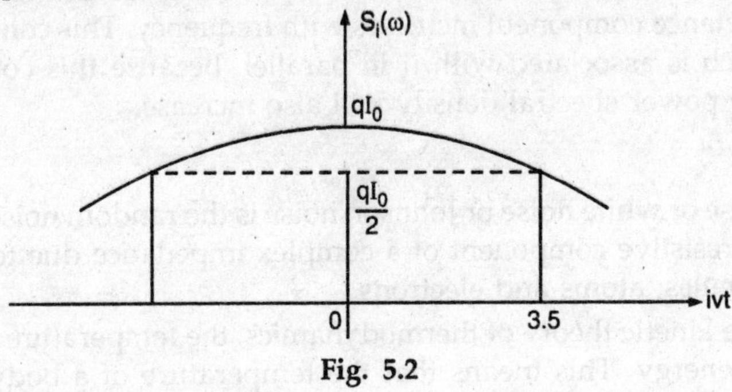

Fig. 5.2

Partition Noise

Partition noise is generated in a circuit when a current has to divide between two or more paths. This means that the partition noise results from the random fluctuations in the division. Hence, it is expected that a diode must be less noisy than a transistor, all else being equal. Due to this reason, the inputs of microwave receivers are generally taken directly to the diode mixers. For partition noise, the spectrum is a flat spectrum. Also, the active three terminal components in which the control terminal draw less current are less noisy. In recent time, Metal-Semiconductor Field Effect Transistors have been developed which draw almost zero gate bias current. Due to this reason, these devices have low partition noise and therefore finds application in low noise microwave amplification.

Flicker Noise, or Low-Frequency Noise

At low frequencies (below few kHz), a particular type of noise appears. The power spectral density of this noise increases as the frequency decreases. This noise is called as flicker noise or ($1/f$) noise. In case of vacuum tubes, the main causes of flicker noise are slow changes which take place in the oxide structure of oxide coated cathode and migration of impurity ions.

In semiconductor devices, flicker noise is generated from the fluctuations in the carrier density and creates more problems in semiconductor amplifying device than in vacuum tubes at low frequencies.

Actually, the fluctuations in the carrier density generates fluctuations in the conductivity of the material. This produces a fluctuating voltage drop when a direct current flows. The fluctuating voltage drop is called as the flicker-noise voltage. The power density spectrum of the flicker noise is inversely proportional to frequency. Mathematically,

$$S(W) \ \alpha \ \frac{1}{f} \qquad ...(5.4)$$

Therefore, the flicker noise becomes significant at very low frequencies, generally below a few kHz.

Transit-Time Noise or High Frequency Noise

It is generally observed in semiconductor devices, when the transit-time of charge-carries crossing a junction is comparable with the time-period of the signal, some charge-carries diffuse back to the source or emitters. This process gives rise to an input admittance in which the conductance component increases with frequency. This conductance has a noise current source which is associated with it in parallel. Because this conductance increases with frequency, the power spectral density will also increase.

Thermal Noise

The thermal noise or white noise or Johnson noise is the random noise which is generated in a resistor or the resistive component of a complex impedance due to rapid and random motion of the molecules, atoms and electrons.

According to the kinetic theory of thermodynamics, the temperature of a particle denotes its internal kinetic energy. This means that the temperature of a body expresses the rms value of the velocity of motion of the particles in body. As per this kinetic theory, the kinetic energy of these particles becomes approximately zero (i.e., zero velocity) at absolute zero.

Therefore, the noise power produced in a resistor is proportional to its absolute temperature. Also the noise power is proportional to the bandwidth over which the noise measured.

Therefore the expression for maximum noise power output of a resistor may be given as

$$P_n \propto T.B$$

or

$$P_n = k.T.B \qquad \text{...(5.5)}$$

where k = Boltzmann's constant

$= 1.38 \times 10^{-23}$ Joule/deg. K

T = absolute temperature

B = Bandwidth of interest in Hz.

5.3 VOLTAGE AND CURRENT MODELS OF A NOISY RESISTOR

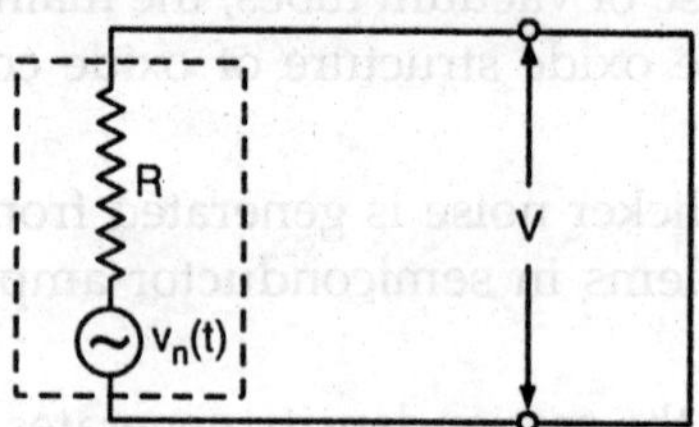

Fig. 5.3 *Voltage Models of a Noisy Resistor*

If we connect a d.c. voltmeter across any given resistor at room temperature i.e. 27°C (or 300 K), then no voltage is displayed by the d.c. voltmeter. However, if a very sensitive electronic voltmeter is used, then it displays some reading. This happens because of the fact that every resistor may be treated as a noise generator and so a sufficiently large voltage may develop across it.

However, the noise voltage so generated is always random and thus it has zero d.c. value but a definite rms value. Therefore only an a.c. meter across the resistor would register a value. This noise voltage is produced by the random movement of electrons within the resistor constituting a current. When measured over a long period of time, as many electrons arrive at one end of the resistor as at the other. But at any particular instant of time, this principle does not hold good i.e., more electrons arrive at one end than at other due to their random movement. This means that the rate of arrival of electrons at either end of the resistor will vary randomly and therefore the potential different between the two ends will also vary randomly. This random voltage across the resistor can be both measured and calculated. The random noise voltage is always expressed in terms of rms value rather than instantaneous value.

Now, we can draw an equivalent circuit of a resistor as a noise voltage generator as shown in Fig. 5.3. This equivalent circuit is also called as **Voltage model of a noisy resistor.**

From this equivalent circuit, we can compute the resistor's equivalent rms noise voltage V_n.

Let us consider that a noiseless load resistor R_L is connected across the noise generator as shown in Fig. 5.3.

According to maximum power transfer theorem, for maximum transfer of power from noise voltage source V_n to load resistor R_L, we must have

$$R_L = R$$

Then the maximum noise power so transferred will be given as

$$P_n = \frac{V^2}{R_L}$$

But $$R_L = R$$

So, $$P_n = \frac{V^2}{R_L}$$

Applying voltage-divider method in Fig. 5.3, we get

$$V = \frac{V_n}{2}$$

So that $$P_n = \frac{V^2}{R}$$

$$P_n = \frac{(V_L/2)}{R} = \frac{V_n^2}{4R}$$

or $$V_n^2 = 4RP_n \quad ...(5.6)$$

But we know that $P_n = kTB$

Putting this value of P_n in Eqn. (5.6), we get

$$V_n^2 = 4R(kTB) = 4R\,kTB$$

or $$V_n = \sqrt{4R\,kTB} \quad ...(5.7)$$

From Eqn. (5.7), we may conclude that the square of the rms noise voltage associated with a resistor is proportional to the absolute temperature T of the resistor, value R of the resistor and the bandwidth B over which noise is being measured. At this point, it may be noted that the noise voltage is independent of the frequency at which it is measured.

This happens because of the fact that the noise is random and so on average is evenly distributed over the frequency spectrum.

Norton's theorem may be used to find an equivalent current generator as shown in Fig. 5.4. This equivalent current generator is called as current model of a noisy resistor.

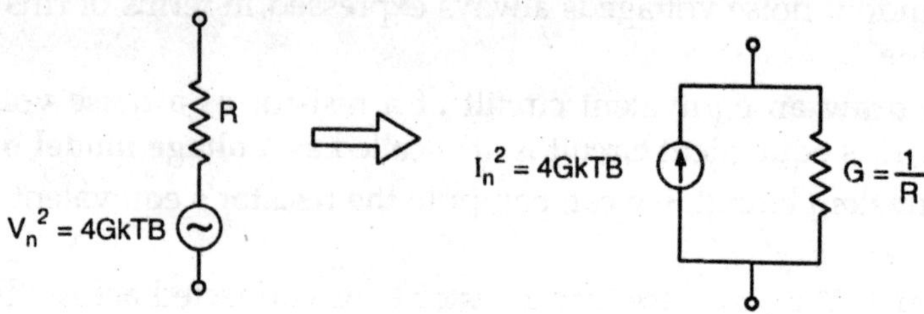

Fig. 5.4

Using conductance $G = \frac{1}{R}$, the rms noise current I_n for current model of a noisy resistor will be expressed as

$$I_n^2 = 4GkTB \quad ...(5.8)$$

5.4 ADDITION OF NOISE DUE TO SEVERAL SOURCES IN SERIES

Let us consider several thermal noise sources i.e., resistors, R_1, R_2, R_3 etc. in series producing noise voltages V_{n1}, V_{n2}, V_{n3} etc. We know that the rms noise voltage produced by resistor R is given as

$$V_n = \sqrt{4kTBR}$$

Therefore,

$$V_{n1} = \sqrt{4kTBR_1}$$

$$V_{n2} = \sqrt{4kTBR_2}$$

and

$$V_{n3} = \sqrt{4kTBR_3} \text{ etc.}$$

Then the resultant noise voltage V_{nr} is given by the square root of the sum of the squares of individual rms noise voltages.

$$V_{nr} = \sqrt{V_{n1}^2 + V_{n2}^2 + V_{n3}^2 + ...}$$

Putting all the values, we get

$$V_{nr} = \sqrt{4kTB(R_1 + R_2 + R_3 + ...)}$$

or

$$V_{nr} = \sqrt{4kTB.R_s} \quad ...(5.9)$$

where R_s is the equivalent series resistance of the individual resistance and is given as

$$R_s = R_1 + R_2 + R_3 + ...$$

5.5 ADDITION OF NOISE DUE TO SEVERAL SOURCES IN PARALLEL

Let us consider several thermal noise sources i.e., resistors R_1, R_2, R_3 etc. in parallel producing rms noise voltages V_{n1}, V_{n2}, V_{n3} etc.

We know that the rms noise voltage produced by resistor R is given as

$$V_n = \sqrt{4kTBR}$$

Therefore, the resultant noise in the case will be

$$V_{nr} = \sqrt{4kTBR_{par}} \quad \text{...(5.10)}$$

where
$$\frac{1}{R_{par}} = \frac{1}{R_1} + \frac{1}{R_2} + \frac{1}{R_3} + \text{....}$$

5.6 ADDITION OF NOISE DUE TO SEVERAL AMPLIFIERS IN SERIES

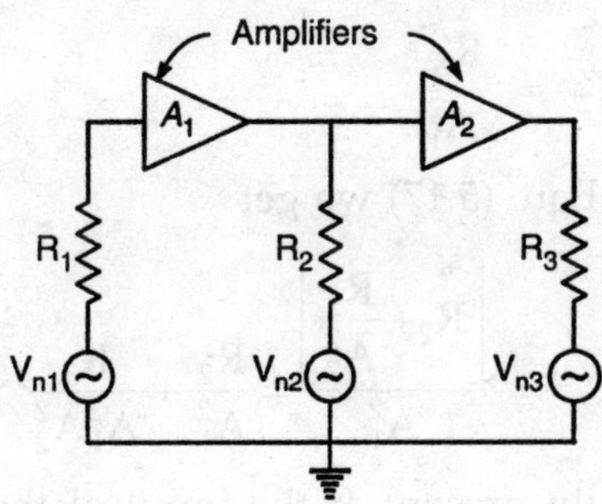

Fig. 5.5 *Calculation of Noise of two Amplifying Stages in Series*

Let us consider a two stage amplifier. Let us assume that gain of the first stage is A_1 and that of second stage is A_2. The first stage has a total input-noise resistance R_1, the second R_2, and the output resistance is taken as R_3. The noise voltage at the output due to resistance R_3 is

$$V_{n3} = \sqrt{4kTBR_3} \quad \text{...(5.11)}$$

The same noise voltage will be present at the output if there was no R_3, but instead R_3' was present at the input of stage 2, such that

$$V_{n3}' = \frac{V_{n3}}{A_2} \quad \text{...(5.12)}$$

Substituting Eqn. (5.11) is Eqn. (5.12), we get

$$V_{n3}' = \sqrt{\frac{4kTBR_3}{A_2}} = \sqrt{4kTB\frac{R_3}{A_2^2}} = \sqrt{4kTBR_3'} \quad \text{...(5.13)}$$

where
$$R_3' = \frac{R_3}{A_2} \quad \text{...(5.14)}$$

R_3' is the resistance which if placed at the input of the second stage would produce the same noise voltage at the output as that produced by R_3. Eqn. (5.14) explains that when a noise resistance is 'transferred' from the output of a stage to its input, it must be divided by the square of the voltage gain of that particular stage. Now the noise resistance actually present at the input of the second stage is R_2, so that the equivalent noise resistance at the input of the second stage, due to the second stage and the output resistance, will be

$$R'_{eq} = R_2 + R_3' \qquad ...(5.15)$$

Substituting Eqn. (5.14) in Eqn. (5.15), we get

$$R'_{eq} = R_2 + \frac{R_3}{A_2^2} \qquad ...(5.16)$$

Similarly, a resistor R_2' can be placed at the input of the first stage to replace R'_{eq} both naturally producing the same noise voltage at the output.

$$R_2' = \frac{R'_{eq}}{A_1^2} \qquad ...(5.17)$$

Substituting Eqn. (5.16) in Eqn. (5.17) we get

$$R_2' = \frac{\left[R_2 + \frac{R_3}{A_2^2}\right]}{A_1^2} = \frac{R_2}{A_1^2} + \frac{R_3}{A_1^2 A_2^2} \qquad ...(5.18)$$

The noise resistance actually present at the input of the first stage is R_1, so that the equivalent noise resistance of the whole cascaded amplifier at the input of the first stage, we will be

$$R_{eq} = R_1 + R_2' \qquad ...(5.19)$$

Substituting Eqn. (5.18) in Eqn. (5.19), we get

$$R_{eq} = R_1 + R_2' = R_1 + \frac{R_2}{A_1^2} + \frac{R_3}{A_1^2 + A_2^2}$$

or

$$R_{eq} = R_1 + \frac{R_2}{A_1^2} + \frac{R_3}{A_1^2 + A_2^2} \qquad ...(5.20)$$

It may be noted that we can extend Eqn. (5.20) to an *n*-stage cascaded amplifier.

5.7 SPECTRAL DENSITIES

We know that periodic waveforms are also known as deterministic waveforms since their values are known at all times. Periodic waveforms are also called as power waveforms. Another type of power waveform encountered in communications engineering is the random waveform, for which the values may not be predicted. Noise waveforms are examples of random waveforms, like all information signals such as audio and video signals. These signals carry finite power, but unlike periodic waves they do not have a harmonic or line spectrum. Fourier methods applied to these random power signals result in a power spectral density function. This is a curve which shows the energy distribution as a continuous function of frequency. Such a typical curve has been illustrated in Fig. 5.6.

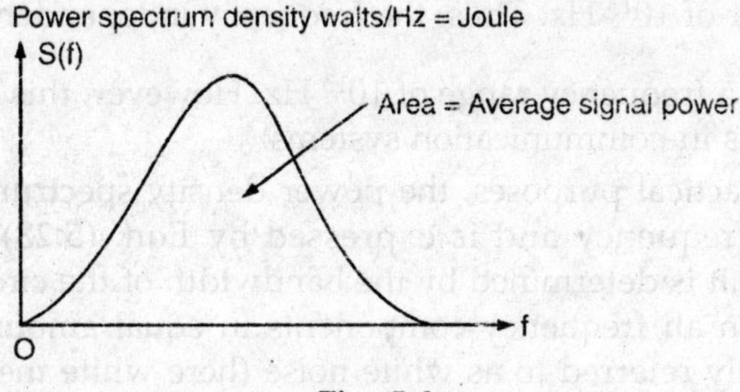

Fig. 5.6

The units for power spectral density are watts per hertz, which are equivalent to joules. The area under the curves has units of joules X hertz, which is dimensionally equivalent to watts. But watts are the units of power. Therefore, the total area under the curve gives the average signal power. This means that the total average power is equal to the product of the power spectral density (Watts/Hz) and bandwidth (Hz).

As discussed in last article, the thermal noise fall into the category of power signals and hence it has a power spectral density. It may be shown that the power density spectrum of the current contributing the thermal noise is expressed as

$$S_i(\omega) = \frac{2kTG}{1+\left(\frac{\omega}{\alpha}\right)^2} \quad ...(5.21)$$

Here, T is ambient temperature in Kelvin.

G is the conductance of the resistor in mhos.

k is the Boltzmann constant.

and α is the average number of collisions per second per electron.

Fig. 5.6 shows the variation of power density spectrum with frequency.

From Fig. 5.6, it may be noted that the power density spectrum can be considered to be flat one for $\frac{\omega}{\alpha} \leq 0.1$. This means that for this range of frequency, the power density spectrum $S_i(\omega)$ is almost constant and is expressed as

$$S_i(\omega) = 2kTG \quad ...(5.22)$$

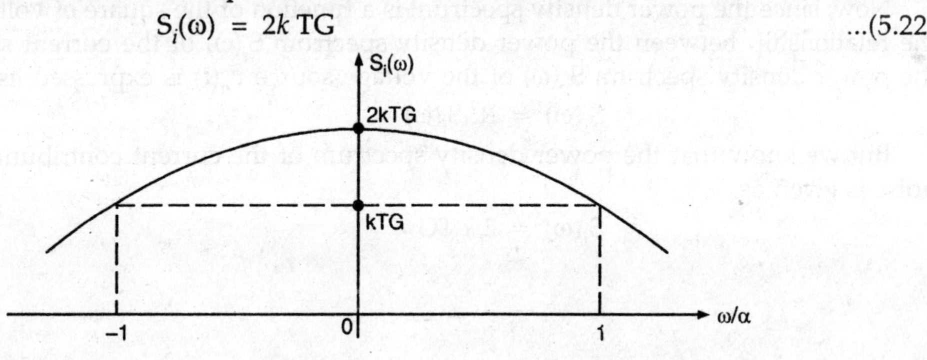

Fig. 5.7

The value of α is generally of the order of 10^{14} and so the frequency range corresponding to $\frac{\omega}{\alpha} < 0.1$ is of the order of 10^{13} Hz. Thus, the frequency independent expression of $S_i(w)$ in Eqn. (5.22) is valid upto a frequency range of 10^{13} Hz. However, this range covers almost all the practical applications in communication systems.

Therefore, for all practical purposes, the power density spectrum $S_i(\omega)$ may be treated to be independent of frequency and is expressed by Eqn. (5.23). This means that the contribution of any circuit is determined by the bandwidth of the circuit. Thermal noise can be considered to contain all frequency components in equal amount. Due to this reason, thermal noise is generally referred to as white noise (here white means the presence of all colours or frequencies).

Thermal noise is also known as Johnson noise after the scientist J.B. Johnson who presented a detailed investigation of it.

5.8 DERIVATION OF THERMAL NOISE EXPRESSIONS USING POWER SPECTRUM DENSITY

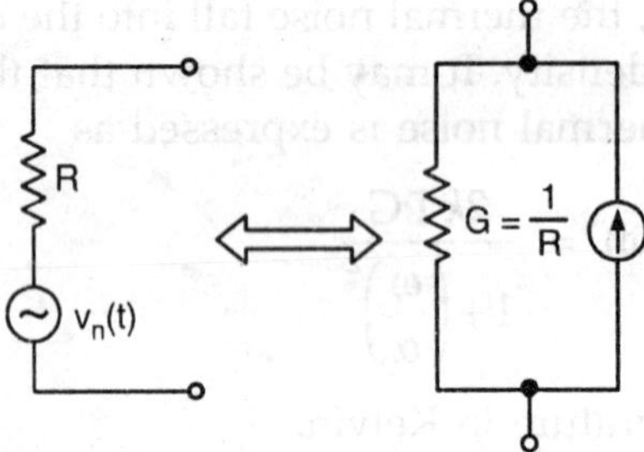

Fig. 5.8

As discussed noisy resistor may be represented as a voltage model and a current model. In voltage model of a noisy resistor, resistor R is connected with a thermal noise voltage source $v_n(t)$ in series. In current model of a noisy resistor, the noise resistor R is represented by a conductance G in parallel with a thermal noise current $i_n(t)$. Fig. 5.8 shows the voltage and current model of a noisy resistor R.

We know that the current $i_n(t)$ and voltage $v_n(t)$ are related as

$$v_n(t) = R.i_n(t) \quad ...(5.23)$$

Now, since the power density spectrum is a function of the square of voltage or current, the relationship between the power density spectrum $S_i(\omega)$ of the current source $i_n(t)$ and the power density spectrum $S_v(\omega)$ of the voltage source $v_n(t)$ is expressed as

$$S_i(\omega) = R^2.S_i(\omega) \quad ...(2.24)$$

But we know that the power density spectrum of the current contributing the thermal noise is given as

$$S_i(\omega) = 2\,k\,TG \quad ...(2.25)$$

Therefore,

$$S_v(\omega) = R^2 S_i(\omega) = R^2.(2kTG)$$

Substituting $G = \frac{1}{R}$, we get

$$S_v(\omega) = R^2.\left(2kT.\frac{1}{R}\right) = 2kTR \quad ...(2.26)$$

Thus, the power density spectrum $S_v(\omega)$ of a thermal noise voltage $v_n(t)$ is independent of frequency. Now, since the power density spectrum is defined as the power per unit bandwidth, therefore the noise power increases with increase in bandwidth.

We know that the average or total power P in terms of power density spectrum $S(\omega)$ is given as

$$P = \frac{1}{2\pi}\int_{-\infty}^{\infty} S(\omega)\, d\omega \quad ...(5.27)$$

Hence, the noise power P_n in terms of power density spectrum is given as

$$P_n = \frac{1}{2\pi}\int_{-\infty}^{\infty} S_v(\omega)\, d\omega \quad ...(5.28)$$

But $S_v(\omega) = 2kTR$

Therefore,

$$P_n = \frac{1}{2\pi}\int_{-\infty}^{\infty} 2kTR\, d\omega \quad ...(5.29)$$

This integral would be in infinite if it is integrated over an infinite bandwidth. But for a finite bandwidth of $2\Delta f$, (from $-\Delta f$ to $+\Delta f$) the noise power is given as

$$P_n = S_v(\omega).2\Delta f = 2kTR.2\Delta f$$

$$P_n = 4kTR(\Delta f) \quad ...(5.30)$$

We know that the average power of a signal is the same as its mean square (ms) value.

Thus $P_n = \overline{v_n^2} = 4kTR(\Delta f)$

Let the rms value of noise voltage v_n be V_n. Then, the rms value is given as

$$V_n = \sqrt{\overline{v_n^2}} = \sqrt{4kTR(\Delta f)} \quad ...(5.31)$$

5.8.1 Generalised Nyquist Theorem for Thermal Noise

With the help of generalised Nyquist theorem, the concept of power density spectrum of a single resistor is extended to a generalised circuit *t*.

Let us consider a generalized linear bilateral passive network (i.e. two port network) consisting of resistance, R capacitance C and inductance L as shown in Fig. 5.9.

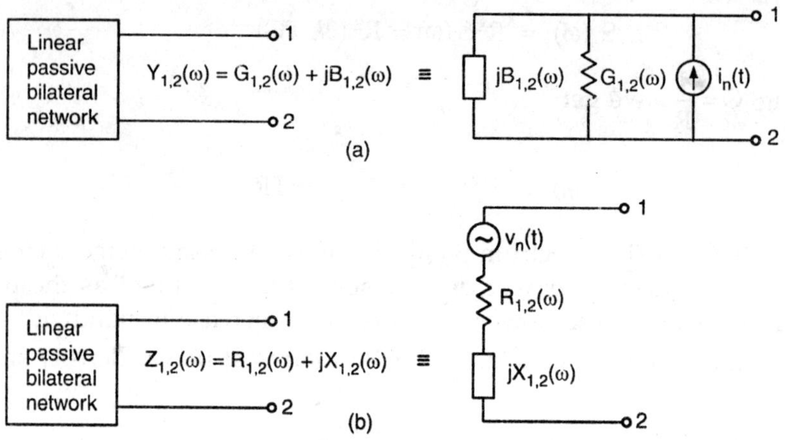

Fig. 5.9

The Norton's equivalent circuit is representing the given network in terms of conductance $G_{1,2}(\omega)$ and susceptance $B_{1,2}(\omega)$ forming an equivalent admittance of the network expressed as

$$Y_{1,2}(\omega) = G_{1,2}(\omega) + jB_{1,2}(\omega) \quad ...(5.32)$$

Here, the source of the thermal noise is conductance $G_{1,2}(\omega)$ and so the given network may be represented by a noiseless admittance $Y_{1,2}(\omega)$ in parallel with a noise current source with power density spectrum expressed as

$$S_i(\omega) = 2k\,TG_{1,2}(\omega) \quad ...(5.33)$$

Similarly, the Thevenin's equivalent circuit is representing the given network in terms of resistance $R_{1,2}(\omega)$ and reactance $X_{1,2}(\omega)$ forming an equivalent impedance of the network expressed as

$$Z_{1,2}(\omega) = R_{1,2}(\omega) + jX_{1,2}(\omega) \quad ...(5.34)$$

Here, the source of the thermal noise is resistance $R_{1,2}(\omega)$ and so the given network may be represented by a noiseless impedance $Z_{1,2}(\omega)$ in series with a noise voltage source with the power density spectrum expressed as

$$S_v(\omega) = 2k\,T.R_{1,2}(\omega) \quad ...(5.35)$$

At this stage, following two points may be noted:

(*i*) In general, $R_{1,2}(\omega)$ and $G_{1,2}(\omega)$ are frequency dependent and therefore the power density spectrum is also frequency-dependent, i.e., a function of frequency, unlike the white noise.

(*ii*) The major limitation of the theorem is that it is applicable to a linear bilateral passive circuit elements in which all of the noise sources are due to resistors only. This means that this theorem cannot be applied to a circuit containing active elements like tube and transistors.

5.8.2 Calculation of Thermal Noise for a Single Noise Source

Let us consider a linear system or circuit consisting of noiseless elements. Let this system be excited by a random noise input voltage $v_{ni}(t)$. We have to determine the rms value of the output noise voltage $v_{no}(t)$.

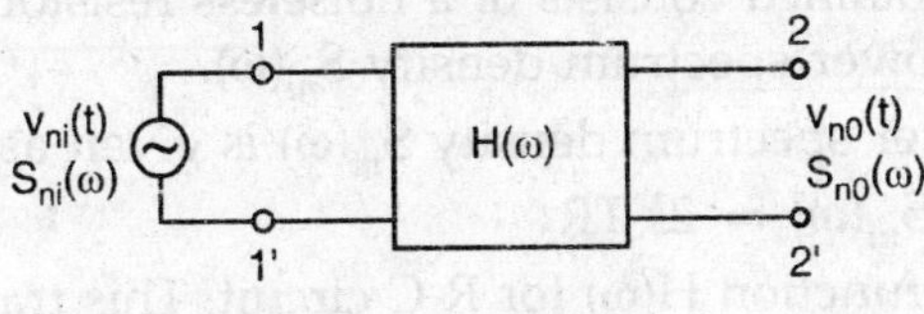

Fig. 5.10

Let H(ω) be the transfer function of the system. Let us consider that $S_{ni}(\omega)$ and $S_{no}(\omega)$ be the power spectrum densities of the noise voltages $v_{ni}(t)$ and $v_{no}(t)$ respectively.

We know that for a linear system, the relationship between input and output power spectrum densities is given as under:

$$S_{no}(\omega) = |H(\omega)|^2 S_{ni}(\omega) \quad ...(5.36)$$

Now, the output noise power P_o may be calculated by integrating output power spectrum density $S_{no}(\omega)$ over the bandwidth under consideration

$$P_o = \text{m.s. value of } v_{no}(t) = \overline{v_{no}^2(t)} \quad ...(5.37)$$

or

$$P_o = \frac{1}{2\pi}\int_{-\infty}^{\infty} S_{no}(\omega)\, d\omega$$

Substituting the values of $S_{no}(\omega)$ in above equation, we get

or

$$P_o = \overline{v_{no}^2} = \frac{1}{2\pi}\int_{-\infty}^{\infty} |H(\omega)|^2 .S_{ni}(\omega)\, d\omega \quad ...(5.38)$$

We know that the rms output noise voltage may be obtained by simply taking square root of mean square (ms) value of output noise voltage, i.e.

$$V_{rms} = \sqrt{\overline{v_{no}^2}} = \left[\frac{1}{2\pi}\int_{-\infty}^{\infty} |H(\omega)|^2 .S_{ni}(\omega)\, d\omega\right]^{1/2} \quad ...(5.39)$$

This is required expression.

5.8.3 Thermal Noise in RC Circuit

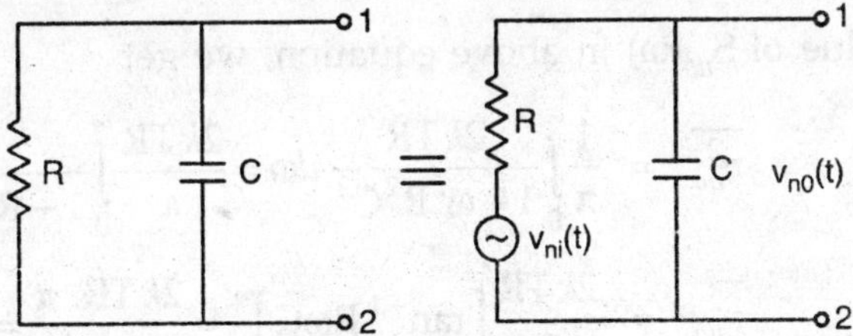

Fig. 5.11

Let us consider an R-C circuit as shown in Fig. 5.11. Let us obtain an equivalent circuit by replacing the thermal resistor with its equivalent thevenin model as shown in Fig. 5.13.

The equivalent circuit so obtained consists of a noiseless resistor and a series noise voltage source of $v_{ni}(t)$ having a power spectrum density $S_{ni}(\omega)$.

We know that the power spectrum density $S_{ni}(\omega)$ is given as

$$S_{ni}(\omega) = 2kTR \quad ...(5.40)$$

We can find a transfer function $H(\omega)$ for R-C circuit. This transfer function $H(\omega)$ relates the output voltage $v_{no}(t)$ with the input noise voltage $v_{ni}(t)$.

Thus,

$$H(\omega) = \frac{\frac{1}{j\omega C}}{R + \frac{1}{j\omega C}} = \frac{1}{1 + j\omega RC}$$

or

$$H(\omega) = \frac{1 - j\omega RC}{1 + \omega^2 R^2 C^2} = \frac{1}{1 + \omega^2 R^2 C^2} - j\frac{\omega RC}{1 + \omega^2 R^2 C^2}$$

or

$$|H(\omega)| = \sqrt{\frac{1}{(1 + \omega^2 R^2 C^2)} + \frac{\omega^2 R^2 C^2}{1 + \omega^2 R^2 C^2}}$$

or

$$|H(\omega)| = \frac{\sqrt{1 + \omega^2 R^2 C^2}}{1 + \omega^2 R^2 C^2} = \frac{1}{\sqrt{1 + \omega^2 R^2 C^2}} \quad ...(5.41)$$

The power density spectrum of the output noise voltage $v_{no}(t)$ is given as

$$S_{no}(\omega) = |H(\omega)|^2 . S_{ni}(\omega) \quad ...(5.42)$$

Substituting the value of $|H(\omega)|$ and $S_{ni}(\omega)$ in above equation, we get

$$S_{no}(\omega) = \frac{1}{1 + \omega^2 R^2 C^2} . 2kTR = \frac{2kTR}{1 + \omega^2 R^2 C^2} \quad ...(5.43)$$

The average power or mean square (ms) value of the output noise voltage $v_{no}(t)$ is given as

$$P_o = \overline{v_{no}^2} = \frac{1}{2\pi} \int_{-\infty}^{\infty} S_{no}(\omega)\, d\omega = \frac{1}{\pi} \int_0^{\infty} S_{no}(\omega)\, d\omega$$

Substituting the value of $S_{no}(\omega)$ in above equation, we get

$$\overline{v_{no}^2} = \frac{1}{\pi} \int_0^{\infty} \frac{2kTR}{1 + \omega^2 R^2 C^2} d\omega = \frac{2kTR}{\pi} \int_0^{\infty} \frac{1}{1 + R^2 \omega^2 C^2} d\omega$$

or

$$\overline{v_{no}^2} = \frac{2kTR}{\pi RC} \left[\tan^{-1} R\omega C\right]_0^{\infty} = \frac{2kTR}{\pi RC} . \frac{\pi}{2} = \frac{kT}{C} \quad ...(5.44)$$

Thus, the output rms noise voltage is expressed as

$$V_{rms} = \sqrt{\overline{v_{no}^2}} = \sqrt{\frac{kT}{C}} \qquad ...(5.45)$$

5.9 CALCULATION OF THERMAL NOISE FOR MULTIPLE NOISE SOURCES

5.9.1 Superposition of Power Spectrum Density

In any electrical circuit, there can be a large number of noise sources due to resistors and electronic elements. All these noise sources are independent sources of random signals.

This theorem states that for any system involving multiple independent sources of noise, the mean square (ms) value of response or output is equal to the sum of the mean square (ms) value of the responses calculated by individual sources at a time. In the same manner, the power spectrum density of the output or response may also be calculated by using the superposition theorem.

Let us consider a linear system with two independent input sources $x_1(t)$ and $x_2(t)$ at two different terminals generating an output or response $y(t)$ as shown in Fig. 5.14. Let us assume that $x_1(t)$ alone produces a response $y_1(t)$ and $x_2(t)$ alone produces a response $y_2(t)$ to this linear system.

Now since $x_1(t)$ and $x_2(t)$ are independent sources, then according to the superposition theorem,

$$\overline{y^2(t)} = \overline{y_1^2(t)} + \overline{y_2^2(t)} \qquad ...(5.46)$$

and

$$S_y(\omega) = S_{y1}(\omega) + S_{y2}(\omega) \qquad ...(5.47)$$

We can extend this result for more than two sources as

$$\overline{y^2(t)} = \overline{y_1^2(t)} + \overline{y_2^2(t)} + \overline{y_3^2(t)} + ... + \overline{y_n^2(t)} \qquad ...(5.48)$$

and

$$S_y(\omega) = S_{y1}(\omega) + S_{y2}(\omega) + S_{y3}(\omega) + ... + S_{yn}(\omega) \qquad ...(5.49)$$

5.9.2 White Noise

Noise an idealized form is known as white noise. This means that in a communication system, the noise analysis is based on an idealized form of noise, i.e., white noise.

As we know that white light consists of all colour frequencies, in the same manner, white noise contains all frequencies in equal amount. Actually, the power density spectrum of a white noise consists of all the frequency components in equal amount. If the probability of occurrence of a white noise is specified by a Gaussian distribution function, it is called as white Gaussian noise. Since the power density spectrum of thermal and shot noise is independent of the operating frequencies, therefore, shot noise and thermal noise can be treated as white Gaussian noise for all practical purposes.

The power spectrum density of white noise is expressed as

$$S_w(\omega) = \frac{N_o}{2} \qquad ...(5.50)$$

Here the factor $\frac{1}{2}$ has been included to show that half of the power is associated with the positive frequencies and remaining half with the negative frequencies. This has been shown in Fig. 5.15 reveals the fact that it has no dc power, i.e., the mean or average value of white noise Fig. 5.14 is zero.

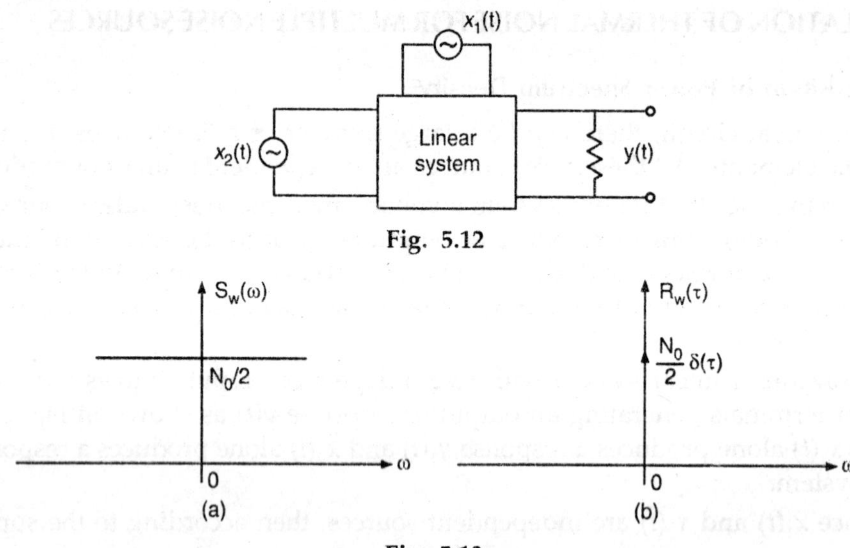

Fig. 5.12

Fig. 5.13

In addition to this, the autocorrelation function of the power spectrum density of white noise may be obtained by simply taking the inverse Fourier transform of both sides of expression.

$$S_w(\omega) = \frac{N_o}{2} \quad \text{...(5.51)}$$

Thus, the autocorrelation function for the white noise may be expressed as

$$R_w(\tau) = \frac{N_o}{2}\delta(\tau) \quad \text{...(5.52)}$$

Eqn. (5.52) reveals the fact that the autocorrelation function of the white noise contains a delta function scaled by a factor $\frac{N_o}{2}$ and situated at $\tau = 0$ as shown in Fig. 5.15.

Also, $$R_w(\tau) = 0 \quad \text{for} \quad \tau \neq 0 \quad \text{...(5.53)}$$

Strictly speaking, white noise has infinite average power and thus is not physically realizable. However, the concept of white noise provides convenient mathematical properties and this is a useful tool for system analysis.

5.9.3 Equivalent Noise Bandwidth

With the help of equivalent noise bandwidth, noise power can be specified at the output of a linear bandpass system.

Let us consider a linear bandpass system as shown in Fig. 5.16. The system is excited by an input noise voltage source $v_{ni}(t)$ producing an output $v_{no}(t)$. Let $H(\omega)$ be the transfer function of linear bandpass system. The parameter $|H(\omega)|^2$, the square of the transfer function is plotted in Fig. 5.17.

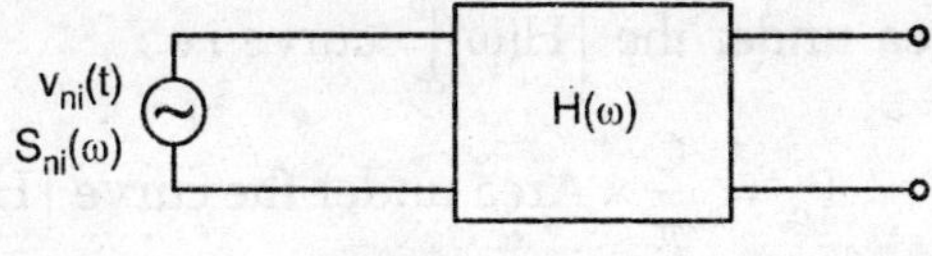

Fig. 5.14

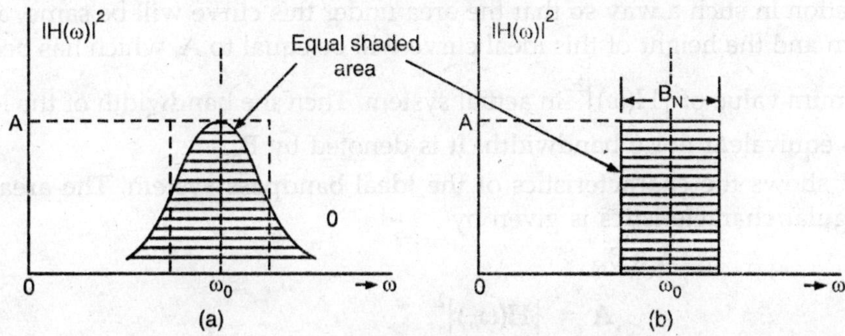

Fig. 5.15

In Fig. 5.17, we have assumed that the negative half of the curve is symmetrical about the vertical axis and hence the positive half of the plot $|H(\omega)|^2$ has been shown.

We know that the power spectrum density of noise voltage at the input and power spectrum density at the output are related as

$$S_{no}(\omega) = S_{ni}(\omega)|H(\omega)|^2 \quad ...(5.54)$$

The mean square (ms) value of output noise signal i.e., output noise power P_o may be evaluated simply by integrating $S_{no}(\omega)$ over the bandwidth under consideration, i.e.,

$$v_{no}^2 = P_o = \frac{1}{2\pi}\int_{-\infty}^{\infty} S_{ni}(\omega)|H(\omega)|^2 \, d\omega \quad ...(5.55)$$

or

$$P_o = \frac{1}{\pi}\int_0^{\infty} S_{ni}(\omega)|H(\omega)|^2 \, d\omega \quad ...(5.56)$$

For the practical purpose, the input noise power spectrum density is taken as constant with frequency. Let this constant value be denoted by 'C'.

Hence, $S_{ni}(\omega) = C$ ...(5.57)

Thus the output noise power is given by

$$P_o = \frac{1}{\pi}\int_0^{\infty} C|H(\omega)|^2 \, d\omega = \frac{C}{\pi}\int_0^{\infty} |H(\omega)|^2 \, d\omega \quad ...(5.58)$$

In Eqn. (5.59), the integral on the right hand side in the area under the curve namely square of transfer function $|H(\omega)|^2$ as shown in Fig. 5.16. This means that the output power P_o is proportional to the area under the $|H(\omega)|^2$ curve i.e.,

$$P_o = \frac{C}{\pi} \times \text{Area under the curve } |H(\omega)|^2$$

Now, we consider an ideal bandpass system with rectangular characteristic of square transfer function in such a way so that the area under this curve will be same, as that of an actual system and the height of this ideal curve will be equal to A, which has been assumed as the maximum value of $|H(\omega)|^2$ in actual system. Then the bandwidth of the ideal system is known as equivalent noise bandwidth. It is denoted by B_N.

Fig. 5.17 shows the characteristics of the ideal bandpass system. The area under this ideal rectangular characteristics is given by

$$A \times B_N$$

Here, $$A = |H(\omega_o)|^2$$

Equating the areas of the actual and ideal systems, we get

$$A \times B_N = \int_0^{\infty} |H(\omega)|^2 \, d\omega$$

From above equation, the expression for equivalent noise bandwidth is given as

$$B_N = \frac{1}{A} \int_0^{\infty} |H(\omega)|^2 \, d\omega \quad \text{...(5.59)}$$

i.e., $$B_N = \frac{\text{Area under the } |H(\omega)|^2 \text{ curve}}{\text{Maximum value of the } |H(\omega)|^2 \text{ curve}}$$

Thus, the output noise power may be obtained ass

$$P_o = \overline{v_{no}^2} \quad \text{...(5.60)}$$

$$= \frac{C}{\pi} \times \left\{ \text{area under the curve } |H(\omega)|^2 \text{ of actual system} \right\}$$

$$P_o = \frac{C}{\pi} \times \left\{ \text{area under the curve } |H(\omega)|^2 \text{ of ideal system} \right\}$$

$$= \frac{C}{\pi} \times \{AB_N\} \quad \text{...(5.61)}$$

Thus, $$P_o = \overline{v_{no}^2} = \frac{CAB_N}{\pi} \quad \text{...(5.62)}$$

where C is a constant equal to the power spectrum density of noise voltage at the input of the system. Hence, equivalent noise bandwidth may be defined as follows:

"Equivalent noise bandwidth may be defined as the bandwidth of an ideal bandpass system which produces the same noise power as the actual system does."

5.9.4 Noise Equivalent Bandwidth for Low-Pass Filter (LPF)

It is worthnoting that the average noise power at the output of an ideal low pass filter (LPF) having bandwidth B and unity passband amplitude response is $N_o B$ when a white-noise source having zero mean and power spectrum density $\frac{N_o}{2}$ is connected across its input terminals.

When this noise source is connected to the input of a practical R-C low-pass filter, it will produce an average output power equal to $\frac{N_o}{4RC}$. This may be verified by finding the autocorrelation function a $\tau = 0$. The filter has a 3 dB bandwidth equal to $\frac{1}{2\pi RC}$.

Hence, we observe that the filtered white noise has a finite average power which is proportional to the bandwidth in both the cases. This is true for all kinds of low-pass filters.

For convenience, to generalize the above observation for all kinds of low-pass filters (LPFs), we define the noise equivalent bandwidth.

Now let us consider an arbitrary low-pass filter (LPF). The transfer function of this low-pass filter (LPF) is $H(f)$ at the input of which a white noise source is connected. This white noise has zero mean and $\frac{N_o}{2}$ power spectrum density.

The power spectrum density (psd) of the output noise is given as

$$S_{no}(f) = |H(f)|^2 \frac{N_o}{2} \quad ...(5.63)$$

here, $|H(f)|^2$ is the squared magnitude of the transfer function of the low-pass filter (LPF). Therefore, the resulting average output noise is given as

$$P_{no}(f) = \int_{-\infty}^{\infty} S_{no}(f)\, df \quad ...(5.64)$$

Substituting Eqn. (5.63) in Eqn. (5.64), we obtain

$$P_{no}(f) = \int_{-\infty}^{\infty} |H(f)|^2 \frac{N_o}{2}\, df$$

$$= \frac{N_o}{2} \int_{-\infty}^{\infty} |H(f)|^2\, df$$

or $$P_{no}(f) = \frac{N_o}{2} \times 2\int_0^{\infty}|H(f)|^2\, df = N_o\int_0^{\infty}|H(f)|^2\, df \quad ...(5.65)$$

The same white noise source is connected to the input of an ideal low-pass filter (LPF). This ideal low-pass filter (LPF) has zero frequency response H(0) and bandwidth B_N.

The average output noise power is given as

$$P_{no}(f) = N_o . B_N |H(0)|^2 \quad ...(5.66)$$

From above equation, it may be observed that the filtered noise power P_{no} is a finite value.

It is also proportional to the bandwidth B_N. This bandwidth B_N is called noise-equivalent bandwidth for a low-pass filter (LPF).

Using Eqns. (5.65) and (5.66), we obtain

$$B_N = \int_0^{\infty}\frac{|H(f)|^2\, df}{|H(0)|^2} \quad ...(5.67)$$

Hence, the noise-equivalent bandwidth may be obtained by replacing the arbitrary low-pass filter of transfer function H(*f*) by an equivalent ideal low-pass filter having zero-frequency response H(0) and bandwidth B_N.

It has been shown in Fig. 5.18.

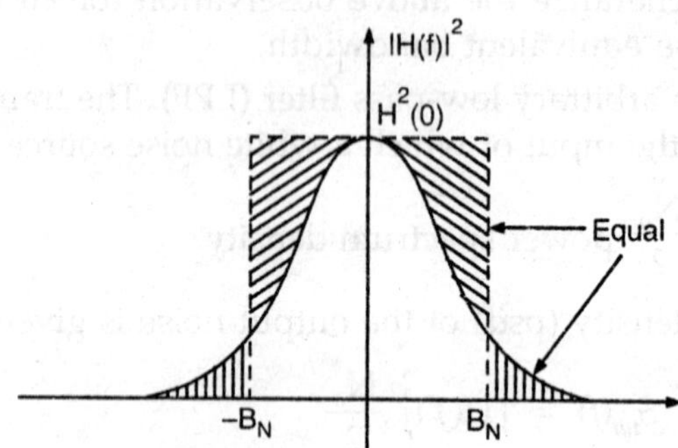

Fig. 5.16 *Illustration of Noise-equivalent Bandwidth of Low-pass Filter (LPF)*

5.9.5 Noise in Reactive Circuits

Fig. 5.19 shows a parallel resonant circuit. In this circuit, the inductor L is assumed to be an ideal one. The resistance R is the resistance associated with the inductor or coil. Thus resistance R is the only source of noise voltage in the circuit. Let v_{ni} be the noise voltage due to resistor R.

The mean square value of noise voltage is given as

$$\overline{v_{ni}^2} = 4k\, RT(\Delta f) \quad ...(5.68)$$

Fig. 5.20 shows the equivalent circuit in which a noise voltage source v_{ni} is connected with the resistor R. Now, our purpose is to find the noise voltage v_{no} developed across the output terminals.

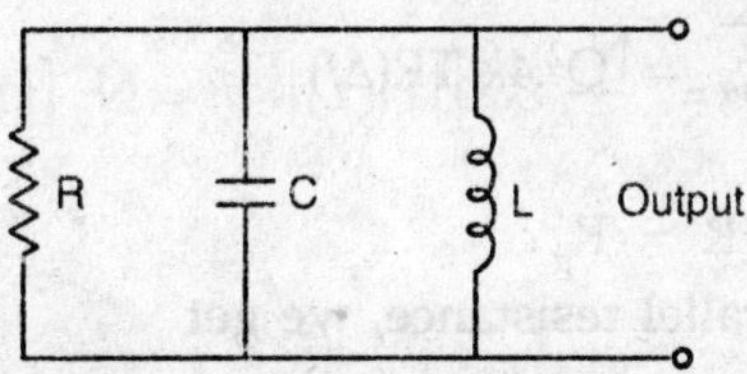

Fig. 5.17 *A Parallel Resonant Circuit*

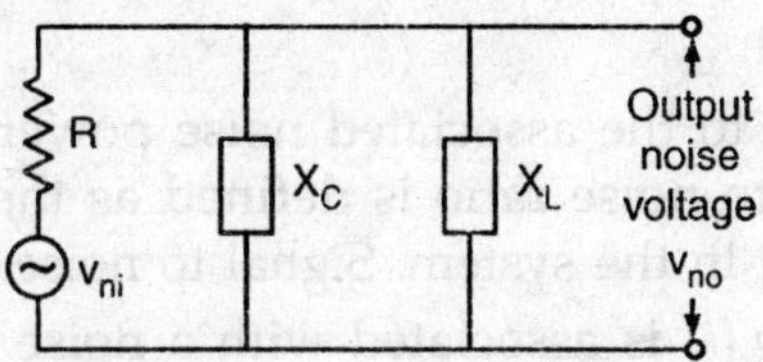

Fig. 5.18 *Equivalent Circuit*

Let V_{ni} be the rms value of noise voltage v_{ni} and Z be the total impedance of the circuit. Let I_{rms} be the rms value of the noise current generated due to noise voltage v_{ni}.

Thus, $$I_{rms} = \frac{V_{ni}}{Z}$$

But at resonance

$$Z = R$$

Hence, $$I_{rms} = \frac{V_{ni}}{Z}$$

Let V_{no} be the rms value of the output noise voltage v_{no}. Thus, the rms value of output noise voltage is given by

$$V_{no} = I_{rms} \cdot X_c$$

where X_c is the reactance of the capacitor *c*.

or $$V_{no} = \frac{V_{ni}}{R} . X_c$$

But $$X_c = Q.R$$

Therefore, $$V_{no} = \frac{V_{ni}}{R} . Q.R = Q.V_{ni} \qquad ...(5.69)$$

Here V_{no} and V_{ni} are the rms value. Since we know that the rms value of a signal is equal to the square root of its mean square (ms) value, therefore,

$$\left(\overline{v_{no}^2}\right)^{1/2} = Q\left(\overline{v_{ni}^2}\right)^{1/2}$$

Squaring both sides, we get

$$\overline{v_{no}^2} = Q^2 . \overline{v_{ni}^2}$$

Putting the value of v_{ni} in above equation from Eqn. (5.69), we get

$$\overline{v_{no}^2} = Q^2.4k\,TR(\Delta f)$$

Substituting

$$Q^2R = R_p$$

where R_p is the equivalent parallel resistance, we get

$$\overline{v_{no}^2} = 4k\,TR_p(\Delta f) \quad ...(5.70)$$

5.9.6 Signal to Noise Ratio

The ratio of signal power to the associated noise power is defined as signal to noise ratio. In other words, signal to noise ratio is defined as the ratio of signal power to the noise power at the same point in the system. Signal to noise ratio is denoted by S/N.

Thus, if a signal voltage $v_s(t)$ is associated with a noise voltage source $v_n(t)$, then the ratio of signal power to the noise power will be

$$\frac{S}{N} = \frac{\overline{v_s^2}}{\overline{v_n^2}} \quad ...(5.71)$$

Because the power spectrum density is power per unit bandwidth, hence the above expression may be expressed as

$$\frac{S}{N} = \frac{S_s(\omega)}{S_n(\omega)}$$

or

$$\frac{S}{N} = \frac{\text{Power spectrum density of signal voltage}}{\text{Power spectrum density of noise voltage}} \quad ...(5.72)$$

The S/N ratio is always a power ratio unless stated otherwise. In some cases, the ratio of a signal voltage (rms) to a noise voltage (rms) is also specified. However, in this case, it is specially stated that S/N voltage ratio is being considering.

The signal to noise ratio (S/N ratio) at the input and output of a noise free two-part network is the same since both signal and noise are treated identically. But, practical two-part networks introduce some additional noise. Thus for a practical two port networks, the S/N ratio deteriorates and we more from input side of the output side. Infact, the comparison of S/N ratio at the input and the output of any two port network gives the noise indication of that network.

5.9.7 Noise Figure

It has been observed that for comparison of receivers or amplifiers working at various impedance levels the use of the equivalent noise resistance is quite misleading. However, instead of equivalent noise resistance, one another quantity known as noise figure is defined and used. Sometimes, this is also called noise-factor.

The noise figure F is defined as the ratio of the signal-to-noise power ratio supplied to the input terminals of a receiver or amplifier to the signal-to-noise power ratio supplied to the output or load resistor.

Hence, mathematically,

$$\text{Noise figure} \qquad F = \frac{\text{Input SNR}}{\text{Output SNR}} = \frac{(\text{SNR})_i}{(\text{SNR})_o}$$

At this point, it may be noted that the noise figure F is one for an ideal receiver which introduces no amount of noise of its own, due to which the SNR does not deteriorate as we pass from the input side to the output side of the receiver. The noise figure for practical receiver can be expressed as an actual ratio or in decibels (dBs).

5.9.8 Noise Temperature

According to the thermodynamic considerations the power spectrum of thermal noise is constant over a very wide frequency range. Noise having such type of spectrum consists of all frequency components in equal proportion and is designated as white noise.

As we know that the available thermal noise power is expressed as

$$P_n = k\,TB \text{ Watts} \qquad ...(5.73)$$

As obvious, the noise depends only upon the value of temperature and bandwidth. Therefore, a thermal resistor delivers a maximum of kT watts per unit bandwidth regardless of the value of R.

Thus, temperature is the fundamental parameter of thermal noise. However, there are other white noise sources which are non-thermal in the sense that noise power is not related to a physical temperature, still we can speak of the noise temperature T_n of any white-noise sources, either thermal or non-thermal, by defining as under:

$$T_n = \frac{P_n}{kB} \qquad ...(5.74)$$

Here P_n is the maximum noise power of the sources delivered over the bandwidth B.

For thermal sources, T_n is a physical temperature and for non-thermal sources, it is a measure of the available noise power. This implies that a non-thermal white-noise source having equivalent internal resistance R_n and noise temperature T_n may be replaced by a thermal resistor $R = R_n$ at physical temperature $T = T_n$ without altering the terminal properties.

Also it may be shown that the equivalent noise temperature T_n may be expressed in terms of noise figure as under:

$$T_n = T_o(F-1) \qquad ...(5.75)$$

where $T_o = 17°C$ i.e., $290°K$

Thus, once the value of noise-figure is known, the equivalent noise temperature can be calculated with the help of above equation.

At this stages, it may be noted that the concept of noise-figure, although frequently used, is not always the most convenient measure of noise particularly in dealing with UHF and microwave low-noise antennas, receivers or devices. Noise temperature is employed extensively for antennas and low-noise microwave amplifiers. Noise temperature is additive like noise power.

It may be observed as under:

We, know that total noise power is given as

$$P_t = k\,T_t.B$$

or $$P_t = P_1 + P_2 = kT_1B + kT_2B$$

or $$T_t = T_1 + T_2 \qquad ...(5.76)$$

where P_1 and P_2 are two individual noise powers (i.e., received by the antenna and generated by antenna respectively) and P_t is the sum of the two powers P_1 and P_2, T_1 and T_2 are the individual noise temperatures,

T_t = the total noise temperature.

One more advantage of the use of noise temperature for low-noise levels is that it exhibits a greater variation for any given noise level. Hence, noise level changes are easier to grasp by using noise temperature concept.

It may also be noted that the concept of equivalent noise resistance is quite fictitious even then it is often employed due to its convenience. Equivalent noise temperature T_{eq} may also be utilized if it proves convenient. In defining the equivalent noise temperature of a receiver or amplifier we have assumed that

$$R_{eq} = R_a, \qquad ...(5.77)$$

Now, if this is to lead to the correct value of the noise output power then clearly R'_{eq} must be at a temperature other than the standard one at which all the components are assumed. Therefore, we can equate noise figure and equivalent noise temperature i.e.,

$$F = 1 + \frac{R'_{eq}}{R_a} \qquad ...(5.78)$$

$$F = \frac{1 + kT_{eq}BR'_{eq}}{kT_o BR_a} = 1 + \frac{T_{eq}}{T_o} \qquad ...(5.79)$$

where $R'_{eq} = R_a$

T_o = 17°C = 17 + 273 = 290°K

T_{eq} = Equivalent noise temperature of the receiver or amplifier whose noise figure is F.

It may be noted that here F is a ratio of powers and is not expressed in decibels. Equivalent noise temperature T_{eq} may be influenced by the actual ambient temperature of the receiver or amplifier. It is also to be noted that the equivalent noise temperature is just a convenient fiction. If all the noise of the receiver were generated by R_a, its temperature would have to be T_{eq}.

Lastly using Eqn. (5.79), we have

$$F = 1 + \frac{T_{eq}}{T_o}$$

or $$T_oF = T_o + T_{eq}$$

or $$T_{eq} = T_o(F - 1) \qquad ...(5.80)$$

Thus, once noise figure F is known, the equivalent noise temperature T_{eq} may be determined from Eqn. (5.80) Graphs of noise temperature of various sources verses frequency and the temperature verses frequency are also available.

5.9.9 Calculation of Noise Figure

To calculate the noise figure, let us consider a network (i.e., receiver). This network has an input impedance R_t, an output impedance R_L and an overall voltage gain A.

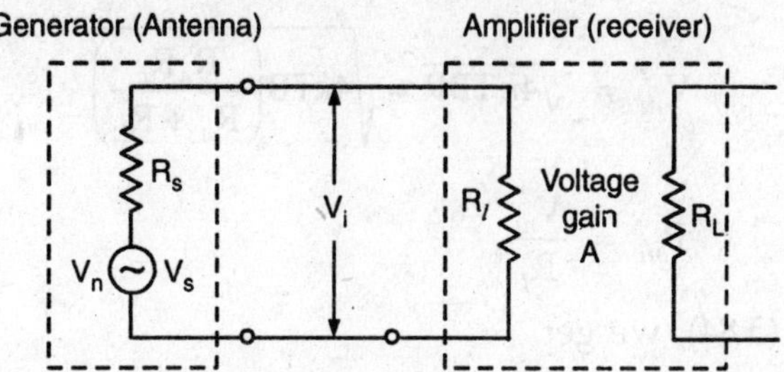

Fig. 5.19 *Block Diagram for Noise Figure Calculation*

It is fed from a source i.e., antenna of internal resistance R_a. The internal resistance R_a may or may not be equal to R_t as the circumstances warrant. Fig. 5.21 shows a block diagram of such a four-terminals network (with the source feeding it.).

The calculation procedure can be divided into a number of steps. These steps are as under:

Step 1: Determination of signal input power P_{si}.

Step 2: Determination of noise input power P_{ni}.

Step 3: Calculation of input signal-to-noise ratio $(SNR)_i$ with the help of P_{si} and P_{ni}.

Step 4: Determination of signal output power P_{so}.

Step 5: Determination of noise output power P_{no}.

Step 6: Calculation of the output signal-to-noise ratio $(SNR)_o$ with the help of P_{so} and P_{no}.

Step 7: Calculation of the generalized form of noise figure with the help of step III and step VI.

Step 8: Calculation of P_{no} from R_{eq} if possible and substitute into the general expression for noise figure F to get the actual formula or determination of P_{no} from measurement and substitute to obtain the formula for noise figure F.

Now, we shall discuss step-by-step as under:

Step 1: From figure we can obtain signal input voltage V_{si} and Power P_{si} as under:

$$V_{si} = \frac{V_s . R_t}{R_a + R_t} \quad ...(5.81)$$

and

$$P_{si} = \frac{V_{si}^2}{R_t} \quad ...(5.82)$$

Substituting the value of V_{si} from Eqn. (5.81) in Eqn. (5.82), we get

$$P_{si} = \frac{V_{si}^2}{R_t} = \frac{\left(\dfrac{V_s . R_t}{R_a + R_t}\right)^2}{R_t}$$

$$P_{si} = \frac{V_s^2 R_t}{(R_a + R_t)^2} \qquad ...(5.83)$$

Step 2: Similarly, we can obtain noise input voltage V_{ni} and power P_{ni} as under:

$$V_{ni} = \sqrt{4k\,TBR} = \sqrt{4k\,TB\left(\frac{R_a R_t}{R_a + R_t}\right)} \qquad ...(5.84)$$

and $$P_{ni} = \frac{V_{ni}^2}{R_t}$$

Using Eqn. (5.84), we get

$$P_{ni} = \frac{4k\,TB\left(\frac{R_a R_i}{R_a + R_t}\right)}{R_t} = \frac{4k\,TBR_a}{R_a + R_t} \qquad ...(5.85)$$

Step 3: The input signal-to-noise ratio will be found as

$$(SNR)_i = \frac{P_{si}}{P_{ni}}$$

Using Eqns. (5.83) and (5.85), we get

$$(SNR)_i = \frac{V_s^2 R_t}{(R_a + R_t)^2} + \frac{4k\,TBR_a}{R_a + R_t}$$

or $$(SNR)_i = \frac{V_s^2 R_t}{4k\,TBR_a(R_a + R_t)} \qquad ...(5.86)$$

Step 4: The output signal power will be given as

$$P_{so} = \frac{V_{so}^2}{R_L} = \frac{(AV_{si})^2}{R_L} \qquad ...(5.87)$$

or $$P_{so} = \frac{A^2 V_{si}^2}{R_L}$$

Substituting Eqn. (5.81) in Eqn. (5.87), we get

$$P_{so} = \frac{A^2\left(\frac{V_s R_t}{R_a + R_t}\right)^2}{R_L} = \frac{A^2 V_s^2 R_t^2}{(R_a + R_t)^2} \qquad ...(5.88)$$

Step 5: The noise output power P_{no} may be quite difficult to calculate. For instance, it can simply be written as

$$P_{no} = \text{Noise output power} \qquad ...(5.89)$$

Step 6: Hence, the output signal-to-noise ratio $(SNR)_o$ will be found as

$$(SNR)_o = \frac{P_{so}}{P_{no}}$$

Using Eqns. (5.89) and (5.89), we get

$$(SNR)_o = \frac{P_{so}}{P_{no}} = \frac{A^2V_s^2R_t^2}{(R_a+R_t)^2 R_L.P_{no}} \qquad ...(5.90)$$

Step 7: Lastly, the general expression for the noise figure is

$$F = \frac{(SNR)_i}{(SNR)_o}$$

Using Eqns. (5.86) and (5.90), we get

$$F = \frac{V_s^2R_t}{4k\,TBR\,(R_a+R_t)} + \frac{A^2V_s^2R_t^2}{(R_a+R_t)\,R_LP_{no}} = \frac{R_LP_{no}(R_a+R_t)}{4kTBA^2R_aR_t}$$

This is required expression.

5.9.10 Noise Figure from Equivalent Noise Resistance

As already discussed that the equivalent noise resistance R_{eq} of an amplifier or receiver is the sum of the input terminating resistance and the equivalent noise resistance of the first stage, together with the noise resistance of the previous stages referred to the input i.e.,

$$R_{eq} = R_1 + \frac{R_2}{A_1^2} + \frac{R_2}{A_1^2A_2^2} + \frac{R_2}{A_1^2A_2^2A_3^2} + ... \qquad ...(5.91)$$

where R_{eq} = Equivalent noise resistance of a multistage amplifier or a receiver.

A_1, A_2, A_3 are gains of first stage, second stage, third stage and fourth stage, respectively of an amplifier or a receiver.

R_1, R_2, R_3, R_4 are the total input-noise resistance of first stage, second stage, third stage, and fourth stage, respectively of an amplifier or a receiver.

Further, we may observe from another angle that all these resistances are added to R_t, giving a lumped resistance. This lumped resistance is thin said to concentrate all the "noise making", of the receiver. The rest of it is now assumed to be noiseless. These noise resistances must now be added to the parallel combination of R_a and R_t.

Now, in order to correlate noise figure and equivalent noise resistance, let us define R'_{eq}.

R'_{eq} is a noise resistance. It is does not incorporate R_t. It is given by

$$R'_{eq} = R_{eq} - R_t \qquad ...(5.92)$$

The total equivalent noise resistance for the amplifier is given by

$$R = R'_{eq} + R_a \| R_t$$

$$R = R'_{eq} + \frac{R_a R_t}{R_a + R_t} \quad \text{...(5.93)}$$

Hence the equivalent noise voltage generated at the input of the receiver will be given as

$$V_{ni} = \sqrt{4k\,TBR} \quad \text{...(5.94)}$$

But, since the amplifier has an overall gain A, therefore, this amplifier is treated as though it were noiseless.

The noise output now will be found as

$$P_{no} = \frac{V_{no}^2}{R_L} = \frac{(AV_{ni})^2}{R_L} = \frac{A^2 V_{ni}^2}{R_L} \quad \text{...(5.95)}$$

Substituting Eqn. (5.94) in Eqn. (5.95), we get

$$P_{no} = \frac{A^2 V_{ni}^2}{R_L} = \frac{A^2\left(\sqrt{4k\text{TBR}}\right)^2}{R_L}$$

or

$$P_{no} = \frac{A^2\,4k\text{TBR}}{R_L} \quad \text{...(5.96)}$$

It may be noted that we have already obtained the expression for noise figure in terms of P_{no} i.e.,

$$F = \frac{R_L P_{no}(R_a + R_t)}{R_L} \quad \text{...(5.97)}$$

It may be noted that we have already obtained the expression for noise figure in terms of P_{no} i.e.,

$$F = \frac{R_L P_{no}(R_a + R_t)}{4k\text{TBA}^2 R_a R_t}$$

Substituting Eqn. (5.96) in Eqn. (5.97), we get

$$F = \frac{R_L P_{no}(R_a + R_t)}{4k\text{TBA}^2 R_a R_t} = \frac{R_L\left[\dfrac{A^2\,4k\text{TBR}}{R_L}\right](R_a + R_t)}{4k\,\text{TBA}^2\,R_a R_t}$$

or

$$F = R\frac{(R_a + R_t)}{R_a R_t} \quad \text{...(5.98)}$$

Substituting Eqn. (5.93) in Eqn. (5.98), we get

$$F = \left\{R'_{eq} + \frac{R_a R_t}{R_a + R_t}\right\}\left(\frac{R_a + R_t}{R_a R_t}\right)$$

or
$$F = 1+R'_{eq}\frac{(R_a+R_t)}{R_aR_t} \quad ...(5.99)$$

This equation is an expression for the noise figure in terms of the equivalent noise resistance.

It may be observed from Eqn. (5.99) that if the noise has to be minimum for any given value of antenna resistance R_a, the ratio $\frac{(R_a+R_t)}{R_t}$ must be a minimum, so that R_t is much larger than R_a. This situation is very much used in practise. Under these mismatched conditions $\frac{(R_a+R_t)}{R_t}$ approaches unity. Thus expression for the noise figure, now, will be reduced to

$$F = 1+\frac{R'_{eq}}{R_a} \quad (5.100)$$

5.9.11 Experimental Determination of Noise Figure

In this article, let us describe a method used for measuring noise figure of an amplifier or a receiver.

Let us assume that amplifier has a source impedance R_a. A diode is connected to the amplifier circuit as shown in Fig. 5.22.

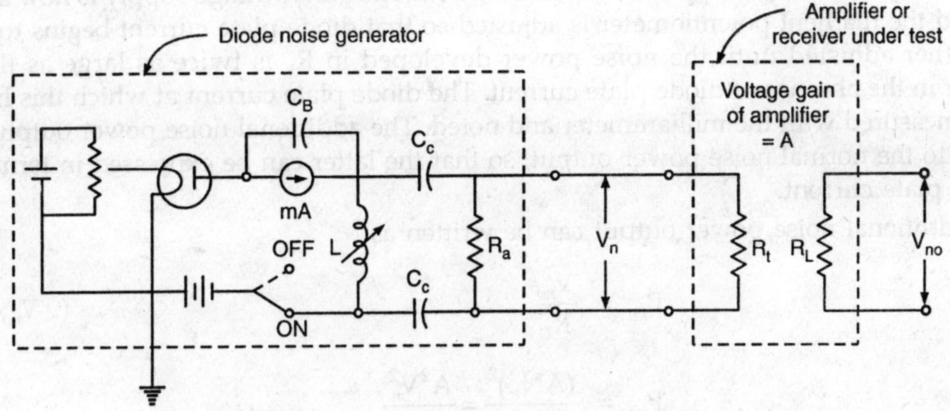

Fig. 5.20 *Experimental Determination of Noise Figure (F)*

The diode is operated in the temperature-limited region so that the effect of space charge can be eliminated and the shot-noise current of the diode is expressed as

$$I_n = \sqrt{2qI_pB} \quad ...(5.101)$$

where I_n = rms shot noise current

q = electronic charge = e = 1.6×10^{-1} coulomb

I_p = direct diode current
B = bandwidth of the interest

Eqn. (5.102) is an expression for the exact plate noise current of a vacuum-tube diode, and this can now be used. Anode current is controlled by means of the potentiometer which varies filament voltage, and that is how shot-noise current is adjusted.

The output capacitance of the diode and its associated circuit is resonated at the operating frequency of the receiver with the help of the variable inductance, and thus it can be ignored. The output impedance of the noise generator or source will now be simply R_a. Hence, the noise voltage supplied to the input of the receiver by the diode is given by

$$V_n = I_n Z_n \quad ...(5.102)$$

where

$$Z_n = R_a \| R_t = \frac{R_a R_t}{R_a + R_t} \quad ...(5.103)$$

Substituting Eqns. (5.101) and (5.103) in Eqn. (5.102), we get

$$V_n = I_n Z_n = \sqrt{2qI_pB}\left(\frac{R_a + R_t}{R_a + R_t}\right)$$

or

$$V_n = \frac{R_a R_t}{R_a + R_t}\sqrt{2qI_pB} \quad ...(5.104)$$

The noise generator or source is connected to the receiver (or amplifier) under test, and the noise output power of the receiver is measured with zero diode plate current, i.e., with the diode plate voltage supply switched off. The diode plate voltage supply is now switched on and the filament potentiometer is adjusted so that diode plate current begins to flow. It is further adjusted until the noise power developed in R_L is twice as large as the noise power in the absence of diode plate current. The diode plate current at which this happens, I_p, is measured with the milliammeter and noted. The additional noise power output is now equal to the normal noise power output, so that the latter can be expressed in terms of the diode plate current.

Additional noise power output can be written as

$$P_{no} = \frac{V_{no}^2}{R_L} \quad (\because V_{no} = AV_n)$$

or

$$P_{no} = \frac{(AN_n)^2}{R_L} = \frac{A^2V_n^2}{R_L} \quad ...(5.105)$$

Substituting Eqn. (5.104) in Eqn. (5.105), we get

$$P_{no} = \frac{A^2V_n^2}{R_L} = \frac{A^2\left\{\dfrac{R_a R_t}{R_a + R_t}\sqrt{2qI_pB}\right\}}{R_L}$$

$$\text{or} \qquad P_{no} = \frac{A^2 R_a^2 R_t^2 2q\, I_p B}{R_L (R_a + R_t)^2} \qquad \text{...(5.106)}$$

As already discussed, the general expression for noise figure is given as

$$F = \frac{R_L P_{no}(R_a + R_t)}{4kTBA^2 R_o R_t} \qquad \text{...(5.107)}$$

Substituting Eqn. (5.106) in Eqn. (5.107), we get

$$F = \frac{R_L P_{no}(R_a + R_t)}{4kTBA^2 R_a R_t} = \frac{R_L \left\{ \dfrac{A^2 R_a^2 R_t^2 2q I_p}{R_L (R_a + R_t)^2} \right\} (R_a + R_t)}{4kTBA^2 R_a R_t}$$

Simplifying, we get

$$F = \frac{q I_p R_a R_t}{2kT(R_a + R_t)} \qquad \text{...(5.108)}$$

If we assume once again that the system is mismatched and $R_t \gg R_a$, then the equation (5.108) can be simplified as

$$F = \frac{R_a q I_p}{2kT} \qquad \text{...(5.109)}$$

If the above procedure is repeated right from the beginning for a system under matched conditions then Eqn. (5.109) applies exactly to such a system instead of being merely a god approximation.

Putting the values of constants in the Eqn. (5.109), we get

$$F = \frac{R_a q I_p}{2kT} = \frac{1.6 \times 10^{-19} R_a I_p}{2 \times 1.38 \times 10^{-23} \times 290} = 20 R_a I_p \qquad \text{...(5.110)}$$

We have assumed that the normal temperature is 17°C or 290°K, where R_a is measured in ohms and I_p is amperes.

5.10 NOISE FIGURE DETERMINATION FOR CASCADED STAGES OF AMPLIFIERS

In this article we shall determine the noise figure (F) of a cascaded amplifier in terms of noise figure of individual stages of amplifiers. It is evident that the noise generated in the first stage is amplified by the subsequent stages. Therefore, the noise figure (F_1) of first stage is much more important compared to the later stages in a multistage amplifier to determine the overall noise figure of the amplifier. Let the two stages of the amplifier have available power gains A_1 and A_2, respectively. Now, the overall noise figure F of the cascaded amplifier can be determined as under.

The total available noise power density S_{no} consists of the total noise power density S_{n1} available at the output due to first stage. It also consists of the total noise power density S_{n2} at the output due to second stage of the amplifier.

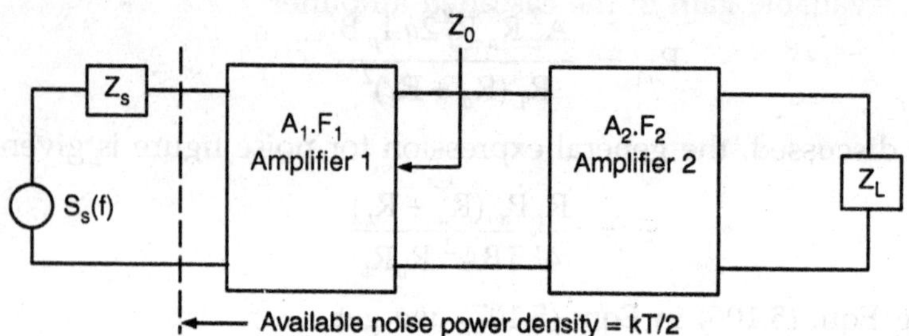

Fig. 5.21 *Determination of Noise Figure of Cascaded Amplifier or a Multistage Cascaded Amplifier*

Let F_1 and F_2 be the noise figures of the first stage and second stage of the amplifier, respectively. Total noise power density S_{n1} available at the output due to the first stage is given as

$$S_{n1} = \frac{kTF_1}{2} A_1 A_2 \qquad ...(5.111)$$

Let S_{n2} is the noise power density available at the output due to the second stage only. The source impedance for the second stage is the impedance seen at the output terminals of the first stage. Impedance seen at the output terminals of first stage is denoted by Z_o. The noise component S_{n2} is due to noise sources inside the second stage amplifier only and it is immaterial whether Z_o is thermal or not. For convenience, we are assuming here that Z_o is thermal.

Hence, the available power density is $kT/2$.

Let the noise figure of amplifier stage 2 be F_2. The available noise power density at the output due to amplifier in second stage will be given by

$$S_{n2} = \frac{kT}{2}(F_2 - 1)A_2 \qquad ...(5.112)$$

Hence, the total noise power density delivered to the load by the cascaded amplifier is given by

$$S_n = S_{n1} + S_{n2} \qquad ...(5.113)$$

Substituting Eqns. (5.111) and (5.112) in Eqn. (5.113), we get

$$S_n = S_{n1} + S_{n2} = \frac{kT}{2} F_1 A_1 A_2 + \frac{kT}{2}(F_2 - 1)A_2$$

or

$$S_n = \frac{kT}{2}\left[F_1 A_1 A_2 + (F_2 - 1)A_2\right] \qquad ...(5.114)$$

Now, we are consider the overall amplifier for which noise power density S_n is expressed as

$$S_n = \frac{kT}{2} FA \qquad ...(5.115)$$

where A = Available gain of the cascaded amplifier

F = Overall spectral noise figure

Because the overall available gain of a cascaded amplifier is the multiplication of the available gain of each stage, therefore,

We have

$$A = A_1A_2 \quad ...(5.116)$$

Substituting Eqn. (5.116) in Eqn. (5.115), we obtain

$$S_n = \frac{kT}{2}FA_1A_2 \quad ...(5.117)$$

On comparing Eqns. (5.114) and (5.117), we get

$$S_n = \frac{kT}{2}[FA_1A_2 + (F_1 - 1)A_2] \quad ...(5.118)$$

$$S_n = \frac{kT}{2}FA_1A_2 \quad ...(5.119)$$

$$F = F_1 + \frac{(F_2 - 1)}{A_1} \quad ...(5.120)$$

It must be remembered that Noise Figure F_2 is the noise figure of the amplifier in second stage under the condition that the source is a positive impedance Z_o equal to the output impedance of the amplifier in first stage.

In general, for a multistage amplifier, the overall spectral noise figure is given by

$$F = F_1 + \frac{F_2 - 1}{A_1} + \frac{F_3 - 1}{A_1A_2} + \frac{F_4 - 1}{A_1A_2A_3} + ... \quad ...(5.121)$$

From Eqn. (5.121), we conclude that the first stage is most significant in determining the overall noise figure of the amplifier. Hence, in low noise amplifiers, the primary consideration is the design of the first stage which must have a low noise figure. This is sometimes obtained even at the cost of gain of the first stage of the multistage amplifier.

SOME SOLVE EXAMPLES

EXAMPLE 1

An amplitude operating over the frequency range from 18 to 20 MHz has a 10 kΩ input resistor. Calculate the rms noise voltage at the input to this amplifier if the ambient temperature is 27°C.

Solution:

The rms noise voltage is given by the expression

$$V_n = \sqrt{4RkTB} \quad ...(i)$$

Here given that

$$R = 10 \text{ k}\Omega$$

$$T = 27 + 273 = 300°K$$

Bandwidth

$$B = 20 - 10 = 2 \text{ MHz}$$

$$k = 1.38 \times 10^{-23} \text{ Joule/deg. K}$$

$$V_n = \sqrt{4 \times 10 \times 10^3 \times 1.38 \times 10^{-23} \times 300 \times 2 \times 10^6}$$

$$= \sqrt{4 \times 1.38 \times 3 \times 2 \times 10^{-11}}$$

$$= 1.82 \times 10^{-5} \text{ Volt}$$

$$= 18.2 \ \mu\text{V}$$

EXAMPLE 2

The first stage of a two-stage amplifier has a voltage gain of 10, a 600 Ω input resistor, a 1600 Ω equivalent noise resistance and a 27 kΩ output resistor. For the second stage, these values are 25, 81 kΩ, 10 kΩ, and 1 MΩ respectively. Find the equivalent input noise resistance of this two-stage amplifier.

Solution:

Here given for two-stage amplifier:

The voltage gain for first stage is $A_1 = 10$

The voltage given for second stage is $A_2 = 25$

Input resistor for first stage is $R_{i_1} = 600\ \Omega$

Equivalent noise resistance for first stage is $R_{eq_1} = 1600\ \Omega$

Output resistor for first stage $R_{o_1} = 27\ \Omega$

Input resistor for second stage $R_{i_2} = 81\ \Omega$

Equivalent noise resistance for second stage is $R_{eq_2} = 10\ \Omega$

Output resistor for second stage is $R_{o_2} = 1\ M\Omega$

$$R_1 = R_{i_1} + R_{eq_1} = 600 + 1600$$

$$= 2200\ \Omega$$

$$R_2 = \frac{R_{o_1} R_{i_2}}{R_{o_1} + R_{i_2}} + R_{eq_2}$$

$$= \frac{27 \times 81}{27 + 81} + 10$$

$$= 30.2\ k\Omega$$

$$R_3 = R_{o_2} = 1\ M\Omega$$

Equivalent input noise resistance can be determined as

$$R_{eq} = R_1 + \frac{R_2}{A_1^2} + \frac{R_3}{A_1^2 A_2^2}$$

Substituting the values, we get

$$R_{eq} = 2200 + \frac{30.2\times10^3}{(10)^2} + \frac{1\times10^6}{(10)^2(25)^2}$$

$$= \mathbf{2518\ W\ Ans.}$$

EXAMPLE 3

Determine the noise equivalent bandwidth of RC low-pass filter, LPF, whose transfer function or frequency response is given by

$$H(f) = \frac{1}{1 + j2\pi fRC}$$

Solution:

Here given that the frequency response

$$H(f) = \frac{1}{1 + j2\pi fRC} \quad ...(i)$$

or

$$H(f) = \frac{1}{1 + j(f/B)} \quad ...(ii)$$

where

$$B = \frac{1}{2\pi RC} \quad ...(iii)$$

Therefore,

$$|H(f)|^2 = \frac{1}{1 + (f/B)^2} \quad ...(iv)$$

For the centre frequency $f = 0$ of the filter, we have

$$|H(o)^2| = \frac{1}{1 + (o/B)^2} = 1 \quad ...(v)$$

We know that the noise-equivalent bandwidth for a low-pass filter is expressed as

$$B_N = \frac{\int_o^\alpha |H(f)|^2\, df}{|H(o)|^2} \quad ...(vi)$$

Substituting equations (*iv*) and (*v*) in equation (*vi*), we get

$$B_N = \frac{\int_o^\alpha \frac{1}{1 + (f/B)} df}{1}$$

$$= \frac{\pi}{2}.B \quad ...(vii)$$

Now substituting equation (*iii*) in equation (*vii*), we get

$$B_N = \frac{\pi}{2}.\frac{1}{2\pi RC} = \frac{1}{4RC} \quad ...(viii)$$

Hence, the noise-equivalent bandwidth decreases with increase in R.

The average noise output power of a low-pass filter, LPF, is given as

$$P_n = N_o B_N |H(o)|^2$$

Now, substituting equations (*v*) and (*viii*) in lost equation, we get

$$P_n = N_o.\frac{1}{4RC} \times 1$$

$$= \frac{N_o}{4RC}$$

But for thermal noise source

$$N_o = 4kTR$$

Therefore,

$$P_n = \frac{N_o}{4RC}$$

$$= \frac{4kTR}{4RC}$$

$$= \frac{kT}{C}$$

EXAMPLE 4

A parallel tuned circuit is made to resonate at a frequency of 100 MHz. The parallel tuned circuit uses a coil having quality factor Q of 10 and a capacitance of 10 pF. The temperature of the circuit is maintained at 17°C. Determine the output voltage across the circuit measured by a wideband voltmeter.

Solution:

We know that the quality factor Q of the coil is given by

$$Q = \frac{f_r}{\Delta f}$$

where f_r is the resonance frequency and Δf is the bandwidth of the tuned circuit. Thus

$$\Delta f = \frac{f_r}{Q} = \frac{100}{10} = 10 \text{ MHz}$$

The quality factor Q is also given by the following expression as

$$Q = \frac{1}{\omega CR}$$

or
$$R = \frac{1}{Q\omega C} = \frac{1}{10 \times \omega \times 10 \times 10^{-12}}$$

or
$$R = \frac{1}{10 \times 2\pi \times 100 \times 10^{6} \times 10 \times 10^{-12}}$$
$$= 16 \text{ ohms}$$

The output voltage, rms value, is given as

$$V_{no} = \left(\overline{v_{no}^2}\right)^{1/2}$$

But mean square, m.s., value v^2 is given by

$$\overline{v_{no}^2} = 4kT(\Delta f)Q^2R$$

Therefore,

$$V_{no} = \sqrt{4kT\,(\Delta f)\,Q^2R}$$
$$= \sqrt{4 \times 1.38 \times 10^{-23} \times (273 + 17) \times 10 \times 10^{6} \times 10^{2} \times 16}$$
$$= 160 \times 10^{-7}$$
$$= \mathbf{16\ mV\ Ans.}$$

EXAMPLE 5

A mixer stage has a noise figure of 20 dB. This mixture stage is preceded by an amplifier which has a noise figure of 9 dB and an available power gain of 15 dB. Find the overall noise figure referred to the input.

Solution:

Here given that

Noise figure an Amplifier, F_1 = 10 dB

Noise figure of Mixer, F_2 = 20 dB

Available power gain of an amplifier, A_1 = 15

This is two-stage system where both stages, amplifier and mixer, are cascaded.

Thus,
$$F_1 \text{ in ratio} = \text{Antilog}_{10}\left(\frac{F_1 \text{ in dB}}{10}\right)$$
$$= \text{Antilog}_{10}\left(\frac{9}{10}\right) = 7.94$$
$$F_2 \text{ in ratio} = \text{Antilog}_{10}\left(\frac{F_2 \text{ in dB}}{10}\right)$$

$$= \text{Antilog}_{10}\left(\frac{20}{10}\right)$$

or $$F_2 \text{ in ratio} = \text{Antilog}_{10}(2) = 100$$

$$A_1 \text{ in ratio} = \text{Antilog}_{10}\left(\frac{A_1 \text{ in dB}}{10}\right)$$

$$= \text{Antilog}_{10}\left(\frac{15}{10}\right)$$

or $$A_1 \text{ in ratio} = \text{Antilog}_{10}(1.5)$$
$$= 31.62$$

Overall noise figure of the cascaded two-stage system referred to input can be found as under.

$$F = F_1 + \frac{F_2 - 1}{A_1}$$

Substituting all the values, we get

$$F = 7.94 + \frac{100 - 1}{31.62}$$

$$= 7.94 + \frac{99}{31.62}$$

$$= 11.07$$

Now, $$F \text{ in dB} = 10 \log_{10} (F_{ratio})$$
$$= 10 \log_{10} (11.07)$$
$$= 10.44 \text{ dB}$$ **Ans.**

EXAMPLE 6

An amplifier operating on a frequency range from 18 to 20 MHz has a 10 kW input resistance. Find the rms noise voltage at the input to this amplifier if the ambient temperature is 27°C.

Solution:

We know that the rms noise voltage of an amplifier is given by

$$V_n = \sqrt{4kTBr} \quad ...(i)$$

Here given that the operating bandwidth of the amplifier is

$$B = (20 - 18) \text{ MHz}$$
$$= 2 \text{ MHz} = 2 \times 10^6 \text{ Hz}$$

$$\text{Temperature, T} = (273°\text{C} + 17)$$
$$= 290 \text{ K}$$

$$\text{Input resistor R} = 10 \text{ k}\Omega$$
$$= 10 \times 10^3 \ \Omega$$

$$= 10^4\ \Omega$$

Substituting all the values in equation (*i*), we get

$$V_n = \sqrt{4kTBR}$$

$$= \sqrt{4 \times 1.38 \times 10^{-23} \times 290 \times 2 \times 10^6 \times 10^4}$$

$$V_n = 1.78 \times 10^{-5} \text{ volt}$$

$$= 17.8 \times 10^{-6} \text{ volt}$$

or $\quad$ $V_n = 17.8\ \mu$ **volt Ans.**

EXAMPLE 7

Two resistors R_1 and R_2 at absolute temperatures T_1 and T_2 are connected in series to form a white noise source. Find the equivalent noise temperature I_{eq}.

Solution:

The equivalent resistance will be

$$R = R_1 + R_2$$

Hence, the available noise power will be

$$P_n = \frac{V_n^2}{4R} = \frac{V_{n1}^2 + V_{n2}^2}{4R}$$

Substituting for V_{n1} and V_{n2}, we get

$$P_n = \frac{4kT_1 BR_1 + 4kT_2 BR_2}{4R}$$

$$= (R_1T_1 + R_2T_2)\frac{kB}{R} \qquad ...(i)$$

Therefore, the equivalent noise resistance R_{eq} is determined by

$$T_{eq} = \frac{P_n}{kB} \qquad ...(ii)$$

Substituting equation (i) in equation (ii), we can have

$$T_{eq} = \frac{(R_1T_1 + R_2T_2)\,kB/R}{kB}$$

or

$$T_{eq} = \frac{(R_1T_1 + R_2T_2)}{R}$$

Because $\quad R = R_1 + R_2$

Hence,

$$T_{eq} = \frac{(R_1T_1 + R_2T_2)}{R_1 + R_2} \qquad ...(iii)$$

If $R_1 = R_2$, the noise resistance is the arithmetic average of the physical temperature, i.e.,

$$T_{eq} = \frac{(T_1 + T_2)}{2}$$

Because T_{eq} is not a physical temperature, the series connection is a non-thermal source, as should be expected since the resistance are not in thermal equilibrium. However, viewed from the terminals, this non-thermal source looks identical to thermal resistance R at temperature T_{eq}.

PROBLEMS

1. Define the term 'noise' used in communication system.
2. What is shot noise? Write expression for the shot noise current in a diode.
3. What is mean by the term 'atmosphere noise'?
4. What are the causes and effects of solar noise?
5. What is meant by the term 'cosmic noise'? Give its salient features.
6. Describe thermal noise.
7. Define transistor signal-to-noise ratio and noise figure of a receiver.
8. What is meant by the transit time noise in a transistor?

❑❑❑

Pulse Modulation 6

6.1 INTRODUCTION

In analog modulation systems, some parameter of sinusoidal carrier is varied in accordance with the instantaneous value of the modulating signal. In pulse modulation systems, the carrier is no longer a continuous signal but consists of a pulse train, some parameter of which is varied in accordance with the instantaneous value of the modulating signal. There are two types of pulse modulation systems:

(*i*) Pulse Amplitude Modulation (PAM)

(*ii*) Pulse Time Modulation (PTM)

In PAM, amplitude of the pulses of the carrier pulse train is varied in accordance with the modulating signal; whereas, in PTM, the timing of the pulses of the carrier pulse train is varied. There are two types of PTM.

(*i*) Pulse Width Modulation (PWM) or
Pulse Duration Modulation (PDM) or
Pulse Length Modulation (PLM)

(*ii*) Pulse Position Modulation (PPM)

In PWM, the width of pulses of the carrier pulse train is varied in accordance with the modulating signal; whereas, in PPM, the position of pulses of the carrier pulse train is varied.

According to the sampling theorem, if a modulating signal is bandlimited to B Hz (i.e., there are no frequency components beyond BHz in the frequency spectrum of the modulating signal), the sampling frequency must be at least 2B Hz and, hence, the frequency of the carrier pulse train must also be at least 2B Hz.

The spectral range occupied by the basic signal is called the baseband frequency range or, simply, baseband. Hence the basic signal is sometimes called as baseband signal. When the basic signal is transmitted without a frequency translation, the system is known as baseband system.

6.2 PULSE AMPLITUDE MODULATION

Fig. 6.11 explains the principle of P.A.M. A baseband signal $f(t)$ is shown in Fig. 6.1(*a*) and carrier pulse train $f_c(t)$ is shown in Fig. 6.1(*b*). The frequency of the carrier pulse train is decided by the sampling theorem. A pulse amplitude modulated signal $f_m(t)$ is shown in Fig. 6.1(*c*). It can be seen that the amplitude of the pulses in Fig. 6.1(*c*) depends on the value of $f(t)$ during the time of the pulse.

In Fig. 6.1(*a*), the baseband signal $f(t)$ is shown to have only a positive polarity. In practise, however, we can have a baseband signal with a positive as well as negative polarity. But, in such a case, the modulated pulses will also be of positive as well as negative polarities. As the transmission of such bipolar pulses is inconvenient; a clamping circuit is used so that we always have a baseband signal with only the positive polarity as shown in Fig. 6.1(*a*).

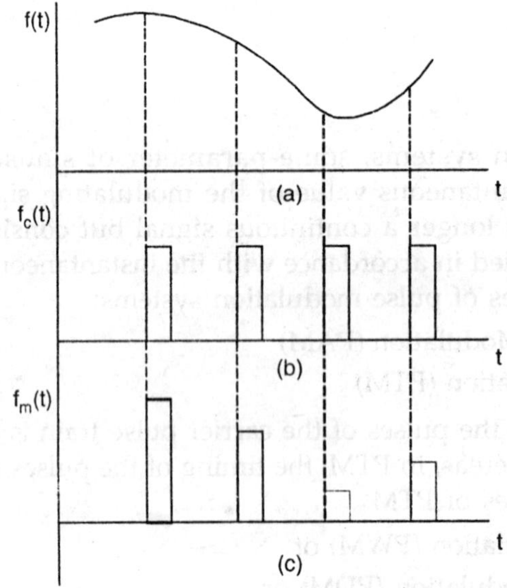

Fig. 6.1 *(a) Pulse Amplitude Modulation-Baseband Signals f(t) (b) Carrier Pulse Train $f_c(t)$ (c) Pulse Amplitude Modulated Signal*

There are two methods of getting the pulse amplitude modulated waveform:

(*i*) Natural sampling or shaped-to sampling

(*ii*) Flat-top sampling

6.2.1 Natural Sampling

This is explained in Fig. 6.2. Here, the amplitude of the carrier pulse train is adjusted to 1, the duration of the pulse is τ, and they are separated by T_s. The PAM signal as shown in Fig. 6.2(*d*) is obtained by multiplying $f(t)$ of Fig. 6.2(*a*) and $f_c(t)$ of Fig. 6.2(*b*) in the multiplier of Fig. 6.2(*c*). It may be seen from Fig. 6.2(*d*) that the tops of the pulse-amplitude modulated pulses are not flat but they follow the natural waveform of the modulating signal $f(t)$ during the respective pulse intervals and, hence, the name natural sampling is given to this method.

The Fourier series of a periodic train of pulses can be given as

$$v(t) = \frac{A\tau}{T_0} + \frac{2A\tau}{T_0}\sum_{n=0}^{\infty} C_n \cos\frac{2\pi nt}{T_0} \quad ...(6.1)$$

where

A is the amplitude of the pulse

τ the duration of the pulse

T_0 the period of the pulse train and

and

C_n is given by

$$C_n = \frac{\sin(n\pi\tau/T_0)}{n\pi\tau/T_0}$$

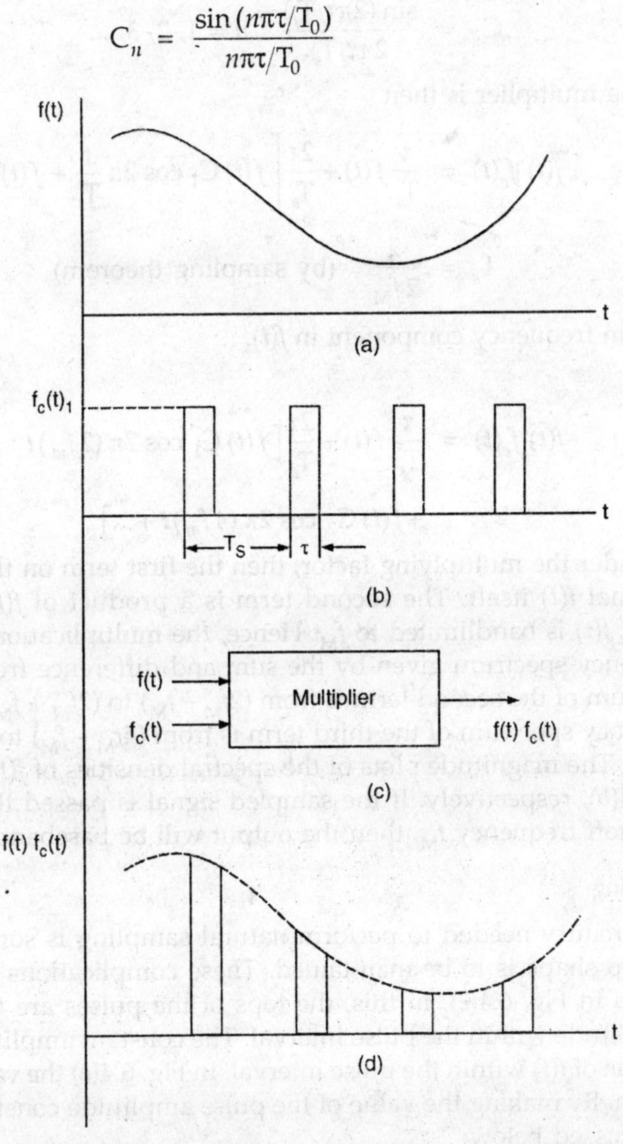

Fig. 6.2 *(a) Natural Sampling-Baseband Signal f(t) (b) Carrier Pulse Train fc(t) (c) Multiplier (d) PAM Signal at Output of Multiplier*

For the carrier pulse train Fig. 6.2(*b*), we have

$$v(t) = f_c(t)$$

$$A = 1$$

and $$T_0 = T_s$$

Therefore Eqn. (6.1) becomes

$$f_c(t) = \frac{\tau}{T_s} + \frac{2\tau}{T_s}\left[C_1 \cos 2\pi \frac{t}{T_s} + C_2 \cos 2 \times 2\pi \frac{t}{T_s} + ...\right] \quad ...(6.2)$$

The constant C_n is given by

$$C_n = \frac{\sin(2\pi\tau/T_s)}{2\pi\tau/T_s} \qquad n = 1, 2, 3, ...$$

The output of the multiplier is then

$$f(t)\, f_c(t) = \frac{\tau}{T_s} f(t) + \frac{2\tau}{T_s}\left[f(t)\, C_1 \cos 2\pi \frac{t}{T_s} + f(t)\, C_2 \cos 2 \times 2\pi \frac{t}{T_s} + ...\right]$$

Let us choose $$T_s = \frac{2}{2f_M} \quad \text{(by sampling theorem)}$$

where f_M = maximum frequency component in $f(t)$.

We then get

$$f(t)\, f_c(t) = \frac{\tau}{T_s} f(t) + \frac{2\tau}{T_s}[f(t)\, C_1 \cos 2\pi (2f_m)t$$

$$+ f(t)\, C_2 \cos 2\pi (4f_m)t + ...] \quad ...(6.3)$$

If we don't consider the multiplying factor, then the first term on the r.h.s. of Eqn. (6.3) is the baseband signal $f(t)$ itself. The second term is a product of $f(t)$ and a sinusoid of frequency $2f_M$. Now, $f(t)$ is bandlimited to f_M. Hence, the multiplication in the second term will yield the frequency spectrum given by the sum and difference frequency terms. Thus the frequency spectrum of the second term is from $(2f_M - f_M)$ to $(2f_M + f_M)$ i.e., from f_M to $3f_M$. Similarly, the frequency spectrum of the third term is from $(4f_M - f_M)$ to $(4f_M + f_M)$, i.e., from $3f_M$ to $5f_M$ and so on. The magnitude plots of the spectral densities of $f(t)$ and $f_c(t)$ are shown in Fig. 6.3(*a*) and 6.3(*b*), respectively. If the sampled signal is passed through an ideal low pass filter with a cutoff frequency f_M, then the output will be baseband signal $f(t)$.

6.2.2 Flat-top Sampling

The electronic circuitry needed to perform natural sampling is somewhat complicated because the pulse-top shape is to be maintained. These complications are reduced by flat-top sampling, shown in Fig. 6.4(*e*). In this, the tops of the pulses are flat. Thus the pulses have a constant amplitude within the pulse interval. The constant amplitude of the pulse can be chosen at any value of $f(t)$ within the pulse interval. In Fig. 6.4(*e*) the value at the beginning of the pulse is chosen. By making the value of the pulse amplitude constant within the pulse interval. This is discussed below:

6.3 SPECTRUM OF A FLAT-TOP SAMPLED SIGNAL

The flat-top sample signal $f_m(t)$ of Fig. 6.4(*e*) may be considered as a convolution of the impulse sampled signal $f_s(t)$ of Fig. 6.4(*a*), and non-periodic pulse $p(t)$ of width τ and height 1 of Fig. 6.4(*c*). The spectrums of $f_s(t)$ and $p(t)$ are shown in Fig. 6.4(*b*) and Fig. 6.4(*d*), respectively. The spectrum of $f_m(t)$ is shown in Fig. 6.4(*f*), which is obtained by multiplying $F_{s(\omega)}$ with $P_{(\omega)}$. As the $P_{(\omega)}$ value is different at different frequencies, the shape of $F_M(\omega)$ is not similar to $F_{s(\omega)}$ which shows that a distortion will be introduced if the signal is recovered by an ideal low pass filter of a cutoff frequency ω_M.

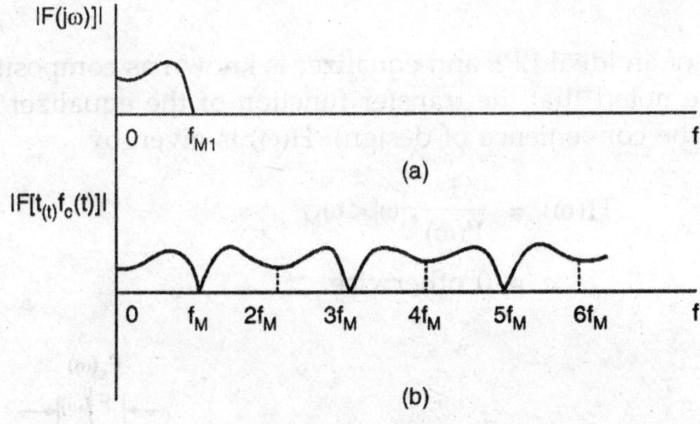

Fig. 6.3 *(a) Magnitude plot of Spectral Density of f(t)*
(b) Magnitude plot of Spectral Density of f(t) and $f_c t$

6.4 A PAM MODULATOR CIRCUIT

A PAM modulator circuit is shown in Fig. 6.5. This circuit is a simple emitter follower. In the absence of the clock signal, the output follows the input. The modulating signal is applied as the input signal. Another input to the base of the transistor is the clock signal. The frequency of the clock signal is made equal to the desired carrier pulse train frequency. The amplitude of the clock signal is so chosen that the high level is at ground (OV), and the low level is at some negative voltage which is sufficient to bring the transistor in the cutoff region. Thus when the clock signal is high, the circuit behaves as an emitter follower, and the output follows the input modulating signal. When the clock signal is low, the transistor is cutoff and the output is zero. Thus the output waveform, shown in Fig. 6.5 is the desired pulse amplitude modulated signal.

6.5 DEMODULATION OF PAM SIGNALS

Demodulation of natural sampled signal can be done with the help of an ideal low pass filter with a cutoff frequency ω_M. But, for this, the pulse-top shape is to be maintained after transmission. This is very difficult due to the transmitter and receiver noise. Therefore, normally, flat-top sampling is preferred over natural sampling.

There are two demodulation methods for the flat-top sampled signal:

6.5.1 Using an Equalizer

If the flat-top sample signal is passed through an ideal low pass filter, the spectrum of the output will be F(ω)P(ω). The time function of the output is somewhat distorted due to the multiplying factor P(ω). If the low pass filter output is passed through a filter having a transfer function 1/P(ω) over the range $0 - \omega_M$, the spectrum at the output of this filter will be $F(\omega)P(\omega).\dfrac{1}{P(\omega)} = F(\omega)$, and, hence, the original time function *f*(*t*) will be recovered. The filter with a transfer function $\dfrac{1}{P(\omega)}$ is known as equalizer. The over-all arrangement is shown in Fig. 6.6(*a*).

The combination of an ideal LPF and equalizer is known as composite filter is shown in Fig. 6.6(*b*). (It may be noted that the transfer function of the equalizer outside ω_M can be chosen according to the convenience of design). H(ω) is given by

$$H(\omega) = \frac{1}{P(\omega)}, |\omega| < \omega_M$$

$$= 0 \text{ otherwise.}$$

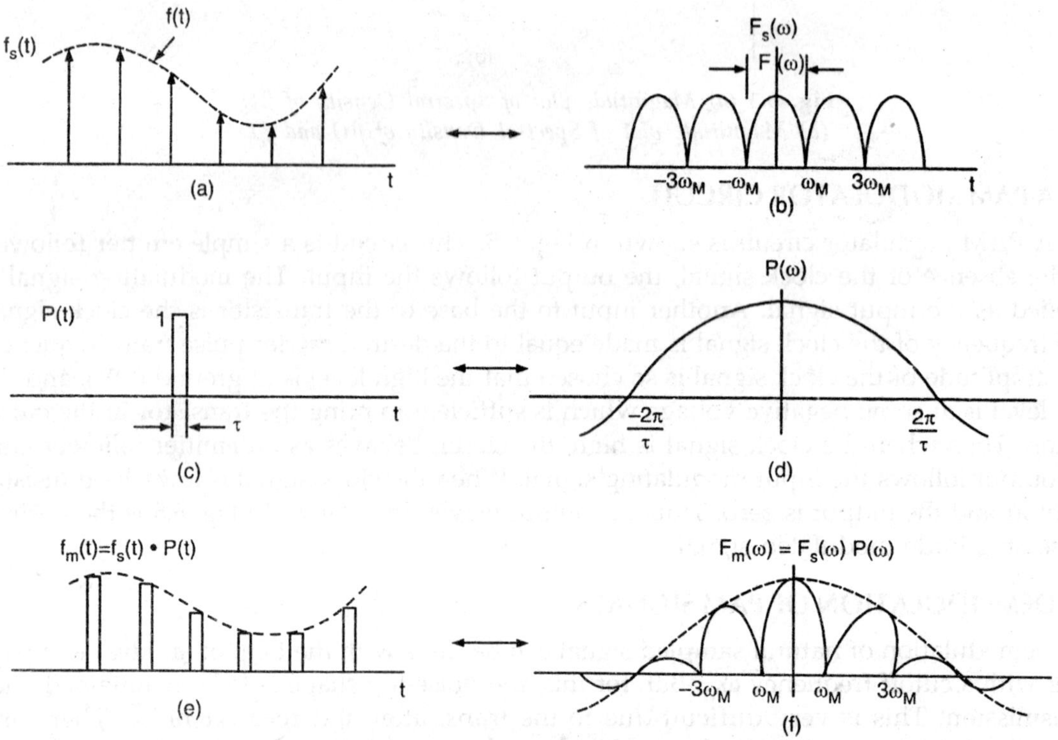

Fig. 6.4 *Flat Top Sampling (a) Impulse Sampled Signal $f_s(t)$ (b) Spectrum of $f_s(t)$ (c) Non-periodic pulse $P_c t$ of Width t and Height 1 (d) Spectrum of p(t) (e) Flat-top Sampled PAM signal $f_m(t)$ (f) Spectrum of $f_m(t)$*

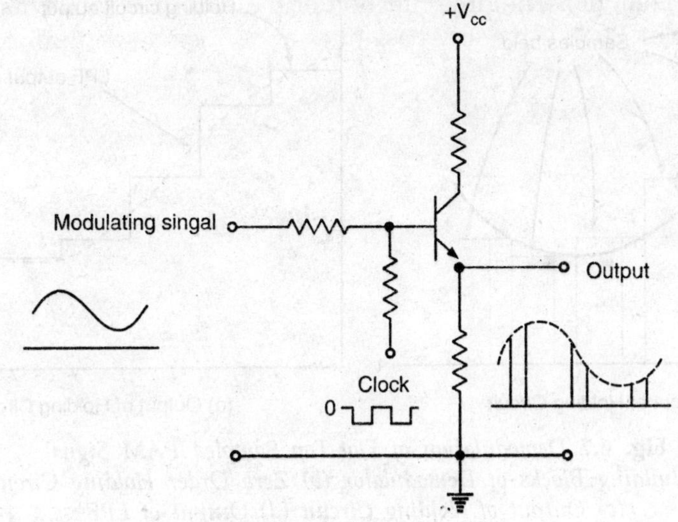

Fig. 6.5 *A PAM Modulator*

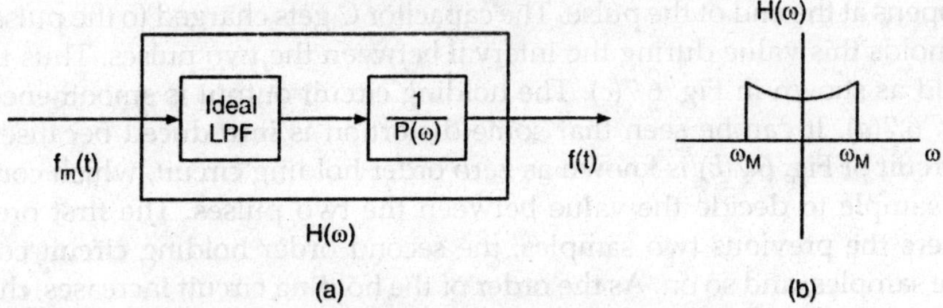

Fig. 6.6 *Demodulation of Flat-Top Sampled PAM Signal Using Equalizer*
(a) Composite Filter with Transfer Function H(ω)
(b) The Desired Characteristics of H(ω)

6.5.2 Using Holding Circuit

In this method, the received signal is passed through a holding circuit and a LPF, shown in Fig. 6.7(*a*).

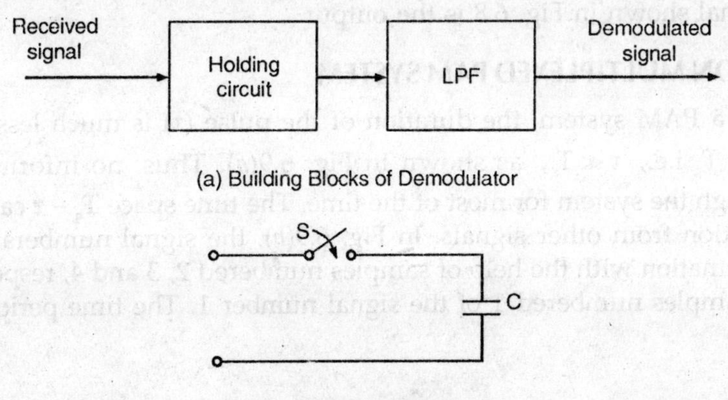

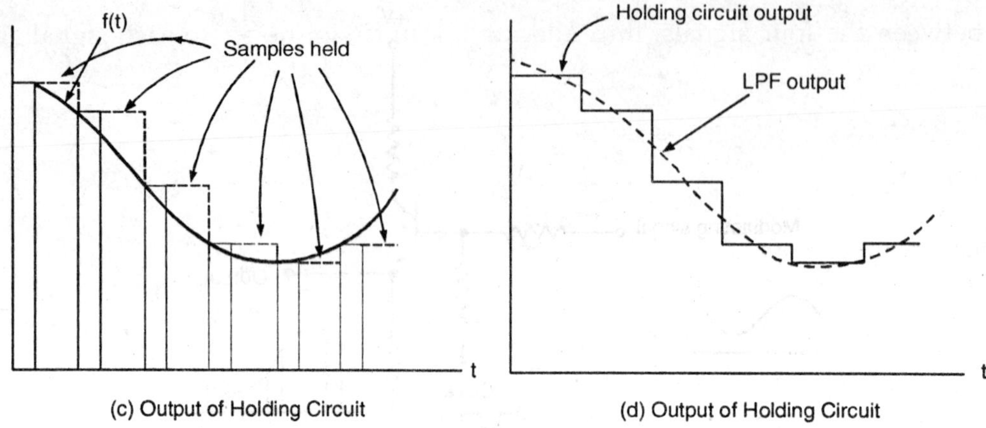

(c) Output of Holding Circuit (d) Output of Holding Circuit

Fig. 6.7 *Demodulation of Flat-Top Sampled PAM Signal (a) Building Blocks of Demodulator (b) Zero Order Holding Circuit (c) Output of Holding Circuit (d) Output of LPF*

Fig. 6.7(*b*) shows a simple holding circuit. The switch S closes after the arrival of the pulse, and it opens at the end of the pulse. The capacitor C gets charged to the pulse amplitude value, and it holds this value during the interval between the two pulses. Thus the sampled values are held as shown in Fig. 6.7(*c*). The holding circuit output is smoothened in LPF as shown in Fig. 6.7(*d*). It can be seen that some distortion is introduced because of holding circuit. The circuit of Fig. 6.7(*b*) is known as zero order holding circuit, which considers only the previous sample to decide the value between the two pulses. The first order holding circuit considers the previous two samples; the second order holding circuit considers the previous three samples; and so on. As the order of the holding circuit increases, the distortion decreases at the cost of the circuit complexity. The amount of permissible distortion decides the order of the holding circuit.

6.6 A PAM DEMODULATOR CIRCUIT

A PAM demodulator circuit is shown in Fig. 6.8. It is just an envelope detector followed by a low pass filter. The diode and R-C combination work as the envelope detector. This is followed by a second order OP-AMP low pass filter to have a good filtering characteristics. Thus, for the received pulse amplitude modulated signal as the input signal, the desired demodulated signal shown in Fig. 6.8 is the output.

6.7 TIME DIVISION MULTIPLEXED PAM SYSTEM

Normally, in a PAM system, the duration of the pulse (τ) is much less than the time period of pulses T_s i.e., $\tau \ll T_s$, as shown in Fig. 6.9(*a*). Thus, no information is being transmitted through the system for most of the time. The time space $T_s - \tau$ can be utilized to transmit information from other signals. In Fig. 6.9(*b*), the signal numbers 2, 3 and 4 are transmitting information with the help of samples numbered 2, 3 and 4, respectively. This is along with the samples numbered 1 of the signal number 1. The time period T_s is equally divided between the four signals, thus allocating a time slot of $\frac{T_s}{4}$ to each signal. The duration of time slot is such that $\frac{T_s}{4} > \tau$. Thus there is a guard time $\frac{T_s}{4} - \tau$ between all successive sampling pulses, ensuring that there is less cross-talk between signals.

The arrangement by which the information from more than more than one signal is transmitted in this manner is known as Time Division Multiplexing (TDM). A Time Division Multiplexed PAM system is shown in Fig. 6.9 which transmits information from n signals.

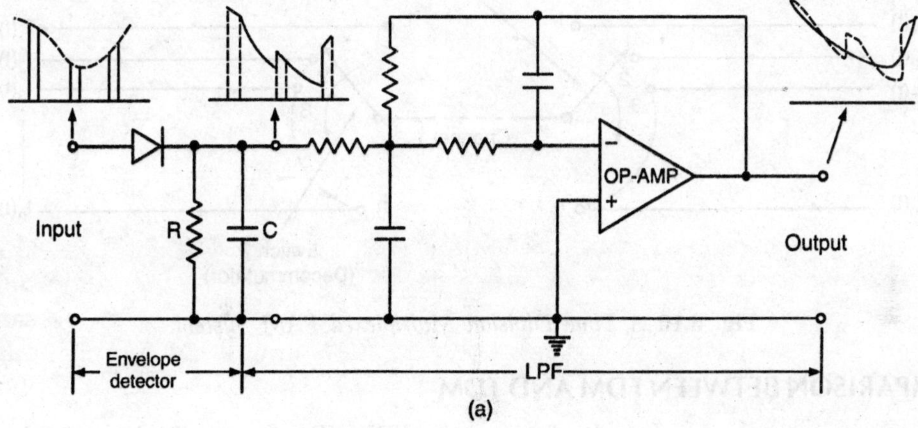

Fig. 6.8 *A PAM Demodulator*

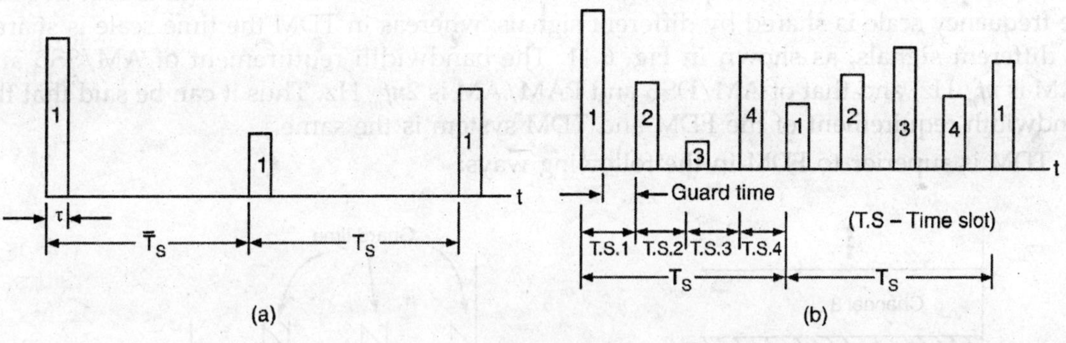

Fig. 6.9

The switch 1 and switch 2, respectively known as commutator and decommutator, are synchronized electronic switches which rotates at the same speed of $2f_M$ rotations per second. The commutator samples and combines the samples. While the decommutator separates the samples belonging to individual signals.

Synchronization is the most crucial operation in any TDM system. Thus, for example, if the commutator is at position 2, the decommutator must also be in position 2. To provide synchronization, a synchronizing pulse is transmitted in every frame (time interval between two successive samples of the same signal, i.e., T_s). Thus to multiplex n channel, n + 1 time slots are provided in a frame; n for channels and 1 for the synchronizing pulse. The synchronizing pulse is chosen in such a way that it is easily distinguishable. For this purpose, one of its properties is adjusted in such a way that it is never attained by the other pulses. In the case of PAM, its amplitude is made larger than the amplitudes of all the other pulses. Fig. 6.10 shows the schematic diagram of a time division multiplexed PAM system.

In the above discussion, it is assumed that all signals are bandlimited to the same frequency, thereby allowing the same sampling frequency for all signals.

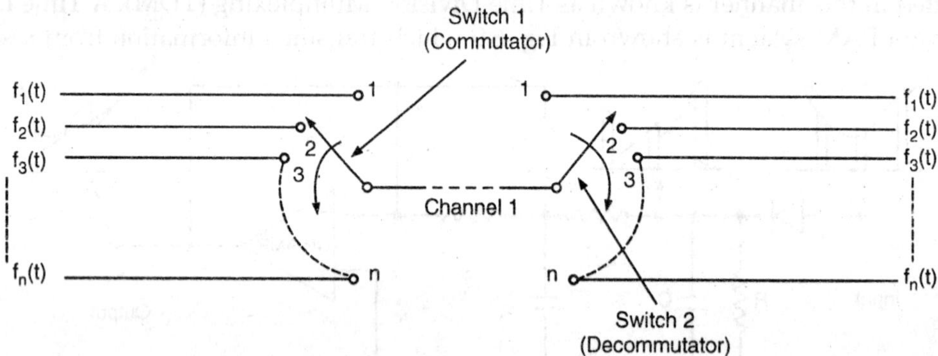

Fig. 6.10 *A Time Division Multiplexed PAM System*

6.8 COMPARISON BETWEEN FDM AND TDM

There are two methods of multiplexing, viz. FDM (Frequency Division Multiplexing) and TDM (Time Division Multiplexing). The basic difference between them is that in FDM the frequency scale is shared by different signals; whereas in TDM the time scale is shared by different signals, as shown in Fig. 6.11. The bandwidth requirement of AM/SSB and PAM is nf_m Hz, and that of AM/DSB and PAM/AM is $2nf_m$ Hz. Thus it can be said that the bandwidth requirement of the FDM and TDM system is the same.

TDM is superior to FDM in the following ways:

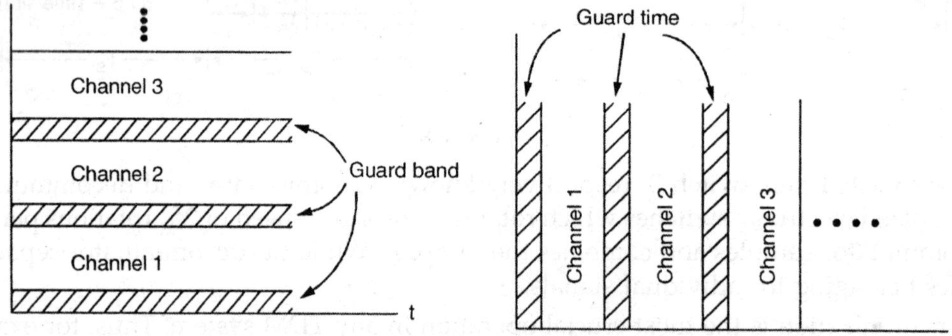

Fig. 6.11 *(a) FDM (b) TDM*

(*i*) In the FDM system, different carriers are to be generated for different channels. Also, as each channel occupies a different frequency band, different bandpass filters are required. On the other hand, in TDM system, all the channels require identical circuit, consisting of simple synchronous switches, gates, and a low pass filter. Thus the circuitry needed in the TDM system is much simple than the one needed in the FDM system.

(*ii*) The nonlinearities in the various amplifiers of an FDM system produce harmonic distortion, hence, they introduce interference within the channels. Therefore, the nonlinearity requirements of an FDM system are much more stringent than those for a single channel. On the other hand, in a TDM system, the signals from different channels are allotted different time slots and they are not applied to the system simultaneously. Hence, the nonlinearity requirements of a TDM system are the same as those for a single channel. Thus the TDM system is relatively immune to interference within the channels (interchannel cross-talk) as compared to the FDM system.

6.9 CROSS-TALKED DUE TO HF CUTOFF OF THE CHANNEL

All communication channels are of limited bandwidths. Hence a communication channel can be represented by an RC lowpass filter as shown in Fig. 6.12(*a*) whose upper cutoff frequency is

$$f_c = \frac{1}{2\pi RC}$$

Now, when a pulse is applied to this channel, the output of the channel will be distorted as shown in Fig. 6.12(*b*). This is due to the HF limitations of the channel. It can be seen that the signal pulse allotted to the time slot 1 extends into the time slot 2, resulting in cross-talk. On an average, we can assume that the multiplexed signals are equally strong. Hence, during the time slot 2, the area corresponding to the signal, i.e., A_2 (not shown in the figure) will be the same as A_1. We can define the cross-talk factor, K, as the ratio of the cross-talk signal to the desired signal. Thus

$$K = \frac{A_{12}}{A_2} = \frac{A_{12}}{A_1} \qquad \text{...(6.4)}$$

The pulse in time slot 1 is almost rectangular (the figure here is exaggerated for charity).

Hence, $$A_1 = V_\tau \qquad \text{...(6.5)}$$

Now,

$$A_{12} = V\tau_c e^{-\tau_g/\tau_c}(1 - e^{-\tau/\tau_c}) \qquad \text{...(6.6)}$$

The time constant τ_c must be much smaller than τ in order to have minimum cross-talk. Therefore, $\tau_c \ll \tau$, and Eqn. (6.5) becomes

$$A_{12} = V\tau_c e^{-\tau_g/\tau_c} \qquad \text{...(6.7)}$$

From Eqn. (6.4) and (6.7), we get

$$K = \frac{\tau_c}{\tau} e^{-\tau_g/\tau_c} \qquad \text{...(6.8)}$$

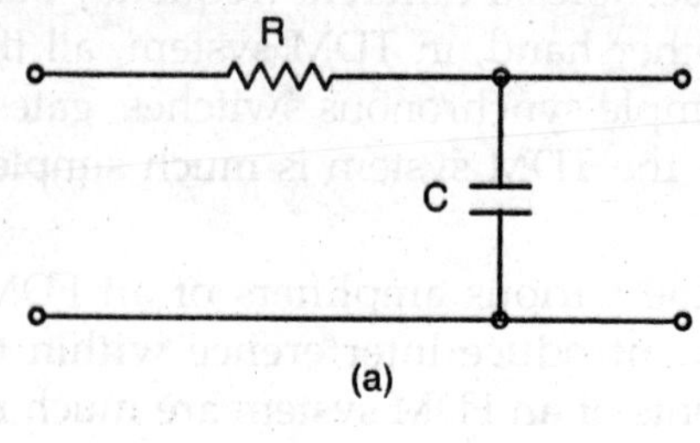

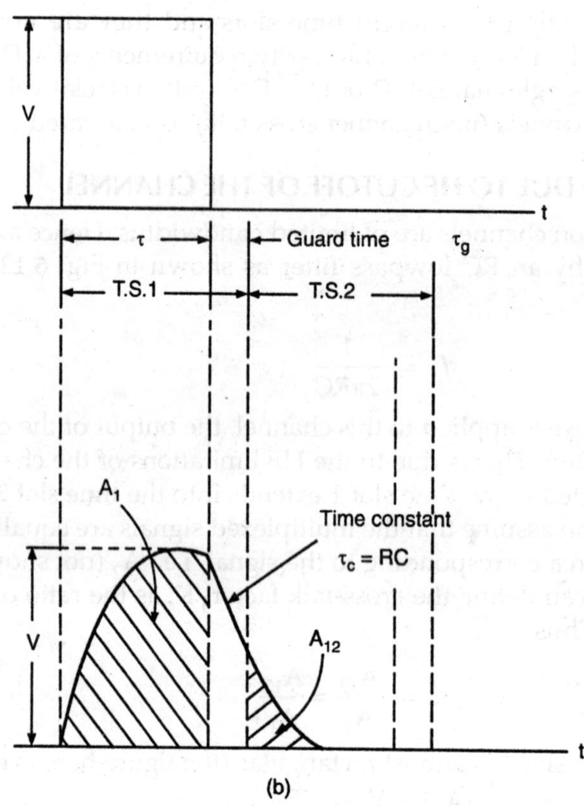

Fig. 6.12 *Cross Talk due to HF Cutoff Channel; (a) Low Pass Filter Representation of the Channel (b) Output of Channel of (a) For a Pulse Input*

The cross-talk factor K should be as low as possible. Eqn. (6.8) suggests that τ_g should be much larger than τ_c. Of course, there are other considerations as well. For example, a larger τ_g will result in less number of channels that can be multiplexed, and/or a reduced signal strength.

This type of cross-talk is restricted to the neighbouring channel because τ_c is very small, and hence the pulse ends in the neighbouring time slot only.

6.10 CROSS-TALK DUE TO LF CUTOFF OF THE CHANNEL

Just as the channel has an upper cutoff frequency, it also has a lower cutoff frequency. Hence the channel can be represented by an RC high pass filter as shown in Fig. 6.13(*a*), whose lower cutoff frequency is

$$f_1 = \frac{1}{2\pi RC}$$

Now, when a pulse is applied to this channel, the output of the channel will be distorted as shown in Fig. 6.13(*b*) due to the LF limitations of the channel. In this case, the time constant τ_c should be much greater, than τ to reduce cross-talk; i.e. $\tau_c \gg \tau$.

Now,

$$\Delta V = V(1 - e^{-\tau/\tau_c}) = \frac{V\tau}{\tau_c} \text{ as } \tau_c \gg \tau$$

Hence,

$$A_{12} = \Delta V\tau = \frac{V\tau^2}{\tau_c}$$

Also, as in the previous case

$$A_2 \approx A_1 \approx V\tau$$

Therefore, the cross-talk factor in this case is

$$K = \frac{A_{12}}{A_2} = \frac{A_{12}}{A_1} = \frac{V\tau^2/\tau_c}{V\tau} = \frac{\tau}{\tau_c}$$

In this case, τ_c is very large and, hence, the pulse extends to many times slots. Thus this type of cross-talk extends to more than one channel. In other words, a channel gets affected not only by the previous channel but also by some more channels prior to that.

6.11 TRANSMISSION OF PAM SIGNALS

If the PAM signals are to be transmitted directly, say over a pair of wires, then no further signal processing is necessary. However, if they are to be transmitted through the space using an antenna, they must first be amplitude or frequency or phase modulated by a high frequency carrier, and only then they can be transmitted. The overall system would be then known as PAM-AM or PAM-FM or PAM-PM respectively. At the receiving side, AM or FM or PM detection is first employed to get the PAM signal, and then the baseband signal is recovered from it.

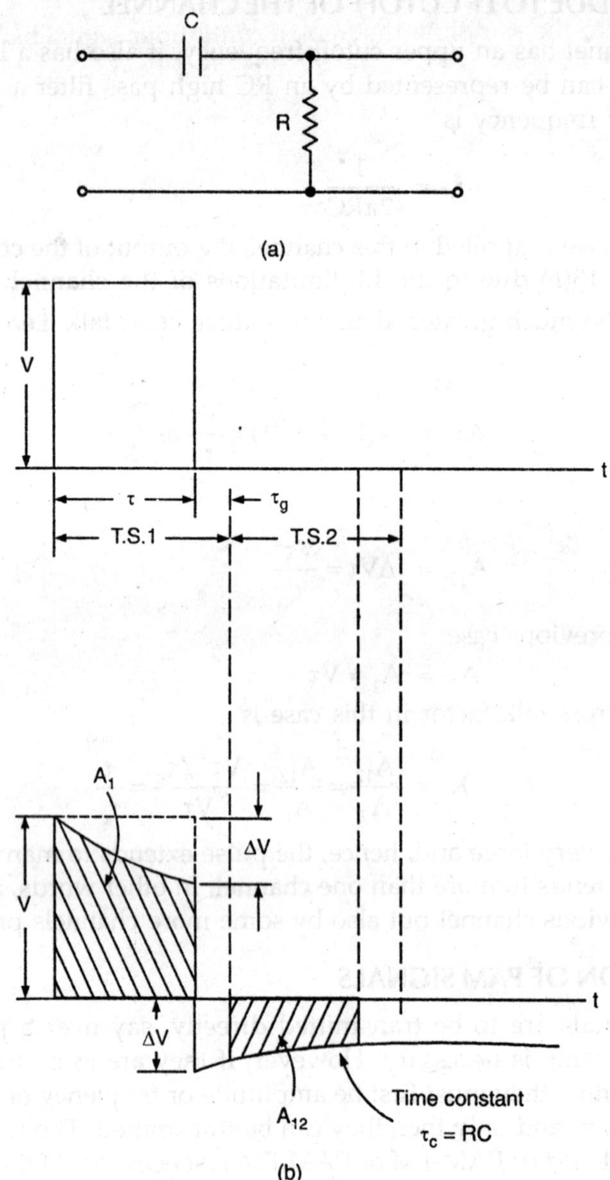

Fig. 6.13 *Cross-talk due to LF Cutoff of the Channel; (a) High Pass Filter Representation of the Channel (b) Output of the Channel of (a) for a Pulse Input*

6.12 BANDWIDTH OF PAM SIGNALS

Let us assume that we have to transmit n signals, each bandlimited to f_M Hz. Then, for each signal, we have to take $2f_M$ samples per second. Thus, in all, we have to transmit $2nf_M$ samples per second.

Now, according to the sampling theorem, a continuous signal band limited to B Hz can be transmitted by 2B samples per second. Conversely, it can be stated that 2B samples per second define a continuous signal band limited to B Hz. Therefore, the bandwidth of the PAM system of n signals, each band limited to f_M Hz, will be nf_M Hz, because we are transmitting $2nf_M$ samples per second.

6.13 S/N RATIO OF PAM SYSTEM

The noise performance of PAM is identical to AM-SC signal. The figure of merit v is unity. It has already been shown that the bandwidth of PAM/AM is nf_M, where n is the number of massages multiplexed. Therefore, the bandwidth per message is f_M. Thus, the transmission of sampled signal message reproduces a bandwidth at the receiver that is equivalent to the message $f(t)$. In other words, the transmission of a sampled signal is equivalent to the direct transmission of $f(t)$.

The signal and noise power at the transmitter and receiver is expected to the identical. It has been shown that the power of a sampled signal.

Hence,

$$S_o = S_i = \overline{f^2(t)}$$

The bandwidth of sampled signal per message is same as that of the baseband signal $f(t)$.

Hence, for white noise

$$N_o = N_i = \eta f_M$$

where f_M is the bandwidth of $f(t)$, and $\eta/2$ is the noise power per unit bandwidth. The figure of merit is given by

$$(v)_{PAM} = \frac{(S_o/N_o)}{(S_i/N_i)} = 1$$

6.14 PULSE TIME MODULATION

The two types of PTM systems, namely PWM and PPM, are shown in Fig. 6.14. The Fig. 6.14(a) is the baseband signal $f(t)$ whereas the Fig. 6.14(b) is the carrier pulse train $f_c(t)$. The Fig. 6.14(c) is the PWM signal where the width of each pulse depends on the instantaneous value of the baseband signal at the sampling instant. The Fig. 6.14(d) is the PPM signal where the shift in the position of each pulse depends on the instantaneous value of the baseband signal at the sampling instant. It can be seen that in PWM, the information about the baseband signal lies in the trailing edge of the pulse, whereas, in PPM, it lies in both the edges of the pulse. Although, basically it lies in the leading edge, but since the width of the pulse is same always, the trailing edge also carriers the same information.

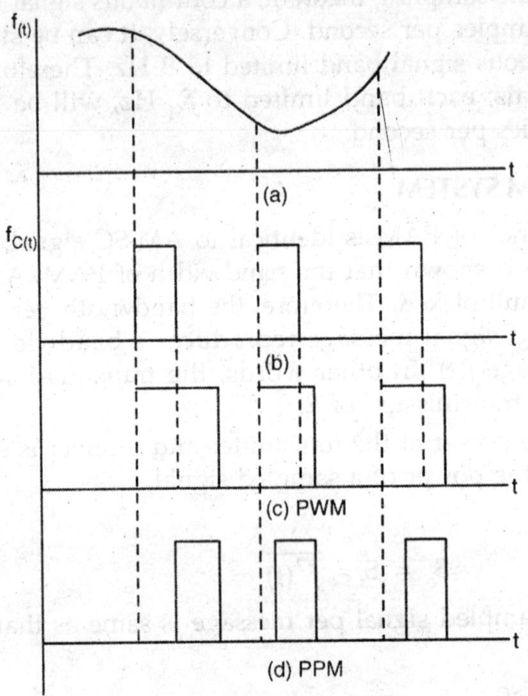

Fig. 6.14 Pulse Time Modulation (*a*) Baseband Signal $f(t)$, (*b*) Carrier Pulse Train $f_c(f)$, (*c*) Pulse Width Modulated Signal, (*d*) Pulse Position Modulation Signal

6.15 GENERATION OF PTM SIGNALS

6.15.1 Indirect Method

The scheme of generation of the PTM signal is shown in Fig. 6.15. Firstly, the flat-topped PAM signals are generated as 6.15(*a*). The synchronoised ramp waveform shown in Fig. 6.15(*b*) is generated during each pulse interval. These two signals are added as shown in Fig. 6.15(*c*), and the sum is applied to a comparator circuit whose reference level is shown by a broken line in Fig. 6.15(*c*).

The second crossing of the comparator reference level by the waveform of Fig. 6.15(*c*) is used to generate the pulses of constant amplitude and width as shown in Fig. 6.15(*d*), giving the desired PPM waveform. The leading edge of the synchronized ramp of Fig. 6.15(*b*) is used to start a pulse, and the trailing edge of the PPM waveform of Fig. 6.15(*d*) is used to terminate the pulse, as shown in Fig. 6.15(*e*), giving the desired PWM waveform.

The ramp amplitude is so adjusted that it is slightly greater that the maximum variation in amplitude of the PAM signals. The comparator reference level is such that it always intersects the slopping portion of the waveform of Fig. 6.15(*c*).

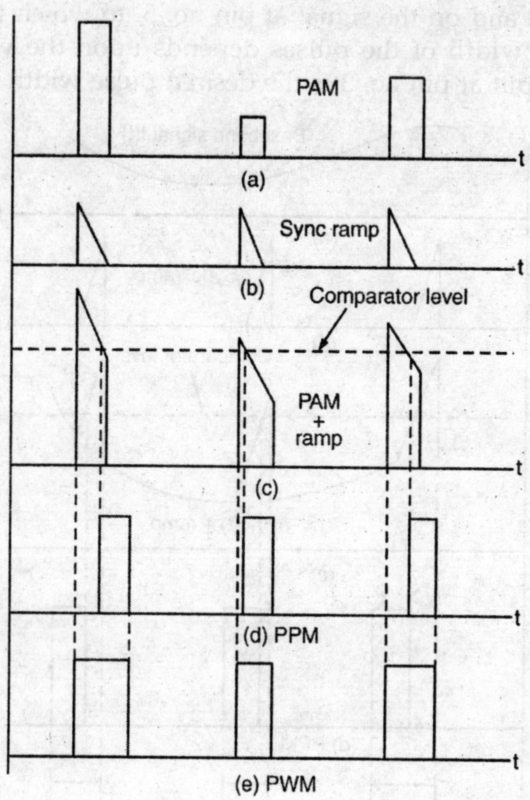

Fig. 6.15 Indirect Method of Generation of PTM Signals (*a*) PAM Signal, (*b*) Synchronized RAMP, (*c*) PAM + Synchronized Ramp, (*d*) PPM Signal, (*e*) PWM Signal

6.15.2 Direct Method

In the direct method, the PTM waveforms are generated without generating the PAM waveform. Here, the baseband signal *f*(*t*) of Fig. 6.16(*a*) and a ramp signal of Fig. 6.15(*b*), occurring at the sampling instants, are added to give the waveform of Fig. 6.16(*c*). This is compared in a comparator whose reference level is shown in Fig. 6.16(*c*) by a broken line.

The PPM, Fig. 6.16(*d*) and PWM Fig. 6.16(*e*) waveforms are then obtained in the same manner as explained in the indirect method of generation of PTM signals.

If we want to modulate the leading edge of pulse in the PWM waveform, the ramp waveform shown in Fig. 6.17(*a*) should be used. For modulating both the edges of the pulses in the PWM waveform, the ramp waveform shown in Fig. 6.17(*b*) needs to be used.

6.16 APWM MODULATOR CIRCUIT

A PWM modulator circuit is shown in Fig. 6.18. The clock signal of the desired frequency is applied as shown, from which the negative trigger pulses are derived with the help of a diode and an $R_1 - C_1$ combination which works as a differentiator. These negative trigger pulses are applied to the pin no. 2 of the 555 timer which is working in the monostable mode. They decide the starting time of the PWM pulses. The end of the pulses depends on an $R_2 - C_2$ combination, and on the signal at pin no. 5, to which the modulating signal is applied. Therefore, the width of the pulses depends upon the value of the modulating signal, and thus the output at pin no. 3 is the desired pulse width modulated signal.

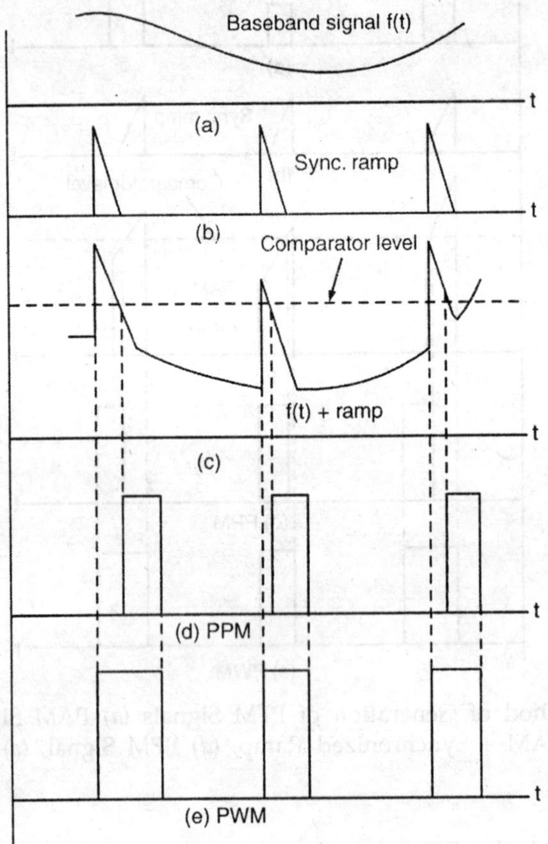

Fig. 6.16 Direct Method Generation of PTM Signals (*a*) Baseband Signal *f*(*t*), (*b*) Synchronized Ramp, (*c*) *f*(*t*) + Synchronized Ramp, (*d*) PPM Signal (*e*) PWM Signal

6.17 A PPM MODULATOR CIRCUIT

A PPM modulator circuit is shown in Fig. 6.19. The PWM signal is applied to pin no. 2 through the diode and $R_1 - C_1$ combination. Thus, the input to pin no. 2 is the negative trigger pulses which correspond to the trailing edges of the PWM waveform. The 555 timer is working in a monostable mode and the width of the pulse is constant as governed by an $R_2 - C_2$ combination.

The negative trigger pulses decide the starting time of the output pulses and, thus, the output at pin no. 3 is the desired pulse position modulated signal.

6.18 DEMODULATION OF PTM SIGNALS

The PWM waveform of Fig. 6.20(*a*) is used to generate a ramp waveform as shown in Fig. 6.20(*b*). The leading edges of the PWM pulses start the ramp of the same slope, and the trailing edges of the PWM pulses terminate the ramp. The height attained by the ramp is, therefore, proportional to the width of the PWM pulses. The height attained by the sustained for same time, thus creating a porch, after which the voltage returns to its initial level.

A similar type of synchronized ramp is shown in Fig. 6.21(*b*). This is generated with the help of the PPM pulses shown in Fig. 6.21(*a*). Here, the ramp is initiated at the beginning of the time slot, and it is terminated by the leading edge of the PPM pulse. Thus, the height attained by the ramp is proportional to the displacement of the leading edge of the PPM pulses from the beginning of the time slot. Here, too, the height attained by the ramp is sustained for some time, thus creating a porch, and then it is returned to the initial level.

The remaining procedure for both PWM and PPM is same. A sequence of locally generated pulses of a fixed amplitude are added to the synchronized ramp on the porch as shown in Fig. 6.15(*c*) and 6.22(*c*). The lower portions in these two waveforms are clipped by a clipping circuit, with the clipping level adjusted in such a way that it never crosses the ramp. The output of the clipper is a PAM wavefront Fig. 6.21(*d*) and 6.21(*d*) from which the baseband signal can be recovered.

6.19 A PWM DEMODULATOR CIRCUIT

A PWM Demodulator circuit is shown in Fig. 6.22. The transistor T_1 works as an invertor. Hence, during the time interval A-B, when the PWM signal is high, the input to the transistor T_2 is low. Therefore, during this time interval, the transistor T_2 is cut-off and the capacitor C gets charged through an R – C combination. During the time interval B – C, when the PWM signal is low, the input to the transistor T_2 is high, and it gets saturated. The capacitor C then discharged very rapidly through T_2. The collector voltage of T_2 during the interval B – C is then low. Thus the waveform at the collector of T_2 is more or less a saw-tooth waveform whose envelope is the modulating signal. When this is passed through a second-order OP-AMP low pass filter, we get the desired demodulated output.

6.20 A PPM DEMODULATOR CIRCUIT

A PPM demodulator circuit is shown in Fig. 6.23. This utilizes the fact that the gaps between the pulses of a PPM signal contain the information regarding the modulating signal. During the gap A – B between the pulses, the transmitter is cut-off, and the capacitor C gets charged through the R – C combination. During the pulse duration B – C, the capacitor discharges through the transistor, and the collector voltage becomes low. Thus, the waveform at the collector is approximately a saw-tooth waveform whose envelope is the modulating signal. When this is passed through a second order OP-AMP low pass filter, we get the desired demodulated output.

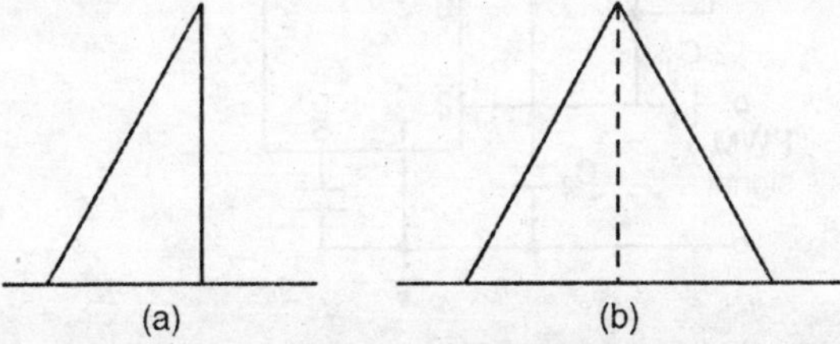

Fig. 6.17 (*a*) Synchronized Ramp for Leading Edge Modulation of PAM Signal (*b*) Synchronized Ramp for Both Edge Modulation of PWM Signal

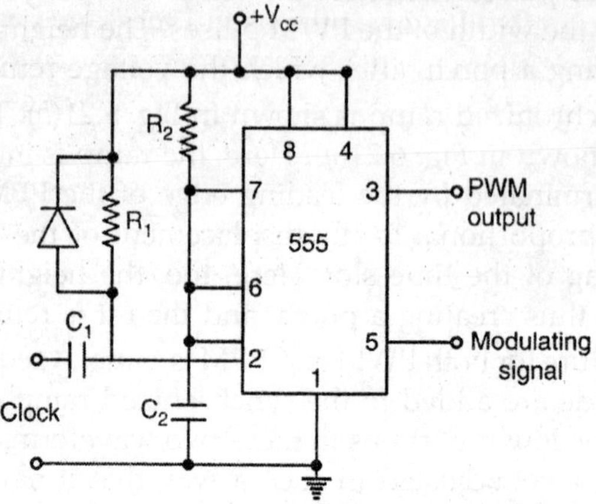

Fig. 6.18 A PWM Modulator

6.21 BANDWIDTH OF PTM SIGNALS

The bandwidth can be estimated by observing the spectrum of the PTM signals. The spectral analysis for such waves is complicated and hence will not be treated here. The one-sided spectrum of a PWM signal is shown in Fig. 6.24. Assume that the modulating signal is a single-tone sinusoid of the frequency ω_m and the sampling frequency is ω_s. The PWM spectrum has the following frequency components:

(*a*) A dc component at $\omega = 0$, which represents the average value of the pulses.

(*b*) The modulating frequency ω_m.

(*c*) The harmonics of the sampling frequency ω_s.

(*d*) Sidebands spaced by ω_m centered around each harmonic of ω_s.

The presence of harmonics of w_s are due to the contribution of the unmodulated pulse train which may be taken as the carrier of the PDM wave. Each harmonic of w_s is associated

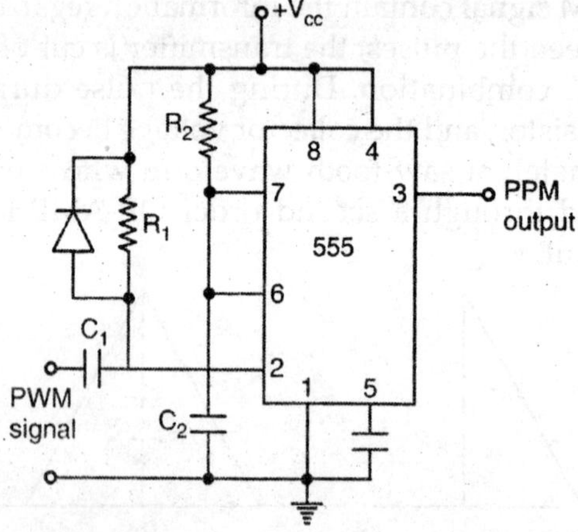

Fig. 6.19 A PPM Modulator

with the sidebands of an FM type. The sidebands of each ω_s extend to infinity outward, but with a decaying magnitude. However, the useful message band is available in a band $0 - \omega_m$ and hence a low pass filter can be used to recover the message from PWM. But the output of LPF is distorted due to the presence of cross-modulation terms that lie in the baseband. The lower sidebands of ω_s may extend to lie in the message baseband to cause distortion. This can be prevented by restricting the maximum excursion of the trailing edge of the PWM pulse.

The spectrum of a naturally sampled PPM wave from a single tone modulating signal has a form similar to that of a PDM wave, with the only difference that it contains a component proportional to the derivative of the modulating signal in place of modulating component itself. Therefore, the PPM detection can be achieved by an LPF followed by an integrator. An alternative detection method is to convert PPM into PWM, and then pass it through an LPF. This provides greater signal amplitude with less distortion in the receiver.

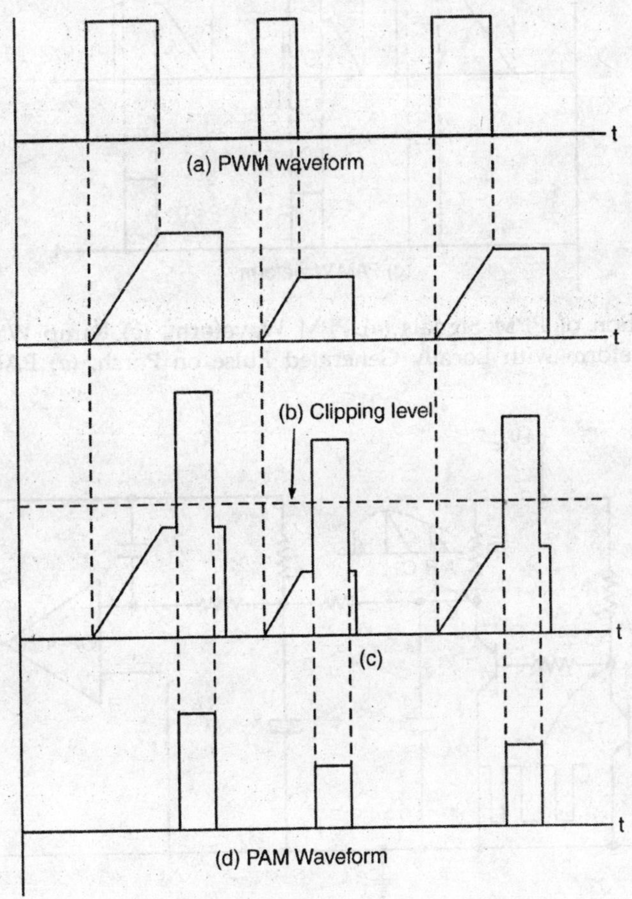

Fig. 6.20 Demodulation of PWM Signals (*a*) PWM Waveform, (*b*) Ramp Waveform with Porch, (*c*) Ram Waveform with locally Generated Pulse on Porch, (*d*) PAM Waveform

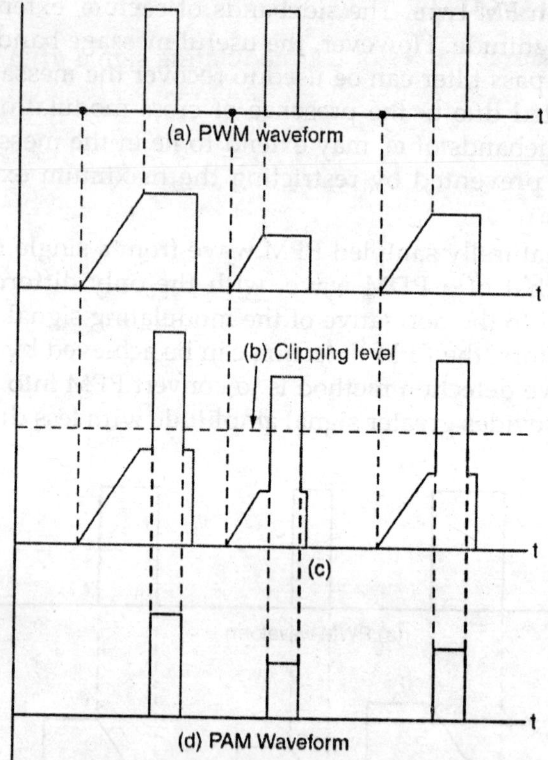

Fig. 6.21 Demodulation of PPM Signals *(a)* PPM Waveform, *(b)* Ramp Waveform with Porch, *(c)* amp Waveform with Locally Generated Pulse on Porch, *(d)* PAM Waveform

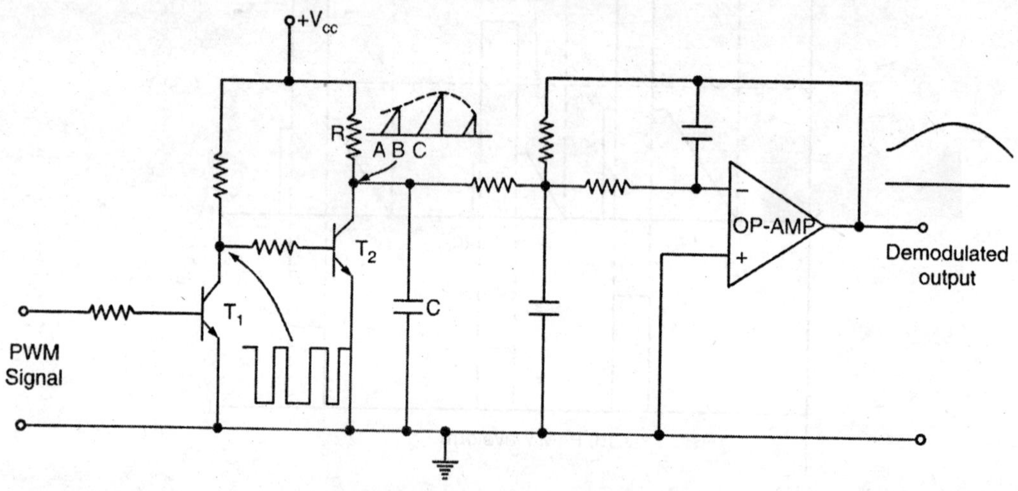

Fig. 6.22 A PWM Demodulator

6.22 S/N RATIO OF PTM SYSTEMS

In PPM signal, the sample values of the modulating signal $f(t)$ are transmitted in terms of the pulse position over a channel of bandwidth BHz. This finite bandwidth of the channel causes dispersion in a received PPM pulse. The resulting trapezoidal pulse is shown in Fig. 6.25(*a*). The rise time of the pulse is given as

$$t_r = \frac{1}{B} \quad \text{...(6.9)}$$

The position of the pulse is sensitive to any additive noise. If a noise signal y is added at any instant, the position is shifted by 'e' as shown in Fig. 6.25(*b*). It is obvious that

$$\frac{A}{y} = \frac{t_r}{e} \quad \text{...(6.10)}$$

As y varies randomly, e also varies randomly. The mean squared (m, s) values of y and e are related as

$$\frac{\overline{e^2}}{\overline{y^2}} = \left(\frac{t_r}{\text{A}}\right)^2 \quad \text{...(6.11)}$$

Therefore, for an additive noise $n(t)$, the m.s. value of e is given as

$$\overline{e^2} = \left(\frac{t_r}{\text{A}}\right)^2 \overline{n^2(t)} \quad \text{...(6.12)}$$

The change in the position of the mth pulse is proportional to have the value of mth sample of modulating signal $f(t)$. Let us denote the change in the position by x_m and the mth sample by $f(m\text{T})$, each sample being spaced by T. Then

$$x_m = \text{K}_1 f(m\text{T}) \quad \text{...(6.13)}$$

where K_1 is a proportionality constant.

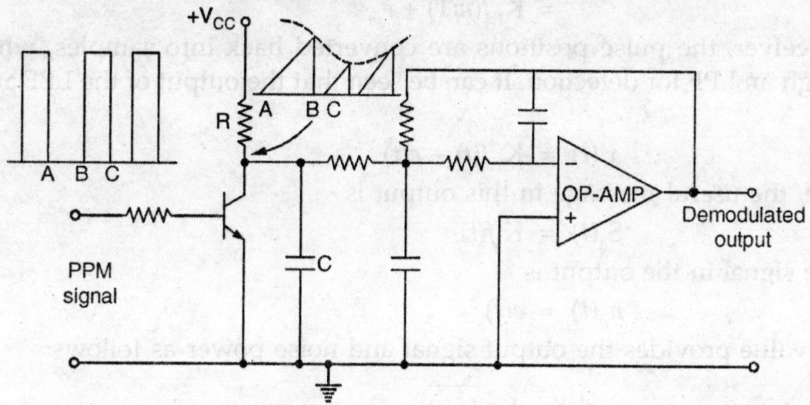

Fig. 6.23 A PPM Demodulator

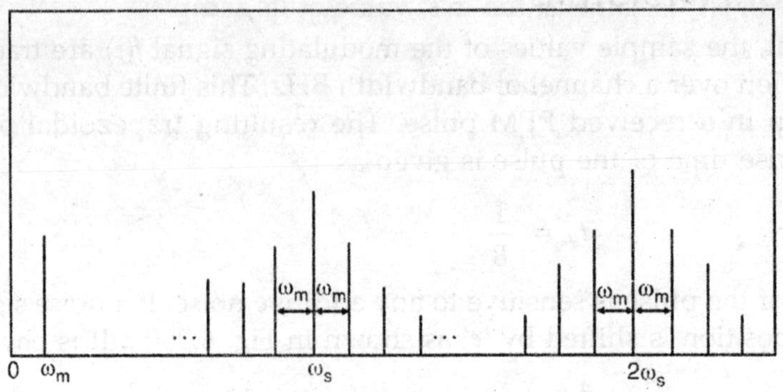

Fig. 6.24 One-sided Spectrum of PWM Signal

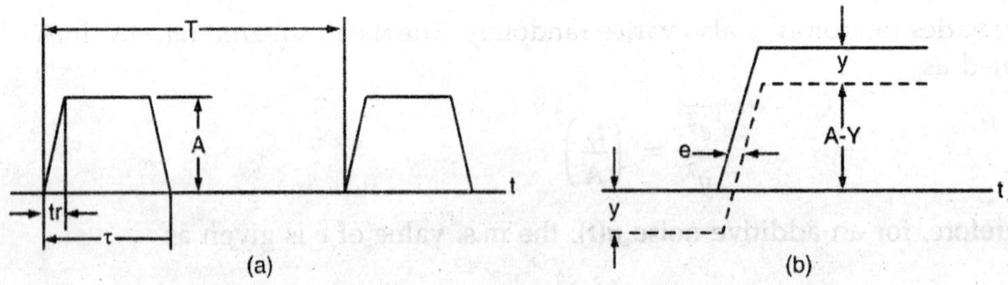

Fig. 6.24 (*a*) Trapezoidal Pulses due to Dispersion

Now, when noise is added, the change in the position of *m*th pulse becomes $x_m + e_m$.

where e_m is random and is defined by Eqn. (6.12). Let us denote this changed position of the *m*th pulse by x'_m, then

$$x'_m = x_m + e_m$$

$$= K_1 f(mT) + e_m \qquad ...(6.14)$$

At the receiver, the pulse positions are converted back into samples, which are then passed through an LPF for detection. It can be seen that the output of the LPF at the receiver is given by

$$x'(t) = K_1 f(t) + e(t)$$

Therefore, the useful message in this output is

$$S_o(t) = K f(t)$$

and the noise signal in the output is

$$n_o(t) = e(t)$$

The m.s. value provides the output signal and noise power as follows:

$$S_o = K_1^2 \overline{f^2(t)} \text{ and}$$

$$N_o = \overline{n_o^2(t)} = \overline{e^2(t)}$$

The m.s. value of $e(t)$ is same as the m.s. value of its samples.

Therefore,

$$\overline{e^2(t)} = \overline{e_m^2} = \left(\frac{t_r}{A}\right)^2 \overline{n^2(t)}$$

and

$$N_o = \left(\frac{t_r}{A}\right)^2 \overline{n^2(t)}$$

Therefore,

$$\frac{S_o}{N_o} = \frac{K_1}{\overline{n^2(t)}}\left(\frac{A}{t_r}\right) \overline{f^2(t)}$$

$$= K_1^2 B^2 \overline{f^2(t)} \frac{A^2}{\overline{n^2(t)}}$$

where

$$B = \frac{1}{t_r}$$

Let the duty ratio 'd' of the pulse be given by

$$d = \frac{\tau}{T}$$

The power contains in the input PPM wave of the amplitude 'A' is given by

$$S_i = dA^2$$

The input noise power is given by,

$$N_i = \overline{n^2(t)}$$

Therefore,

$$\frac{S_o/N_o}{S_i/N_i} = \frac{K_1^2}{d} \overline{f^2(t)} B^2 \qquad ...(6.15)$$

Thus, the S/N ratio improvement is proportional to the square of the channel bandwidth. A similar result can be drawn for the PWM signal after evaluating its noise figure. It has been observed that the noise performance of PWM is inferior to PPM. This is true because the PPM system preserves all the signal information at the terminating instant of the pulses, and yet it avoids a considerable loss of power which a PWM system expands during the pulse. PWM, is, thus, less efficient than PPM with regard to the transmitter power utilization.

PROBLEMS

1. What is pulse-width modulation? What other names does it have? How is it demodulated?
2. Explain the methods for demodulation of PAM signals.
3. Define and describe pulse-position modulation, and explain with waveform how it is derived from PWM.

4. Explain the cross-talk in PAM due to the HF and LF limitations of the channel. Which one of the two affects more than one channel and why?
5. Explain how PPM and PWM signals are generated:
   (*a*) from PAM signals
   (*b*) directly.
   How are these detected?
6. What are FDM and TDM? Compare these two.
7. Draw and explain the following circuits:
   (*a*) PPM modulator and demodulator
   (*b*) PWM modulator and demodulator
   (*c*) PPM modulator and demodulator

❑❑❑

Digital Modulation 7

7.1 INTRODUCTION

As we know digital representation of signal and certain techniques useful for baseband pulse transmission. Here first of all we shall discuss passband transmission of digital signal and then take up inter channel and inter symbol interference, way to reduce bandwidth of digital signal. Nyquist criterion and related issues. We have already discussed advantage of modulating a carrier in passband transmission of analog signal that efficiently utilizes frequency spectrum through FDM. For digital signal transmission instead of modulation; the term 'Iceying', i.e., switching is used. In Amplitude slift keying (ASK), amplitude of the carrier is switched from one value to other depending on modulating digital input. Similarly, we have frequency swift keying (FSK). Plase slift keying (PSK) etc. in digital passband transmission.

7.2 TYPES OF DIGITAL MODULATION

Digital modulation may be classified into two sections as coherent technique and non colerent technique depending on whether the receiver is equipped with a plase recovery circuit or not. The phase recovery circuit ensures that the oscillator supplying the locally generated carrier wave receiver is synchronized to the oscillator supplying the carrier wave used to originally modulate the incoming date stream in the transmitter.

7.2.1 Coherent Digital Modulation

Those techniques which employ coherent detection are knows as soherent digital modulation. In this modulation the local carrier generated at the receiver is phase locked with the carrier at the transmitter. Thus, the detection is done by correlating received noisy signal and locally generated carrier. The colerent detection is a synchronous detection.

7.2.2 Non-Coherent Digital Modulation

The technique in which the detection process does not need receiver carrier to be phase locked with transmitter carrier are known as non-coherent digital modulation. The advantage of such type of modulation is that the system becomes simple. But over the drawback, error probability may increase.

In fact, the different digital modulation are used for various specific application areas. The binary i.e. digital modulation has three basis forms amplitude shift keying (ASK), plase shift keying (PSK) and frequency shift iceying (FSK).

7.3 COHERENT BINARY AMPLITUDE SHIFT KEYING

Amplitude shift keying (ASK) which is also known as On-off keying (OOK) is the simplest digital modulation. In this method, there is only are unit energy carrier and it is switched on or off technique depending upon the input binary sequence. The ask waveform may be represented as,

$$s(t) = \sqrt{2P_s} \cos 2\pi f_c t \qquad ...(7.1)$$

To transmit symbol 'O', the signal $s(t) = 0$ ie. no signal is transmitted. Signal $s(t)$ contains some, complete cycles of carrier frequency 'f_c'.

Hence, the ASK waveform looks like ON-OFF of the signal. Therefore it is also known as the ON-OFF keying (OOK). Fig. Shows the ASK waveform.

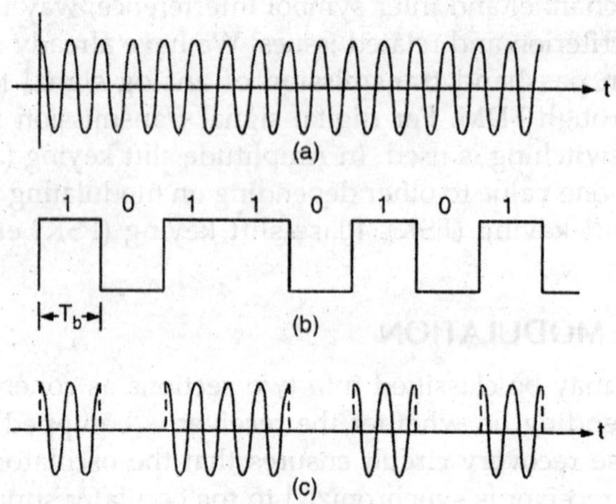

Fig. 7.1 *Amplitude-shift keying waveforms, (a) Unmodulated carrier, (b) Unipolar bit sequence, (c) ASK waveform*

7.4 SIGNAL SPACE DIAGRAM OF ASK

The ASK waveform of equation (7.1) for symbol, '1' can be represented as,

$$s(t) = \sqrt{P_s T_b} \cdot \sqrt{2/T_b} \cos(2\pi f_c t)$$

or

$$s(t) = \sqrt{P_s T_b \phi_1(t)} \qquad ...(7.2)$$

This means that there is only one carrier function $\phi_1(t)$. The signal space diagram will have two points on $\phi_1(t)$. One will be at zero and other will be at $\sqrt{P_s T_b}$. Figure 7.2 shows this as

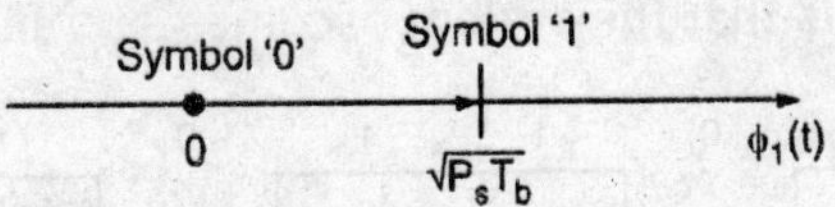

Fig. 7.2 *Signal space diagram of ASK*

Thus, the distance between the two signal points is

$$d = \sqrt{P_s T_b} = \sqrt{E_b} \qquad ...(7.3)$$

7.5 BINARY PHASE SHIFT KEYING (BPSK)

In binary phase-shift keying (BPSK) the transmitted signal is a sinusoid of fixed amplitude. If has one fixed phase when the data is at one level and when the data is at the other level the phase is different by 180°. In BPSK the binary symbol '1' and '0' modulate the phase of the carrier. Let us assume that the carrier is given as

$$s(t) = A \cos(2\pi f_c t) \qquad ...(7.4)$$

Here 'A' represents peak value of sinusoidal carrier. For the standard 1 Ω load resistor, the power dissipated would be

$$P = \frac{1}{2} A^2$$

or

$$A = \sqrt{2P} \qquad ...(7.5)$$

Now, when the symbol is changed, then the phase of the carrier will also be changed by an amount of 180°.

For example, let us consider

For symbol '1'. We have

$$S_1(t) = \sqrt{2P} \cos(2\pi f_c t) \qquad ...(7.6)$$

If next symbol is '0', then we have

$$S_2(t) = \sqrt{2P} \cos(2\pi f_c t + \pi) \qquad ...(7.7)$$

Because we know that

$$\cos(\theta + \pi) = -\cos\theta,$$

then Eq. (7.7) can be written as

$$S_2(t) = -\sqrt{2P} \cos(2\pi f_c t) \qquad ...(7.8)$$

with the above equation, we can define

BPSK signal in combined way as

$$s(t) = b(t) \sqrt{2P} \cos(2\pi f_c t) \qquad ...(7.9)$$

Where

$b(t) = +1$ When binary '1' is to be transmitted

$\quad\;\; -1$ When binary '0' is to be transmitted

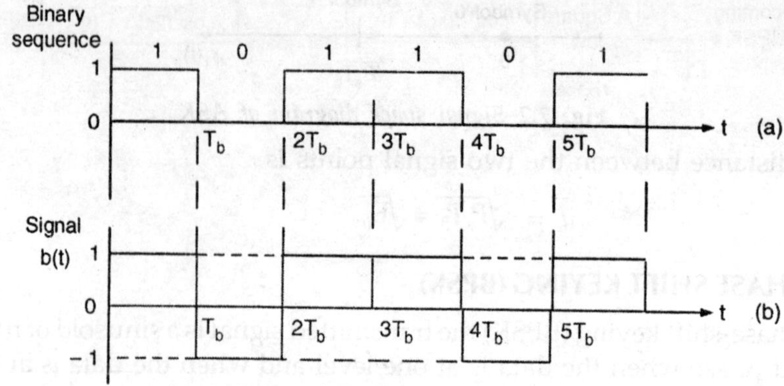

Fig. 7.3 *(a) Binary sequence, (b) The corresponding bipolar signal b(t)*

Figure 7.3 illustrates binary signal and its equivalent signal $b(t)$.

7.6 GENERATION OF BPSK SIGNAL

The BPSK signal may be generated by applying carrier signal to a balanced modulator. Here, the bipolar signal $b(t)$ is applied as a modulating signal to the balanced modulator.

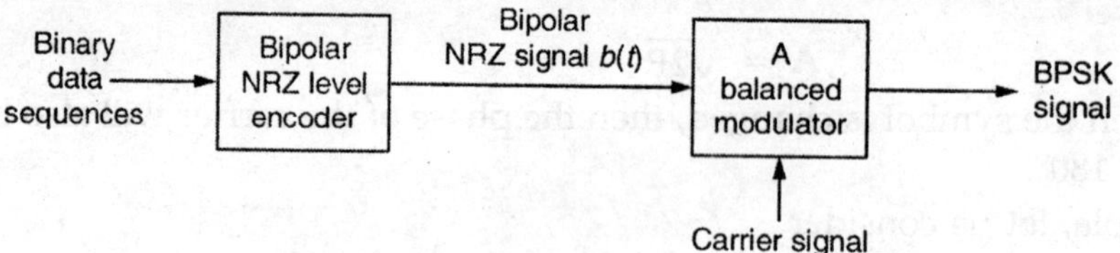

Fig. 7.4 *Generation of BPSK*

Figure 7.4 shows the block diagram of a BPSK signal generator. ANRZ level encoder converts the binary data sequence into bipolar NRZ signal.

7.7 PRECEPTION OF BPSK SIGNAL

Figure 7.5 shows the block diagram of the scheme to recover baseband signal from BPSK signal. The transmitted BPSK signal is given as

$$s(t) = b(t)\sqrt{2P}\cos(2\pi f_c t)$$

This signal undergoes the phase change depending upon the time delay from transmitter end to receiver end. This phase change is, usually, a fixed phase shift in the transmitted signal.

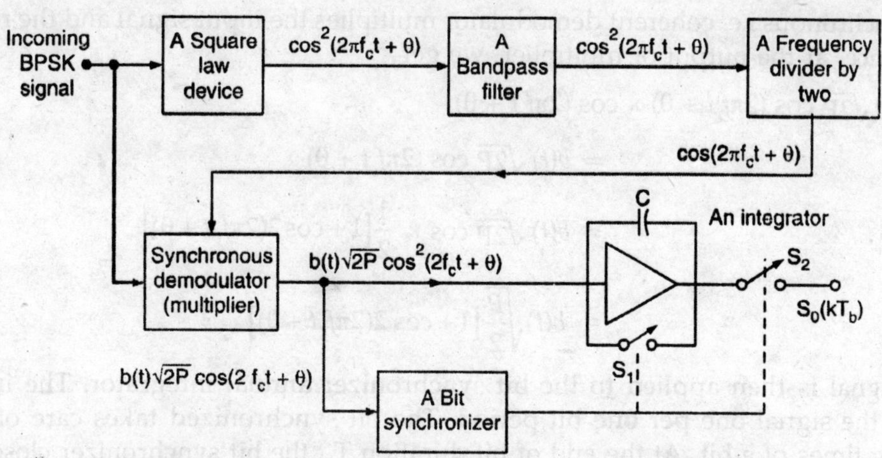

Fig. 7.5 *Reception of baseband signal in BPSK signal*

Let us consider that this phase slift is θ. Because of this, the signal at the input of the receiver can be written as

$$s(t) = b(t) \sqrt{2P} \cos(2\pi f_c t + \theta) \quad \text{...(7.10)}$$

Now, from this received signal, a carrier is repeated because this is coherent detection. As shown in Figure 7.5. The received signal is allowed to pass through a square law device. At the output of the square law device, we get a signal which is given as

$$\cos^2(2\pi f_c t + \theta)$$

It may be noted that we have neglected the amplitude here. Since we are only interested in the carrier of the signal.

Also, we know that

$$\cos^2\theta = \frac{1 + \cos 2\theta}{2}$$

Therefore, we have

$$\cos^2(2\pi f_c t + \theta) = \frac{1 + \cos 2(2\pi f_c t + \theta)}{2}$$

$$= \frac{1}{2} + \frac{1}{2} \cos 2(2\pi f_c t + \theta)$$

Here, $\frac{1}{2}$ represents a DC level. This signal is then allowed to pass through a bandpass fitter (BPF) whose pass and is centered around $2f_c$. Bandpass fitter removes the DC level of 1/2 and at the output. We obtain,

$$\cos 2(2\pi f_c t + \theta)$$

This signal is having frequency equal to $2f_c$. hence, it is passed through a frequency divider by two. Thus, at the output of frequency divider. We get a carrier signal whose frequency is f_c i.e.,

$$\cos 2(2\pi f_c t + \theta).$$

The synchronous i.e. coherent demodulator multiplies the input signal and the recovered carrier. Hence, at the output of multiplier, we get

$$b(t)\sqrt{2P}\cos(2\pi f_c t+\theta)\times\cos(2\pi f_c t+\theta)$$

$$= b(t)\sqrt{2P}\cos(2\pi f_c t+\theta)$$

$$= b(t)\sqrt{2P}\cos\times\frac{1}{2}[1+\cos 2(2\pi f_c t+\theta)]$$

$$= b(t)\sqrt{\frac{P}{2}}[1+\cos 2(2\pi f_c t+\theta)] \qquad ...(7.11)$$

This signal is then applied to the bit synchronizer annual integrator. The integrator integrates the signal one per one bit period. The bit synchronized takes care of starting and ending times of a bit. At the end of bit duration T_b, the bit synchronizer closes switch S_2 temporarily. This concepts the output of an integrator to the decision device. In fact, it is equivalent the sampling the output the integrator. The synchronozer then opens switch S_2 and switch S_1 is closed temporarily. This resets the integrator voltage to zero. The inegrator then integrates next bit. Let us assume that one bit period 'T_b' contains integral number of cycles of the carrier. This means that the phase change occurs in the carrier only at zero crossing. This has been shown in Fig. 7.6. This BPSK waveform has full cycles of sinusoidal carrier.

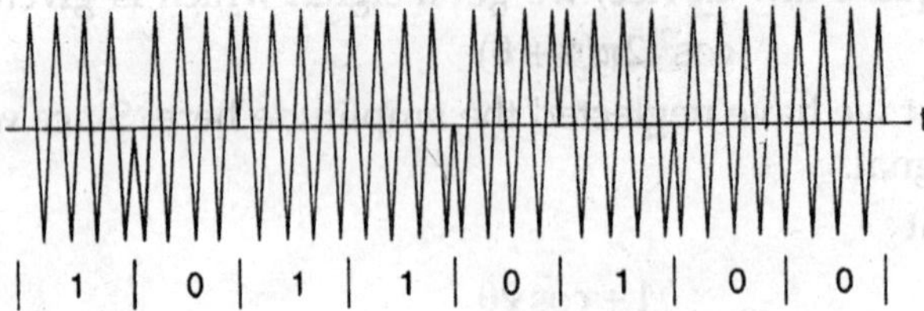

Fig. 7.6 *The BPSK waveform*

Also, in the kth bit interval, we can write output signal ss under :

$$s_0(kT_b) = b(k\,T_b)\int_{(k-a)T_b}^{k\,T_b}[1+\cos(2\pi f_c t+\theta)]\,dt$$

The equation gives the output of an interval for k^{th} bit. Hence, integration is performed from $(k-1)\,T_b$ to $k\,T_b$. Here, T_b is the one bit period. Now, above equation can be written as :

$$S_0(k\,T_b) = b(k\,T_b)\sqrt{\frac{P}{2}}\left[\int_{(k-1)T_b}^{k\,T_b}1dt+\int_{(k-1)T_b}^{k\,T_b}\cos(2\pi f_c t+\theta)dt\right]$$

Where $\int_{(k-1)T_b}^{k\,T_b}\cos 2(2\pi f_c t+\theta)dt = 0$, since

average value of sinusoidal waveform is integration is done over full cycles. Hence, we can have

$$S_0(k\,T_b) = b(k\,T_b)\sqrt{\frac{P}{2}}\int_{(k-1)T_b}^{k\,T_b} 1\,dt$$

or
$$S_0(k\,T_b) = b(k\,T_b)\sqrt{\frac{P}{2}}\,[t]_{(k-1)T_b}^{k\,T_b}$$

or
$$S_0(k\,T_b) = b(k\,T_b)\sqrt{\frac{P}{2}}\{k\,T_b - (k-1T_b\}$$

or
$$S_0(k\,T_b) = b(k\,T_b)\sqrt{\frac{P}{2}}T_b \quad \text{...(7.12)}$$

The last equation shows that the output of the receiver depends on input. Thus,

$$S_0(k\,T_b) \propto b(k\,T_b)$$

Depending upon the value of $b(k\,T_b)$, the output $S_0(k\,T_b)$ is generated in receiver.

This signal is then applied to a decision device which decides whether transmitted symbol was zero or one.

7.8 THE SPECTRUM OF BPSK SIGNALS

We know that the waveform $b(t)$ is a NRZ binary waveform. In this waveform there are rectangular pulses of amplitude $\pm V_b$. If we assume that each pulse is $\pm \frac{T_b}{2}$ around its centre, then it becomes easy to find Fourier transform of such pulse. The Fourier transform of this type of pulse is given as,

$$X(f) = V_b\,T_b\frac{\sin(\pi f\,T_b)}{(\pi f\,T_b)} \quad \text{...(7.13)}$$

For a large number of such positive and negative pulses, the power spectral density S(*f*) is expressed as

$$S(f) = \frac{\overline{|X(f)|^2}}{T_S} \quad \text{...(7.14)}$$

Here $\overline{X(f)}$ represents average value of X(*f*) due to all the pulse in $b(t)$. And T_S is symbol duration. Substituting value of X(*f*) from equation

$$\int_0^T S_2(t)S_k(t)dt = S_i^T\,S_k$$

Which states that the inner product of the signal $S_i(t)$ and $S_k(t)$ over the interval [0, 1], using their time domain representations, is equal to the inner product of their respective vector representations S_i and S_k and in equation (7.14). We get

$$S(f) = \frac{V_b^2 T_b^2}{T_S}\left[\frac{\sin(\pi f T_b)}{\pi f T_b}\right]^2$$

For BPSK, because only one bit is transmitted at a time, therefore, symbol and bit durations are same i.e., $T_b = T_S$. Now, the above equation becomes,

$$S(f) = V_b^2 T_b \left[\frac{\sin(\pi f T_b)}{\pi f T_b}\right]^2 \quad ...(7.15)$$

Equation (7.15) gives the power spectral density (psd) of basband signal $b(t)$. The BPSK signal is generated by modulating a carrier by the baseband signal $b(t)$. Due to modulation of the carrier of frequency f_c, the spectral components are translated from f to $f_c + f$ and $f - f$. The magnitude of these components is divided by half.

Therefore, from equation (7.15) we can write the power spectral density of BPSK signal as

$$S_{BPSK}(f) = V_b^2 T_b \left\{\left[\frac{1}{2}\frac{\sin \pi(f_c - f)T_b}{\pi(f_c - f)T_b}\right]^2 + \frac{1}{2}\left[\frac{\sin \pi(f_c - f)T_b}{\pi(f_c - f)T_b}\right]^2\right\}$$

This equation consists of two half magnitude spectral components if same frequency 'f' above and below f_c. Let us assume that the value of

$$\pm V_b = \pm\sqrt{P}$$

This means that the NRZ signal is having amplitude of $\pm\sqrt{P}$ and $-\sqrt{P}$. Then the above equation becomes

$$S_{BPSK}(f) = \frac{P T_b}{2}\left\{\left[\frac{\sin \pi(f - f_c)T_b}{\pi(f - f_c)T_b}\right]^2 + \frac{1}{2}\left[\frac{\sin \pi(f_c - f)T_b}{\pi(f_c - f)T_b}\right]^2\right\} \quad ...(7.17)$$

Equation (7.16) gives power spectral density (psd) of BPSK signal for modulating signal $b(t)$ having amplitudes equal to $\pm\sqrt{P}$.

Further, we know that the modulated signal is given as

$$s(t) = \pm\sqrt{2P}\cos(2\pi f_c t)$$

because $A = \sqrt{2P}$

If $b(t) = \pm\sqrt{P}$, then the carrier becomes,

$$\phi(t) = \sqrt{2}\cos(2\pi f_c t) \quad ...(7.17)$$

Equation (7.15) describes power spectral density (psd) of the NRZ waveform. For one rectangular pulse, the shape of S(f) will be a since pulse as shown in Fig. 7.7.

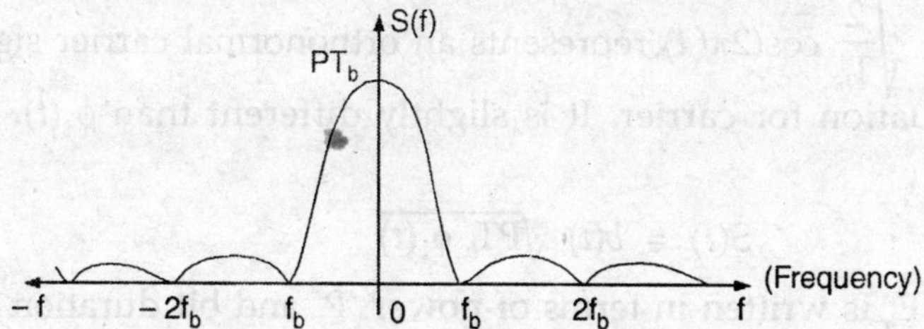

Fig. 7.7 *Plot of power spectral density (psd) of NRZ baseband signal*

From the figure it is clear that the main lobe ranges from $-f_b$ to $+f_b$. As we have

$$\pm V_b = \pm \sqrt{P}$$

so, the peak value of the main lobe is P T_b. Now, let us consider the power spectral (psd) of BPSK signal expressed by equation (7.16)

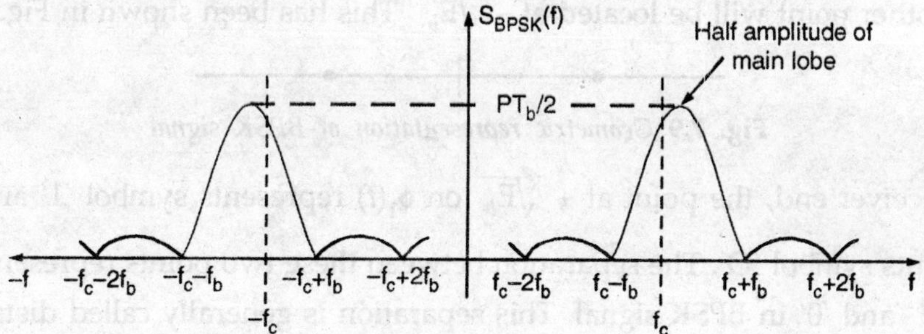

Fig. 7.8 *Plot of power spectral density of BPSK signal*

Figure 7.8 clearly shows that there are two lobes, one at f_c and other at $-f_c$. The same spectrum of figure 7.7 has been placed at $+f_c$ and $-f_c$. However, the amplitude of main lobes are $\frac{PT_b}{2}$ in figure 7.8.

Hence they are reduced to half. The spectrum of S(*f*) as well as $S_{BPSK}(f)$ extends overall the frequencies.

7.9 A GEOMETRICAL REPRESENTATION FOR BPSK SIGNALS

As we know that BPSK signal carries the information about two symbols. These symbols are symbol '1' and symbol '0'. We can represent BPSK signal geometrically to show these two symbols.

From above equation, we know that BPSK signal is expressed as,

$$s(t) = b(t)\sqrt{2P}\cos(2\pi f_c(t)) \qquad ...(7.18)$$

Now, let us rearrange the last equation as

$$s(t) = b(t)\sqrt{PT_b}\cdot\sqrt{\frac{2}{T_b}}\cos(2\pi f_c t) \qquad ...(7.19)$$

Now, let $f_1(t) = \sqrt{\frac{2}{T_b}}\cos(2\pi f_c t)$ represents an orthonormal carrier signa. Equation (7.14) also represents equation for carrier. It is slightly different than $\phi_1(t)$. So, equation (7.19) may be written as

$$S(t) = b(t)\sqrt{PT_b}\,\phi_1(t) \quad ...(7.20)$$

The bit energy E_b is written in terms of power 'P' and bit duration T_b as

$$E_b = PT_b \quad ...(7.21)$$

Hence, equation (7.20) becomes

$$S(t) = \pm\sqrt{E_b}\,\phi_1(t) \quad ...(7.22)$$

Here, $b(t)$ is simply ± 1.

Thus, on the single axis of $\phi_1(t)$m there will be two points. One point will be located at $+\sqrt{E_b}$ and other point will be located at $-\sqrt{E_b}$. This has been shown in Fig. 7.9.

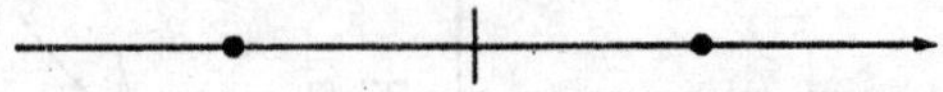

Fig. 7.9 *Geometric representation of BPSK signal*

At the receiver end, the point at $+\sqrt{E_b}$ on $\phi_1(t)$ represents symbol '1' and point at $-\sqrt{E_b}$ represents symbol 'O'. The separation between these two points represents is isolation in symbols '1' and '0' in BPSK signal. This separation is generally called distance 'd' From Fig. 7.9 it is obvious that the distance between the two point is.

$$d = +\sqrt{E_b} - (-\sqrt{E_b})$$

or

$$d = 2\sqrt{E_b} \quad ...(7.23)$$

If this distance 'd' increases, the isolation between the symbols in BPSK signal is more. So, probability of error reduces.

7.10 BANDWIDTH FOR BPSK SIGNAL

We have seen that the spectrium of the BPSK signal is centred around the carrier frequency f_c.

If $f_b = \frac{1}{T_b}$, then for BPSK, the maximum frequency in the baseband signal will be f_b as shown in Fig. 7.10. In this figure, the main lobe is centred around carrier frequency f_c and extends from $f_c - f_b$ to $f_c + f_b$.

Therefore bandwidth of BPSK signal will be

$$BW = \text{Highest frequency} - \text{Lowest frequency in the main lobe}$$

$$BW = f_c + f_b - (f_c - f_b)$$

or

$$BW = 2f_b \quad ...(7.24)$$

Hence, the minimum bandwidth of BPSK signal is equal to twice of the highest frequency contained in basband signal.

7.11 ERROR PROBABILITY OF BPSK SIGNAL

7.11.1 Employing Coherent

Reception or Metched Filter

The BPSK receiver shown in Fig. 7.5 is equivalent to the coherent receiver. If is nothing but coherent receiver. Let us say that the signal $x(t)$ represents $S_1(t)$ and $S_2(t)$. Then the BPSK receiver of Fig. 7.5. Can be approximated to coherent as shown in Fig. 7.10

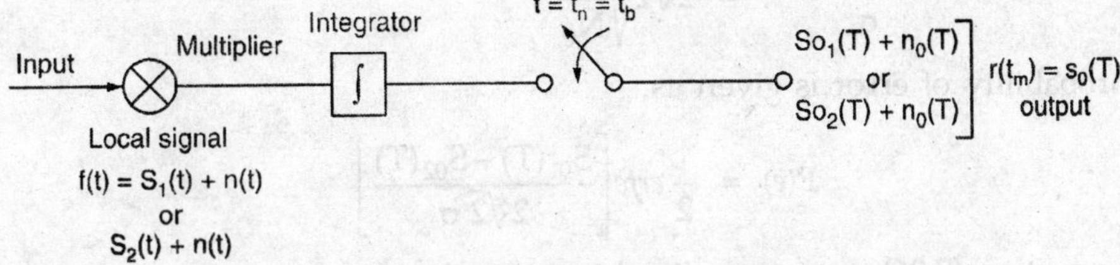

Fig. 7.10 *Coherent reception system*

Because, the correlator of Fig. 7.10 is equivalent to matched fitter, we can apply the analysis of metch fitter detection. We IC now that signal to noise ratio ρ_{max} at the output of receiver is given by

$$\rho_{max} = \frac{2}{N_0} \int_{-\infty}^{\infty} x^2(t)dt \quad ...(7.25)$$

Here, $x(t) = S_1(t) - S_2(t)$

i.e., input difference signal, then we have

$$\rho_{max} = \frac{2}{N_0} \int_{-\infty}^{\infty} S_1(t) - S_2(t)^2 dt$$

or

$$\rho_{max} = \frac{2}{N_0} \left[\int_{-\infty}^{\infty} S_1^2(t)dt + \int_{-\infty}^{\infty} S_2^2(t)^2 dt - \int_{-\infty}^{\infty} S_1(t)S_2(t)dt \right]$$

Since

$S_1(t) = S_2(t)$, we have

$$\int_{-\infty}^{\infty} S_1(t)S_2(t)dt = -E_b$$

and

$$\int_{-\infty}^{\infty} S_1^2(t)dt = \int_{-\infty}^{\infty} S_2^2(t)dt = E_b$$

Hence,

$$\rho_{max} = \frac{2}{N_0} [E_b + E_b + 2E_b]$$

or $$\rho_{max} = \frac{8\,E_b}{N_0}$$

For the correlator, we can write

$$\left[\frac{S_{01}(t) - S_{02}(t)}{\sigma}\right]^2 = \frac{8\,E_b}{N_0}$$

or $$\frac{S_{01}(t) - S_{02}(t)}{\sigma} = 2\sqrt{2}\sqrt{\frac{E_b}{N_0}} \qquad ...(7.26)$$

The probability of error is given as,

$$P(e) = \frac{1}{2}\,erfc\left[\frac{S_{01}(T) - S_{02}(T)}{2\sqrt{2}\,\sigma}\right]$$

From equation (7.26), we can write above equation as

$$P(e) = \frac{1}{2}\,erfc\left[\frac{1}{2\sqrt{2}}\cdot 2\sqrt{2}\sqrt{\frac{E_b}{N_0}}\right]$$

or $$P(e) = \frac{1}{2}\,erfc\sqrt{\frac{E_b}{N_0}} \qquad ...(7.27)$$

This is the required expression for error probability of BPSK reception using coherent or matched filter detection.

7.12 DRAWBACK OF BPSK

Fig. 7.5 shows the block diagram of BPSK receiver. To regenerate the carrier in the receiver. We start by squaring $b(t)\sqrt{2P}\cos(2\pi f_c t + \theta)$. If the received signal is $-\,b(t)\sqrt{2P}\cos(2pf_c t + \theta)$ then the squared signal remains same as before. Hence, the recovered carrier is unchanged even if the input signal was changed its sign. Therefore, it is not possible to determine whether the received signal is equal to $b(t)$ or $-\,b(t)$. Infact, this is the result in ambiguity in the output signal.

This problem can be removed if we use differential plase shift keying (DPSK). However, differential phase shift keying (DPSK) also has some other problems. Other problems of BPSK are ISI and Inter-Channel Interference. However, these problems can be reduced to some extent by making use of filters.

7.13 BINARY FREQUENCY SHIFT KEYING (BFSK)

In binary frequency shift keying (BFSK), the frequency of the carrier is shifted according to the binary symbol. However, the phase of the carrier is unaffected. This means that we have two different frequency signals according to binary symbols. Let us consider a frequency shift represented by Ω. Then we can write following equations :

If $$b(t) = '1'$$

then $$S_H(t) = \sqrt{2P_S}\cos(2\pi f_c + \Omega)t \quad ...(7.28)$$

Again, if

$$b(t) = \text{'0'}$$

then $$S_L(t) = \sqrt{2P_S}\cos(2\pi f_c + \Omega)t \quad ...(7.29)$$

Hence, there is increase or decrease in frequency by Ω. Let us use the following conversion table to combine above two FSK equations :

Conversion Table for **BPSK** representation

Table 7.1

$b(t)$ Input	$d(t)$	$P_H(t)$	$P_L(t)$
1	+ 1 V	+ 1 V	0 V
0	− 1 V	0 V	+ 1 V

The equations (7.28) and (7.29) combinely may be written as

$$s(t) = \sqrt{2P_S}\cos[(2\pi f_C + d(t)\Omega)t] \quad ...(7.30)$$

Hence, if symbol '1' is to be transmitted, the carrier frequency will be $f_C + \left(\frac{\Omega}{2\pi}\right)$. If symbol '0' is to be transmitted, then the carrier frequency will be $f_C - \left(\frac{\Omega}{2\pi}\right)$.

Therefore, we have

$$f_H = f_C + \frac{\Omega}{2\pi} \quad \text{for symbol '1'} \quad ...(7.31)$$

$$f_L = f_C - \frac{\Omega}{2\pi} \quad \text{for symbol '0'} \quad ...(7.32)$$

7.14 GENERATION OF BFSK

From Table 7.1 it is observed that $P_H(t)$ same as $b(t)$ and also $P_L(t)$ is inverted version of $b(t)$. The block diagram for BFSK generation is shown in Fig. 7.11.

We know that input sequence $b(t)$ is same as $P_H(t)$. An inverter is added after $b(t)$ to get $P_L(t)$. An inverter is added after $b(t)$ to get $P_L(t)$. The level shifter converts the '+ 1' level to $\sqrt{P_S T_b}$. Zero level is unaffected. Thus, the output of the level shifters will be either $\sqrt{P_S T_b}$ (if '+ 1') or zero (if input is zero). Further, there are product modulators after level slifter. The two carrier signals $\phi_1(t)$ and $\phi_2(t)$ are used. $\phi_1(t)$ and $\phi_2(t)$ are orthogonal to each other. In one bit period of input signal i.e. T_b, $\phi_1(t)$ or $\phi_2(t)$ have integral number of cycles.

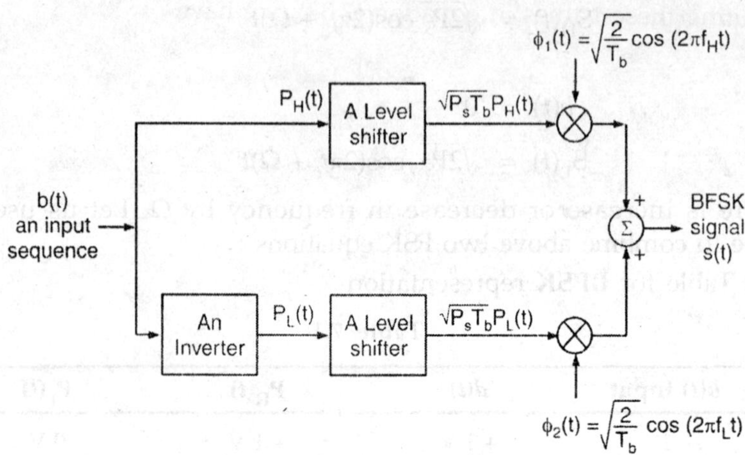

Fig. 7.11 *Block diagram for BFSK generation*

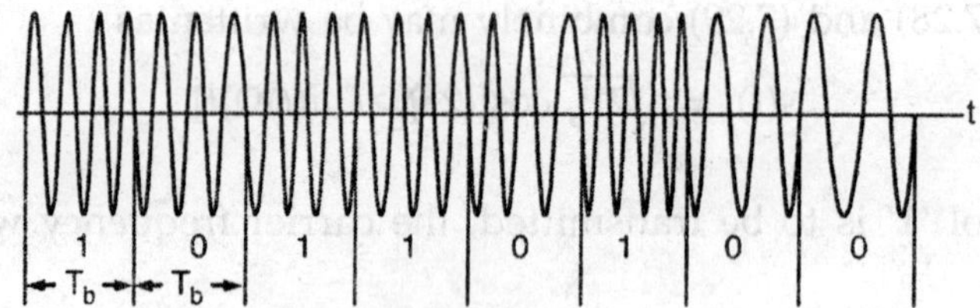

Fig. 7.12 *The BFSK signal*

Thus, the modulated signal is having continuous phase. Fig. 7.12 shows such type of BFSK signal. The adder then adds the two signals.

7.15 THE SPECTRUM OF BFSK SIGNAL

From Fig. 7.12, the BFSK signal $s(t)$ may be written as,

$$S(t) = \sqrt{2P_S}\ P_H(t) \cos(2\pi f_H t) + \sqrt{2P_S}\ P_L(t) \cos(2\pi f_L t) \quad ...(7.33)$$

Equation (7.33) represents the expression for BFSK signal. Now, let us compare this equation with BPSK equation which is written as :

$$S_{BPSK}(t) = b(t)\sqrt{2P}\ \cos(2\pi f_C\ t) \quad ...(7.34)$$

It may be pointed out that this equation is identical to BFSK equation. In BPSK equation, $b(t)$ is a bipolar signal where as in BFSK the similar coefficients $P_H(t)$ or $P_L(t)$ are unipolar. Hence, let us convert these coefficient in bipolar form as :

$$P_H(t) = \frac{1}{2} + \frac{1}{2} P_H(t) \quad ...(7.35)$$

and

$$P_L(t) = \frac{1}{2} + \frac{1}{2} P_L'(t) \quad ...(7.35)$$

Where $P_H'(t)$ and $P_L'(t)$ will be bipolar

i.e. + 1 or − 1

Substituting these values in equation (7.33), we have

$$s(t) = \sqrt{2P_S}\left[\frac{1}{2}+\frac{1}{2}P_H(t)\right]\cos(2pf_H t)$$

$$+ \sqrt{2P_S}\left[\frac{1}{2}+\frac{1}{2}P_L'(t)\right]\cos(2\pi f_L t)$$

or

$$s(t) = \sqrt{\frac{P_S}{2}}\cos(2\pi f_H t) + \sqrt{\frac{P_S}{2}}\cos(2\pi f_L t) + \sqrt{\frac{P_S}{2}}P_H'(t)\cos(2\pi f_H t)$$

$$+ \sqrt{\frac{P_S}{2}}P_L'(t)\cos(2\pi f_L t)\ \cos(2pf_L t) \quad \text{...(7.37)}$$

The first term of this equation represents the single frequency impulse situated at frequency f_H. The second term represents the impulse of f_L. These are constant amplitude pulses. The last two terms are intentical to BPSK equation of equation (7.34)

Here $P_H'(t)$ and $P_L'(t)$ are equivalent to $b(t)$. Therefore, these last two terms in equation (7.37) produce the spectrum which are similar to that of BPSK. One spectrum is located at f_H and other at f_L. Hence, we can write the power spectral density of BFSK as

$$S(f) = \sqrt{\frac{P_S}{2}}\left[\int(f-f_H) + \int(f-f_L) + \frac{P_S T_b}{2}\right.$$

$$\left.\left\{\frac{\sin(\pi f_H T_b}{\pi f_H T_b}\right\}^2 + \frac{P_S T_b}{2}\left\{\frac{\sin(\pi f_L T_b}{\pi f_L T_b}\right\}\right\} \quad \text{...(7.38)}$$

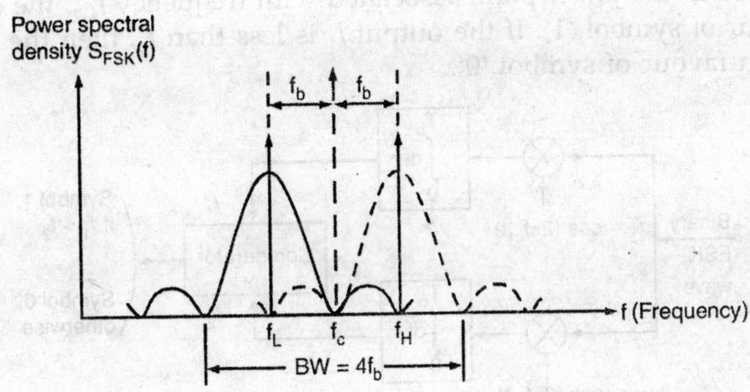

Fig. 7.13 *Illustration of Power spectral density (psd) of a BFSK signal*

Fig. 7.13 represents the plot of power spectral density of BFSK signal expressed by equation (7.38)

Also, f_H and f_L are elected such that

$$f_H - f_L = 2f_b \quad \text{...(7.39)}$$

With such types of selection, it is obvious from the spectrums in the above Fig. That the two frequencies f_H and f_L may be identified properly. The interference between the spectrums is not much with the assumptions above.

7.16 BANDWIDTH OF BFSK SIGNAL

It is bovious from fig. 7.12 that width of one lobe is $2f_b$. The two main lobe due to f_H and f_L are placed such that the total width due to both main lobes is $4\ f_b$.

Therefore, we have

$$\text{Bandwidth of BFSK} = 2f_b + 2f_b$$

or

$$BW = 4f_b \quad \text{...(7.40)}$$

Now, if we compare this band width with that of BPSK, we note that

$$BW(BFSK) = 2 \times BW\ (BPSK) \quad \text{...(7.41)}$$

7.17 DETECTION OF BFSK

The Block diagram of a scheme for demodulation of BFSK wave using coherent detection technique is shown in Fig. 7.14. The detector consists of two correlatorss that are individually tuned to two different carrier frequencies to represent symbols '1' and '0'. A correlator consists of a multiplier followed by an integrator. Then, the received binary FSK signal is applied to the multipliers of both the correlators. To the other input of the multipliers, carriers with frequency f_{C1} and f_{C2} are applied as shown in Fig. 7.14. The multiplied output of each multiplier is subsequently passed through integrators generating output l_1 and l_2 in the two paths. The output of the two integrators are then fed to the decision making device. The decision making device is essentially a comparator which compares the output l_1 in the upper path and output l_2, in the lower path. If the output l_1 produced in the upper path associated with frequency f_{C1} is greater than the output l_2 produced in the lower path, associated with frequency f_{C2}, the detector makes a decision in favour of symbol '1'. If the output l_1 is less than l_2, then the decision making device decides in favour of symbol '0'.

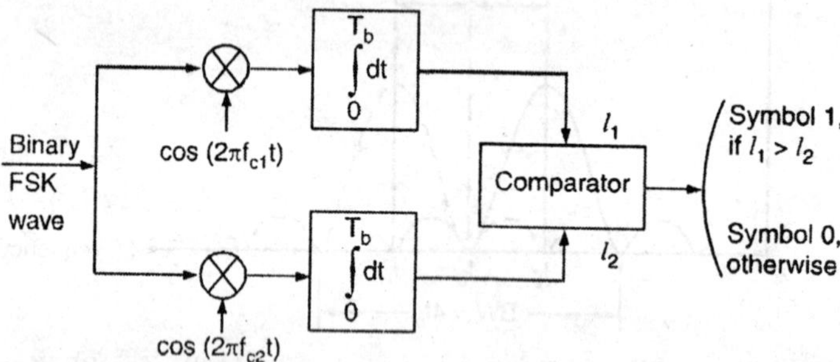

Fig. 7.14 *Block diagram of BFSK receiver (detection of BFSK)*

This type of digital communication receivers are also called correlation receiver. As we know the delector based upon coherent detection requires phase and timing synchronisation.

7.18 GEOMETRICAL REPRESENTATION ORTHOGONAL BFSK

It is fact that orthogonal carriers are used for M-ary PSK and QASK. The different signal points are represented geometrically in ϕ_1, ϕ_2-plane. For geometrical representation of BFSK signals, such orthogonal carrier are required. From Fig. 7.15. We know that two carriers $\phi_1(t)$ and $\phi_2(t)$ of two different frequencies f_H and f_L are used for modulation. To make $\phi_1(t)$ and $\phi_2(t)$ orthogonal, the frequencies f_H and f_L must be some integer multiple of band frequency 'f_b'.

Thus,
$$f_H = mf_b \quad \text{...(7.42)}$$
and
$$f_L = nf_b \quad \text{...(7.43)}$$

Here, $f_b = \dfrac{1}{T_b}$, then the carrirs would be

$$\phi_1(t) = \sqrt{\frac{2}{T_b}} \cos(2\pi\, mf_b\, t) \quad \text{...(7.44)}$$

and
$$\phi_2(t) = \sqrt{\frac{2}{T_b}} \cos(2\pi nf_b\, t)$$

The carriers $\phi_1(t)$ and $\phi_2(t)$ are orthogonal over the period T_b. Now the equations (7.28) and (7.29) can be written as

$$S_H(t) = \sqrt{P_S\, T_b}\sqrt{\frac{2}{T_b}} \cos(2\pi f_H\, t)$$

$$S_L(t) = \sqrt{P_S\, T_b}\sqrt{\frac{2}{T_b}} \cos(2\pi f_L\, t)$$

Here
$$f_H = f_C + \frac{\Omega}{2\pi} \quad \text{...(7.45)}$$

and
$$f_L = f_C - \frac{\Omega}{2\pi}$$

Using the relations in equations (7.42) to (7.45), the above equations can be written as

$$S_H(t) = \sqrt{P_S\, T_b} \cdot \phi_1(t) \quad \text{...(7.46)}$$

and
$$S_L(t) = \sqrt{P_S\, T_b} \cdot \phi_2(t) \quad \text{...(7.47)}$$

Thus, based on the above two equations we can draw the signal space diagram as shown in Fig. (7.19)

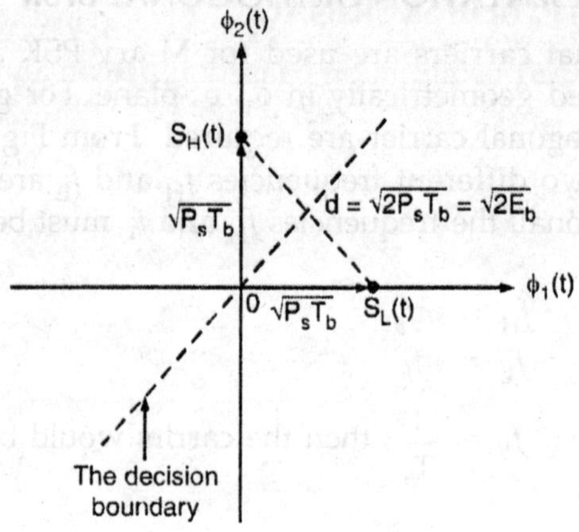

Fig. 7.15 *Illustration of signal space representation of orthogonal BFSK signal*

7.19 DISTANCE BETWEEN SIGNAL POINTS

We know that there are two signal points in the signal space. The distance between these two points may be evaluated as

$$d^2 = \left(\sqrt{P_S T_b}\right)^2 + \left(\sqrt{P_S T_b}\right)^2$$

or

$$d^2 = 2P_S T_b$$

or

$$d = \sqrt{P_S T_b} \qquad ...(7.49)$$

Since $P_S T_b = E_b$, the above equation can be written as

$$d = \sqrt{2E_b} \qquad ...(7.49)$$

With the comparision to the distance of BPSK it may be observe that this distance is smaller than BPSK.

7.20 GEOMETRICAL REPRESENTATION OF NON-ORTHOGONAL BFSK SIGNALS

It is fact, whenever the carriers $\phi_1(t)$ and $\phi_2(t)$ are non-orthogonal, then the signal points $S_H(t)$ or $S_L(t)$ would not lie exactly on the axes $\phi_1(t)$ and $\phi_2(t)$. Such a representation has been shown in Fig. 7.16.

The distance *'d'* for non-orthogonal signal shown in figure may be written approximately as :

$$d^2 = 2E_0\left[1 - \frac{\sin 2\pi(f_H - f_L)T_b}{2\pi(f_H - f_L)T_b}\right] \qquad ...(7.50)$$

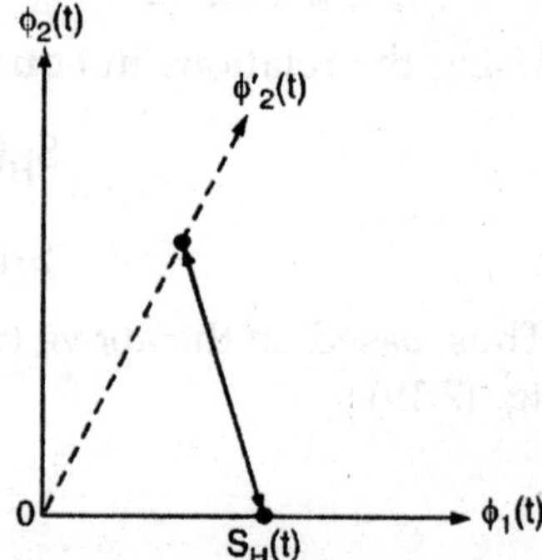

Fig. 7.16 *Geometrical representation of non-orthogonal BFSK signals*

7.21 PROBABILITY OF ERROR, P_e of BFSK SIGNALS

The generated equation derived using union baased approximation gives probability of error as

$$P(e) = \frac{1}{2}\sum_{\substack{k=1 \\ k \neq 1}}^{m} erfc\left(\frac{d_{ik}}{2\sqrt{N_0}}\right) \quad \text{for all } i \qquad \text{...(7.50a)}$$

Here, we have only two points in BFSK. Let us consider that $S_H(t)$ be $S_1(t)$ and $S_L(t)$ be $S_2(t)$. Then, distance between these two points can be written as,

$$d = d_{12} = \sqrt{2E_b}$$

Since there are only two points i.e., M = 2 take $i = 2$ and $k = 1$ then equation (7.50*a*) will be

$$P(e) = \frac{1}{2} erfc\left(\frac{d_{12}}{2\sqrt{N_0}}\right)$$

or

$$P(e) = \frac{1}{2} erfc\left(\frac{\sqrt{2E_b}}{2\sqrt{N_0}}\right)$$

or

$$P(e) = \frac{1}{2} erfc\sqrt{\frac{E_b}{2N_0}}$$

This equation shows that error probability of BFSK is higher compared to that of BPSK.

7.22 ADVANTAGES AND DISADVANTAGES OF BFSK SIGNALS

Although the generation of BFSK is essier, it has many disadvantages compared to BPSK signal. Its bandwidth is greater than $4t_b$ which is almost double the bandwidth of BPSK. Also, the distance between the signal points is less in case of BFSK. Therefore, the error rate of BFSK is more compared to BPSK.

7.23 NON-COHERENT BINARY MODULATION DETECTION

As coherent detection exploits knowledge of the carrier wave's phase reference, thus providing the optimum error performance attainable with a digital modulation format of interest. However, it is impractical to have knowledge of the carrier phase at the receiver, we make use of non-colerent detection. Thus in the case of phase-shift keying (PSK), we can hot have "non-coherent PSK since non-coherent means doing without phase information. However, there is a pseudo PSK technique known as differential phase slift keying (DPSK) which can be viewed as the non-coherent form of PSK.

7.24 AMPLITITUDE SHIFT KEYING (ASK)

In the binary ASK case, the transmitted signal is expressed as :

$$s(t) = \sqrt{2P_S}\cos(2\pi f_C t)$$

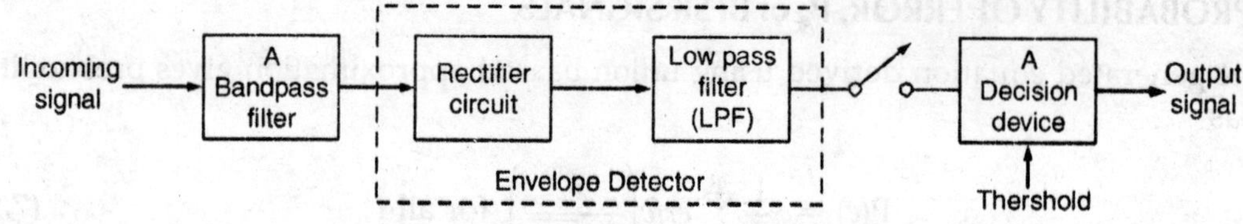

Fig. 7.17 *Non-coherent ASK detector*

Binary ASK signal can also be demodulated non-coherently using envelop detector. This simplifies the design consideration required in synchronous detection. Non-coherent detection schemes do not require a phase-coherent local oscillator. This method involves some form of rectification and low pass filtering at the receiver. The block diagram of a non-coherent receiver for ASK signal has been shown in Fig. 7.17.

7.25 NON-COHERENT DETECTION OF FSK

Binary FSK waves may be demodulated non-coherently using envelope detector. The received FSK signal is applied to a bank of two bandages filters, one tuned to frequency f_{C1} and the other tuned of f_{C2}. Each filter is followed by an envelope detector. The resulting outputs of the two envelope detectors are sampled and then compared with each other.

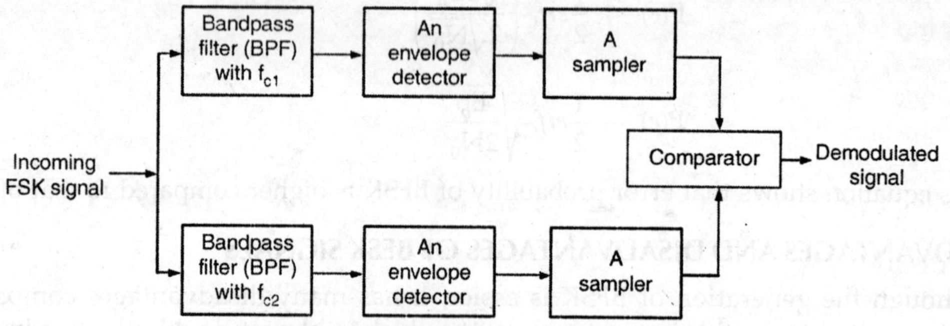

Fig. 7.18 *Non-coherent detection of FSK binary signals*

The arrangement for non-coherent detection of FSK signal has been shown in Fig. 7.18.

A decision is made in favour of symbol '1' if the envelope detector output derived from the fitter tuned to frequency f_{C1} is larger than that derived from the second filter. Otherwise, a decision is made in favour of the symbol '0'.

7.26 GENERATION OF DPSK

In order to eliminate the need for phase synchronisation of coherent receiver with PSK, a differential encoding system can be used with PSK. The digital information content of the binary data is encoded in therms of signal transitions. Let us consider, the symbol '0' may be used to represent transition in a given binary sequence and symbol '1' to indicate no transition. This new signaling technique which combines differential encoding wilt phase-shift keying. PSK is known as differential phase shift keying (DPSK).

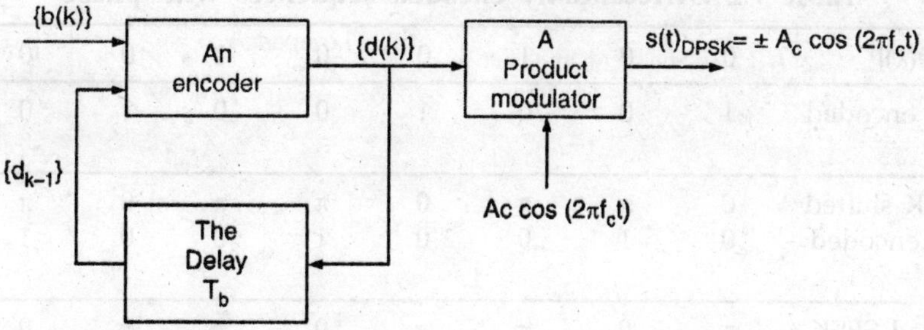

Fig. 7.19 *Illustration of the scheme to generate DPSK signals*

A schematic arrangement for generating DPSK signal has been shown in Fig. 7.19. The data stream $b(t)$ is applied to the input of the encoder. The output of the encoder is applied to one input of the product modulator. To the other input of this product modulator a sinusoidal carrier of fixed amplitude and frequency is applied.

7.27 DETECTION OF DPSK

For detection of the differentially encoded DPSK. The receiver arrangement as shown in Fig. 7.20 may be use. The received DPSK signal is applied to one input of the multiplier. To the other of the multiplier, a deloyed version of the received DPSK signal by the time interval T_b is applied. The delayed version of the received DPSK signal, in the absence of channel noise, has been shown in the 4th row of the table. The output of the difference is proportional to cos (ϕ), where ϕ is the difference between the carrier phase angle of the received DPSK signal and its delayed version, measured in the same bit interval. The phase angle of the DPSK signal and its delayed version have been shown in 3rd and 5th rows respectively. The phase difference between the two sequences for each bit interval is used to determine the sign of the phase comparator output. When $\phi = 0$, the integrator output is positive whereas when $\phi = \pi$, the integrator output is negative. By comparing the integrator output with a decision level of zero volt, the decision device can reconstrivet the binary sequence by assigning a symbol '0' for negative output and a symbol '1' for positive output. The reconstructed binary data is shown in the last row of the table. It seen that in the absence of noise, the receiver can reconstrout the transmitted binary data exactly. DPSK

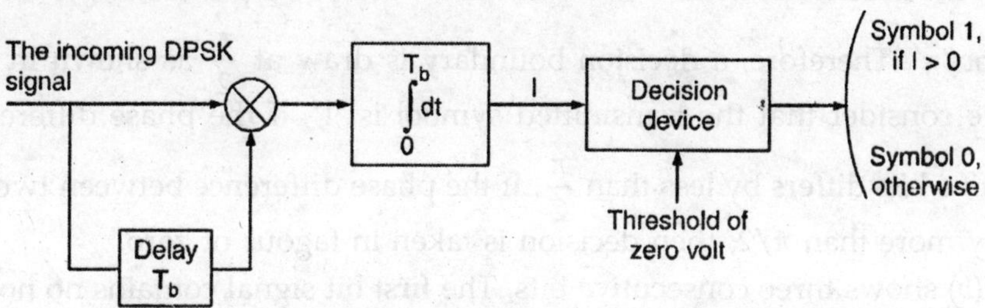

Fig. 7.20 *Receiver for the detection of DPSK signals*

may be viewed as a non-coherent version of PSK. It may be noted that the reconstruction is invariant with the choice of the initial bit in the encoded data.

Table 7.2 Differentially encoded sequences with phase

Binary data $\{b(k)\}$	0	0	1	0	0	1	0	0	1	1
Differentially encoded data $\{d(k)\}$	1	0	0	1	0	0	1	0	0	0
Phase of DPSK shifted differentially encoded data $\{d_{k-1}\}$	0 0	π 1	π 0	0 0	π 1	π 0	0 0	π 1	π 0	π 0
Phase of shifted SPSK	π	0	π	π	0	π	π	0	π	π
Phase comparison output	–	–	)	–	–	+	–	–	+	+
Detected binary sequence Starting reference bit	0	0	1	0	0	1	0	0	1	1

7.28 EVALUATION OF BANDWIDTH OF DPSK SIGNAL

As we have seen that one previous bit is used to decide the phase shift of next bit. Thus, change in $b(t)$ occurs only if input bit is at level '1'. No change happens if output bit is at level '0'. Because, one previous bit is always used to define the phase shift in next bit, therefore, the symbol can be said to have two bits. Hence one symbol duration (T) is equivalent to two bits duration ($2T_b$) i.e.,

$$\text{Symbol duration} \quad T = 2\,T_b \qquad \text{...(7.51)}$$

$$\text{Bandwidth is given as} \quad BW = \frac{2}{T} = \frac{1}{T_b} = f_b \qquad \text{...(7.52)}$$

Hence, the minimum bandwidth in DPSK is equal to f_{b1} i.e. maximum baseband signal frequency.

7.29 ERROR PROBABILITY OF DPSK

The phase diagram of DPSK signal in Fig. 7.21 shows when no noise is present. This means that in the absence of noise and transmission delay, the phase shift of the DPSk signal is either '0' or 'π'. Therefore, a decision boundary is draw at $\frac{\pi}{2}$ as shown in Fig. 7.21 (*a*). Therefore, we consider that the transmitted symbol is '1', if the phase difference between two consecutive bits differs by less than $\frac{\pi}{2}$. If the phase difference between two consecutive bits differs by more than $\pi/2$, then decision is taken in fagour of zero.

Fig. 7.21 (*b*) shows three consecutive bits. The first bit signal contains no noise, hence its phasor is along the horizontal line. Therefore, the symbol transmitted in first bit is assumed to be '1'. Because of noise, there is a phase difference of 'θ_1' between first and second bit. Since $\theta_1 < \frac{\pi}{2}$, second bit is also taken as symbol '1'. The phase difference between second and third bit θ_2.

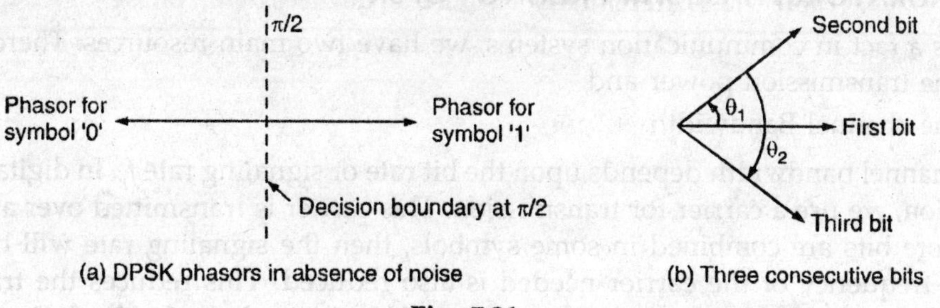

Fig. 7.21

From figure, it is clear that $\theta_2 > \pi/2$, hence third bit is taken as symbol zero. Since the phase differences are not exact between two successive bits, some error is introduced in the decision. Therefore, DPSK system is called 'suboptimum' in nature. If the synchronization is used to make phase differences exact, then it becomes PSK system.

Because of the suboptimal nature of DPSK, the error probability is higher than that of BPSK. The average probability of error of non-coherent receiver is given as :

$$P(e) = \frac{1}{2}e^{-E/2N_0}$$

Here E = the energy per symbol and

$\frac{N_0}{2}$ = spectrual density of white Gaussian noise.

We know that symbol duration $T = 2T_b$. Hence, energy of the symbol will be

$$E = 2E_b$$

Now, expression for $P(e)$ becomes,

$$P(e) = \frac{1}{2}e^{-E_b/N_0}$$

This is called average probility of error or bit error, BER of DPSK system.

7.30 ADVANTAGES AND DISADVANTAGES OF DPSK

Overall observations and discussions there are some advantages as well as some disadvantages of DPSK over BPSK.

7.30.1 Advantages

(*a*) DPSK does not need carrier at the receiver end. This means that the complicated circuiting for generation of local carrier is not required.

(*b*) The bandwidth requirement of DPSK is reduced as compared to that of BPSK.

7.30.2 Disadvntages

(*a*) The probability or error *i.e.*, bit error rate, of DPSK is higher than that of BPSK.

(*b*) Because DPSK uses two successive bits for its reception, error in the first bit creates error in the second bit. Therefore, error propagation in DPSK is more. On the other hand, in BPSK single bit can go in error since detection of each bit is independent.

(*c*) Noise inteference in DPSK is more.

7.31 QUADRATURE PHASE SHIFT KEYING—QPSK

This is a fact in communication systems, we have two main resources. There are :

(*i*) The transmission power and

(*ii*) The channel Bandwidth.

The channel bandwidth depends upon the bit rate or signaling rate f_b. In digital bandpass transmission, we use a carrier for transmission. This carrier is transmitted over a channel. If two or more bits are combined in some symbols, then the signaling rate will be reduced. Thus, the frequency of the carrier needed is also reduced. This reduces the transmission channel bandwidth. Hence, because of grouping of bits in symbols, the transmission channel bandwidth can be reduced. In quadrature phase shift keying. QPSK, two successive bits in the data requence are grouped together. This reduces the bits rate or signaling rate, i.e., f_b and thus reduces the bandwidth of the channel.

Table 7.3 Symbol and corresponding phase shifts in QPSK

S. No.	Input successive bits		Symbol	Phase shift in carrer
$i = 1$	1(1 V)	0(– 1 V)	S_1	$\pi/4$
$i = 2$	1(– 1 V)	0(– 1 V)	S_2	$3\pi/4$
$i = 3$	1(– 1 V)	1(1 V)	S_3	$5\pi/4$
$i = 4$	1(1 V)	1(1 V)	S_4	$7\pi/4$

In case of BPSK, we know that when symbol changes the level, the phase of the carrier is changes by 180°. Because, there were only two symbols in BPSK, the phase shift occurs in two levels only. However, in QPSK, two successive bits are combined. In fact, this combination of two bits forms four distinct symbols. When the symbol is changed to next symbol, then the phase of the carrier is changed by 45°, $\pi/4$ radius. Table 7.3 shows these symbols and their phase shifts.

7.32 GENERATION OF QPSK

The block diagram of QPSK transmitter is shown in Fig. 7.23. Here, the input binary sequence is first converted to a bipolar NRZ type of signa. This signal is denoted by $b(t)$. It represents binary '1' by + 1 V and binary '0' by – 1 V. This signal has been shown in Fig. 7.23(*a*). The demultiplexer divides $b(t)$ into two separate bit streaws of the odd numbered and even numbered bits.

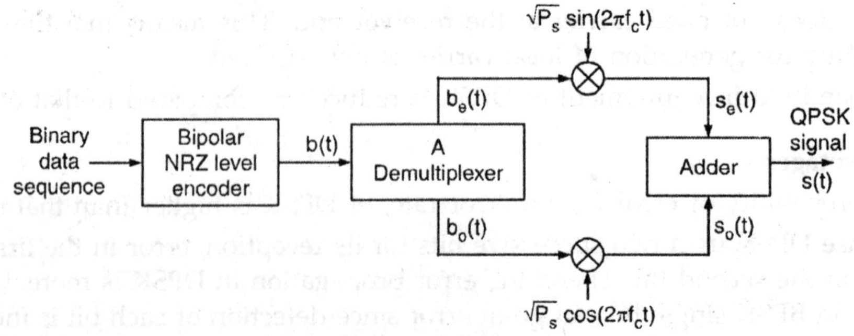

Fig. 7.22 *Generation of QPSK*

Here, $b_e(t)$ represents even numbered sequence and $b_0(t)$ represents odd numbered sequence. The symbol duration of both of these odd and even numbered sequence is $2T_b$. Hence, each symbol consists of two bits. Fig. 7.23 (*b*) and (*c*) illustrates the waveform of $b_e(t)$ and $b_0(t)$.

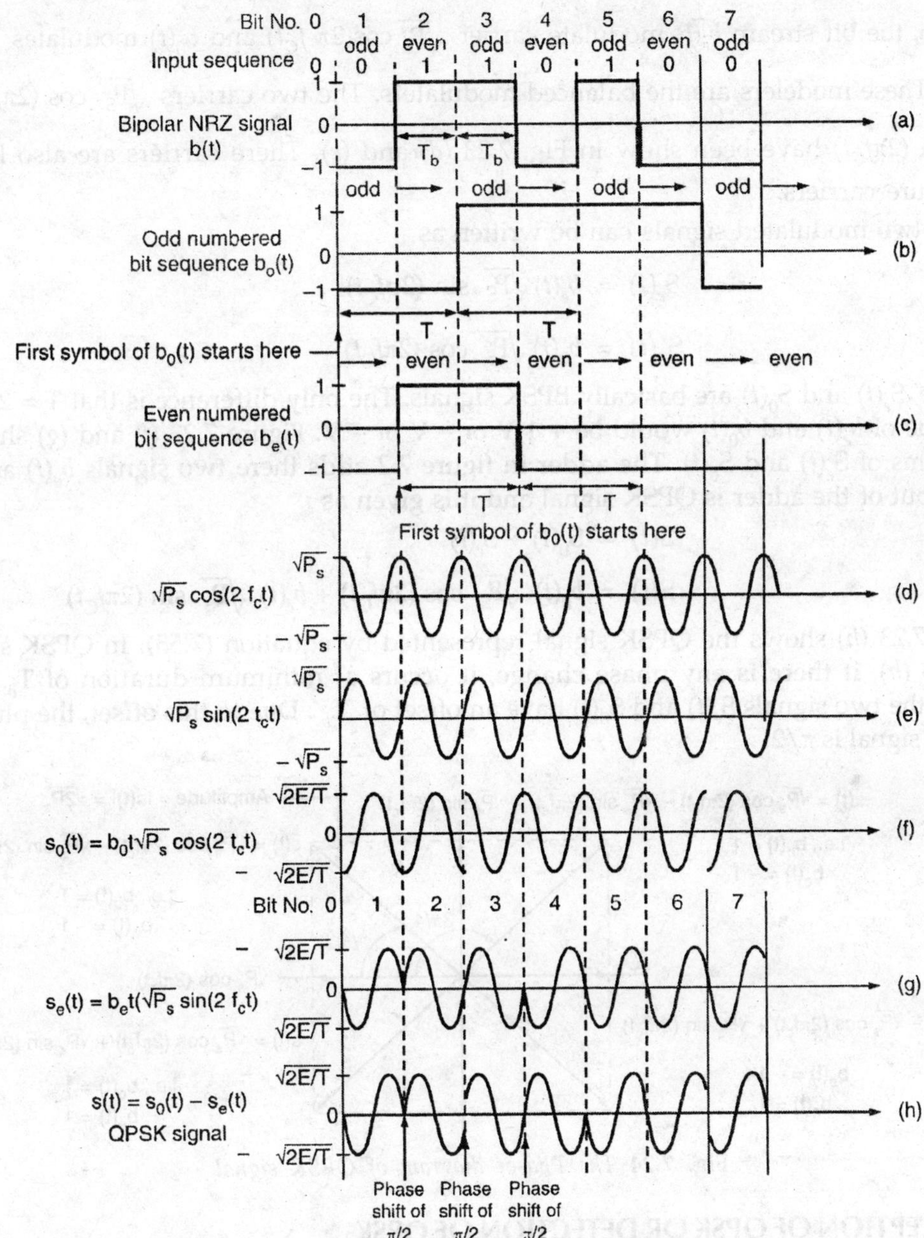

Fig. 7.23 *QPSK waveforms, (a) Input sequence and its corresponding NRZ waveform, (b) Odd numbered bit sequence and its corresponding waveform (c) Even numbered bit sequence and its NRZ waveform (d) Basis function $\phi_1(t)$ (e) Basis function $\phi_2(t)$ (f) Binary PSK waveform for odd numbered channel (g) Binary PSK waveform for even numbered channel (h) Final QPSK waveform*

It may be observed that the first even bit occurs after the first odd bit. Hence, even numbered bit sequence $b_e(t)$ starts with the delay of one bit period due to first odd bit. Thus, first symbol of $b_e(t)$ is delayed by one bit period T_b' with respect to first symbol of $b_0(t)$. This delay of T_b is known as offset. This shows that the change in levels of $b_e(t)$ and $b_0(t)$ cannot occur at the same time due to offset or staggering.

Also, the bit stream $b_e(t)$ modulate carrier $\sqrt{P_S}\cos(2\pi f_C t)$ and $b_0(t)$ modulates $\sqrt{P_S}$ sin $(2\pi f_C t)$. These modelers are the balanced modulaters. The two carriers $\sqrt{P_S}\cos(2\pi f_C t)$ and $\sqrt{P_S}\sin(2\pi f_C t)$ have been show in Fig. 7.23 (*d*) and (*e*). There carriers are also know as quadrature carriers.

The two modulated signals can be written as

$$S_e(t) = b_e(t)\sqrt{P_S}\sin(2pf_C t) \quad ...(7.54)$$

and

$$S_0(t) = b_0(t)\sqrt{P_S}\cos(2\pi f_C t) \quad ...(7.54)$$

Here, $S_e(t)$ and $S_0(t)$ are basically BPSK signals. The only difference is that $T = 2T_b$ here. The value of $b_e(t)$ and $b_0(t)$ would be + 1 V or – V or – V. Figure 7.23 (*f*) and (*g*) shows the waveforms of $S_e(t)$ and $S_0(t)$. The adder in figure 7.2 adds there two signals $b_e(t)$ and $b_0(t)$. The output of the adder is QPSK signal and it is given as :

$$S(t) = S_0(t) + S_e(t)$$

or

$$S(t) = b_0(t)\sqrt{P_S}\cos(2\pi f_C t) + b_e(t)\sqrt{P_S}\sin(2\pi f_C t) \quad ...(7.55)$$

Fig. 7.23 (*h*) shows the QPSK signal represented by equation (7.55). In QPSK signal in Fig. 7.23 (*h*), if there is any phase change, it occurs at minimum duration of T_b. This is because the two signals $S_e(t)$ and $S_0(t)$ have an offset of 'T_b'. Due to this offset, the phase slift in QPSK signal is $\pi/2$.

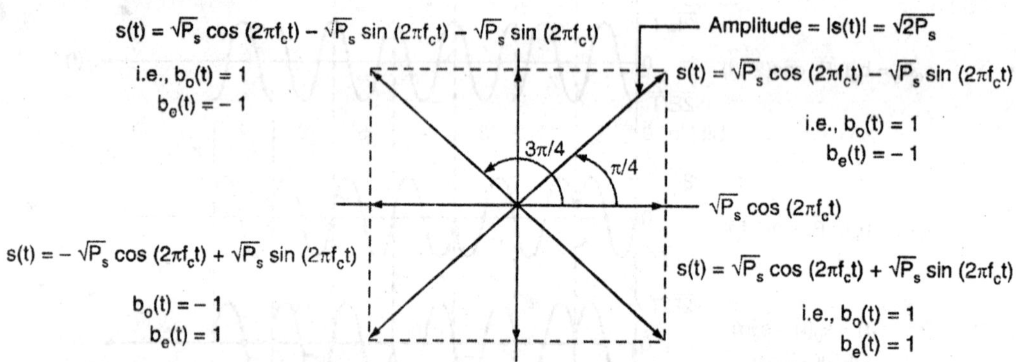

Fig. 7.24 *The Phasor diagram of QPSK signal*

7.33 RECEPTION OF QPSK OR DETECTION OF QPSK

Fig. 7.25 shows the QPSK receiver. This is synchronous reception. Hence, the coherent carrier is to be recovered from the received signal *s*(*t*). The received signal *s*(*t*) is first raised to its 4th power, i.e., $S^4(t)$. After that, it is allowed to pass through a bandpass filter, BPF, which is centred around $4f_C$. The output of the bardpass filter is a coherent carrier of frequency $4f_C$. This is divided by 4 and it provides two coherent quadrature carriers, i.e.,

$$\cos(2\pi f_C t)$$

and

$$\sin(2\pi f_C t)$$

These coherent carriers are applied to two synchronous demodulators. These synchronous demodulators consist of multiplier and an integrator.

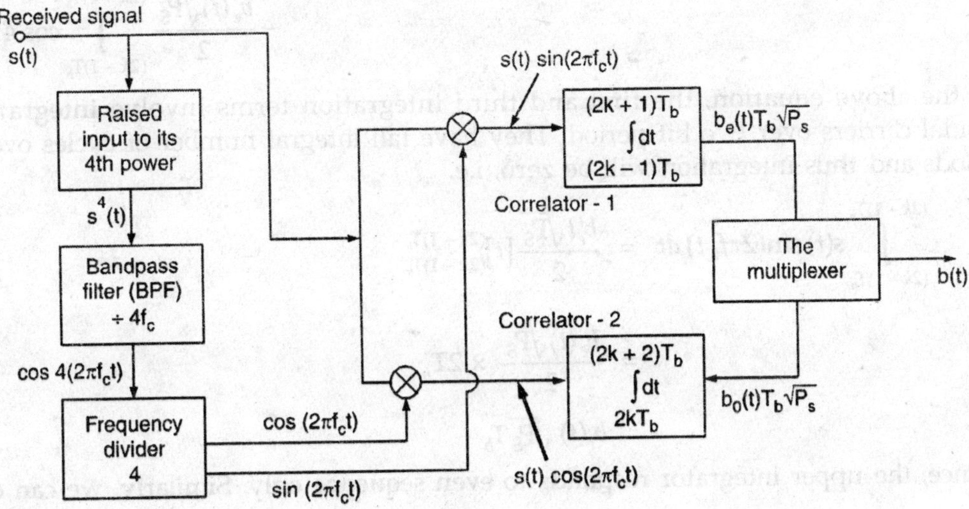

Fig. 7.25 *Reception of QPSK*

The incoming signal is applied to both the multipliers. Here, the integrator integrates the product signal over two bit interval, i.e., $T_S = 2\,T_b$. At the end of this period, the output of integrator is sampled. The outputs of the two integrators are sampled at the offset of one bit period T_b. Hence, the output of the multiplexer is the signal $b(t)$. This means that the odd and even sequences are combined by multiplexer.

Now, set us consider the product signal at the output of upper multiplier, i.e.,

$$S(t) = \sin(2\pi f_C t) = b_0(t)\sqrt{P_S}\cos(2\pi f_C t)\sin(2\pi f_C t) + b_e(t)\sqrt{P_S}\sin^2(2\pi f_C t) \quad ...(7.56)$$

This signal in integrated by the upper integrator in Fig. 7.25.

Therefore, we have

$$\int_{(2k-1)}^{(2k+1)} s(t)\sin(2\pi f_C t) = b_0(t)\sqrt{P_S}\int_{(2k-1)T_b}^{(2k+1)T_b}\cos(2\pi f_C t)\sin(2\pi f_C t)\,dt$$

$$+\, b_e(t)\sqrt{P_S}\int_{(2k-1)T_b}^{(2k+1)T_b}\sin^2(2\pi f_C t)\,dt$$

Because $\frac{1}{2}(\sin 2x) = \sin x \cdot \cos x$

and $\sin^2(x) = \frac{1}{2}[1 - \cos(2x)]$

Hence, with the help of these two trignometric identities in equation (7.57), we have

$$\int_{(2k-1)T_b}^{(2k+1)T_b} \sin(2\pi f_C t)\, dt = \frac{b_0(t)\sqrt{P_S}}{2}\int_{(2k-1)T_b}^{(2k+1)T_b} \sin 4\pi f_C t\, dt + \frac{b_e(t)\sqrt{P_S}}{2}\int_{(2k-1)T_b}^{(2k+1)T_b} 1\cdot dt$$

$$- \frac{b_e(t)\sqrt{P_S}}{2}\int_{(2k-1)T_b}^{(2k+1)T_b} \cos 4\pi f_C t\, dt$$

In the above equation, the first and third integration terms involve integration of sinusoidal carriers over two bit period. They have fall integral number of cycles over two bitperiods and thus integration will be zero, i.e.

$$\int_{(2k-1)T_b}^{(2k+1)T_b} s(t)\sin(2\pi f_C t)\, dt = \frac{b_e t\sqrt{P_S}}{2}[t]_{(2k-1)T_b}^{(2k+1)T_b}$$

$$= \frac{b_e(t)\sqrt{P_S}}{2} \times 2T_b$$

$$= b_e(t)\sqrt{P_S T_b} \qquad \text{...(7.58)}$$

Hence, the upper integrator responds to even sequence only. Similarly, we can obtain the output of lower integrator as $b_0(t)\sqrt{P_S T_b}$.

7.34 CONCEPT OF CARRIER SYNCHRONIZATION IN QPSK

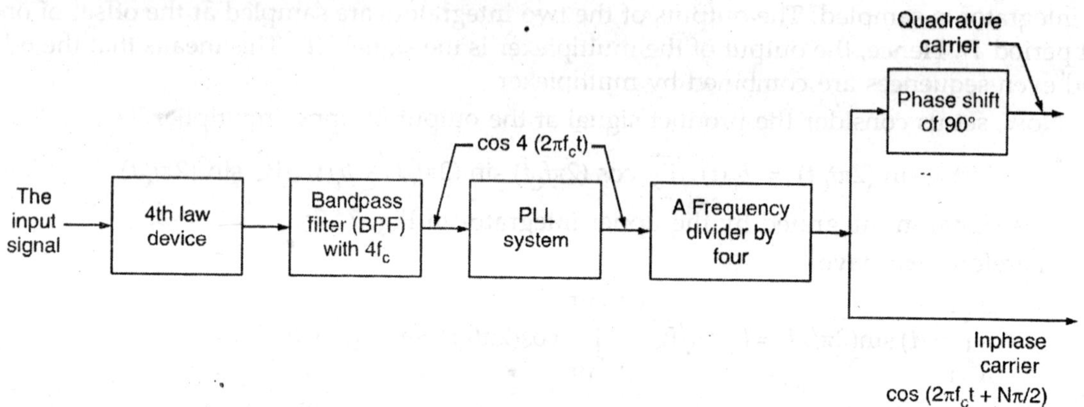

Fig. 7.26 *PLL system used for carrier synchronization in QPSK*

Both the carriers are to be synchronized properly in coherent detection in QPSK. Fig. 7.26 shows the PLL system for carrier synchronization in QPSK.

The fourth power of the input signal consists of discrete frequency component at $4f_C$.

We know that

$$\cos 4(2\pi f_C t) = \cos(8\pi f_C t + 2\pi N) \quad ...(7.59)$$

Where 'N' is the number of cycles over the bit period. It is always an integer value. When the frequency division by four takes place, the right hard side, R.H.S. of this equation becomes

$$\cos\left(2\pi f_C t + \frac{N\pi}{2}\right)$$

This indicates that the output has a fixed phase error of $\frac{N\pi}{2}$.

Differential encoding can be used to pullify the phase error events. The PLL remains locked with the phase of $4f_C$ and then output of PLL is divided by 4. This provides a coherent carrier. A 90° phase shift is added to this carrier to produce a quadrature carrier.

7.35 SIGNAL SPACE REPRESENTATION IN QPSK SIGNALS

Fig. 7.27 shows the phasoe diagram of QPSK signal. Depending upon the combination of two successive bits, the phase shift occurs in carrier. This means that the QPSK signal in equation (7.55) can be written as

$$S(t) = \sqrt{2P_S}\cos\left[2\pi f_C t + (2m+1)\frac{\pi}{4}\right]$$

where $m = 0, 1, 2, 3$ ...(7.60)

This equation takes four value and they represent the phases of Fig. 14.31. This equation can be expanted as :

$$s(t) = \sqrt{2P_S}\cos(2\pi f_C t)\cos\left[(2m+1)\frac{\pi}{4}\right] - \sqrt{2P_S}\sin(2\pi f_C t)\sin\left[(2m+1)\frac{\pi}{4}\right]$$

Let us rerrange the above equation as

$$s(t) = \left\{\sqrt{P_S T_S}\cos\left[(2m+1)\frac{\pi}{4}\right]\right\}\sqrt{\frac{2}{T_S}}\cos(2\pi f_C t) - \left\{\sqrt{P_S T_S}\sin\left[(2m+1)\frac{\pi}{4}\right]\right\}\sqrt{\frac{2}{T_S}}\sin(2\pi f_C t) \quad ...(7.61)$$

Again, let

$$\phi_1(t) = \sqrt{\frac{2}{T_S}}\cos(2\pi f_C t) \quad ...(7.62)$$

and

$$\phi_1(t) = \sqrt{\frac{2}{T_S}}\sin(2\pi f_C t) \quad ...(7.63)$$

These two signals are known as orthogonal signals and they are used as carriers in QPSK modulator.

Let $$b_0(t) = \sqrt{2}\cos\left[(2m+1)\frac{\pi}{4}\right] \quad ...(7.64)$$

and $$b_e(t) = -\sqrt{2}\sin\left[(2m+1)\frac{\pi}{4}\right] \quad ...(7.65)$$

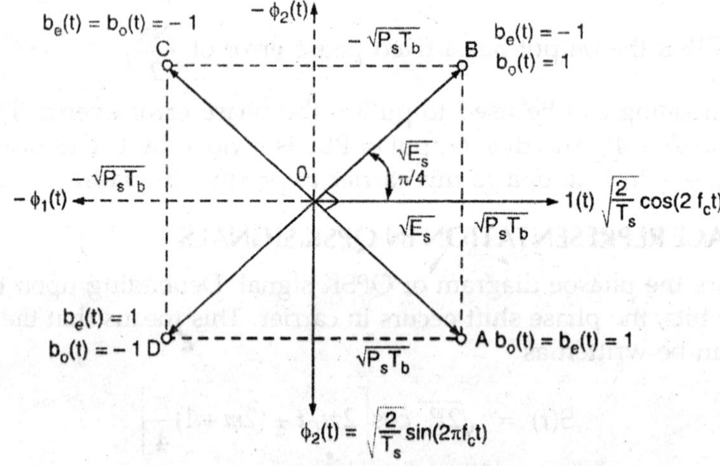

Fig. 7.27 *The signal space representation for QPSK signals*

With the use of equations (7.61) to (7.64) we can have from equation (7.60) as

$$S(t) = \sqrt{P_S T_S}\cdot\frac{1}{\sqrt{2}}\, b_0(t)\,\phi_1(t) + \sqrt{P_S T_S}\cdot\frac{1}{\sqrt{2}}\, b_C(t)\,\phi_2(t)$$

or $$S(t) = \sqrt{P_S\cdot\frac{T_S}{2}}\, b_0(t)\,\phi_1(t) + \sqrt{P_S\cdot\frac{T_S}{2}}\, b_e(t)\,\phi_2(t)$$

$$T_S = \text{symbol duration}$$

and $$T_S = 2T_b$$

or $$T_b = \frac{T_S}{2} \quad ...(7.66)$$

Then, the above equation can be written as

$$S(t) = \sqrt{P_S T_b}\, b_0(t)\,\phi_1(t) + \sqrt{P_S T_b}\, b_e(t)\,\phi_2(t) \quad ...(7.67)$$

Because bit energy

$$E_b = P_S\, T_b$$

Therefore,

$$S(t) = \sqrt{E_b}\, b_0(t)\,\phi_1(t) + \sqrt{E_0}\, b_0(t)\,\phi_2(t) \quad ...(7.68\_$$

This equation gives signal space representation of QPSK signal. The two orthogonal signals $\phi_1(t)$ and $\phi_2(t)$ form the two axes of the signal space. Fig. 7.27 shows the signal space representation. The possible 4 signal points have been shown by small circles on ϕ_1, ϕ_2 plane.

From each signal points, we obtain two bits. For example, from point 'A', we obtain two bit as (1, 1) and from 'B' we obtain as (– 1, 1). The distance of any signal point from origin '0' given as,

$$\text{Distance} \quad \text{'OB'} = \sqrt{P_S T_b + P_S T_b}$$

$$= \sqrt{2P_S T_b}$$

$$= \sqrt{P_S T_S} \qquad [\because\ 2T_b = T_S] \quad \text{...(7.69)}$$

$$\text{'OB'} = \sqrt{E_S} \qquad [\because\ P_S T_S = E_S] \quad \text{...(7.70)}$$

Hence, the length of each signal point origin is $\sqrt{E_S}$. As we know that $b_e(t)$ and $b_0(t)$ represents two successive bits. There is of offset of 'T_b' between $b_e(t)$ and $b_0(t)$. Therefore, $b_e(t)$ and $b_0(t)$ both cannot change their levels simultaneously. Hence, either $b_e(t)$ or $b_0(t)$ can change at a time.

Suppose $\quad b_e(t) = b_0(t) = 1$ is

representating signal point 'A' in Fig. 7.27. In the next bit interval, if $b_0(t) = -1$, then signal point will be 'D'. Otherwise, if $b_e(t)$ changes its level, i.e., $b_e(t) = -1$, then next signal point will be 'B'. Hence, from signal point 'A', then next signal point will be either 'D' or 'B'.

7.36 DISTANCE BETWEEN SIGNAL POINTS

This is a fact, the ability to determine a bit without error is measured by the distance between two nearest possible signal points in the signal space. Such points differed in signal bit. For, example, signal points 'A' and 'B' are two nearest points since they differe by a signal bit $b_e(t)$. As 'A' and 'B' become closer to each other, the possibility of error increases. Therefore, this distance must be as large as possible. This distance is denoted by 'd'. In figure 14.31, the distance between signal points 'A' and 'B' can be written as :

$$d^2 = \left(\sqrt{E_S}\right)^2 + \left(\sqrt{E_S}\right)^2 = \sqrt{2E_S} \quad \text{...(7.71)}$$

or

$$d = 2\sqrt{P_S T_b}$$

$$= 2\sqrt{E_b} \quad \text{...(7.72)}$$

7.37 SPECTRUM OF QPSK SIGNAL

The input sequence $b(t)$ is of bit duration T_b. And also, it is a NRZ bipolar waveform. We know that the power spectrul density of such waveform can be given as :

$$S(f) = V_b^2 T_b \left[\frac{\sin \pi f\, T_b}{\pi f\, T_b}\right]^2$$

Also, $$V_b = \sqrt{P_S}$$

then, $$S(f) = P_S T_b \left[\frac{\sin(\pi f\, T_b)}{\pi f\, T_b}\right]^2 \qquad \text{...(7.73)}$$

This equation gives power spectral density, psd, of signal $b(t)$. This signal is divided into $b_e(t)$ and $b_0(t)$ each of bit period $2T_b$. If we consider that symbols 1 and 0 are equally likely, then we can write power spectral densities, psds, of $b_e(t)$ and $b_0(t)$ as :

$$S_e(f) = P_S T_S \left[\frac{\sin \pi f\, T_S}{\pi f\, T_S}\right]^2 \qquad \text{...(7.74)}$$

and $$S_0(f) = P_S T_S \left[\frac{\sin \pi f\, T_S}{\pi f\, T_S}\right]^2 \qquad \text{...(7.75)}$$

In these two equations, we have just replaced T_b by T_S and T_S is the period of bit in $b_e(t)$ and $b_0(t)$. Because in phase and quadrature components $b_e(t)$ and $b_0(t)$ are statistically independent, the baseband power spectral density of QPSK signal equals the sum of the individual power spectral densities of $b_e(t)$ and $b_0(t)$, i.e.,

$$S_B(f) = S_e(f) + S_0(f)$$

or $$S_B(f) = 2P_S T_S \left[\frac{\sin(\pi f\, T_S)}{\pi f\, T_S}\right]^2 \qquad \text{...(7.76)}$$

This equation gives baseband power spectral density of QPSK signal. Upon modulation of carrier of frequency f_c, the spectral density given by above equation is shifted at $\pm f_C$. Thus, plots of power spectral density of QPSK will be similar to that of BPSK.

7.38 BANDWIDTH OF QPSK SIGNAL

We know the bandwidth of BPSK signal is equal fo $2f_b$. Here,

$$T_b = \frac{1}{f_b} \text{ is the one bit period.}$$

In QPSK, the two waveforms $b_e(t)$ and $b_0(t)$ forms the baseband signals. One bit period for both of these signals is equal to $2T_b$. Therefore, bandwidth of QPSK signal will be

$$BW = 2 \times \frac{1}{2T_b} = f_b \qquad \text{...(7.77)}$$

Hence, the bandwidth of QPSK signal is half of the bandwidth of BPSK signal. Earlier, we have observed that noise immunity of QPSK and BPSK is same. This shows that inspite of the reduction in bandwidth in QPSK, the noise immunity remains same as compared to BPSK. BW of WPSK can also be obtained by plotting equation (7.71) as shown in Fig. 7.28 below.

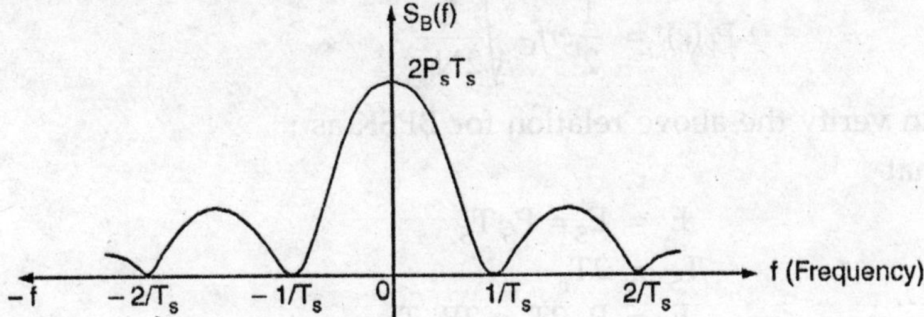

Fig. 7.28 *Plot of power spectral density (psd) of QPSK signal*

$$BW = \text{Highest frequency} - \text{Lowest frequency in main lobe}$$

$$BW = \frac{1}{T} - \left(-\frac{1}{T_S}\right) \quad \text{since carrier frequency } f_c \text{ cancels out}$$

$$BW = \frac{2}{T_S}$$

We know that $$T_S = 2\,T_b$$

or $$BW = \frac{2}{2T_b} = \frac{1}{T_b}$$

or $$BW = f_b \qquad ...(7.78)$$

7.39 PROBABILITY OF ERROR OF QPSK SYSTEM

From Fig. 7.28 between signal phasors 'OA' and 'OB' which can be observed carefully, the axis $\phi_1(t)$ may be called decision boundary. Let us consider signal vector 'OB' is present, but because of imperfect phase synchronization, it is detected as 'OA'. For happening of this, the phase shift must be at least $\pi/4$. The same thing can be happen in case of other phasors also.

There are two correlators in the QPSK receiver. One correlator is used to detect even hits and other detects odd bits. Thus, any correlator, can make a mistake if phase slift of $\pi/4$ occurs in the corresponding carrier. Therefore, the probability that correlator 1 or correlator 2 will make a mistake is given as,

$$P_1'(e) = P_2'(e) = \frac{1}{2} erfc \sqrt{\frac{E\cos^2\theta}{N_0}}$$

This equation has been written because each correlator is independent PSK receiver. 'E_b' is replaced by 'E' in above equation, since energy of one bit in $b_e(t)$ or $b_0(t)$ is E also written as E_S. Putting phase shift $\theta = \pi/4$ in above equation, we have

$$P_1'(e) = P_2' = \frac{1}{2} erfc \sqrt{\frac{E\cos^2\frac{\pi}{4}}{N_0}}$$

or $$P_1'(e) = \frac{1}{2} erfc \sqrt{\frac{E}{2N_0}}$$

Now, we can verify the above relation for BPSK as :

We know that

$$E = E_S = P_S T_S$$

and $$T_S = 2T_b$$

Therefore, $$E = P_S 2T_b = 2P_S T_b$$

$$E_b = P_S T_b$$

$$E = 2E_b$$

By using the above relation. We obtain error probability of BPSK i.e.,

$$P_1'(e) = P_2'(e)$$

$$= \frac{1}{2} erfc \sqrt{\frac{2E_b}{2N_0}}$$

or $$P_1'(e) = \frac{1}{2} erfc \sqrt{\frac{E_b}{N_0}}$$

The above equation gives bit error probability of QPSK. Thus, bit error probability of QPSK and BPSK is same.

The probability P(e) that the QPSK receiver will correctly detect the transmitted signal is equal to product of probabilities that both correlator 1 and correlator 2 will receive their bits correctly. The probability of correct reception of correlators 1 and 2 is,

$$P_1'(e) = 1 - P_1(e)$$

and $$P_2(e) = 1 - P_2'(e)$$

Therefore, $$P(e) = P_1'(e) \times P_2'(e)$$

$$= \left[1 - P_1'(e)\right] \times \left[1 - P_2'(e)\right]$$

$$= 1 - P_1'(e) - P_2'(e) + P_1'(e) \times P_2'(e)$$

Therefore, $$P_1'(e) = P_2'(e)$$

We can write abover equation as

$$P(e) = 1 - 2P'^1(e) + P'^2(e)$$

Here, $$P_1'(e) = P_2'(e)\ P'^2(e)$$

The term $P'^2(e)$ will be very small and can be neglected. Hence,

$$P(e) = 1 - 2P'(e)$$

Therefore, probability of error of QPSK system is

$$P(e) = 1 - P'(e)$$

or
$$P(e) = 1 - 1 + 2P'(e)$$

or
$$P(e) = 2P'(e)$$

Putting value of $P'(e) = \frac{1}{2} erfc \sqrt{\frac{E}{2N_0}}$

We have
$$P(e) = erfc \sqrt{\frac{E}{2N_0}} = erfc \sqrt{\frac{E_b}{N_0}} \quad \text{because } E = 2E_b$$

The individual probabilities $P_1'(e)$ and $P_2'(e)$ correlators are sometimes called as bit error probabilities or Bit Effort Rate BER. The bit error rate of QPSK is given as

$$BER = \frac{1}{2} erfc \sqrt{\frac{E_b}{N_0}}$$

7.40 ADVANTAGES OF QPSK

QPSK has some certain advantages as compared to BPSK and DPSK as :

(*a*) For the same bit error rate, the bandwidth required by QPSK is reduced to half as compared to BPSK.

(*b*) Because of reduced bandwidth, the information transmission rate of QPSK is higher.

7.41 MINIMUM SHIFT KEYING, MSK

We know that the bandwidth requirement of QPSK is high. Filters or other methods can overcome these problems, but they have other side effects i.e., filters alter the amplitude of the waveform.

Minimum shift iceying, MSK overcomes these problems. In MSK, the output waveform is continuous in phase, hence there are no absupt changes in amplitude. The side lobes of MSK are very small hence bandpass filtering is not required to avoid interchannel in terference Fig. 7.29 (*a*) shows the waveform of MSK. The binary bit sequence is shown at the top. Figure 7.29 (*a*) shows the corresponding NRZ waveform $b(t)$. From $b(t)$, two waveform are generated for odd and even bits. $b_0(t)$ represents odd bits and $b_e(t)$ represents even bits. Fig. 7.29 (*b*) and (*c*) shows the waveform of $b_0(t)$ and $b_e(t)$. As shown in those waveforms b_1, b_3, b_5 etc. are represented by odd waveform i.e., $b_0(t)$.

The duration of each bit in $b_0(t)$ or $b_e(t)$ is $2T_b$. Whereas it is T_b in $b(t)$ i.e.,

$$T_S = 2T_b \qquad ...(7.79)$$

the waveforms $b_0(t)$ and $b_e(t)$ have an offset of T_b. This offset is essential in NSK. Two waveforms sin $2\pi(t/4T_b)$ and cos $2p(t/4T_b)$ are generated as shown in Fig. 7.29 (*d*). The waveform of sin $2\pi(t4\ T_b)$ passes through zero at the end of symbol time in $b_0(t)$. Hence, one symbol duration of $b_0(t)$ consists of complete half cycle of cos $2\pi(t/4T_b)$. this means that similarly, one symbol duration of $b_e(t)$ contains half cycle of sin $2\pi(t/T_b)$. Thus, there is a phase shift of 'T_b' in sine and consine waveforms. $b_e(t)$ is multiplied by sin $2\pi(t4T_b)$ and $b_0(t)$

is multiplied by cos $2\pi(t/4T_b)$. These product waveforms are shown in Fig. 7.29 (*e*) and (*f*). The transmitted MSK signal is represented as :

$$s(t) = \sqrt{2P_S}\left[b_e(t)\sin(2\pi t/4T_b)\right]\cos(2\pi f_C t)$$

$$+\sqrt{2P_S}\left[b_e(t)\cos(2\pi t/4T_b)\right]\sin 2\pi f_C t \quad ...(7.80)$$

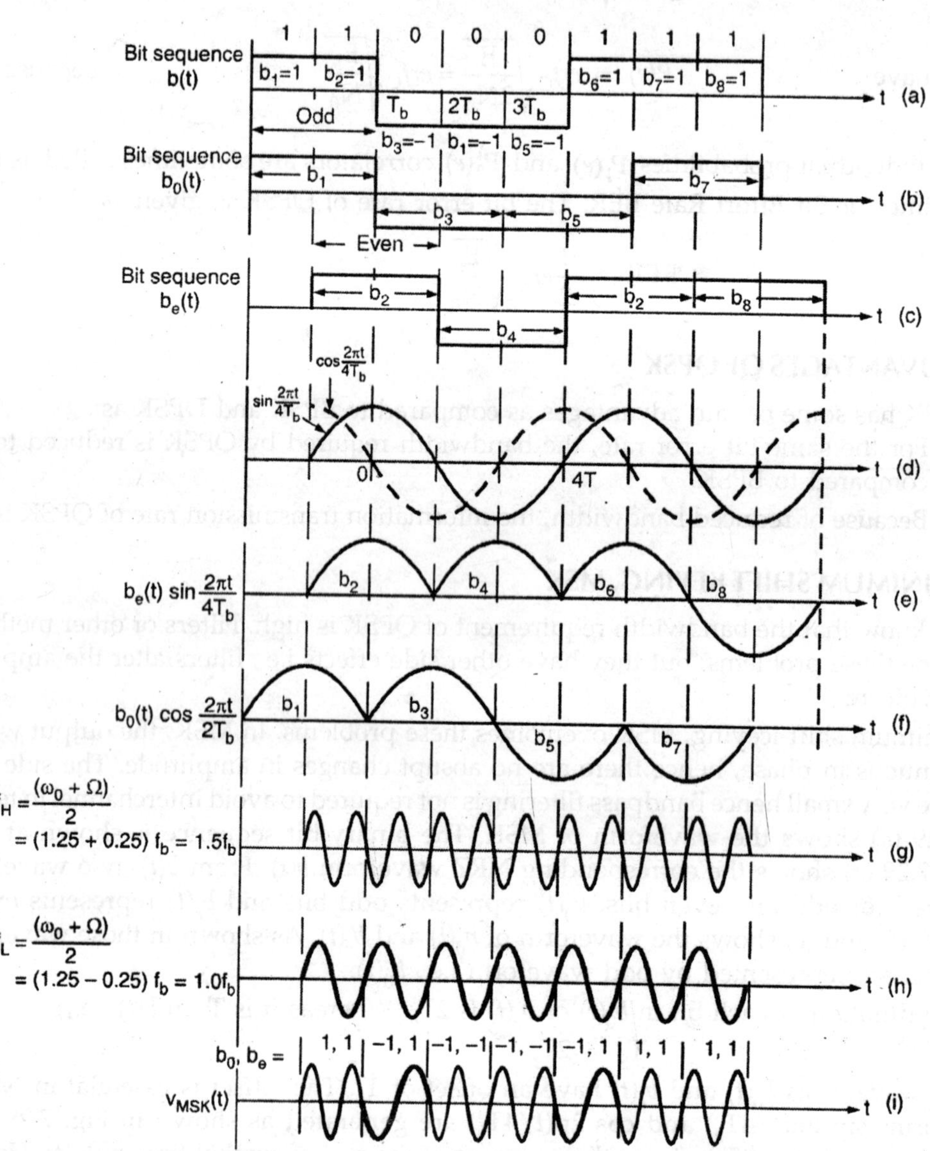

Fig. 7.29 *(a) Bipolar NRZ waveform representing bit sequence (b) Odd bit sequence waveforms $b_0(t)$ (c) Even bit sequence waveform $b_e(t)$ (d) Waveforms of frequency $f_b/4$ used for smoothing of $b_e(t)$ and $b_0(t)$ (e) Modulating waveform of even sequence (f) Modulating waveform of odd sequence (g) Waveform of frequency f_H (h) Waveform of frequency f_L(i) MSK waveform.*

This means that the product signal $b_e(t)\sin(2\pi t/4T_b)$ and $b_0(t)\cos(2\pi t/4T_b)$ madulate the quadrature carriers of frequency f_C. We can write last equation as :

$$S(t) = \sqrt{2P_S}\left[\frac{b_0(t)+b_e(t)}{2}\right]\sin 2\pi\left(f_C+\frac{1}{4T_b}\right)t$$

$$+\sqrt{2P_S}\left[\frac{b_0(t)+b_e(t)}{2}\right]\sin 2\pi\left(f_C-\frac{1}{4T_b}\right)t \quad ...(7.81)$$

We know that $f_b = \frac{1}{T_b}$, then the last equation (7.81) becomes,

$$S(t) = \sqrt{2P_S}\left[\frac{b_0(t)+b_e(t)}{2}\right]\sin 2\pi\left(f_C+\frac{f_b}{4}\right)t$$

$$+\sqrt{2P_S}\left[\frac{b_0(t)-b_e(t)}{2}\right]\sin 2\pi\left(f_C-\frac{1}{4}\right)t \quad ...(7.81)$$

Let
$$C_H(t) = \left[\frac{b_0(t)+b_e(t)}{2}\right]$$

and
$$C_L(t) = \frac{b_0(t)+b_e(t)}{2}$$

and let
$$f_H = f_e + \frac{f_b}{4} \quad ...(7.83)$$

and
$$f_L = f_C - \frac{f_b}{4} \quad ...(7.84)$$

With these substitutions, equation (7.82) becomes,

$$S(t) = \sqrt{2P_S}\ C_H(t)\sin(2\pi f_H t) + \sqrt{2P_S}\ C_L(t)\sin(2\pi f_L t) \quad ...(7.85)$$

If $b_0(t) = b_e(t)$ then $C_L(t) = 0$ and $C_H(t) = \pm 1$ then least equation becomes

$$S(t) = \sqrt{2P_S}\ C_H(t)\sin(2\pi f_H t) \quad ...(7.86)$$

Hence, the transmitted frequency is f_H.

If $b_0(t) = -b_e(t)$, then $C_H(t) = 0$ and $C_L(t) = \pm 1$.

Then equation 7.86 can be written as

$$S(t) = \sqrt{2P_S}\ C_H(t)\sin(2\pi f_L t) \quad ...(7.87)$$

Hence, the transmitted frequency is f_L.

The frequencies f_H and f_L are chosen such that $\cos(2\pi f_H t)$ and $\sin(2\pi f_L t)$ are orthogonal over the interval T_b. For orthogonality

$$\int_0^{T_b}\sin(2\pi f_H t)\sin(2\pi f_L t)\,dt = 0 \quad ...(7.88)$$

This relation will be satisfied if we have integers 'm' and 'n' such that,

$$2\pi(f_H - f_L)T_b = n\pi \quad ...(7.89)$$

and

$$2\pi(f_H + f_L)T_b = m\pi \quad ...(7.90)$$

Now, Let us substitute the values of f_H and f_L from equations (7.81) and (7.82) in above relations. From equation (7.87), we get

$$2\pi\left(f_C + \frac{f_b}{4} - f_C + \frac{f_b}{4}\right)T_b = n\pi$$

or

$$f_b\, T_b = n$$

or

$$f_b \times \frac{1}{f_b} = n \Rightarrow n = 1 \quad ...(7.91)$$

Similarly from equation (7.90), we get

$$2\pi\left(f_C + \frac{f_b}{4} - f_C - \frac{f_b}{4}\right)T_b = m\pi$$

or

$$4f_b\, T_b = m$$

or

$$4f_b \times \frac{1}{f_b} = m \Rightarrow f_C = \frac{m}{4}f_b \quad ...(7.91)$$

With $n = 1$ inequation (14.89), we get

$$2p(f_H - f_L)T_b = 1 \times \pi$$

or

$$(f_H - f_L) = \frac{1}{2T_b} = \frac{f_b}{2} \quad ...(7.93)$$

Here $n = 1$ means the difference between f_H and f_L is minimum and at the same time, MSK, they are orthogonal. Therefore, this technique is called minimum shift keying. MSK. This minimum difference is given by equation (7.93). From equation (7.92). We know that

$$f_C = \frac{m}{4}f_b.$$

This show that carrier frequency 'f_C' is integer multiple of $f_b/4$. Substituting this value of f_C in equation (7.83), We get

$$f_L = (m-1)\frac{f_b}{4} \quad ...(7.94)$$

Similarly, substituting $f_b = m\frac{f_b}{4}$ in equation (7.84), We get

$$f_L = (m-1)\frac{f_b}{4} \quad ...(7.95)$$

Fig. 7.29 (g) and (h) shows the waveforms of $\sin(2\pi f_H t)$ and $\sin(2\pi f_L t)$. For these waveforms $m = 5$. Using equations (7.94) and (7.90), f_H and f_L are calculated with $m = 5$. Fig. 7.38 (i) shows the final MSK waveform. From equation (7.86), We know that if $b_0(t) = b_e(t)$, then transmitted waveform is of frequency f_H. And if $b_0(t) = -b_e(t)$ then the transmitted waveform is given by equation (7.82), which has frequency of f_L. This shows that MSK is basically FSK with reduced bandwidth and continuous phase.

7.42 SIGNAL SPACE REPRESENTATION OF MSK AND DISTANCE BETWEEN THE SIGNAL POINTS OR GEOMETRICAL REPRESENTATION OF MSK

Let us rearrange equation (7.80) as follows

$$S(t) = C_H(t)\sqrt{P_S T_S}\cdot\sqrt{\frac{2}{T_S}}\sin(2\pi f_H t)$$

$$+ C_L(t)\sqrt{P_S T_S}\cdot\sqrt{\frac{2}{T_S}}\sin(2\pi f_L t) \quad ...(7.96)$$

Let

$$\phi_H(t) = \sqrt{P/T_S}\sin(2\pi f_H t) \quad ...(7.97)$$

$$\phi_L(t) = \sqrt{P/T_S}\sin(2\pi f_L t) \quad ...(7.97)$$

The carriers $\phi_H(t)$ and $\phi_L(t)$ are in quadrature. They are in quadrature because their frequencies are in quadrature. In QPSK the carriers are in quadrature because of phase shift. Depending on the values of $C_H(t)$ and $C_L(t)$, there will be four signal points in $\phi_H\,\phi_L$ plane. This has been illustrated in Fig. 7.30. The distance of each signal point from the origin is $\sqrt{P_S T_S}$.

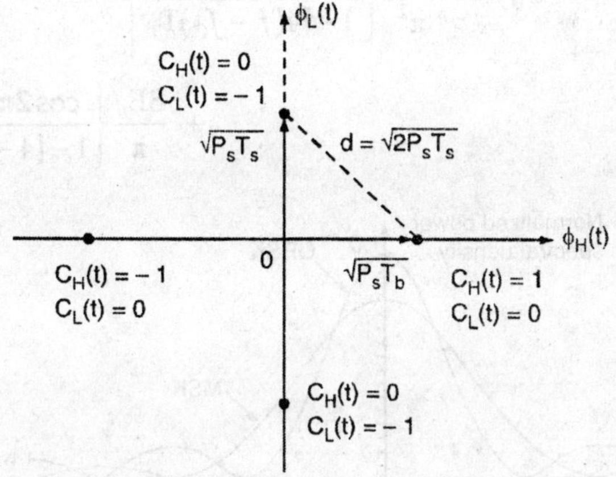

Fig. 7.30 *Geometrical (Signal Space) representation of MSK signals*

7.43 DISTANCE BETWEEN SIGNAL POINTS

Because the Points are symmetric, the distance between any two nearest points is same, i.e.,

$$d^2 = \left(\sqrt{P_S T_S}\right)^2 + \left(\sqrt{P_S T_S}\right)^2$$

or $$d = \sqrt{2P_S T_S} \quad ...(7.99)$$

or $$d = \sqrt{2E_S} \quad \text{Since } P_S T_S = E_S \quad ...(7.100)$$

or $$d = \sqrt{4E_b} \quad ...(7.101)$$

Since $$E_S = 2E_b = 2\sqrt{E_b} \quad ...(7.102)$$

These relations give distance between signal points in MSK. This distance is same as in QPSK.

7.44 POWER SPECTRAL DENSITY, PSD, AND BAND WIDTH OF MSK

Let us consider the baseband signal of equation (7.85). The waveform which medulates $\sin(2\pi f_C t)$ is,

$$p(t) = [b_0(t) \cos(2\pi t/4T_b)] \quad ...(7.103)$$

$$= \sqrt{2P_S}\ b_0(t) \cos(2pf_b\, t/2) \quad ...(7.104)$$

The power spectral density, psd, of above waveform is expressed as,

$$S_p(f) = \frac{32E_b}{\pi^2}\left[\frac{\cos(2\pi f\, T_b)}{1-(4f\, T_b)^2}\right]^2 \quad ...(7.105)$$

When this signal modulates the carrier 'f_C' then total power spectral density, pds, of baseband signal is divided by '4' and is placed at $\pm f_C$ i.e.,

$$S(f) = \frac{8E_b}{\pi^2}\left\{\frac{\cos 2\pi(f - f_C)T_b}{1-[4(f - f_C)T_b}\right\}^2 + \frac{8E_b}{\pi}\left\{\frac{\cos 2\pi(f + f_C)T_b}{1-[4-(f - f_C)T_b}\right\}^2 \quad ...(7.106)$$

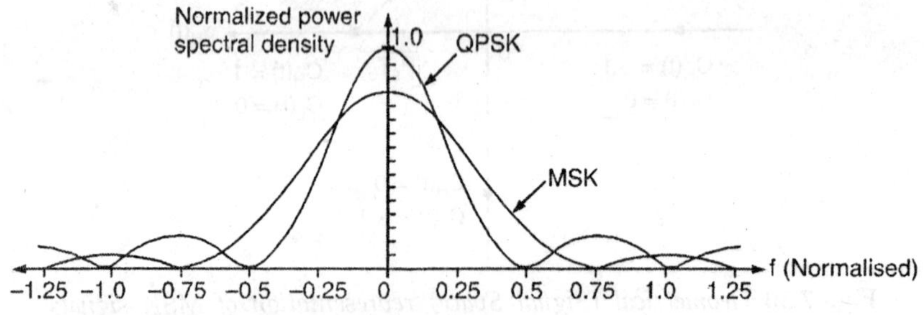

Fig. 7.31 *Power spectral densitities (psd) of MSK and QPSK*

The above equation gives power spectral density, psd, of MSK signal. Figure 7.30 shows the normalized spectral densities of MSK and QPSK. Normalization means maximum amplitudes of signals are scaled with respect to '1'.

The above plots show that the main lobe in MSK is wider than QPSK. The sides lobes in MSK are very small compared to QPSK.

7.45 BANDWIDTH CALCULATION OF MSK

From Fig. 7.31.. We observe that the width of main lobe in MSK is ± 0.75, i.e.

$$f\,T_b = \pm\, 0.75$$

or

$$f = \pm\, 0.75\, f_b$$

Hence, bandwidth will be equal to width of the main lobe, i.e.,

$$BW = 0.75\, f_b - (-\,0.75\, f_b)$$

$$= 1.5\, f_b \qquad ...(7.107)$$

Thus, the BW of MSK is higher than that of QPSK.

7.46 GENERATION OF MSK

Fig. 7.32 shows the block diagram of MSK transmitter. The two sinusoidal signals $\sin(2\pi f_C t)$ and $\cos(2\pi t/4T_b)$ are mixed, i.e., multiplied. The bandpass filters then pass only sum and difference components

$$f_C + \frac{f_b}{4}$$

and

$$f_C - \frac{f_b}{4}$$

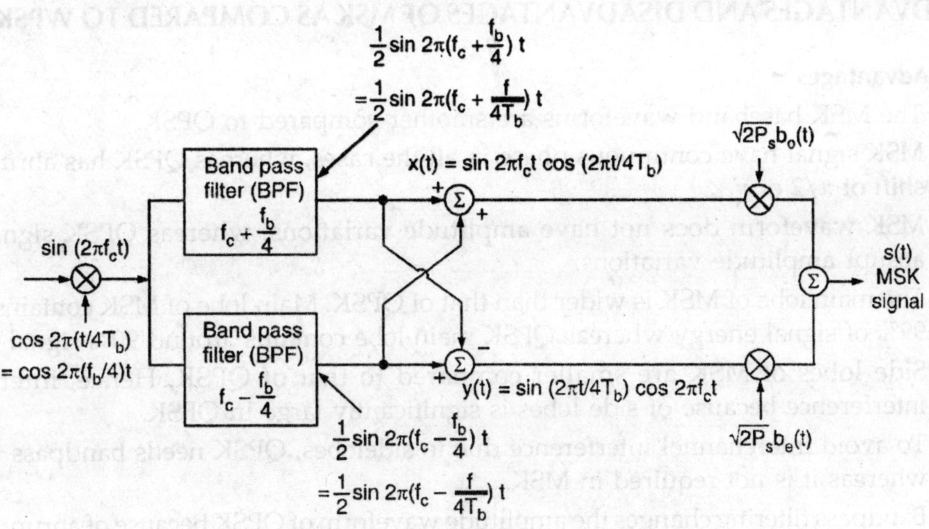

Fig. 7.32 *MSK transmitter block diagram*

The outputs of bandpass filters, BPF_S, are then added and subtracted such that two signals $x(t)$ and $y(t)$ are generated. Signal $x(t)$ is multiplied by $\sqrt{2P_S}\; b_e(t)$. The outputs of the multiplies are then added to give final MSK signal. Thus the block diagram of Fig. 7.31 is the step to step implementation of equation 7.80.

7.47 RECEPTION OF MSK OR DETECTION OF MSK

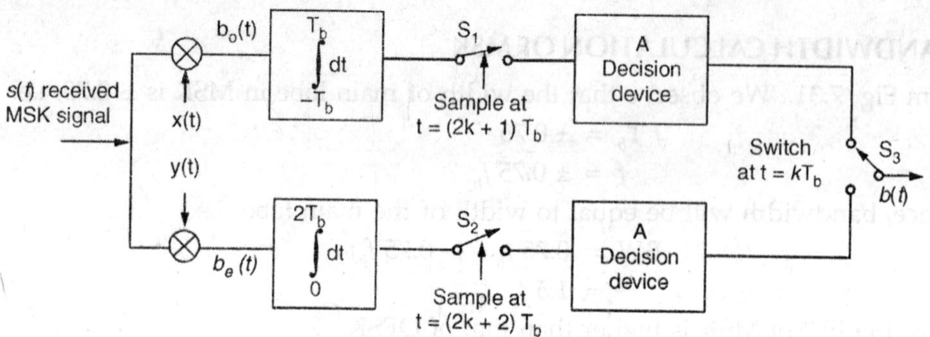

Fig. 7.33 *MSK receiver block diagram*

Fig. 7.33 shows the block diagram of MSK receiver. MSK uses synchronous detection. The signals $x(t)$ and $y(t)$ are multiplies with the received MSK signal. Here $x(t)$ and $y(t)$ have same values as shown in transmitter block diagram of figure 7.32. The outputs of the multipliers are $b_e(t)$ and $b_0(t)$. The integrators integrates over the period of $2T_b$. For the upper correlator, the sampling switch samples output of integrator at $t = (2k + 1)T_b$.

Then the decision device decides whether $b_0(t)$ is + 1 or – 1. Similarly, lower correlator output is $b_e(t)$. The outputs of two decision devices are staggered by T_b. The switch S_3 operates at $t = k\ T_b$ and simply multiplexes the two correlator outputs.

7.48 ADVANTAGES AND DISADVANTAGES OF MSK AS COMPARED TO WPSK

7.48.1 Advantages

1. The MSK baseband waveforms are smother compared to QPSK.
2. MSK signal have continuous phase in all the cases, whereas QPSK has abrupt phase shift of $\pi/2$ or π.
3. MSK waveform does not have amplitude variations, whereas QPSK signals have absupt amplitude variations.
4. The main lobe of MSK is wider than that of QPSK. Main lobe of MSK contains around 99% of signal energy whereas QPSK main lobe contains around 90% signal energy.
5. Side lobes of MSK are smaller compared to that of QPSK. Hence, interchannel interference because of side lobes is significantly large in QPSK.
6. To avoid interchannel interference due to sidelobes, QPSK needs bandpass filtering, whereas it is not required in MSK.
7. Bandpass filtering changes the amplitude waveform of QPSK because of abrupt changes in phases. This problem does not exist in MSK.

The distance between signal points is same in QPSK as well as in MSK. Hence, there are some drowback of MSK.

7.48.2 Disadvantages

1. The bandwidth requirement of MSK is 1.5 f_b. Whereas it is f_b in QPSK. Actually this cannot be said serious drawback of MSK. Because power to bandwidth ratio of MSK is more. In fact, 99% of signal power can be transmitted within the bandwidth of 1.2 f_b in MSK. While QPSK needs around $8f_b$ to transmit the same powers.
2. The generation and detection of MSK is slightly complex. Because of in correct synchronization. Phase jitter can be present in MSK. This degrades the performance of MSK.

PROBLEMS

1. Describe coherent Binary FSK.
2. Explain coherent Binary ASK.
3. Explain the difference between different modulation techniques.
4. Draw the block diagram of QPSK system and explain its working.
5. Explain the geometrical representation of non-orthogonal BFSK.
6. Derive the expression for the spectum of BPSK and sketch the same.
7. What is the advantage of gray coding of the QPSK system? What is the advantage of differential encoding of the input to the QPSK system?
8. What type of synchronization is used in QPSK system?
9. Draw the block diagram of DPSK modulator and explain how synchronization problem is avoided for its detection.
10. What is the ambiguity in the decodes output in case of PSK systems? How it is corrected?
11. Write the power spectral density of BPSK and QPSK signals and draw the power spectrum of each.
12. How is phase continuity achieved in MSK system?

INDEX

A

a harmonic generator 118
a phase-coherent 368
allocated 25
AM detection 112
AM generation 109
AM receiver 128
AM signals 111
AM wave 111
ambiguity 360
amplification 3
amplify 117
amplitude fluctuations 197
analog communication 19
analog information sources 20
analyzer 30
aperiodic 11
arbitrary 255
articulation 8
asynchronous 21
attenuated 67
attenuation 79
audio frequencies 14
audio frequencies 3
audio signal 15
audio signal 189
automatic volume control (AVC) 129

B

Balanced Modulator 57
bandpass 10
bandpass 298
bandpass 52
bandpass filter (BPF) 61
bandpass filters 32
baseband 79
binary digit 20
binary sequence 21
binary symbols 360
bipolar transistor 186
bitperiods 376
bits 21

C

carrier amplitude 139
causal 245
center-tapping 193
chopper modulator 103
clipper 341
codeword 21
colerent 349
collector modulator 103
compatible single sideband (CSSB) 87
complexity 22
Complexity 23
computer communication 1
conjugate 279
continuous wave 31
conventional 227
conventional AM 64
convolution theorem 90
correlators 364
crystal 118

D

...ng factor 267
...bels 305
deciphered 4
decommutator 331
decommutator 331
de-emphasis 215
Delta falls 268
demodulator 354
demodulator circuit 83
demolition 4
Demoulation 51
detection 346
Detection 51
deviation ratio 177
diagonal clipping 114
dielectric waveguide 8
digital communication 19
discrete information sources 20
discrete symbols 20
distorted 3
distorted 4
distortion 114
distortion 144
distortion 234
distortion 24
distortion 67
Distortion 83
distortion 96
Double-sideband 54
double-sideband suppressed carrier (DSB-SC) 70
drain modulator 103
DSB-SC signal 55

E

electromagnetic 120
encoder 21
enormous 8
envelope detector 138
envelope detector 54
extent 223
extraterrestrial 282

F

F(i) Frequency division multiplexing (FDM) 31
fact-acting 234
fascimile 6
fidelity 25
field-effect transistor 186
filter method 80
filtering changes 390
filtering property 237
Flicker Noise 283
flicker noise 285
Fourier method 246
Fourier series 257
Fourier series 26
Frequency Modulation (FM) 163
Frequency Modulation (FM) 166
Frequency Modulation 14
frequency modulation 161
frequency voltages 194

G

galactic noise 283
ganged capacitors 127
geostationary 9
geostationary 9
geosynchronous 9
glottis 5
Graphical synthesis 30
Guassian 297

H

harmonics 30
heterodyning 121
Hilbert Transform 72

I

IF amplifier 128
IF frequency 122
impedance 118
impedance 50
Impulse 252
inductively coupled 193
inevitable 25
infinitesimal 267
information 347
information rate 20
in-phase 91
input frequencies 145
instantaneous 323
instantaneous angular frequency 162
instantaneous angular velocity 162
instantaneous input power 107
instantaneous power 107
integrator 354
intercepted 120
Inter-Channel Interference 360
interference 333
interlaced 5
intermediate frequency transformer (IFT) 125
Intermediate Power Amplifier (IPA) 119
interplay 15
invariant 10
invention 2

L

linear modulation 100
lower sideband (LSB) 37
low-pass filter (LPF) 62

M

Metal-Semiconductor 285
microcontrollers 22
microphone 3
microwave 305
mobile communication, 1
modulating quadrature 68
modulating signal 18
modulation indexes 47
modulation inherently 161
multiple-tone modulation 47
Multiplexing 332

N

narrowband 131
Negative spikes 213
Neutralisation 120
noise component 139
noise-figure 305
noncausal 245
non-coherent 367
Non-linear distortion 145
non-linear modulator 108
non-orthogonal 366
nonperiodic 264
non-repetitive pulse 28
non-sinusoidal 26

O

optical fibres 3
optimum 114
orthogonal 256
orthogonal 256
oscillator circuit 186
oscillator circuit 186
oscillators 64
overmodulated signal 96

P

PAM, PWM, PCM, DM 31
parameter 305
passband 8
peak-separation range 196
perceives 6

phase angle 163
phase locked loop (PLL) 69
Phase Modulation (PM) 163
Phase Modulation (PM) 166
phase opposition 194
phase shifting 73
phase-modulation 183
phase-shift keying 367
phase-shifter 80
phasor diagram 194
phasor diagram 85
phasor OC 98
photocathode 5
photosensitive 5
picture signals 88
piecewise-linear phase 231
potentiometer 312
probabilistic 251
product modulator 57
propagate 282
propagates 5
propagation 13
pulse train 119
pulse-amplitude 14
pulse-duration 14
pulse-position 14
push-pull power 129

Q

QAM transmitter 65
quadrature-phase 91
quantization 17
quantization 248
quantizing 20
quarter-wavelength 14

R

radar communication 1
radiating 15
radio telemetry 1
raster scan 6
reaquaint 26
resonant 302
Ring Modulator 57
ruggedness 8

S

sampling theorem 17
scaling property 263
selecting fading 144
signal-to-noise 15
significantly 390
single-sideband suppressed carrier (SSB-SC) 70
sinusoidal 12
sinusoidal 22
sinusoidal 245
sinusoidally 230
spatially 5
spectral 211
spectral density 388
spectrum 55
spectrums 364
spike 211
spike path 208
sub-harmonic 117
subharmonic 118
superheterodyne 121
superheterodyne 123
superposition 101
synchronization 331
synchronous 21
synchronous 349
synchronous 374
synchronous detection 97
Synchronous Dtection 61

T

T(ii) Time-division multiplexing (TDM) 31
telegraphy 1
tensile 8
terminated 341
threshold effect 143
threshold reduction 219
time-domain 255
time-instants 17
tolerated 24
transconductance 118
transducer 2
transistor 2
Transit-time Noise 283
transmission 1
transmission 117
transmission bandwidth 64
transmitter 2
transmitter 281
transmitter 347
transmitters 117
transponder 10
TRF receiver 123
tuned ratio frequency (TRF) 120

U

unit-impulse 253
unmodulated 96
unpredictable 281
upper sideband (USB) 37
utilization 347

V

vestigial sideband modulation (VSB) 88
video shooting 18
vocal 5

W

waveform 14
waveform 245
waveform 354
wideband FM 172
wiggling 284

□□□